1st International Symposium on Physics, Engineering and Technologies for Bio-Medicine 2016

Journal of Physics: Conference Series Volume 784

Moscow, Russia
18 - 23 October 2016

ISBN: 978-1-5108-3603-7
ISSN: 1742-6588

Printed from e-media with permission by:

Curran Associates, Inc.
57 Morehouse Lane
Red Hook, NY 12571

Some format issues inherent in the e-media version may also appear in this print version.

Printed by Curran Associates, Inc. (2017)

For permission requests, please contact the Institute of Physics
at the address below.

Institute of Physics
Dirac House, Temple Back
Bristol BS1 6BE UK

Phone: 44 1 17 929 7481
Fax: 44 1 17 920 0979

techtracking@iop.org

Additional copies of this publication are available from:

Curran Associates, Inc.
57 Morehouse Lane
Red Hook, NY 12571 USA
Phone: 845-758-0400
Fax: 845-758-2633
Email: curran@proceedings.com
Web: www.proceedings.com

Table of contents

Volume 784

1st International Symposium 'Physics, Engineering and Technologies for Bio-Medicine'

18–23 October 2016, Moscow, Russia

Accepted papers received: 6 December 2016
Published online: 25 January 2017

Preface

Papers

IOP Conf. Series: Journal of Physics: Conf. Series **784** (2017) 011001 doi:10.1088/1742-6596/784/1/011001

The 1st International Symposium «Physics, Engineering and Technologies for Bio-Medicine» October 18-23, 2016, Moscow (Russia)

The 1st International Symposium "Physics, Engineering and Technologies for BioMedicine" was held in Moscow at the occasion of the foundation of the new Institute PhysBio at MEPHI (Russia) on October 20-23, 2016.

Under the auspices of the Russian Ministry of Science and Education, the Ministry of Health and the State Company Rosatom, the Symposium is organized by the Institute of Engineering Physics for Biomedicine (PhysBio), which has recently been established at National Research Nuclear University MEPhI (Moscow Engineering Physics Institute).

The PhysBio's goal is to train highly-skilled personnel through the research and development in engineering physics for biomedicine, including nuclear medicine, material science, laser physics and biophotonic technologies.

The Symposium was aimed to meet the leading scientists and experts in nuclear medicine, biophysics, biophotonics, and emerging fields to present their works and to have invited lectures.

1. Organizers

- The Ministry of Education and Science of the Russian Federation
- The Ministry of Health of Russian Federation
- The State Company Rosatom
- The Institute of Engineering Physics for Biomedicine (PhysBio), which has recently been established at National Research Nuclear University MEPhI (Moscow Engineering Physics Institute)

2. Committees

Symposium Co-Chairs

- Mikhail N. Strikhanov, MEPhI, Russia
- Paras Prasad, University of Buffalo, USA

Program-Committee Co-Chairs

- Andrei Kabashin, CNRS, Aix-Marseille University, France
- Oleg B. Loran, S.P. Botkin City Clinical Hospital, Department of Urology and Surgical Andrology of the Russian Medical Academy of Post-Graduate Education, Moscow, Russia.

Organizing-Committee Chair

- Irina N. Zavestovskaya, PhysBio MEPhI, LPI of RAS, Russia

International Program Advisory Committee

- Walter Blondel, Université de Lorraine, France

- Vladimir I. Chernov, Department of Radiologic Diagnosis of the Tomsk Research Institute of Oncology, Tomsk, Russia
- Mikhail I. Davydov, N.N. Blokhin Russian Cancer Research Center, Moscow, Russia
- Boris I. Dolgushin, Research Institute of Clinical Oncology of N.N. Blokhin Cancer Research Center, Moscow, Russia
- Dmitry K. Fomin, Nuclear Medicine Clinic of the Russian Scientific Center of Radiology and Nuclear Medicine of the RF Ministry of Health, Head of the Society of Nuclear Medicine, Moscow, Russia
- Hans H. Gutbrod, GSI Helmholtzzentrum für Schwerionen forschung, Darmstadt, Hesse, Germany
- Andrey D.Kaprin, Herzen Moscow Oncology Research Institute, Moscow, Russia
- Anatoly V. Karalkin, Nuclear Medicine Clinic of N.I. Pirogov State Clinical Hospital No. 1, Moscow, Russia
- Aleksandr A.Khasin, NRSKurchatov Institute, director of "LLC Nuclear Medicine Development Center"
- Galina E. Kodina,A.I. Burnazyan Federal Medical BiophysicsCenter, Department of Pharmaceutical and Radiopharmaceutical Chemistry, Russia
- Vitali I. Konov, Prokhorov General Physics Institute of RAS, PhysBio, MEPhI, Russia
- Eduard V. Kotlyarov, FACNM, FACR, International Radiology and Nuclear Medicine, USA
- Oleg N.Krokhin, LPI, MEPhi, Russia
- Yurii B. Lishmanov, Research Institute of Cardiology, Laboratory of Radionuclide Research Techniques, Russia
- Hans-PeterMeinzer, Department Medical and Biological Informatics E130, Germany
- Igor R. Nabiev, University of Reims Champagne-Ardenne, France, LNBI, MEPhI, Moscow, Russia
- Rainer Rienmuelle, Radiology of Medical University of Graz, European Society of Cardiac Radiology, Austria
- Sergey A. Rumiantsev, Department of Innovation development and Scientific Engineering , Ministry of Healthcare of the Russian Federation
- Nidal Salim, Center of beam therapy of The European Medical Center
- Marc Santis, CNRS, Aix-Marseille University, France
- Sergey V. Shestakov, Academician of RAS, member of the Genetics Department of the M.V. Lomonosov Moscow State University
- Rudolf Steiner, Institut für Lasertechnologien in der Medizin (ILM), Universität Ulm, Germany
- GlebSukhorukov, Queen Mary University of London, UK

Organizing Committee

- Yulia A. Akmalova, PhysBio of MEPhI
- Yurii A. Aleshchenko, PhysBio of MEPHI, Lebedev Physical Institute of RAS, Russia
- Yuri N. Anokhin, MEPhI, Russia
- Vladimir N.Belyaev, MEPhi, Russia
- Vladimir G.Bychenko, Research Centerof Obstetrics, Gynecology and Perinatology, Ministry of Healthcare of the Russian Federation
- Marco Durante, Ph.D. TIFPA-INFN Director Department of Physics, University of Trento, Italy
- Natalia B. Epshtein, PhysBio of MEPhI, Russia
- Elena Gromushkina, MEPhI, Russia
- Igor A. Gulidov, Department of beam therapy of the Radiological Medical Scientific Center A.F.Tsyba

IOP Conf. Series: Journal of Physics: Conf. Series **784** (2017) 011001 doi:10.1088/1742-6596/784/1/011001

- Evgeniy V. Khmelevskiy, Department of beam therapy of Herzen Moscow Oncology Research Institute, Ministry of Healthcare of the Russian Federation
- Liudmila Komarova, professor of biology, Head of Department of Biology, Obninsk
- Youlia B. Kurashvili, Herzen Moscow Oncology Research Institute, MEPhI, Russia
- Victor B. Loschenov, Prokhorov General Physics Institute of RAS, PhysBio, MEPhI, Russia
- Vladimir Lysenko, CNRS, INL, INSA de Lyon, France
- Valentin G. Nikitaev, MEPhI, Russia
- Petriev V.M., laboratory of experimental nuclear medicine, Medical Radiological Research Center, Ministry of Health, Russia
- Sergey V. Shiryaev, Deparment of Radiologic Diagnosis of the N.N. Blokhin Cancer Research Center, President of the Oncological Society of Molecular Imaging, Moscow, Russia
- Victor Yu. Timoshenko, MSU, PhysBio of MEPhI, Lebedev Physical Institute of RAS, Russia
- Igor E. Tyurin, Department of Radiodiagnostics and Medical Physics of the Russian Academy of Post-qualifying Education, Moscow, Russia

Peer review statement

All papers published in this volume of *Journal of Physics: Conference Series* have been peer reviewed through processes administered by the proceedings Editors. Reviews were conducted by expert referees to the professional and scientific standards expected of a proceedings journal published by IOP Publishing.

Stress-induced molecules MICA as potential target for radioimmunotherapy of cancer

E V Abakushina[1, 2], Yu N Anokhin[2], D N Abakushin[3] and A D Kaprin[4]

[1] A Tsyb Medical Radiological Research Centre – branch of the National Medical Research Radiological Centre of the Ministry of Health of the Russian Federation, 4 Koroleva, Obninsk, 249036, Russia
[2] National Research Nuclear University MEPhI, 31 Kashirskoe shosse, Moscow, 115409, Russia
[3] LLC "Clinic №1", 6A Aksenova, Obninsk, 249039, Russia
[4] National Medical Research Radiological Centre of the Ministry of Health of the Russian Federation, 10 Zhukova, Obninsk, 249031, Russia

E-mail: abakushina@mail.ru

Abstract. Improving the treatment of cancer, increasing their effectiveness and safety is the main objective in the medicine. Molecular nuclear medicine plays an important role in the therapy of cancer. Radioimmunotherapy (RIT) involves the use of antibodies conjugated with therapeutic radionuclides. More often for RIT use the radiolabeled monoclonal antibodies against tumor-associated antigens. Encouraging results have been achieved with this technology in the management of hematologic malignancies. On the contrary, solid tumors have been less responsive. Despite these encouraging results, new potential target for radioimmunodetection and RIT should be found. It was found to increase the level of tumor-associated molecules MICA in the serum of cancer patients. Use of anti-MICA monoclonal antibodies capable a specifically attach to cancer cell via NKG2D ligands and destroy it, is a very promising direction, both therapeutic and diagnostic standpoint.

1. Introduction

One of the most urgent problems of modern oncology is early diagnosis and more effective treatment of patients with malignant tumors. Currently, methods of radionuclide therapy are developing rapidly around the world, which has become an effective tool, both independently and combined treatment. Radioimmunotherapy (RIT) - a promising trend that combines advances in modern nuclear medicine, immunology and biotechnology. Molecular nuclear medicine plays an important role in the diagnosis and therapy of cancer and infectious diseases. It is based on targeted therapy when the therapeutic monoclonal antibody (mAb) coupled to radionuclides. Antibodies bind specifically to antigenic determinants on the tumor or other cells and destroys these cells radionuclide by topical exposure [1].

Receptor NKG2D is expressed by all NK cells but is not limited to NK cells, as it is also expressed by many T cells, including all $CD8^+$ T cells in humans, subsets of $\gamma\delta$ T cells, and subsets of NKT cells. In humans, the NKG2D ligands include MICA and MICB (MHC class I chain–related proteins A and B). These ligands are typically induced in response to stress, commonly found within cancers. However, some cancers are known to develop escape mechanisms that reduce the levels of these ligands on their surface, including down regulation of expression or internalization of the ligands; or shedding of their soluble extracellular domains (sMICA, sMICB). Recently, we demonstrated that MICA is released as a soluble form from the cell surface of tumor cells and can be detected at high levels in sera of patients with malignancies but not in healthy donors [2].

A combination of measurements of circulating MICA/B domains and tumor volume can be combined to determine the rate of shedding, and that this data, could be used to determine likely response to RIT or cell therapy. Immunizing animals with recombinant human MICA protein can be obtained anti-MICA antibodies that will serve as a basis for joining radionuclide and subsequent RIT. The developed method of radioimmunotherapy can improve the treatment of cancer patients, as well as help to reduce the risk of metastasis and recurrence of tumor growth [1].

Purpose to this work was to justification find surface molecules of tumor cells as targets for radioimmunotherapy.

2. Material and methods

In the study were involved 80 patients with a variety cancer: melanoma (n=44), prostate cancer (n=16), colon cancer (n=20) and healthy volunteers (n=20). For the collection of biological material used was taken a peripheral blood samples. Before treatment, all patients performed an enzyme immunoassay for the presence of serum label of soluble forms of molecules MICA. A concentration of sMICA protein in sera of patients were determined by ELISA and reagent MICA Human ELISA Kit according to manufacturer's instructions (Abcam, USA).

3. Result and discussion

In Russia the first studies using radionuclide-labeled anti-tumor antibodies for the diagnosis and therapy of tumors were performed by Anokhin Yu N in 1985 year. The results of experimental studies on animals with transplants of solid tumors have been encouraging and have shown high specificity of ^{131}I-labeled anti-tumor antibodies to target antigens in comparison with non-specific immunoglobulins, labeled with ^{125}I. The concentration of the antitumor antibodies in tumor tissue is 7-10 times greater than the content of these antibodies in normal tissues [3, 4].

In this study we revealed the presence of soluble forms of highly polymorphic stress-induced MICA molecules in a number of cancers. The significant increase of level sMICA in the amount of 4.6 times in serum in patients with melanoma, colon cancer and prostate were found compared with the control group (figure 1).

These molecules, related to MHC class I antigens, have been described in the late '90s by American scientists [5]. It is shown that in neoplasia expression of these molecules is one of the earliest events that are manifested in the process of malignant transformation of cells. On the surface of normal cells they are absent or are contained in the insignificant amounts, but their expression can be induced under conditions of cellular stress, viral or bacterial infections and in tumor growth.

The adapter molecule DAP10 accept impulse occurring upon binding of NKG2D surface receptor on lymphocytes with stress-induced molecules on the surface of target cells, and carry out further transfer a signal that induces lymphocyte cytotoxic effect. Thus, expression of stress-induced molecules should lead to immunological destruction of the infected or transformed cells by the virus.

Therapy approaches are tailored to enhance or redirect NK cell– and/or T cell–mediated cytotoxicity by utilizing agents that may increase NKG2D ligand expression or bifunctional fusion proteins including an NKG2D ligand and a single-chain Ab fragment (scFv) targeting a specific tumor cell surface marker. Among pharmaceuticals tested for upregulation of NKG2D ligands on tumor lines in vitro are demethylating agents, proteasomal inhibitors, and genotoxic drugs used for chemotherapy such as histone deacetylase inhibitors [6]. Objectives for the bifunctional fusion proteins are to target NK cells and T cells onto tumor cells and engage NKG2D.

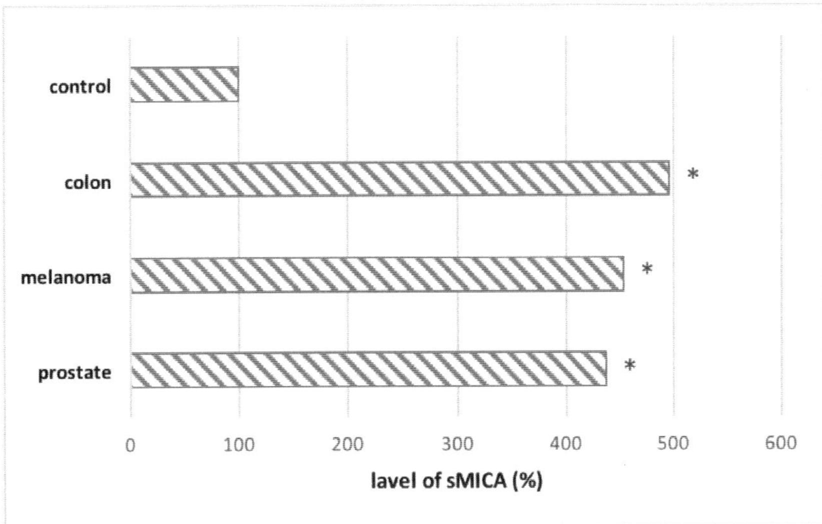

Figure 1. The level of sMICA in patients with prostate cancer, melanoma and colon cancer in comparison with healthy volunteers (100% control, * $p<0.05$).

Another type of bispecific T cell and NK cell engager, thus far tested in mice only, combines the extracellular domain of NKG2D with either CD3ε or the Fc portion of IgG2a for binding to CD16. In an in vivo tumor model, administration of the NKG2D–CD3ε fusion protein reduced growth of tumors bearing NKG2D ligands and promoted T cell infiltration. Likewise, the NKG2D–Fc region fusion protein triggered lysis and reduced growth of ligand-positive lymphomas [7, 8].

This study confirms our earlier results and gives reason to conclude that in the serum of cancer patients with solid tumors multiply exceeded the levels of soluble forms of the NKG2D ligands [9]. This claim suggests that stress-induced molecules MICA may serve as markers of cancer.

4. Conclusion

We can assume that NKG2D ligands the antigenic determinants MICA or MICB will serve as an attractive target for radioimmunotherapy [10]. Conjugation of anti-MICA mAbs to these stress-induced molecules with a specific radioisotope that may be broadly used in the treatment of many tumor types. Radioimmunotherapy can be used in chemoresistant patients and patients are unaffected to the action unlabeled antibody. Also RIT can be effective for elimination of multiple-drug-resistant tumor cells. Main goal of such drugs is associated with the purposive destruction certain cells, such as tumor. NKG2D ligands may present a useful target for immunotherapeutic approaches in cancer. A monoclonal antibodies NKG2D with radionuclide capable of specifically accede to the cancer cell and destroy it. Broad applicability of these therapies to many types of tumors offer great promise for the both treatment and diagnostic of cancer.

Acknowledgments
This work was supported by a grant from the Russian Scientific Foundation (grant № 14-35-00105) and the Russian Federation President Grant Scientific School-9069.2016.4.

References

[1] Tsyb A F, Abakushina E V, Abakushin D N, Romanko Yu S 2013 Present and future radioimmunotherapy *Medical Radiology and Radiation Safety* **58**(3) 48-55
[2] Abakushina E V, Klinkova A V, Kanevskiy L M, Kovalenko E I 2014 *Molecular medicine (Molekuliarnaya medicina)* Elevation of serum levels of soluble forms of stress-induced

molecules MICA in oncological diseases **3** 34-8

[3] Anokhin Yu N, Norec T A, Dedenkov A N 1986 *Medical radiology (Medicinskaja Radiologija)* Monoclonal antibodies in radio nuclide diagnostics and therapy of cancer diseases **6** 72-8

[4] Anokhin Yu N, Norec T A, Petrova G A, Endolov V V, Trofimova M V, Novikov V V 1988 *Medical radiology (Medicinskaja Radiologija)* Opportunity in vivo radionuclide diagnostics of hematologic tumors by means of marked monoclonal antibodies **1** 31-3

[5] Groh V, Bahram S, Bauer S, Herman A, Beauchamp M, Spies T 1996 *Proc. Natl. Acad. Sci.* Cell stress-regulated human major histocompatibility complex class I gene expressed in gastrointestinal epithelium **93** 12445-50

[6] Krieg S, Ullrich E 2012 *Front Immunol* Novel immune modulators used in hematology: impact on NK cells **3** 388-97

[7] Zhang T, Sentman C L 2011 *Cancer Res* Cancer immunotherapy using a bispecific NK receptor fusion protein that engages both T cells and tumor cells **71** 2066-76

[8] Zhang B, Kracker S, Yasuda T, Casola S, Vanneman M, Hömig-Hölzel C, Wang Z, Derudder E, Li S, Chakraborty T, Cotter S E, Koyama S, Currie T, Freeman G J, Kutok J L, Rodig S J, Dranoff G, Rajewsky K 2012 *Cell* Immune surveillance and therapy of lymphomas driven by Epstein-Barr virus protein LMP1 in a mouse model **148** 739-51

[9] Abakushina E V, Abakushin D N, Neprina G S, Pasova I A, Berdov B A, Klinkova A V, Kovalenko E I, Kaprin A D 2015 *Cytokines and Inflammation* Elevation of serum levels of cytokines and stress-induced molecules MICA in patients with gaster and colon cancer **14**(1) 63-7

[10] Abakushina E V, Abakushin D N, Anokhin Yu N 2014 *Siberian Journal of Oncology (Sibirskiy onkologicheskiy zhurnal)* Experimental and clinical approaches to a radioimmunotherapy in oncology **63**(3) 56-64

Design study of Thomson Laser-Electron X-ray Generator (LEX) for Millisecond Angiography

I A Artyukov[1], E G Bessonov[1], R M Feshchenko[1], M V Gorbunkov[1], Yu Ya Maslova[1], N L Popov[1], N V Dyachkov[1], A A Postnov[2], S L Vinogradov[1] and A V Vinogradov[1]

[1]P.N.Lebedev Physical Institute, 53 Leninsky Prospekt, 119991 Moscow, Russia,
[2]National Research Nuclear University MEPhI, 31 Kashirskoye shosse,115409 Moscow, Russia

E-mail: postnov.email@gmail.com

Abstract. In this concept study a laser-electron X-ray generator (LEX) is considered for the medical imaging of the inner vessel structure. It is demonstrated that the modern lasers and linear electron accelerators are suitable for the design of the new generation of angiography medical equipment combining higher spatial and time resolution with the reduced patient dose. Angiography setup based on LEXG can make use of different contrast media (iodine, gadolinium) working on absorption edge due to the narrow tuneable spectrum which is not possible with conventional X-ray tubes. In the present study all estimations are made for iodine-based contrast agents. The conclusion is that modern technologies allow practical implementation of LEX for angiography based on multibunch linear accelerator and photon storage device.

1. Introduction

Initially the idea and the implementation of a laser-electron X-ray generator (LEX) based on Thomson scattering was proposed by F.E.Carroll [1,2]. Currently two LEXG are fully operational [3,4] and even commercially available [5]. In all cases these are multipurpose setups shared by many investigators. Whereas LEX setups are 10 to 30 times bigger than the conventional X-ray sources based on X-ray tubes they can provide several orders of magnitude higher brilliance which stimulates applications in traditional X-ray technologies like security [6], therapy [7] and medical diagnostics [8,9]. The present paper discusses the requirements to LEX and its components to apply it to vessels visualization with 1 ms exposure time and the spatial resolution of 0.1 mm which exceeds the current image quality in clinical practice.

Angiography in medical practice is performed by the radiography of the vessels filled with a contrast agent. Let us estimate the required number of photons for one projection of an object with the blood vessel thickness δ (figure 1). The condition of blood vessel visualization in a bulk of non-contrasted media is:

$$\sqrt{N_{\delta}e^{-\mu L}} < c\left[N_{\delta}e^{-\mu L} - N_{\delta}e^{-\mu L-(k-\mu)\delta}\right] \quad (1)$$

Therefore the minimal exposure for the one picture element is:

$$N_\delta(E) = e^{\mu L}\left\{c\left[1 - e^{-(k-\mu)\delta}\right]\right\}^{-2}, \tag{2}$$

where attenuation coefficients μ (outside the vessel) and κ (in the vessel) depend on an X-ray photon energy E, c is a contrast ratio defined by the particular task and by the detector, L is the body thickness. Hence the number of photons required for the 2D detector M by M pixels is:

$$N(E) = M^2 e^{\mu L}\left\{c\left[1 - e^{-(k-\mu)\delta}\right]\right\}^{-2}. \tag{3}$$

This formula does not contain any information about the source itself. It estimates the necessary number of photons at the detector side to be able to extract small contrasted detail surrounded by attenuating material.

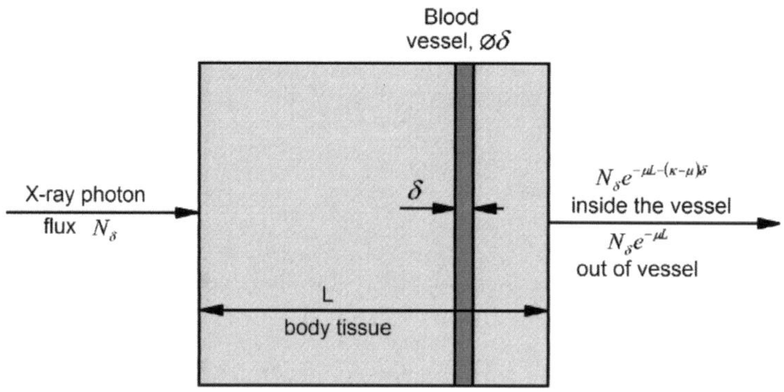

Figure 1. Exposure dose in angiography: L – total object thickness, N_δ – number of incident photons for the square δXδ required for one image, μ - linear attenuation coefficient of the body, κ- linear attenuation coefficient of the vessel with the contrast agent.

2. Coronary angiography. Exposure and radiation dose

For the last decades the "golden standard" for the heart vessel tree diagnostics remains a selective coronary angiography – an invasive procedure which requires artery catheterization with the subsequent injection of a contrast agent next to the investigated region. The risk of complications remains high. Noninvasive CT methods have been developed in the last 15 years [10] with the contrast agent injected intravenously prior to the study. At the same time there were attempts to improve image quality using monochromatic synchrotron radiation [11], X-ray tubes with rotating anode [12] and LEX [13]. However the time and spatial resolution of non-invasive techniques remained worse than those of the "golden standard" method [14-16]. Later in the text we present the LEX design that allows to overcome these limitations.

Let us first define the required dose i.e. the number of photons N(E) demanded for one-projection angiographic image. For that let's use formula (5) with the values typical for medical radiography patient thickness L=20 cm , vessel diameter δ = 1 mm. Attenuation coefficient μ is taken for water [16], attenuation coefficients for iodine contrasted vessels κ are expressed through the mass iodine concentration η related to Hounsfield units CT_{Number} [17], which is standard in radiology diagnostics. $N(E)$ for different iodine mass concentrations η in the energy range 30-120 keV is presented on figure 2 assuming that the imaging matrix contains 0.06 megapixel (256*256) and contrast ratio is c = 0,6. Data for the calculated iodine mass concentrations correspond to $CT_{Number} = 50 \div 500$ measured by CT angiography [19, 20].

Figure 2 demonstrates that the exposure dose is minimal at the energies of 35-45 keV . These doses correspond to $\sim 10^{11} \div 10^{13}$ photons for the clinically approved iodine concentrations (several mSv) and do not exceed approved radiation exposure for angiography and coronography [21]. The optimal energy estimated is above the absorption K-edge of iodine and it is confirmed experimentally [30].
Let us consider the requirements to LEX capable of generating the required number of photons in 1 ms of exposure time.

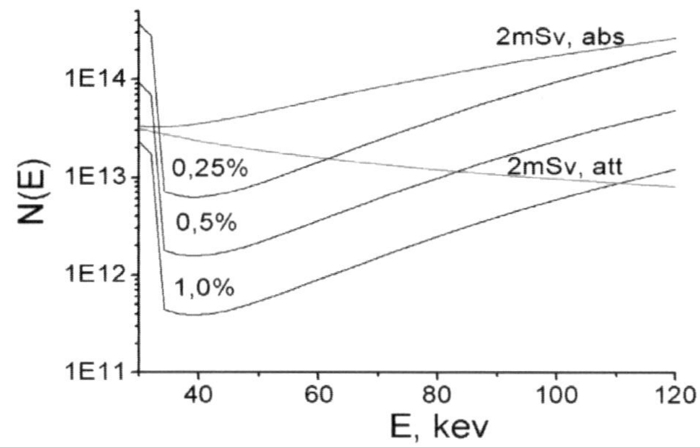

Figure 2. Number of X-ray photons required for one angiography image obtained in 1 ms (see (5)) as a function of photon energy E. Vessel thickness δ = 1 mm, body thickness L = 20 cm, contrast ratio c = 0,6, matrix size 256*256 (M =256). Iodine mass concentration in the vessel η is indicated in percent above the curves. Red curves correspond to the dose of 2 mSv for the 75 kg men: abs – one scattering event, att – multiple scattering, photons are totally absorbed.

3. Principles of Laser-Electron X-ray Generator

The basic scheme of LEX is presented on figure3. The X-rays are generated in the photon storage as a result of Thomson scattering of laser pulses on the relativistic electron bunches produced by the linear accelerator. Electron trains have a frequency of ν_e and contain n_e bunches N_e electrons each. Photon device amplifies (resonator [22]) or stores (circulator [23-25]) laser radiation. The number of stored photons is $N_L^s = \eta N_L$, where N_L is the number of laser photons and η is the photon device efficiency. Further in our estimations we are using $\eta = 10^3$ for the resonator and $\eta = 0,25$ for the circulator, which are feasible at the current technological level [26,27]. The energy and the X-ray flux emitted by the Thomson scattering of the laser photons on electrons depend on the geometry of the colliding beams. For the head-on collision the following formula can be used [28]:

$$E \approx 4\gamma^2 E_L,\ \Phi = n_e \nu_e N = n_e \nu_e N_e N_L^s W = \frac{1}{eE_L} I \Sigma^s W, \quad \text{where } N = N_e N_L^s W, \tag{4}$$

Σ^s – laser pulse energy accumulated in the optical storage device, e – electron charge, I – accelerator current, E_L - laser photon energy, W – the number of X-ray photons produced at one collision. W is used for the total flux calculations as well as for the spectral and angular distributions. For the radiography the key parameter is the total flux. In this case $W \approx \sigma_T / s$, where $\sigma_T = 6.6 \cdot 10^{-25} cm^2$ – Thomson cross-section, s – interaction area. For instance for s=30 μm^2:

$$W \approx \sigma_T / s = 0.93 \cdot 10^{-19}. \tag{5}$$

The general picture of Thomson scattering of a laser pulse on a relativistic electron bunch is different both from scattering on a bunch in a rest condition as well as from a scattering of a single photon on a single relativistic electron. Scattering diagram is elongated along the electron beam direction and concentrated in narrow angle $\delta\theta \sim 1/\gamma \approx 10$ mrad, where $\gamma = E_e / m_e c^2$ – relativistic factor and the spectrum width is 15-20%. Angular and spectral distributions are defined by the momentum distribution functions of the colliding beams. Exact formulas for W and for the LEX spectrum can be found in [28] and the previously cited literature.

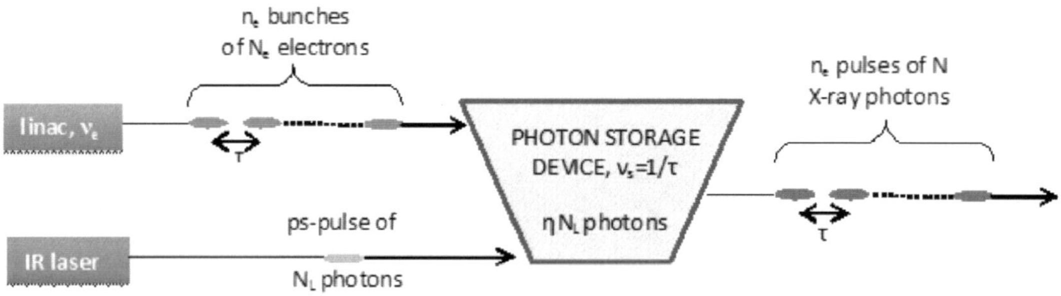

Figure 3. Basic scheme of LEX.

4. Laser-electron X-ray generator for angiography

Due to the narrow angular distribution and the spectrum almost a significant part of the flux can be used for imaging. Using (4) the total number of X-ray photons in one LEX pulse can be calculated:

$$N = \frac{Q}{e}\frac{\Sigma^s}{E_L}W, \tag{6}$$

where Q is the full charge in one bunch. According to the basic scheme (figure 3) one LEX pulse contains n_e X-ray pulses corresponding to n_e electronic bunches generated in one accelerator cycle with the duration not more than 1 ms (temporal resolution of the imaging setup). Formula (5) remains valid because all generated photons are used in the imaging. Finally, accounting for the photon-electron interaction cross-section area the total number of photons for one LEX pulse is:

$$N = \sigma_T \frac{Q}{es}\frac{\Sigma^s}{E_L}. \tag{7}$$

Therefore the LEX X-ray output is the product of the two factors: the first one is related to the electron beam, the second one to the number of laser photons accumulated in the optical storage device. Formula (7) can be used in the development of LEX for the laser and the accelerator optimization in order to achieve the exposure of $N \sim 10^{11} \div 10^{13}$ required for coronal angiography (see section II). From this point of view we analyzed the experimental setups build so far and current projects for the photon energies of 12-30 keV. Estimations were performed for the IR lasers with the pump wavelength 1.03 μm (E_L = 1.2 eV). The laser pulse energy accumulated in the photon storage device was taken as Σ^S = 10 mJ.. Two LEX systems with the close parameters (electron bunch charge, interaction cross-section area) are presented in table 1. demonstrating that the accelerators currently under construction provide output close to the required one for the millisecond angiography.

Table1. Total X-ray flux in 1 ms for the accelerators under development in [27,29].

Linear accelerator	s, μ^2	Q, μC	N	
			calculated see. (13)	required see section 2
[27]	3.6	0.01	$6 \cdot 10^{10}$	10^{11} - 10^{13}
[29]	100	10	$2 \cdot 10^{12}$	10^{11} - 10^{13}

4. Discussion and conclusions

Modern angiography instruments use X-ray tubes for illumination allowing to dynamically observe heart vessel tree condition with 0.01 second temporal resolution and 0.15-0.2 mm spatial resolution [15]. Based on the simplified model developed we demonstrated that LEX can outperform these parameters. In particular the LEX exposure time for one image does not exceed 1 μsec and the spatial resolution is much less than the diameter of the electron beam (10-100 μm). Another key feature of PLEX is the tunable spectrum of X-ray illumination which allows to benefit from another contrast media like gadolinium. Table 2 summarizes some of the main differences between X-ray tube and LEX regarding angiography applications.

Table 2. X-ray sources for coronal angiography.

	Source	
	X-ray tube	LEX
Directionality of illumination	-	10 mrad
Temporal resolution	10 ms	< 1 ms
Spatial resolution	0,15-0,20 mm	<0,1mm
Dynamic imaging	+	-
Safety	Low	High
Spectrum optimization for the contrast agent used	-	+

Acknowledgments

This work was supported by the program of fundamental investigations of the Russian Academy of Science Presidium "Fundamental and applied problems of photonics and physics of new optical materials".

This work was supported by Competitiveness Program of National Research Nuclear University MEPhI.

References

[1] Carroll, F.E., Waters, J.W., Price, R.R., Brau, C.A., Roos, C.F., Tolk, N.H., Pickens, D.R., Stephens W.H. 1990 Near-monochromatic X-ray beams produced by the free electron laser and Compton backscatter, Invest. Radiol. 25, 465.

[2] Carroll, F.E 2003 Tunable, monochromatic X-rays: An enabling technology for molecular/cellular imaging and therapy, Cell J. Biochem., 90, 502.

[3] Pogorelsky, I.V. 2016, Progress and prospects of a Compton X-ray source driven by a high-power CO2 laser, X-Ray Lasers 2014, Ed.: Jorge Rocca, Carmen Menoni, Mario Marconi, Springer, 133-138.

[4] The "Munich Compact Light Source", https://www.munich-photonics.de

[5] http://www.lynceantech.com/

[6] Albert, F., Anderson, S.G., Anderson, G.A., Betts, S.M., Gibson, D.J., Hagmann, C.A., Hall, J., Johnson, M.S., Messerly, M.J., Semenov, V.A., Shverdin, M.Y., Tremaine, A.M., Hartemann, F.V., Siders, C.W., McNabb, D.P., Barty, C.P.J. 2010 Isotope-specific detection of low-density materials with laser-based monoenergetic gamma-rays, Optics Letters, 35(3), 354-356.

[7] Jacquet, M. 2014 High intensity compact Compton X-ray sources: Challenges and potential of applications, NIM B, 331, 1-5.

[8] Bessonov, E.G., Vinogradov, A.V., Gorbunkov, M.V., Tur'yanski, A.G., Feshchenko, R.M., Shabalin, Yu.V. 2003 Laser electron-beam X-ray source for medical applications, Physics-Uspekhi, 46(8), 872.

[9] Bessonov, E.G., Gorbunkov, M.V., Kostryukov, P.V., Maslova, Yu.Ya., Tunkin, V.G., Postnov, A.A., Mikhailichenko, A.A., Shvedunov, V.I., Ishkhanov, B.S., Vinogradov, A.V. 2009 Design study of compact Thomson X-ray sources for material and life sciences applications, JINST, 4, P07017, 1-14.

[10] Mastouri, R., Sawada, S.G., Mahenthiran, J. 2010 Current Noninvasive Imaging Techniqes for Detection of Coronary Artery Disease,. Expert Rev. Cardiovasc Ther, 8(1), 77-91.

[11] Dix, W.-R., Kupper, W., Dill, T., Hamm, C.W., Job, H., Lohmann, M., Reime, B., Ventura, R. 2003 Comparison of intravenous coronary angiography using synchrotron radiation with selective coronary angiography, J. Synchrotron Rad., 10, 219–27.

[12] Sarnelli, A., Taibi, A., Tuffanelli, A., Baldazzi, G., Bollini, D., Cabal Rodriguez, A.E., Gombia, M., Prino, F., Ramello, L., Tomassi, E., Gambaccini M. 2004 K-edge digital subtraction imaging based on a dichromatic and compact x-ray source, Phys. Med. Biol., 49, 3291–3305.

[13] Bessonov, E.G., Gorbunkov, M.V., Kostryukov, P.V., Maslova, Yu.Ya., Tunkin, V. G., Postnov, A.A., Mikhailichenko, A.A., Shvedunov, V.I., Ishkhanov, B.S., Vinogradov, A.V. 2008 Design Study of Compact Thomson X-Ray Sources for Material and Life Sciences Applications, Springer Proceedings in Physics, 130, 521-535

[14] Achenbach, S. 2015 Can Coronary Computed Tomography Angiography Replace Invasive Angiography?, Circulation, 131, 410-417.

[15] Stefanini, G.G., Windecker, S., 2015 "Can Coronary Computed Tomography Angiography Replace Invasive Angiography? , Circulation, 131, 418-426,.

[16] Khovrin, V.V. 2016 "X-ray and MR tomography of aorta in diagnostics, planning and evaluation of surgical results" (in Russian), Doctoral Thesis, *National Center of Surgery named after academician B.V. Petrovsky, Russian Academy of Medical Sciences, Moscow, Russian Federation*

[17] http://physics.nist.gov/PhysRefData/Xcom

[18] Bryant, J.A., Drage, N.A., Richmond, S. 2012 CT number definition, Radiation Physics and Chemistry, 81, 358–361.

[19] Cademartiri, F.,de Monye, C., Pugliese, F., Mollet, N.R., Runza, G., van der Lugt, A., Midiri, M., de Feyter, P.J., Lagalla, R., Krestin, G. 2006 High Iodine Concentration Contrast Material for Noninvasive Multislice Computed Tomography Coronary Angiography: Iopromide 370 Versus Iomeprol 400, Invest Radiol., 41(3), 349-53.

[20] Komatsu, S., Kamata, T., Imai A., Ohara, T., Takewa, M., Miyaji, K., Yoshida, J., Kodama, K. 2013 Controlling intracoronary CT number for coronary CT angiography, Journal of Cardiology, 61, 155–161,.

[21] Komarova, M.A. 2016 Image quality and radiation exposure optimization in CT coronography (in Russian), PhD Thesis, Moscow State University, Faculty of Fundamental Medicine, Moscow, Russian Federation.
[22] Pupeza, I. 2012 Power Scaling of Enhancement Cavities for Nonlinear Optics, Springer Theses, Springer-Verlag New York,.
[23] Bessonov, E.G., Gorbunkov, M.V., Maslova, Y.Y., Kostryukov, P.V., Tunkin, V.G., Ishkhanov, B.S., Shvedunov, V.I., Vinogradov, A.V. 2007 Relativistic Thomson scattering in compact linacs and storage rings: a route to quasi-monochromatic tunable laboratory-scale X-ray sources, in Soft X-Ray Lasers and Applications VII; edited by G. J. Tallents and J. Dunn (SPIE, San Diego, CA, 2007), Proceedings SPIE 6702.
[24] Bessonov, E.G., Gorbunkov, M.V., Ishkhanov, B.S., Kostryukov, P.V., Maslova, Yu.Ya., Shvedunov, V.I., Tunkin, V.G., Vinogradov, A.V. 2008 Laser-electron generator for X-ray applications in science and technology, Laser and Particle Beams, 26, No. 3, 489-495,.
[25] Gorbunkov, M.V., Maslova, Yu.Ya., Vinogradov, A.V. 2009 Optical unit of Laser-Electron X-ray Generator designed for medical applications, NIM A 608, S32-S35,.
[26] Carstens, H., Lilienfein, N., Holzberger, S., Jocher, C., Eidam, T., Limpert, J., Tünnermann, A., Weitenberg, J., Yost, D.C., Alghamdi, A., Alahmed, Z., Azzeer, A., Apolonski, A., Fill, E., Krausz, F., Pupeza I. 2014 Megawatt-scale average-power ultrashort pulses in an enhancement cavity, Optics Letters, 39, 9, 2595-2598,.
[27] Graves, W.S., Bessuille, J., Brown, P., Carbajo, S., Dolgashev, V., Hong, K. H., Ihloff, E., Khaykovich, B., Lin, H., Murari, K., Nanni, E.A., Resta, G., Tantawi, S., Zapata, L.E., Kärtner, F.X., Moncton, D.E. 2014 Compact x-ray source based on burst-mode inverse Compton scattering at 100 kHz, Phys. Rev. ST Accel. Beams, 17, 120701,.
[28] Artyukov, I.A., Bessonov, E.G., Gorbunkov, M.V., Maslova, Yu.Ya., Popov, N.L., Vinogradov, A.V. 2016 Thomson linac-based X-ray generator: a primer for theory and design, Laser and Particle Beams, pp. 1–8
[29] Shimizu, H. et al, 2015 X-ray generation by inverse Compton scattering at the superconducting RF test facility, NIM A, 772, 26-33,.
[30] Meyer, M., Haubenreisser, H., Schoepf, U.J., Vliegenthart, R., Leidecker C., Allmendinger, T., Lehmann, R., Sudarski, S., Borggrefe, M., Schoenberg, S.O., Henzler, T. 2014 Closing in on the K Edge: Coronary CT Angiography at 100, 80, and 70 kV—Initial Comparison of a Second- versus a Third-Generation Dual-Source CT System, Radiology, 273, 373–382,.

The use of optical microscope equipped with multispectral detector to distinguish different types of acute lymphoblastic leukemia

A N Pronichev[1], E V Polyakov[1], N N Tupitsyn[2], M A Frenkel[2] and A V Mozhenkova[2]

[1]National Research Nuclear University MEPhI (Moscow Engineering Physics Institute), Kashirskoe shosse 31, 115409, Moscow, Russia.
[2]N.N. Blokhin Russian Cancer Research Center, Kashirskoe shosse 23, 115478, Moscow, Russia

E-mail: EVPolyakov@mephi.ru

Abstract. The article describes the use of a computer optical microscopy with multispectral camera to characterize the texture of blasts bone marrow of patients with different variants of acute lymphoblastic leukemia: B- and T- types. Specific characteristics of the chromatin of the nuclei of blasts for different types of acute lymphoblastic leukemia were obtained.

1. Introduction

Diagnosis of acute lymphoblastic leukemia (ALL) and their variants is based on morphological, cytochemical and immunophenotypic characterization of leukemic cells pool. In light microscopy an important role plays accurate assessment of some parameters of blasts: cell shape and size, shape of the nuclei, features of the chromatin structure, taking into consideration thin loop texture, lumpy texture, dense structure. Polymorphism of blasts associated with a variety of phenotypic, genetic and molecular biological characteristics. So far, it failed to show a significant correlation between the structure of blasts and immunophenotypical status. In conventional light microscopy heterogeneity of the blast population is ascertained in cases of T- and B- types of acute lymphoblastic leukemia [1-3]. Computer optical microscopy using multispectral camera has more features for the study of the structure of nuclear chromatin fibers and allows to objectify data in the form of a numeric indexes. Comparative study of blasts at different variants of acute lymphoblastic leukemia with the help of modern high technologies will expand our knowledge of the biology of leukemic cells and to identify possible patterns in the structure of the nuclear blasts differences in different types of ALL [4-10].

The aim of this work is to study features of nuclear chromatin structure of bone marrow blasts T- and B- types of acute lymphoblastic leukemia cases the using computer optical microscopy with multispectral camera and formalization of the resulting figures for comparative analysis.

2. Materials and methods

Computer microscopy was performed on stained bone marrow smears of 26 ALL patients, 11 of which have been set T-ALL, 15 of which have been set B-ALL. Morphocytochemical and immunophenotypic studies were conducted in the laboratory of immunology hematopoietic of N.N. Blokhin Russian Cancer Scientific Center (Head of Laboratory, Professor Dr. med Tupitsyn N.N.).

Images of cells were obtained in the RGB color model based on measurements of the function of brightness in three spectral ranges corresponding to red (R), green (G) and blue (B) regions of spectrum of visible electromagnetic radiation (automated microscope Olympus BX43 with a multispectral camera Imperx IPX-4M1ST-GCFB). Images were saved in BMP format, color-coded RGB24 (over 16 million colors).

4866 images of bone marrow lymphoblasts were received, 2164 of these were from patients with T-ALL, 2702 cells with B-ALL.

Investigation of the nuclei chromatin structure was performed with data transformation into the components of color models RGB, XYZ, HSL, Lab, Luv, LHC, HLS, HSV, YUV, YIQ, YCbCr, CMY[8]. Texture features were based on spatial adjacency matrices and matrices of the lengths of the series. These features were considered separately. Averages on group characteristics of blasts nuclei were calculated for T-ALL and B-ALL. The support vector machine and Bayes classifier were used with Euclidean, Manhattan and Chebyshev distance functions in two-dimensional feature space to separate the data with different acute lymphoblastic leukemia variants.

3. Results

Characteristics of the structure of blasts were calculated in a 360 features specific to each of the 26 cases. Among them a couple of features was identified that characterized all the events with minimal error. First feature P_{IPS} is the value of the texture characteristic "fraction of image in the series", it was calculated according to the matrix of the series in the HSV color space for color tone H. Second feature P_{OMS} is the value of the texture characteristic "reverse moment of series", it was calculated from the matrix of series length in LAB color space for the component B. The variability characteristics of the cells was estimated within each type. The statistical characteristics P_{IPS} and P_{OMS} for T- and B-cells (Table 1): $(\bar{x}, S, D_{\bar{x}}, Min, Max, t)$ were calculated. Here $\bar{x}$ – arithmetic mean of characteristic value, $D_{\bar{x}}$ – estimation of the mean arithmetic value variance of the characteristic, Min – minimum value of the characteristic, Max - maximum value of characteristic for the sample corresponding variant T-ALL cells and B-ALL cells, t – student's t-criterion to test the hypothesis of equality of sample averages.

Table 1. Statistical characteristics of features P_{IPS} and P_{OMS} for T- and B- blast cells

	number of cells	P_{IPS}						P_{OMS}					
		$\bar{x}$	S	$S_{\bar{x}}$	Min	Max	t	$\bar{x}*10$	S	$S_{\bar{x}}$	Min	Max	t
T-cells	2164	3,909	0,007	2,96	5,52	2164	3	6,165	0,008	0,48	0,73	6,165	15
B-cells	2702	3,880	0,006	2,90	6,52	2702		6,327	0,007	0,45	0,79	6,327	

Student's t-criterion was calculated according to the formula $t = \frac{\bar{x}_T - \bar{x}_B}{\sqrt{D_{\bar{x}_T} + D_{\bar{x}_B}}}$ to test the hypothesis about equality of means for samples of T-blasts and B-blasts from the experimental data. Here the indexes T and B define cell types. For a significance level of 0.05 t-test values greater than 1.96 indicate statistical significance of differences in mean values for the T- and B- blast cells.

Table 2. Generalized characteristics M_{IPS} andM_{OMS} for groupT- and B-cells patients with acute lymphoblastic leukemia.

The patient (type of blasts)	Quantity of cells	M_{IPS}	M_{OMS}	The patient (type of blasts)	Quantity of cells	M_{IPS}	M_{IPS}
P1(B)	222	4,16	0,643	P 14(B)	202	3,89	0,626
P 2(B)	181	3,93	0,626	P 15(B)	205	3,71	0,633
P 3(B)	202	3,68	0,648	P 16(T)	206	4,14	0,623
P 4(B)	122	3,76	0,642	P 17(T)	209	3,91	0,600
P 5(B)	210	3,91	0,643	P 18(T)	204	4,04	0,624
P 6(B)	204	3,83	0,671	P 19(T)	200	3,77	0,579
P 7(B)	226	3,90	0,642	P 20(T)	169	3,74	0,595
P 8(B)	193	4,19	0,582	P 21(T)	208	3,76	0,601
P 9(B)	125	3,85	0,625	P 22(T)	161	3,87	0,604
P 10(B)	200	3,48	0,629	P 23(T)	193	4,32	0,635
P 11(B)	125	3,73	0,628	P 24(T)	200	3,82	0,652
P 12(B)	204	3,86	0,640	P 25(T)	206	3,64	0,626
P 13(B)	81	4,16	0,646	P 26(T)	208	3,83	0,601

Thus on the average values of these features can be judged on ALL version (T or B-type). At the same time ranges of attribute values for the T blasts completely cover the range of values of B-blasts, that does not allow determine the type of individual cells for the values of the considered features.

In the next phase of the study generalized characteristics M_{IPS} and M_{OMS} were examined, the average values of characteristics P_{IPS} и P_{OMS} were defined for a group of cells of the same patient. The values for the generalized characteristics are presented in table 2.

A graphical representation of the results of table 2 is shown in Figure 1.

As you can see, most of the features of the B-blast cells are located in the area above the line segment with the coordinates M0(3,4;0,598) and M1(4,4;0,649), while T-blast cells are concentrated in areas below the specified line segment. The exceptions were in two cases of T-ALL for P24 - P25 patients and one case of B-ALL for patient P8. Consequently, in 23 (87%) of the 26 patients had a certain distribution of the data pattern. In three cases this association was absent. Therefore, the study of the texture of nuclei blasts allows in most cases to establish a definite link between research results and ALL variant. Thus, the data obtained with the use of optical microscopy with multispectral camera can be used for the determination of objectifying study of acute lymphoblastic leukemia variants.

The present observation is that the characteristic of blast chromatin texture is essential biological criterion that gives grounds to assume the relationship between the orientation of cell differentiation and the nuclei chromatin structure. Further refinement of research methods and the extension of the sample of patients will allow to characterize in depth this problem.

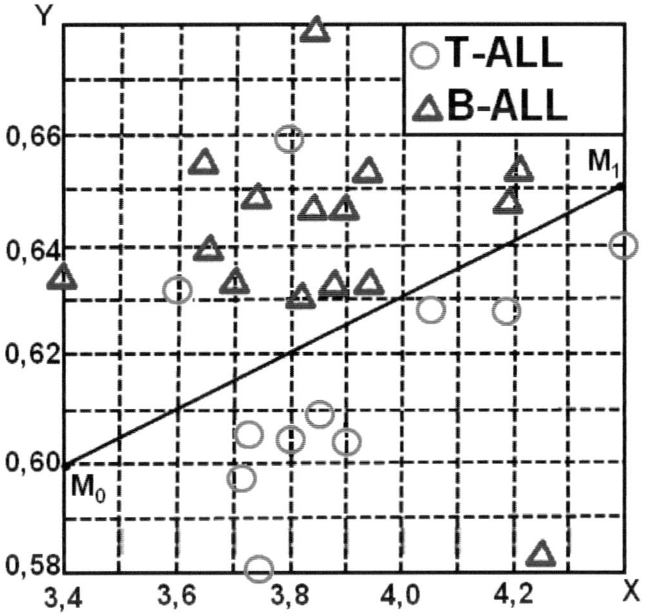

Figure 1. The distribution of the two classes (T-ALL and B-ALL) in two-dimensional feature space. Features are represented according to the coordinate axes: first feature plotted along the Y-axis $Y = M_{OMS}$, : second feature plotted along the X-axis $X= M_{IPS}$.

4. Conclusion

The study showed the possibility of using a computer optical microscopy with multispectral camera to establish T - or B - variants of acute lymphoblastic leukemia.

Planned step for further research is analysis of possibilities of application of computer optical microscopy with multispectral camera to define subtypes within T-ALL and B-ALL.

Acknowledgments

This work was supported by Competitiveness Program of National Research Nuclear University MEPhI.

References

[1] Frenkel M A 2014 (in Russia) Haematopoiesis immunology 12(1-2) 18
[2] Swerdlow S H *et al.* 2008 WHO classification of tumours of haematopoietic and lymphoid tissues, (IARC: Lyon)
[3] Nikitaev V G *et al.* 2015 Measurement Techniques, **57**(10) 1203
[4] Azam B *et al.* 2014 Research Journal of Recent Sciences **3**(4) 34
[5] Mohapatra S *et al.* 2012 ISRN Artificial Intelligence 2012 1
[6] Shivhare S and Shrivastava R 2012 International Journal of Scientific & Technology Research **1**(4) 125
[7] Nikitaev V G *et al. 2015* TRANSACTIONS on BIOLOGY and BIOMEDICINE 12 16
[8] Nikitaev V G *et al.* 2015 Proceedings of the 6th International Confererence on Bioscience and Bioinformatics (ICBB`15) Duba United Emirates February 22-24 p 120
[9] Nikitaev V G *et al. 2015* Proceedings of the 15th International Conference on Signal Processing, Computational Geometry and Artificial Vision (ISCGAV`15) Sliema Malta August 17-19 p 49
[10] Nikitaev V G *et al. 2016* INTERNATIONAL JOURNAL OF BIOLOGY AND BIOMEDICAL

ENGINEERING 10 2016 p 109

Cytology 3D structure formation based on optical microscopy images

A N Pronichev[1], E V Polyakov[1], I P Shabalova[2], T V Djangirova[2], S M Zaitsev[1]

[1]National Research Nuclear University MEPhI (Moscow Engineering Physics Institute), Kashirskoe shosse 31, 115409, Moscow, Russia
[2]Russian Medical Academy of Post-Graduate Education of Ministry of Health of Russian Federation, Barrikadnaya str., 2/1, 125993, Moscow, Russia

E-mail: EVPolyakov@mephi.ru

Abstract. The article is devoted to optimization of the parameters of imaging of biological preparations in optical microscopy using a multispectral camera in visible range of electromagnetic radiation. A model for the image forming of virtual preparations was proposed. The optimum number of layers was determined for the object scan in depth and holistic perception of its switching according to the results of the experiment.

1. Introduction

Cytological examination is a highly specialized, complete, world-renowned method of morphological verification of diagnosis and at the same time it is one of the least automated and standardized types of laboratory diagnostics. It is well known that in the cytological diagnosis as in any other form of laboratory research, dominated by the subjective factor - the cytologic conclusion essentially depends on the skill and experience of the doctor-cytologist. But even experienced professionals sometimes have objective difficulties: in rare or complex in diagnosis tumors and tumor-like lesions; tumors which require additional diagnostic methods to specify histological form; non-neoplastic lesions, morphologically similar to tumors; borderline states, etc. At the same time, cytologic diagnosis is often decisive in determining the nature of the pathological process and the selection of treatment, so its importance for the life and health of the patient can not be overestimated.

Various additional methods of improving the quality of cytological diagnosis are developed to overcome the objective and subjective difficulties encountered in cytology. The doctor-cytologist information technology are actively developed and are successfully putted into practice. The rapid development and distribution of computer information systems, as well as the appearance of scanning microscopes contributes to this [1].

Scanning microscopy solves many problems: raises to a new level capabilities of telepathology (consultations using the Internet), archiving of preparations, improves the quality of intralaboratory and external quality control, cytology training and self-learning [2].

Scanning microscopes allow to automatically input digital image of cytological preparation in the computer. Preparation is scanned at different magnifications, image areas are automatically stitched together, you can view all of the preparation on the screen in the usual form of conventional microscopy, moving in the right areas to a greater magnification [3].

Currently, an increasing number of laboratories apply the preparations made by the method of "liquid-based cytology". These preparations are characterized by a multi-layer structure. In such cases scanning microscopes can not always ensure the completeness of the information of the images if they form an image of cytological preparations using conventional single layer scanning method. In this regard, the development of scanning system of preparations prepared with "liquid-based cytology" is important. This system uses the method based on three-dimensional structure of the objects. [1].

Complexity of preparations prepared by the method of liquid-based cytology lies in the fact, that the investigated cytological structure located in different layers and the focus of the microscope at one of these objects, in other planes are somewhat out of focus (blurred).

In cytological diagnosis color of researched structures is important, so multispectral camera is commonly used for registering microscopic images of cytological preparations. That camera records the electromagnetic radiation in three ranges of the visible spectrum (550-950nm) (400-650nm), (350-550 nm). Accordingly, specified range of recorded signals are indicated by the symbols of RGB (R - Red, G - Green, B - blue). A modern computer monitors (with appropriate color calibration) provided acceptable reproduction color of objects in cytological preparations to diagnose. Determination of the necessary layers of scanning and the distance between the layers remains a problem area [4].

The purpose of this study was to determine the optimum number of scanning layers and the distance between them for preparations prepared by the method of liquid-based cytology, for example preparations of the cervix.

An image forming model on example of one spectral zone for a microscope object located in one layer of the preparation, can be represented by the following equation:

$$I(x,y)=\iint_{-\infty}^{+\infty} p(w,h)g(x-w,y-h)dwdh,$$

where $I(x,y)-$ illumination function on the photosensitive surface of the camera matrix (recording device), depending on the spatial coordinates (x, y), formed by the optical microscope system for the preparation installed on the objective table;

$g(x,y)-$ illumination function on the photosensitive surface of the camera matrix formed by the optical system of the "ideal" microscope for preparation mounted on the objective table (here an ideal microscope is the microscope, without distorting the formed image, and obtained image is an accurate likeness of the image of the preparation with an appropriate scaling factor - excluding the diffraction effects);

$p(w,h)-$ point spread function (PSF) in the plane of the formed image corresponding to the optical system of the "real" microscope, w и h – the spatial coordinates in the plane of the camera recorded image (for the "ideal" microscope $p(w,h)=\delta(w,h)$, where $\delta(w,h)-$ delta function).

For preparations in which objects can be in different layers in height, model given above is not acceptable, since different layers correspond to the different shape of the PSF (figure 1). The difference between the PSF for the different layers allows changing the focus to observe the microscopic image of the objects located in different layers in height and having the same coordinates (x,y), while for a fixed position of focusing can be observed only one of those objects.

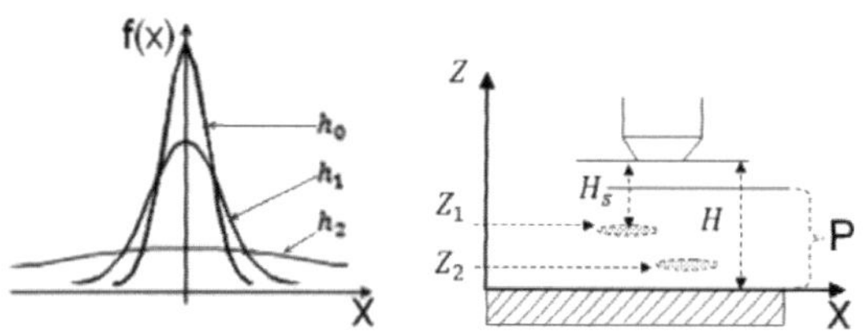

Figure 1. Point Spread Function for different layers of the preparation

In the case of an image forming in the preparation of multilayer objects we propose the following model:

$$\text{Im}\ (H,x,y)=\iiint_{-\infty}^{+\infty}p(Z-H+H_s,w,h)g(Z,x-w,y-h)dZdxdy.$$

Here function $g(Z,x,y)$ represents luminance distribution in the image plane (space coordinates (x,y)) for the "ideal" microscope due to the location of objects in the layer of preparation with the vertical coordinate Z. H - the distance between the lens and the surface of the microscope stage, H_s - the distance between the lens and the plane, objects in which form a sharp image. Here, the use of characters Z, H, Hs in the capital form reflects the fact, they are measured in the space of the preparation (in the thickness of the preparation), while the lower case characters x,y,w,h correspond to the spatial coordinates in the image plane formed by the microscope on the photosensitive matrix of electronic camera, recording this image. Layer with the coordinate Z = H $-H_s$ corresponds to layer, which forms in the "real" microscope maximally sharp image, the other layers are formed blurry images of objects in these layers. At the same time, the larger the difference Z from (H-H_s), the greater the degree of blur images of objects in the corresponding layer.

Thus, an infinite number of different images Im (H, x, y) can be obtained for preparations with a multilayer structure by changing the value of H. The researcher has interest for those images that contain sharp images of objects that represent multi-layered structure. With automatic scanning of the preparation is unknown in advance, which layers contain objects, so before you start scanning, you need to determine how many layers you want to scan and set the step size between the scanned layers so as to "not lose" important objects for the diagnostic conclusion. To solve this problem was an experimental study on real preparations from the cervix, prepared by the method of liquid-based cytology.

Determination of the number of layers to scan carried out on the basis of recording in each position (a step for moving the axis Z equal to 60 nm) for 20 objects in 5 preparations diagnostically significant objects. Then scanning step was selected so, that registration of the most important objects was provided and holistic perception of diagnostically significant objects was remained when viewing layers.

As a result of experiments for cervical preparations prepared by liquid-based cytology, the distance was determined as of not less than 0.6 micrometers, which provides capture of diagnostically significant objects between layers, and the distance between the objects during the experiment was 12 micrometers, which corresponds to 20 layers.

2. Conclusion

The study determined the distance between the layers of automatic microscopic scanning of preparations prepared by the method of liquid-based cytology, and containing multilayer objects (for example, cervical preparations). It was proposed analytical model describing the formation of a microscopic image when observing biological preparations with a multilayer structure in the visible range of the electromagnetic radiation. Method of experimental research scan modes has been described. In results of experiment the distance between the diagnostically relevant objects for automatic scanning cytology preparations prepared by the method of liquid-based cytology (the example of cervical preparations) was determined. Application of the results of the study will allow to form a virtual preparations for multilayer objects without losing the informative part of these objects. This results of work, described in the article, are important both in remote cytological diagnosis and in assessing the quality of cytology laboratories.

Acknowledgments
This work was supported by Competitiveness Program of National Research Nuclear University MEPhI.

References

[1] Djangirova T V *et al.* 2015 *Klinicheskaia laboratornaia diagnostika* **60**(8) 29
[2] Es Van S.L. *et al.* 2016 *Diagnostic pathology* **11**(1) 1
[3] Zang Yi *et al.* 2016 *6th International Congress on Image and Signal Processing (CISP)* (IEEE, Hangzhou, China) **2** p. 927
[4] Nozaka H., Miura T and Zheng Z 2013 *IEICE Transactions on Information and Systems* **96**(4) 856

Wireless system for monitoring Intra-abdominal pressure in patient with severe abdominal pathology

S S Sokolovskiy[1], Y V Shtotskiy[1] and A D Leljanov[1,2]

[1]National Research Nuclear University "MEPhI", Moscow, Russian Federation
[2] CJSC "SPE"Cyclone-Device", Fryazino, Moscow region, Russian Federation

E-mail: sokol-forward-9@mail.ru

Abstract. The paper discusses an experimental design of the wireless system for monitoring intra-abdominal pressure (IAP) using Bluetooth Low Energy technology. The possibility of measuring IAP via the bladder using a wireless pressure sensor with a hydrophobic bacteria filter between the liquid transmitting medium and the sensor element is grounded.

1. Introduction

The intra-abdominal pressure (IAP) measurement is fundamental for the evaluation of patients with severe abdominal surgical pathology at risk of intra-abdominal hypertension [1]. IAP is the pressure concealed within the abdominal cavity as defined superiorly by the diaphragm, anteriorly and posteriorly by the abdominal wall, and inferiorly by the pelvic floor. The World Society of the Abdominal Compartment Syndrome (WSACS, www.wsacs.org) has outlined definitions and recommendations relating to the measurement of IAP [2].

Special interest deserves continuous IAP monitoring allowing to improve assessment of the effects of medical management. A continuous trend of 24-h IAP recordings can obtained with a 3-way Foley catheter or a balloon-tipped catheter placed in the stomach [3]. During continuous IAP monitoring respiratory variations can be more easily identified which is very important for diagnostic purposes [5, 6]. Multiple assembled kits for IAP monitoring are commercially available. They are attached to the monitor unit by wires or directly [2]. Newer techniques including wireless transducers for IAP monitoring are being continuously developed [7, 8, 9]. At the same time this techniques, do not allow to use IAP monitoring for a long term, due to low battery capacity.

The goal of the project described in this paper was to design an experimental system for wireless IAP monitoring with the help of intravesical technique and to research dynamic characteristics of its catheter-sensor part.

The following problems were solved:

1) implementation of wireless transmission channel with high noise immunity and security;
2) elimination of sensor contamination;
3) registration of respiratory variations.

The design of the wireless IAP monitoring system, the theoretical model and identification results of its catheter-sensor part were described. The analysis of experimental data was presented.

2. The design of the wireless pressure monitoring system

The wireless pressure monitoring system consisted of wireless pressure sensor (WPS), USB Bluetooth modem and a computer with installed software. In figure 1 the system's block diagram is shown.

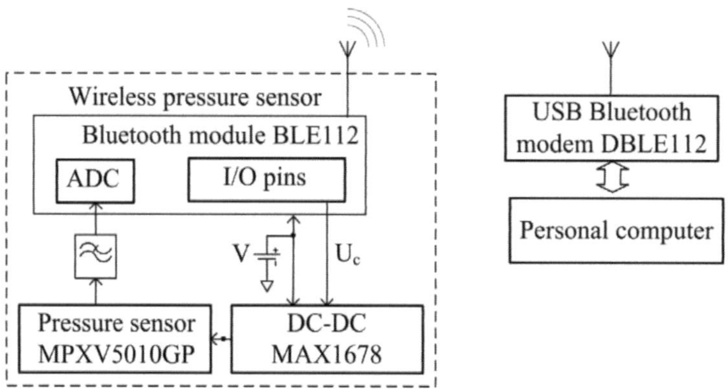

Figure 1. The block diagram of the wireless pressure monitoring system.

Wireless communication was achieved by utilizing the Bluetooth Low Energy wireless channel that is implemented on the basis of BLE112 module and BLED112 USB modem.

The MPXV5010GP transducer was used as differential pressure sensor in range 0 to 10 kPa. The WPS was powered with a pair of AAA batteries. The pressure sensor was powered with a DC-DC voltage converter made on the basis of MAX1678. Suppression of high frequency components in the sensor output signal was provided by an analog second order low-pass filter with bandwidth of 5 Hz, which is implemented using OPA376. The 12 bits analog-to-digital conversion, formation of data packets and transmission via wireless channel with a frequency of 1 Hz were made by BLE112 module. In addition the BLE112 module controlled the DC-DC converter by means of U_c signal in order to save energy.

Electronic components of WPS were placed in a small polymer casing as shown in figure 2. Measurement results were displayed on the computer using a pressure monitoring software.

Figure 2. Internal placement of the WPS components.

3. The mathematical model of the catheter-sensor system

An air barrier with a hydrophobic anti-bacterial filter was placed between the liquid pressure transmission medium and sensor to eliminate contamination of the pressure sensor. This design required optimization of dynamic characteristics of the catheter-sensor system.

The physical model of the system consisting of the sensor and the catheter can be represented as second order model with lumped parameters [10]. Figure 3 shows an electric analog of the simplified pressure-measuring system taking into account the air barrier. The substitutes for hydraulic inertance, resistance and compliance are electric inductance L_c, resistance R_c and capacitance C_c, respectively. The substitutes for the compliance of the sensor diaphragm and the compliance of the air barrier are electric capacitance C_d and capacitance C_b. It is assumed that the compliance of the sensor diaphragm is much larger than that of the liquid-filled catheter and the catheter material is noncompliant. The inertance L_b and the resistance R_b of the air barrier are neglected.

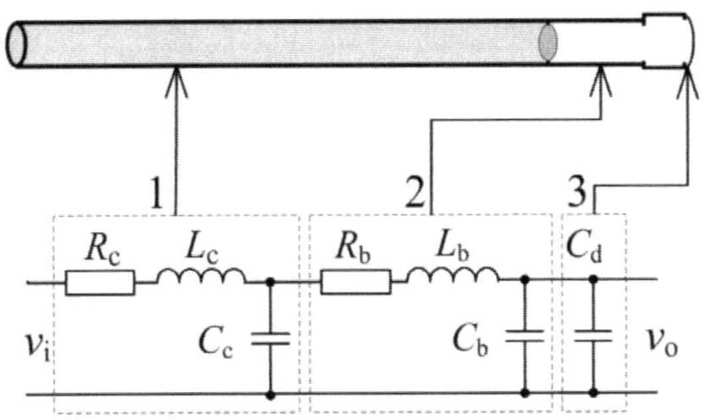

Figure 3. Electric analog of the simplified pressure-measuring system: 1 - partially filled fluid catheter, 2 - air barrier, 3 - pressure sensor membrane.

The relation between the input voltage v_i, analogous to applied pressure, and the output voltage v_0, analogous to pressure at the sensor membrane is determined by the following second order differential equation

$$v_i(t) = \frac{L_c(C_d + C_b)\,d^2 v_0(t)}{dt^2} + \frac{R_c(C_d + C_b)\,d v_0(t)}{dt} + v_0(t).$$

The liquid resistance in the catheter is due to viscous friction of its layers moving along the catheter at different speeds. Assuming that the motion of the liquid with viscosity η in the catheter with length L and radius r is laminar, it is possible, using the Poiseuille's equation, to determine the value of resistance R_c as

$$R_c = \frac{8\eta L}{\pi r^4}.$$

The inertance L_c of the liquid with density ρ in the catheter with radius r is due to the mass of the liquid. It can be estimated by the following equation

$$L_c = \frac{\rho L}{\pi r^2}.$$

The compliance C_d of the sensor diaphragm with the volume modulus of elasticity Ed is determined by the equation

$$C_d = \frac{1}{E_d}.$$

The compliance C_b caused by the isothermal compression of the air barrier is calculated by equation

$$C_b = \frac{\Delta V}{\Delta P}.$$

The values of the natural frequency ω_n and damping ratio ξ can be calculated as

$$\omega_n = \frac{1}{(L_c(C_d + C_b))^{1/2}},$$

$$\xi = \frac{R_c(C_d + C_b)^{1/2}}{2L_c^{1/2}}.$$

4. Experimental part

The goal of the experiment was to record transient responses of the catheter-sensor system using an experimental kit (figure 4) and a setup for the abdominal pressure simulation (figure 5).

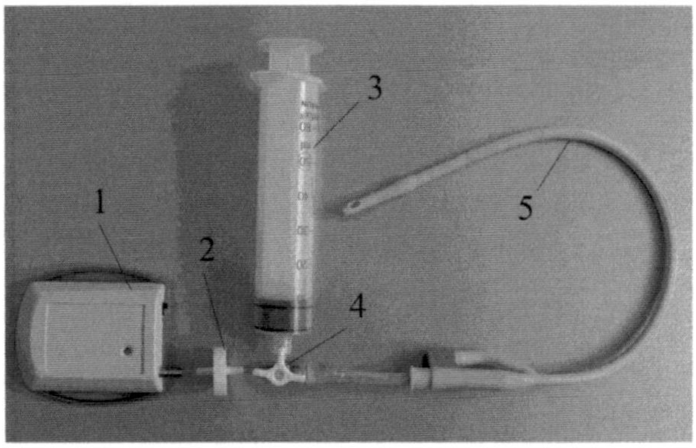

Figure 4. A kit for wireless IAP measurement: 1 – WPS; 2 – hydrophobic anti-bacterial air filter; 3 – syringe with 50 ml volume; 4 – three-way tap; 5 – Foley catheter.

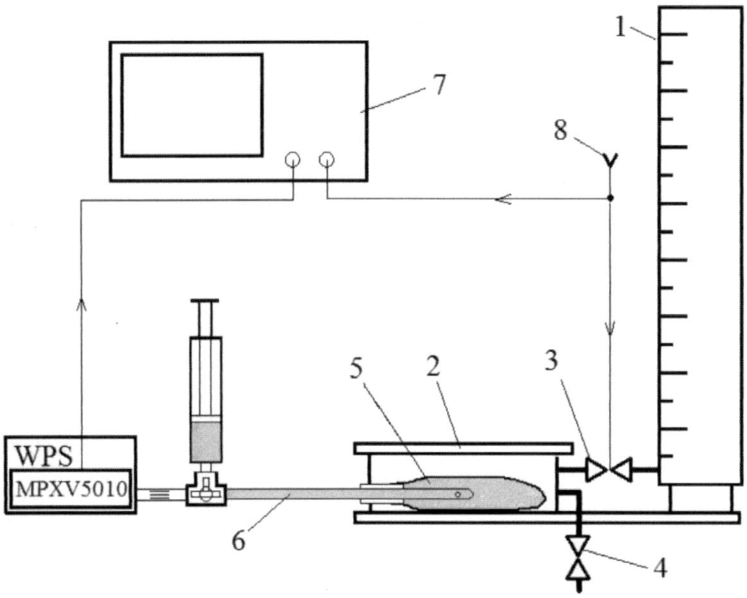

Figure 5. Block diagram of the setup for the abdominal pressure simulation: 1 – 3 l opened container; 2 – 1 l sealed container; 3 – solenoid valve; 4 – water tap; 6 – the kit for wireless IAP measurement; 7 – digital oscilloscope; 8 – synchronization input.

The abdominal model was implemented in the form of two containers which are connected to each other with a solenoid valve. A thin-walled silicone bag 200 ml volume modeling the bladder was placed at the bottom of the sealed container. The silicone bag was connected to the polymer adapter installed in its wall. The Foley catheter was placed in the silicone bag through the adapter.

Both containers were filled with 3 l of water through a water tap located at the bottom of the sealed container. The Foley catheter and the silicone bag were filled with 50 ml of water using a syringe. After filling the system the solenoid valve was closed and only then the water tap was closed.

The transition process was simulated by opening the solenoid valve. Transient responses were recorded from the output of the pressure sensor using a digital oscilloscope which is synchronized with the solenoid valve. The transitional characteristics, obtained in a series of experiments, were processed using the MATLAB System Identification Toolbox.

Assessment of stability of the wireless channel was made by recording cases of communication failures between WPS and the computer. The WPS was placed on the patient's thigh (without being connected to the Foley catheter). The patient was allowed arbitrary physical activity within a range of no more than 10 m from the computer.

5. Results and discussion

The time constant T_{MD}, the natural frequency f_{MD} and the damping ratio ξ_{MD} of the catheter-sensor system were calculated using the mathematical model and presuming the volume of the air barrier equal to 630 mm^3 (the air barrier of the kit).

Following results were obtained: T_{MD} = 0,0159 s; f_{MD} = 9,99 Hz; ξ_{MD} = 0,016.

Figure 6 shows the normalized experimental transient response of the catheter-sensor system and the result of the identification.

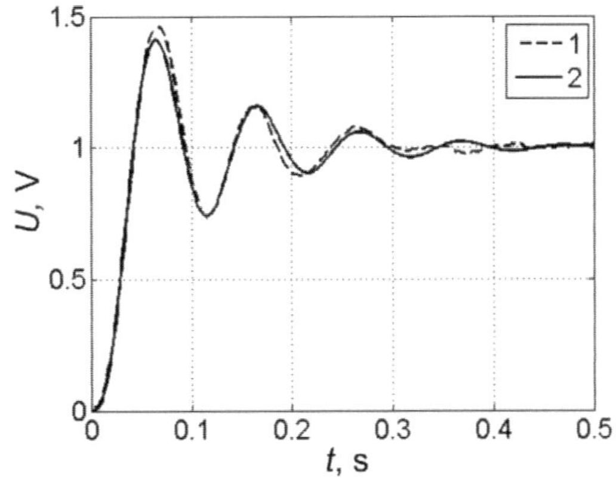

Figure 6. Normalized transient responses: 1 - the experimental catheter-sensor system; 2 - the result of the identification.

The structure and parameters of the model, produced by the result of the identification, were defined as

$$W(p) = \frac{1}{T_{ID}^2 p^2 + 2T_{ID}\xi_{ID} p + 1},$$

where T_{ID} is the time constant and ξ_{ID} is the damping ratio, which amounted to: T_{ID} = (0,0156 ± 0,0006) s; ξ_{ID} = 0,183 ± 0,005.

It was concluded that the structure of models and the values of their natural frequencies are in good agreement with each other. Damping ratios of models were different. It was due to the design features of the actual measuring system (the presence of antibacterial filter and a three-way tap).

The maximum value of the air barrier to register respiratory variations was determined on the basis of the mathematical model. This value was about 100 mm^3.

Oscillations in the natural frequency of the catheter-sensor system can serve as an indicator of the artifact caused by the patient's motor activity.

WPS demonstrated stable operation for a long term without loss of communication.

6. Conclusion

The results of this study demonstrate the ability of the wireless continuous IAP monitoring with the help of the intravesical technique and the Bluetooth Low Energy technology.

The possibility of applying a hydrophobic anti-bacterial filter, placed between the liquid pressure transmission medium and sensor to eliminate the contamination of the pressure sensor, was approved. Maximum air barrier volume of the measuring path was defined, to ensure the registration of respiratory variations. Mathematical and identificational catheter-sensor models demonstrated good coherence.

Wireless technology allows free movement within the laboratory setting which simplifies experimental research activity. The developed wireless system for IAP monitoring will be used to create a non-invasive method for estimating IAP variations.

Acknowledgments
This work was supported by Competitiveness Program of National Research Nuclear University MEPhI.

References

[1] Cheatham ML, Safcsak K 2011 *Intra-abdominal hypertension and abdominal compartment syndrome: the journey forward.* Am Surg 77: pp 1−5

[2] Kirkpatrick AW, Roberts DJ, De Waele J et al 2013 *Intra-abdominal hypertension and the abdominal compartment syndrome: updated consensus definitions and clinical practice guidelines from the World Society of the Abdominal Compartment Syndrome*. Intensive Care Med 39: pp 1190−1206. doi: 10.1007/s00134-013-2906-z

[3] Wauters J, Spincemaille L, Dieudonne AS, Van Zwam K, Wilmer A, Malbrain ML 2012 *A Novel Method (CiMON) for continuous Intra-abdominal pressure monitoring: pilot test in a pig model.* Crit Care Res Pract 181563. doi: 10.1155/2012/181563

[4] Kron IL 1989 *A simple technique to accurately determine Intra-abdominal pressure.* Crit Care Med 17: pp 714−715

[5] Sturini E, Saporito A, Sugrue M, Parr MJ, Bishop G, Braschi A 2008 *Respiratory variation of Intra-abdominal pressure: indirect indicator of abdominal compliance?* Intensive Care Med 34: pp 1632−1637

[6] Ahmadi-Noorbakhsh S, Malbrain ML 2012 *Integration of inspiratory and expiratory Intra-abdominal pressure: a novel concept looking at mean Intra-abdominal pressure.* Ann Intensive Care 2 (Suppl 1): p 18

[7] Coleman TJ, Thomsen JC, Maass SD, Hsu Y, Nygaard IE, Hitchcock RW 2012 *Development of a wireless intra-vaginal transducer for monitoring Intra-abdominal pressure in women.* Biomed Microdevices 14: pp 347−355

[8] Nordin M, Elfstrom G, Dahlquist P 1984 I*ntra-abdominal pressure measurements using a wireless radio pressure pill and two wire connected pressure transducers: a comparison.* Scand J Rehabil Med 16: pp 139−146

[9] Rauch S, Johannes A, Zollhofer B, Muellenbach RM 2012 *Evaluating Intra-abdominal pressures in a porcine model of acute lung injury by using a wireless motility capsule.* Med Sci Monit 18: pp 163−166

[10] Webster, John G 2009 Medical Instrumentation Application and Design 4. Edition. p 720

Synthesis of polystyrene core/SiO2 shell composite particles and fabrication of SiO2 capsules out of them

V Terekhin [1,2]

[1] Laboratory of Nano-Bioengineering, National Research Nuclear University MEPhI (Moscow Engineering Physics Institute), Moscow, Russian Federation
[2] Frumkin Institute of Physical Chemistry and Electrochemistry, Russian Academy of Sciences, bld 4, 31 Leninsky prospekt, 119071 Moscow, Russian Federation

E-mail: mendeleeveckm@mail.ru

Abstract. Systemic studies on the dependence of the morphology of polystyrene core/SiO_2 shell composite particles on the conditions of their fabrication have been performed. Spherical polystyrene particles synthesized in the presence of a cationic initiator of polymerization were used as cores. SiO_2 shells were formed by modified Stober's method. Exposure of the synthesized composite particles to high temperatures has been shown to cause destruction of their polystyrene core, thereby allowing the formation of mesoporous SiO_2 capsules with a mean pore diameter of ~3 nm and specific surface area of ~270 m^2/g. Model experiments on loading the SiO_2 capsules with amoxicillin have been carried out. Spectrophotometry in the UV and visible spectral regions has been used to estimate the kinetics of amoxicillin release from the SiO_2 capsules.

1. Introduction

Hollow inorganic microbeads are used in various fields of science and technology, e.g., as containers for drugs [1–4], absorbents [5], and catalyzers [6]. An effective method for obtaining these capsules is removal of the cores from composite particles (CPs) of the organic core/inorganic shell type, polystyrene core/SiO_2 shell CPs being the most commonly used. In this case, the morphology of the resultant capsules is determined by the CP structure.

The methods for obtaining these CPs may be divided into two main groups: one-stage and two-stage. In the methods of the former group, CPs are synthesized without separation of the target product (the CPs) from the template (polymer particles). In general, polymerization reactions involving different type of initiators (neutral [7], cationic [7], or anionic [8]) are used to synthesize polystyrene (PS) particles and simultaneously form SiO_2 shells by adsorbing preliminarily synthesized SiO_2 nanoparticles (NPs) onto the PS particle surface [9, 10]. However, the resultant CPs are usually polydisperse with respect to size and have a heterogeneous surface.

Two-stage methods of synthesis are used more frequently than one-stage ones. These methods consist in successive obtaining PS particles and formation of the SiO_2 shell (and subsequent fabrication of SiO_2 capsules out of these CPs). Two-stage methods are advantageous in that they yield monodisperse CPs with controllable morphology. For example, in [11], CPs were formed by successive deposition of alternately charged polyelectrolyte and SiO_2 NP layers onto PS particles.

Stober's method [12] is one of the most common ones used to form SiO_2 shells on PS particles. This method, based on the reaction of tetraethyl orthosilicate (TEOS) hydrolysis, is reasonably simple and yields SiO_2 shells of uniform thickness [13]. In Ref. 13, it was used to form a SiO_2 layer on PS particles whose surface carried NH_2 groups. However, obtaining the particles with surface amino groups is a rather sophisticated procedure.

It should be noted that, despite common interest in obtaining PS core/SiO_2 shell CPs (and SiO_2 capsules out of these particles), most researchers confine themselves to studying the effect of one or two parameters of synthesis on the particle morphology. For example, the effects of the dichloromethane concentration and the PS particle to TEOS ratio on the structure of the SiO_2 shells formed [14] and the effects of the polarity of the medium and the TEOS concentration (C(TEOS)) on the structure of the SiO_2 layer on the PS particles [15] have been estimated. In another study, the type of precursor in the synthesis of PS core/SiO_2 shell CPs was varied (either TEOS or a mixture of TEOS with vinyl triethoxysilane was used) [16]. In one of the most comprehensive studies [17], the effects of three parameters of the process, namely, C(TEOS), pH of the reaction medium, and reaction temperature, on the structure of the SiO_2 shells formed on PS particles have been used. Note that the reaction was carried out at a low pH (~2.5), relatively high temperature (60–70°C), and prolonged reaction time (5 h).

Polyvinylpyrrolidone (PVP) is rather frequently used for forming pores in the SiO_2 layer on PS particles [18]. For example, it has been shown that SiO_2 shells with controllable porosity and specific surface can be obtained by varying the PVP concentration C(PVP) in the solution [19, 20]. The use of PVP also allows PS particles of controllable size to be made [21].

Here, we propose a method for forming homogeneous SiO_2 shells over preliminarily PVP-modified positively charged PS particles synthesized in the presence of a cationic polymerization initiator. These approaches to obtaining PS core/SiO_2 shell CPs have been used separately in different studies, but we suggest that they be combined. We also plan integrated systemic studies on the dependence of the morphology of SiO_2 shells formed on PS particles on the synthesis conditions (the time of synthesis of SiO_2 shells, pH of the reaction medium, and the concentrations of the precursor and pore-forming agent). The CPs can be used to obtain hollow SiO_2 capsules and carry out model experiments on loading them with a drug (amoxicillin) and subsequent release of the drug from the capsules incubated in physiological saline.

2. Materials and methods

2.1 Materials

We used styrene monomer (99.5%), 2,2'-azobis(2-methylpropionamidine) dihydrochloride (ABMPAD) (97%), PVPs with molecular weights (MWs) of 10, 30, 55, and 340 kDa, TEOS (98%), aqueous solution of ammonium hydroxide (30 wt %), and amoxicillin from Sigma-Aldrich (United States). Simulated body fluid (SBF) was prepared according to Ref. 22. The salts for preparing SBF were high-grade (Khimreaktiv, Russia).

2.2 Synthesis of PS Beads

PS beads with a positively charged surface were synthesized by the surfactant-free emulsion polymerization method. ABMPAD served as a cationic initiator of the reaction. The necessary amount of styrene (12.5 g) was added to 250 g of deionized water (resistivity, 18.2 MΩ·cm; Millipore Direct-Q, France), after which oxygen was removed from the reaction mixture by bubbling gaseous nitrogen for 30 min. The reaction mixture was heated to 65°C, and the necessary amount of ABMPAD (0.35 g) was added. The reaction was performed at a temperature of 65°C for 24 h with permanent nitrogen bubbling. The reaction products were sedimented by centrifugation at 8000 rpm for 10 min and repeatedly washed with methanol and deionized water. The precipitate was air-dried.

2.3 Synthesis of PS Core/SiO_2 Shell CPs

100 mg of PS was dispersed in 200 ml of a water–ethanol mixture (volume ratio, 35:165). PVP was gradually added to the dispersion while stirring to a final concentration of 3 to 10 g/l. The pH of the mixture was controlled by adding ammonium hydroxide. Then, the necessary amount of TEOS was added to the reaction mixture while intensely stirring. The resultant mixture was heated to 50°C and stirred for 6 h. The conditions of the synthesis of different samples are shown in the Table 1. After the reaction was completed, CPs were washed from excess PVP and TEOS in four cycles of sedimentation by centrifugation and resuspension in water. The precipitate obtained at the final stage was dried at room temperature.

Table 1. Conditions of the synthesis of PS Core/SiO2 Shell composite particles.

Sample	TEOS concentration in ethanol–water mixture, M	Reaction time, h	pH
1	0.2	3	10.0
2	0.2	3	11.0
3	0.2	3	12.5
4	0.2	1	11.0
5	0.2	6	11.0
6	0.05	3	11.0
7	0.1	3	11.0
8	0.4	3	11.0

2.4 Formation of SiO_2 Capsules

The dried CPs were exposed to a temperature of 400°C in the air for 6 h.In estimating the effect of the thermal treatment on the specific surface area of SiO_2 capsules and pore size, the temperature was varied from 200 to 1200°C.

2.5 Loading of the SiO_2 Capsules with Amoxicillin

The necessary amount of preliminarily dried SiO_2 capsules was dispersed in 50 ml of a 20 mg/l aqueous solution of amoxicillin (with the resultant amoxicillin to SiO_2 weight ratio of 2:1). The dispersion was intensely stirred for 48 h at room temperature. Then, the SiO_2 capsules were sedimented by centrifugation (12,000 rpm for 15 min), washed with acetone three times as described in Ref. 23, and vacuum-dried at 35°C. The amount of the drug loaded into the capsules was monitored by measuring the light absorbance of the supernatant at a wavelength of 245 nm using a Cary 50 spectrophotometer (Varian, United States).

Amoxicillin release was induced by dispersing the loaded SiO_2 capsules in physiological saline and stirring the dispersion at 37°C for 20 h. At predetermined intervals, the amoxicillin concentration was estimated by measuring the light absorbance of the supernatant obtained after centrifugation of the dispersion (12,000 rpm for 15 min) at 245 nm.

2.6 Instrumental techniques

The morphology of the particles was studied using the transmission (TEM) and scanning (SEM) electron microscopy methods. The TEM measurements of CPs were performed using a Technai G2 20 FEG microscope (FEI, Netherlnds). Before the measurements, preliminarily dried CPs were thermally treated (400°C for 6 h). The residue was redispersed in 500 μl of ethanol. A drop of the dispersion was applied onto the surface of carbon-coated 300-mesh copper grids for TEM 3 mm in diameter.

SEM studies were performed using a Quanta 200 microscope (FEI, Netherlands). A drop of the dispersion was applied onto a silicon substrate. After evaporation of the dispersion medium, a layer of gold was applied onto the sample surface by means of magnetron spluttering.

In thermogravimetric analysis (TGA) (TGA/STDA 851, Mettler-Toledo, United States), approximately 10-mg samples were used. The heating mode was 20°C/min in a nitrogen atmosphere (flow rate, 50 ml/min).

The specific surface area (S) and volumes of pores of CPs and SiO_2 capsules were determined by the method of low-temperature nitrogen adsorption (ASAP-2010N, Micromeritics, United States). The S values were calculated by the Barrett–Joyner–Halenda (BJH) method [24]. The S and size distribution of the pores were calculated using the desorption arm of the adsorption–desorption isotherm.

The ξ-potential of the particles and their size (determined by the dynamic light scattering (DLS) method) were measured by means of a ZetaSizer Nano device (Malvern, United States).

For measurements by the method of Fourier transform infrared spectrometry (MAGNA 550, Nicolet, United States), the necessary amount of dried particles was mixed with KBr powder, and the mixture was pressed into a tablet.

3. Results and discussion

Figure 1a shows a microphotograph of the PS particles synthesized. The particles are spherical, with a smooth surface, 120 ± 10 nm in diameter. A cationic initiator was used in the styrene polymerization reaction. As a result, the surface of the PS particles was positively charged (at pH 7.0), with a ξ-potential of about +74 mV.

At the next stage, SiO_2 shells were formed on the PS particle surface by Stober's method. We found that, at pH << 11.0 (at C(TEOS) = 0.2 M and reaction time of 3 h), the TEOS hydrolysis and condensation reaction rates were too low to form a SiO_2 layer on the particle surface (Figure 1b). According to TGA data, the weight of the solid residue of the CPs synthesized under these conditions was as small as 0.55% of the initial amount. An increase in the pH of the reaction mixture to 11.0 led to an increase in the TEOS hydrolysis rate and formation of a continuous, smooth SiO_2 shell ~9 nm in thickness (Figure 1c). Further increase in pH (to 12.5) caused not only an increase in the SiO_2 shell thickness (to ~13 nm), but also an undesirable increase in its roughness and heterogeneity compared to the shell obtained at pH 11.0 (Figure 1d). Apparently, the nucleation of SiO_2 NPs occurred not only on the surface of the PS particles and the SiO_2 layer formed at early reaction stages, but also in the bulk reaction medium. We believe that this is related to, first, an increase in the number of nucleation centers of SiO_2 particles (due to a high TEOS hydrolysis rate) and, second, adsorption of SiO_2 NPs formed in the reaction medium with pH 12.5 onto the positively charged surface of PS particles, followed by their incorporation into the SiO_2 shell during its formation, which decreases the shell heterogeneity. Subsequent experiments were performed in the reaction mixture with pH 11.0.

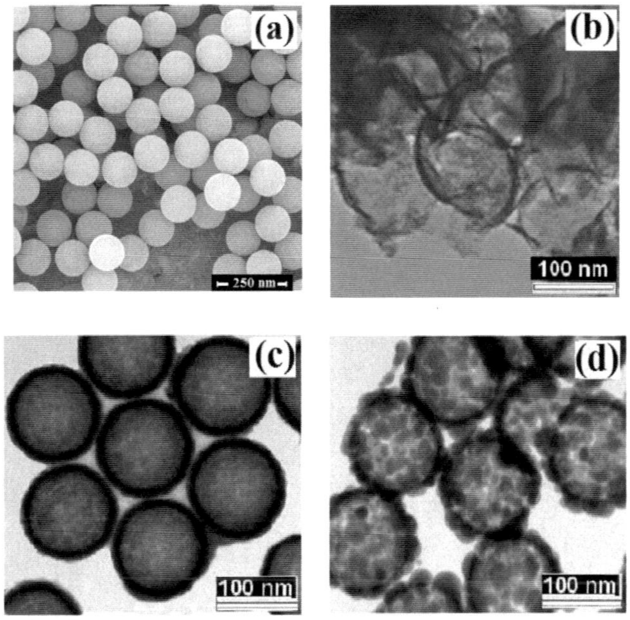

Figure 1. (a) SEM image of the as-synthesized PS particles and TEM photographs of SiO_2 shells formed on the PS particle surface at pH values of (b) 10.0, (c) 11.0, and (d) 12.5 after thermal treatment of the composite particles. In all cases, the reaction time was 3 h and the TEOS concentration was 0.2 M.

Note that, to perform TEM measurements, the CPs synthesized were dried and exposed to a temperature of 400°C in the air for 6 h. This resulted in the destruction and elimination of the PS core, as confirmed by IR spectrometry (Figure 2). The spectrum of the original PS particles (after the addition of PVP before the SiO_2 shell formation) contained absorption bands at 3100–2900 (aliphatic C–H bonds), 1600–1500 (aromatic C=C bonds), and 800–700 cm^{-1} (aromatic C–H bonds). After the formation of the SiO_2 shells, the intensities of these bands were considerably decreased. In addition, an intense band with a peak at 1090 cm^{-1} corresponding to Si–O–Si bonds appeared in the spectrum, which confirmed the formation of a SiO_2 layer on the surface of PS particles. After thermal treatment of the CPs, the absorption bands characteristic of PS disappeared, which indicated its complete elimination. The elimination of the PS core allowed the contrast of TEM images to be considerably increased and the analysis of the SiO_2 shell morphology to be improved. In addition, bands characteristic of the C=O and C–N bonds (1660–1680 cm^{-1}) can be seen in Figures 2a and 2b, which is explained by the presence of PVP molecules on the PS particle surface and in the CPs. After thermal treatment, these bands disappeared, which indicated thermal destruction of PVP.

We established that the thickness of the synthesized SiO_2 shells increased with increasing TEOS hydrolysis reaction time (at pH 11.0 and C(TEOS) = 0.2 M). Non-continuous SiO_2 shells were formed on PS particles within 1 h; continuous shells ~9 nm in thickness, within 3 h, and 14-nm shells, within 6 h. Further increase in the reaction time did not lead to thickening of the SiO_2 shells. Note that the thickness of the synthesized SiO_2 shells was smaller than the calculated value. This was related to the formation of small SiO_2 NPs in the reaction mixture in the course of shell growth. However, subsequent washing entirely eliminated these NPs, which was confirmed by SEM and TEM data.

The results of TGA showed that the weight content (ω) of SiO_2 in CPs increased with increasing reaction time. The ω(SiO_2) was approximately 2, 11, and 27% after 1, 3, and 6 h of the reaction, respectively. These data reasonably agree with the thicknesses of SiO_2 shells indicated above.

We also found that the SiO_2 content of the CPs increased with increasing C(TEOS) (Figure 3a). The TGA curves for CPs synthesized at different C(TEOS) values show a drastic decrease in the sample weight at a temperature of ~420°C and a plateau at temperatures higher than 500°C.

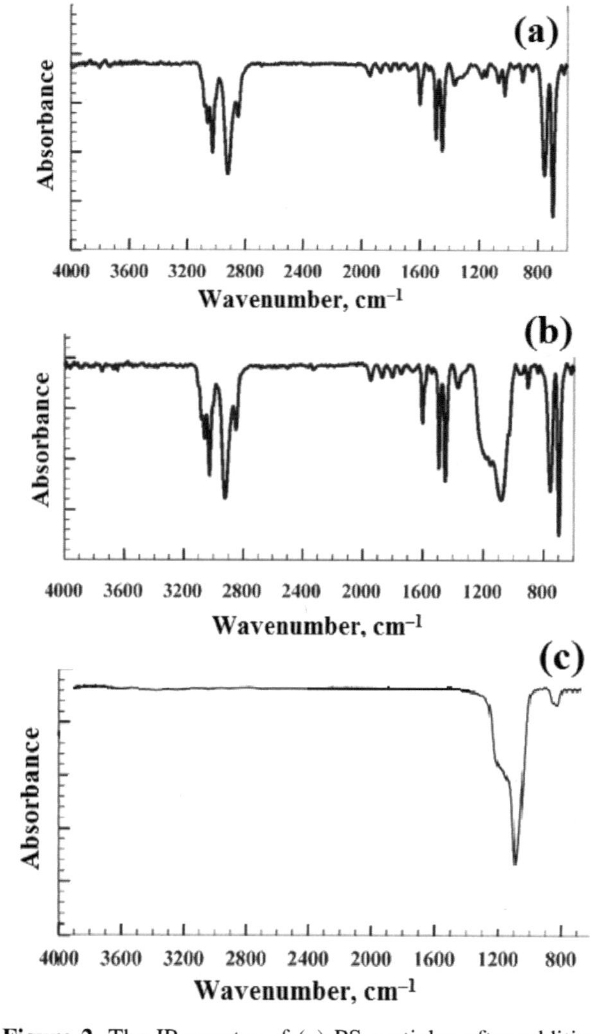

Figure 2. The IR spectra of (a) PS particles after addition of PVP to the reaction mixture and (b, c) PS core/SiO_2 shell composite particles (b) before and (c) after thermal treatment.

As seen from TEM images (Figures 3b-3d), continuous, almost defect-free SiO_2 shells from ~4 to 17 nm in thickness formed on PS particles at C(TEOS) from 0.05 to 0.4 M, respectively (reaction time, 3 h; pH 11.0). At the C(TEOS) of 0.05 and 0.1 M, the SiO_2 layers had about the same thickness of 4–5 nm, whereas an increase in C(TEOS) to 0.2 and 0.4 M led to their thickening to ~9 and 17 nm, respectively. Note that it was earlier shown [21] that the diameter of SiO_2 NPs increased with increasing C(TEOS) in the reaction mixture. This agrees with the dependence of the SiO_2 shell thickness on C(TEOS) observed in our study. Therefore, an increase in the SiO_2 content of the samples was entirely accounted for by an increase in the SiO_2 shell thickness. Figure 4a shows the dependence of the S of CPs exposed to temperatures from 200 to 1200°C calculated using the BJH model, as well as the pore size distribution (at a thickness of the SiO_2 shell of ~17 nm). All the nitrogen adsorption–desorption isotherms belong to type IV, which is characteristic of mesoporous materials (see the inset in Figure 4b). The mean size of the pores was ~3 nm. Note that the pore size distribution curve of none of the samples has a second maximum in the region of larger sizes, which indicates the absence of pores larger than 3–4 nm.

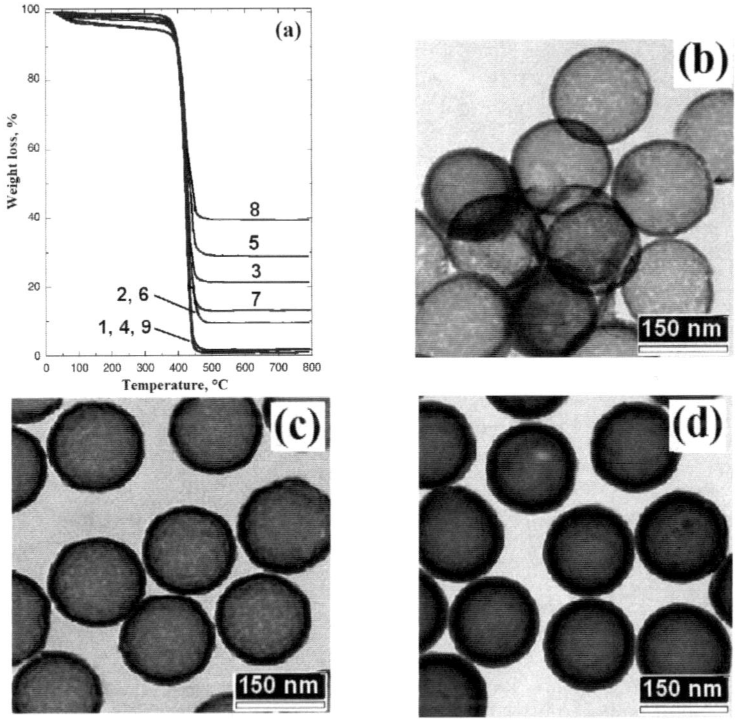

Figure 3. (a) TGA curves of PS core/SiO_2 shell composite particles (the numbers at the curves correspond to the ordinal numbers of the samples in the table; see the "Materials and methods") and original PS particles. (b–d) TEM images of SiO_2 shells obtained after thermal treatment of the composite particles synthesized at different TEOS concentrations: (b) 0.05 M, (c) 0.2 M, and (d) 0.4 M.

Obviously, the pores were formed by PVP molecules adsorbed onto the PS particles and incorporated in the SiO_2 shells. PVP added to the reaction medium was adsorbed onto the positively charged surface of the particles, a certain amount of PVP remaining in the solution. During the formation of a thin SiO_2 layer (at short TEOS hydrolysis reaction times), PVP molecules were adsorbed on this layer due to hydrogen bonding between the carbonyl groups of PVP and OH groups formed on the SiO_2 surface [25]. Further growth of the shell (with increasing reaction time) was accompanied by incorporation of more PVP (apart from that adsorbed on the PS particles) into the SiO_2 shells.

TGA showed that the temperature of the thermal destruction of PVP, as well as that of PS, was ~400°C. The PVP elimination upon thermal treatment at 400°C resulted in the formation of pores. The C(PVP) in the reaction mixture was 4 mg/ml and its MW was 55 kDa in all samples.

With increasing temperature of the thermal treatment from 200 to 400°C, the S of the CPs increased from 15 to 270 m^2/g. The reason was that exposure to a temperature of 200°C did not cause destruction of the PS core or the PVP molecules incorporated in the SiO_2 shell, whereas heating to 400°C resulted in their complete elimination and formation of pores and a cavity within the SiO_2 shell, thereby increasing S. Exposure to temperatures of 600°C and higher caused a considerable decrease in S because of sintering of the pores.

An increase in the concentration of PVP (MW = 55 kDa) introduced into the system before the formation of SiO_2 shells from 1 to 7 mg/ml led to an increase in S from ~40 to 270 m^2/g (Figure 4c). At C(PVP) = 3 mg/ml, it reached the maximum value and did not change with further increase in C(PVP) to 7 mg/ml. This indicated that the surface of the PS particle was completely saturated with PVP at C(PVP) = 3 mg/ml. The pore diameter remained unchanged (~3 nm) in the entire C(PVP) range studied.

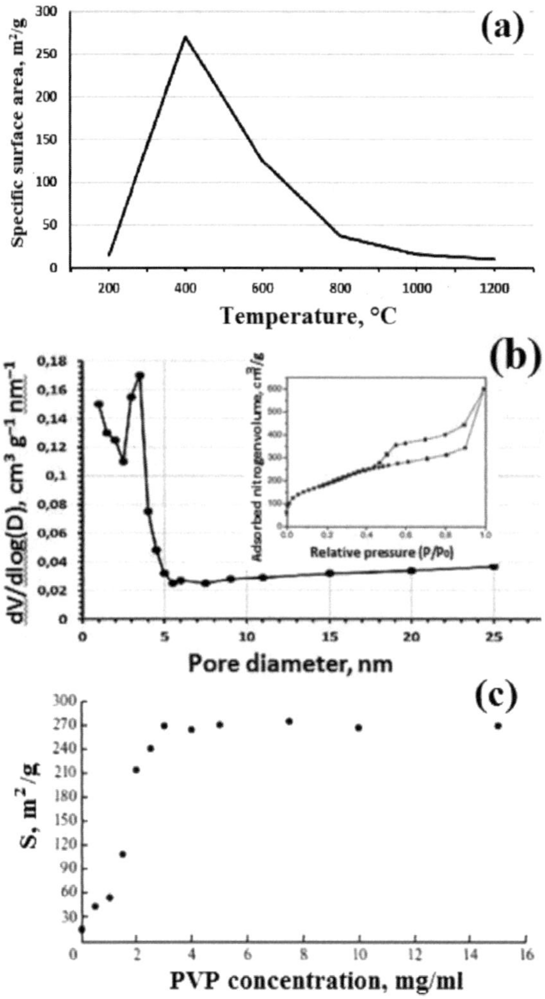

Figure 4. (a) Dependence of the specific surface area (S) on the temperature of treatment of PS core/SiO_2 shell composite particles. (b) Size distribution of the pores in the SiO_2 capsules obtained by treatment of the composite particles at 400°C. The inset shows a typical adsorption–desorption isotherm of a sample of these SiO_2 capsules. (c) Dependence of the S value of the composite particles exposed to a temperature of 400°C on the PVP concentration in the reaction mixture.

Porometry data showed that an increase in the PVP MW from 10 to 340 kDa only slightly affected the pore diameter (Figure 5a). For example, the mean pore diameter was ~3 nm at PVP MW = 10 kDa and ~4 nm at PVP MW = 340 kDa. In all cases, C(PVP) was 7 mg/ml; i.e., it was considerably higher than the concentration at the saturation of the PS surface with PVP molecules. These data agree with the results of other studies [20, 26]; however, the authors of those studies do not suggest any explanation to this effect.

It was demonstrated that the amounts of the polymer adsorbed onto TiO_2 [27] and PS [28] particles are independent of the PVP MW (in the range from 10 to 360 kDa). Photon correlation spectroscopy has shown that the thickness of the adsorbed PVP layer is considerably smaller than the diameter of a globule of the polymer with the corresponding MW [27]. This has led to the conclusion that the most part of each PVP polymeric chain (irrespective of the MW) is adsorbed on the PS surface [27]. The other part of the molecule, which does not interact with the PS particle, is oriented perpendicular to the surface. Apparently, this latter part of the molecule is involved in the formation of pores in the SiO_2 layer. This explains the independence of the diameter of pores formed in the SiO_2 layer on the PVP MW observed in our study. It should be noted that additional studies are necessary to reliably determine the cause of this nontrivial phenomenon.

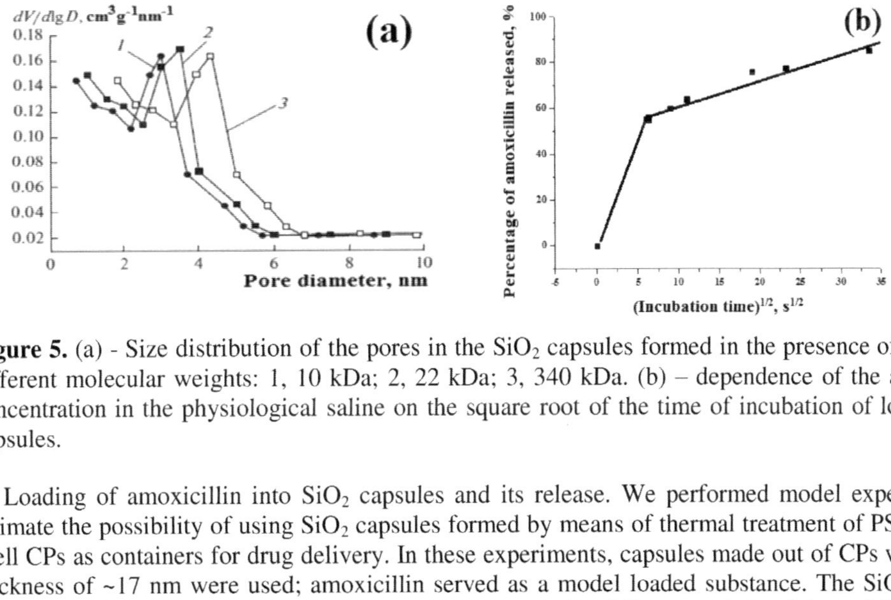

Figure 5. (a) - Size distribution of the pores in the SiO_2 capsules formed in the presence of PVP with different molecular weights: 1, 10 kDa; 2, 22 kDa; 3, 340 kDa. (b) – dependence of the amoxicillin concentration in the physiological saline on the square root of the time of incubation of loaded SiO_2 capsules.

Loading of amoxicillin into SiO_2 capsules and its release. We performed model experiments to estimate the possibility of using SiO_2 capsules formed by means of thermal treatment of PS core/SiO_2 shell CPs as containers for drug delivery. In these experiments, capsules made out of CPs with a shell thickness of ~17 nm were used; amoxicillin served as a model loaded substance. The SiO_2 capsules were dispersed in SBF and incubated at room temperature for different periods of time while intensely stirring. After different periods of incubation in physiological saline (0.5, 1, 2, 6, 9, and 24 h), the SiO_2 capsules were sedimented by centrifugation. The degree of amoxicillin release was determined from the light absorbance of the supernatant at the wavelength corresponding to the peak of amoxicillin light absorbance (245 nm). The temperature and pH of the physiological saline were maintained at the levels close to physiologically normal ones (37°C, pH 7.4) to simulate the release of amoxicillin from SiO_2 capsules in the human body.

With increasing time of incubation of amoxicillin-loaded SiO_2 capsules in SBF, the light absorbance of the supernatant increased due to an increase in the amoxicillin concentration in SBF. Approximately 50% of the amoxicillin loaded into the capsules was released within 30 min of incubation. As the incubation time increased to 24 h, another ~30% of the drug was released into the solution. The release of another 15% of amoxicillin took substantially more time (about five days).

Figure 5b shows the dependence of the amount of amoxicillin released into SBF on the square root of the incubation time. As seen in the figure, this dependence is close to linear, which suggests that the release of the drug from SiO_2 capsules is a diffusion-controlled process and can be satisfactorily approximated by Higuchi's diffusion model [29, 30]. Note that this dependence would have been linear in the case of a purely diffusion process. The deviation from the linear pattern in our case may have resulted from partial dissolution of SiO_2 in the physiological saline [31] and change in the diffusion coefficient in the course of the release of the amoxicillin load.

4. Conclusions

Stober's method has been used to synthesize PS core/SiO_2 shell composite particles. The dependences of the SiO_2 shell structure (its thickness, roughness, homogeneity, continuity, and porosity) on the conditions of its formation, including pH, TEOS concentration, reaction time, and PVP concentration and molecular weight, have been determined. It has been demonstrated that variation of these conditions makes it possible to obtain SiO_2 shells with thicknesses from ~4 to 17 nm and different morphologies on the surface of PS particles.

The PVP molecules preliminarily adsorbed on the PS particles have been shown to be incorporated into the SiO_2 shell in the course of its growth, which results in the formation of pores ~3 nm in diameter. Variation of the PVP concentration in the reaction medium allows the number of pores and the specific surface area of both PS and SiO_2 capsules made out of them to be changed within wide ranges.

We have demonstrated that the resultant mesoporous SiO_2 capsules can be used as containers for delivery of drugs (as exemplified by amoxicillin). The kinetics of the release of amoxicillin loaded into SiO_2 capsules under simulated physiological conditions (capsules incubated in physiological saline) has been studied. The characteristic stages of this process have been determined.

Acknowledgments

I am grateful to O. V. Dementieva and V. M. Rudy (Frumkin Institute of Physical Chemistry and Electrochemistry, Russian Academy of Sciences) for valuable comments during the preparation of the manuscript.

This study was supported by the Federal Targeted Program for Research and Development of the Ministry of Education and Science of Russian Federation (Grant 14.578.21.0054, contract No. RFMEFI57814X0054).

References

[1] Lee J E, Lee N and Kim T 2011 Acc. Chem. Res. 44 893
[2] Slowing I I, Vivero-Escoto J L, Wu C W and Lin V S-Y 2008 Adv. Drug Deliv. Rev. 60 1278
[3] Popat A, Hartono S B, Stahr F, Liu J, Qiao S Z, Qing M and Lu G 2011 Nanoscale 3 2801
[4] Li Z X, Barnes J C, Bosoy A, Stoddart J F and Zink J I 2012 Chem. Soc. Rev. 41 2590
[5] Zhai Z, Chen Y, Wang Y J and Luo G S 2009 Chirality 21 760
[6] Zhang Y, Xin H, Meng Q, Fang Y, Fi L and Li X 2015 Mater. Focus 4 4
[7] Schmid A, Fujii S and Armes S P 2006 Langmuir 22 4923
[8] Lee J, Hong C K, Choe S and Shim S E 2007 J. Colloid Interf. Sci. 310 112
[9] Tissot I, Reymond J P, Lefebvre F and Bourgeat-Lami E 2002 Chem. Mater. 14 1325
[10] Ding X, Jiang Y, Yu K, Hari B, Tao N, Zhao J and Wang Z 2004 Mater. Lett. 58 1722
[11] Caruso F, Caruso R A and Möhwald H 1998 Science 282 1111
[12] Stöber W, Fink A and Bohn E 1968 J. Colloid Interf. Sci. 26 62
[13] Lu Y, McLellan J and Xia Y 2004 Langmuir 20 3464
[14] Cao S, Jin X, Yuan X, Wu W W, Hu J and Sheng W C 2010 J. Polym. Sci. Pol. Chem. **48** 1332
[15] Zhen L and Wang S B 2011 J. Inorg. Mater. 8 885
[16] Bao Y, Yang Y, Shi C and Ma J 2014 J. Mater Sci. 49 8215
[17] Chen M, Zhou S, Wu L, Xie S and Chen Y 2005 Chem. Phys. 206 1896
[18] Graf C, Vossen D L J, Imhof A and van Blaaderen A 2003 Langmuir 19 6693
[19] Chou K S and Chen C C 2007 Micropor. Mesopor. Mat. 98 208

[20] Pang J, Li X and Zhou G 2015 RSC Adv. 5 6599
[21] Deng Z, Chen M, Zhou S, You B and Wu L 2006 Langmuir 22 6403
[22] Kovalev E U, Veresov A G and Soin A B 2007 Surface (in Russian) 11 72
[23] Vafayi L and Gharibe S 2013 Bioinorg. Chem. Appl. 2013 541030
[24] Barrett E P, Joyner L G and Halenda P P 1951 J. Am. Chem. Soc. 73 373
[25] Saegusa T 1995 Pure Appl. Chem. 67 1965
[26] Castillo S I R, Ouhajji S, Fokker S, Ben H, Erné C, Schneijdenberg T M M, Dominique M E and Philipse A P 2014 Micropor. Mesopor. Mat. 195 75
[27] Smith J N, Meadows J and Williams P A 1996 Langmuir 12 3773
[28] Sato T and Kohnosu S 1994 Colloids Surf. A. 88 197
[29] Higuchi T 1961 J. Pharm. Sci. 50 874
[30] Higuchi T 1963 J. Pharm. Sci. 52 1145
[31] Andersson J, Rosenholm J and Lindén M 2004 Chem. Mater. 16 4160

Personalized radiosynoviorthese: dosimetry treatment planning

A A Trukhin[1,2], V G Nikitaev[1], L Yu Dubov[1] and Yu V Shtotsky[1]

[1] National Research Nuclear University «MEPhI»
[2] Endocrinology Research Centre

E-mail: Alexey.trukhin12@gmail.com

Abstract. This work is devoted to the quantitative dosimetric support method for endoradyotherapy of inflammatory joints diseases. To reduce inflammation in synovium radiopharmaceutical is injected intraarticularly and provides local ablation of excessive synovial tissue. Present-day guidelines give empirical approaches to therapeutic activity selection. Such approaches lead to 20% untreated patients. Article shows way to improvement of cure level by introduction of dosimetric support.

1. Introduction

For over 40 years radiosynoviorthese (RSO) successfully used for treatment of arthritis [1], inter alia rheumatoid arthritis (RA). This approach had an extensive development in a recent decade. Clinical evaluation of RSO effectiveness and clinical statistic of RSO application to arthritis and osteoarthritis are described in works [2-4].

Main role in the development of inflammation and soreness is played by lining cells of the synovial membrane, that lost ability of contact inhibition. Normally, these cells produce synovial fluid, ensuring the wettability and the restoration of hyaline cartilage surface. Excess of lining cells leads to joint effusion and loss of movement function.

During RSO microparticles with β-radioactive isotope, ranging from 0.05 to 2 mm in diameter, injects in the cavity of immobilized joint. The microparticles are captured by a lining cells of the synovial membrane, as a result β-particles destroy layer of lining cells that produces excess of lubricant and its vascular supply layer. It is known that after administration of standard activity ^{90}Y (185-220 MBq) disease remission is observed in 80% of clinical cases [4]. The value of required dose D for target tissue is defined by the radiologist on the basis of synovial tissue radiosensitivity.

When there are several organs of known activity, according to MIRD-method a dose in the target organ can be represented as the sum of the individual components:

$$D_i = \sum_j \tilde{A}_j \times S(i \leftarrow j) \quad (1)$$

where i, j – the indexes of targets and sources respectively. Each component directly proportional to the cumulated activity in the source organ $\tilde{A}_j = \int A_j(t)dt$. The proportionality factor (S-factor) depends on the properties of radionuclide radiation and on the relative geometrical characteristics of the source and target. In the case of RSO there are one target (intima) and two source regions (intima and sinovial fluid):

$$D_{int} = \tilde{A}_{int} \times S_{int} + \tilde{A}_{sf} \times S_{sf} \quad (2)$$

RSO as any other therapeutic technique of nuclear medicine requires dosimetry treatment planning that takes into account the structural features of the particular joint and the stage of the disease. The existing "EANM Procedure Guidelines for Radiosynovectomy" contains only limited list of radionuclides - ^{90}Y (large joints), ^{186}Re (medium joints), ^{169}Er (small joints) with fixed activities. The further development of the method requires the expansion of available radionuclide list and optimization of assigning activities of radiopharmaceutical depending on the individual characteristics of the patient.

Dosimetric support is divided into two related tasks: dosimetric planning and dosimetric control of side irradiation. Dosimetry planning is based on the MIRD-formalism principles [5]. According to MIRD-formalism medical physicist should take into account the pharmacokinetics of the radiopharmaceutical and anatomic parameters of joint. Current model of synovial joint does not consider other structures of organism. We believe that the entire radiopharmaceutical localized in synovial joint. Important role here plays a pharmaceutical form containing the radionuclide. It must be microparticles, which is not eliminated from joint capsule during treatment and effectively captured by macrophages. Range of beta-particles has to be about 2-4 mm.

In the current paper we propose the method of calculating administered activity of ^{188}Re-colloid for RSO of the knee joint, developed on the basis of dose simulation by toolkit GATE/Gean4.

2. Materials and methods

For absorbed dose calculation open-source software GATE/GEANT4 [7-8] based on Monte-Carlo technique was used. Model of the joint (see Figure 1) consists of ranked cylindrical disks, representing the main elements of the synovial joint: bone, articular cartilage, synovium, layer of lining cells (intima), synovial fluid (cavity). The synovium is divided into two zones: the intima (0.3 ÷ 0.6 mm), which has a contact with synovial fluid, and sub-intima (1 ÷ 9mm). Diameter R equals to intercondylar distance (2 ÷ 9 cm).

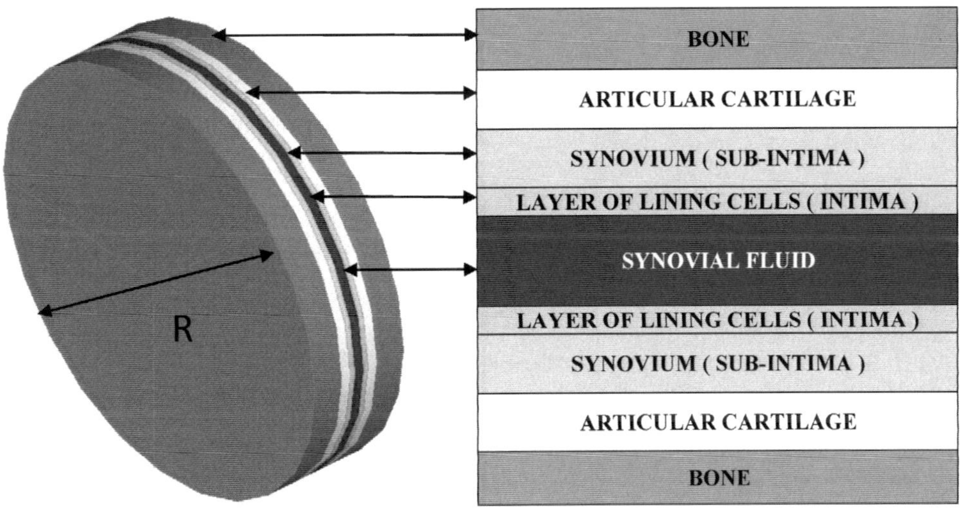

Figure 1. Geometrical model of synovial joint.

The elemental composition of [9] synovial joint tissues applied in this work is shown in Table 1.

Table 1. Synovial joint tissue elemental composition

Element	H	C	N	O	Na	Mg	P	S	Cs	Cl	ρ (g/cm³)
Bone	3,4	15,5	4,2	43,5	0,1	0,2	10,3	0,3	22,5	-	1,92
Articular cartilage	9,6	9,9	2,2	74,4	0,5	-	2,2	0,9	-	0,3	1,1
Soft tissues	10,0	14,9	3,5	71,6	-	-	-	-	-	-	1,0

If the radioisotope is localized in the layer of lining cell (intima) and synovial fluid dose in the intima according to Eq.2 can be expressed as

$$D_{\text{int}} = \int A_{\text{int}}(t)dt \cdot S_{\text{int}} + \int A_{SF}(t)dt \cdot S_{SF} \tag{3}$$

where $A_{\text{int}}(t)$, $A_{SF}(t)$ – activities localized in intima and synovial fluid respectively.

$$A_{\text{int}}(t) = A_0^{\text{int}} \cdot e^{-\frac{t \cdot \ln 2}{T_{1/2}}}, \quad A_{SF}(t) = A_0^{SF} \cdot e^{-\frac{t \cdot \ln 2}{T_{1/2}}}$$

Effective elimination half-life is considered here as isotope decay half-life $T_{1/2}$. Leakage of pharmaceutical during the exposure is considered negligible. Considering that $A_0^{\text{int}} + A_0^{SF} = A_0$ initial activities in intima and in synovial fluid can be expressed as:

$$A_0^{\text{int}} = A_0 \cdot \delta_{\text{int}}, \quad A_0^{SF} = A_0 \cdot \delta_{SF} = A_0 \cdot (1 - \delta_{\text{int}}) \tag{4}$$

Where δ_{int} – share of total activity captured by intima (int); δ_{SF} – share of total activity captured by synovial fluid (SF). Then

$$D_{\text{int}} = A_0 \cdot \delta_{\text{int}} \cdot K \cdot S_{\text{int}} + A_0 \cdot (1 - \delta_{\text{int}}) \cdot K \cdot S_{SF} \tag{5}$$

where K – exposition coefficient, which describes reduction of activity during the exposure due to decay of radionuclide:

$$K = \int_0^{T_{\text{exp}}} e^{-\frac{t \cdot \ln 2}{T_{1/2}}} dt = \frac{T_{1/2}}{\ln 2}\left(1 - e^{-\frac{T_{\text{exp}} \cdot \ln 2}{T_{1/2}}}\right) \tag{6}$$

Where T_{exp} –time of exposure. Immobilization of the joint ensures that leakage is insignificant and the radioisotope remains inside the joint. Therefore the exposure time T_{exp} is considered as the time of immobilization.

To estimate S-factors absorbed dose distribution were simulated in GATE for two different cases: 1) radionuclide is uniformly distributed only in the intima; 2) radionuclide is uniformly distributed only in the synovial fluid (SF). Then S-factors are calculated according to formula (2):

$$S_{\text{int}} = \frac{D_{\text{int}}^1}{\int A_{\text{int}}(t)dt} = \frac{D_{\text{int}}^1}{A_0^{\text{int}} \cdot K}, \quad S_{SF} = \frac{D_{\text{int}}^2}{\int A_{SF}(t)dt} = \frac{D_{\text{int}}^2}{A_0^{SF} \cdot K} \tag{7}$$

Where D_{int} – dose in intima [Gy], A_{int} – activity of radionuclide in intima [MBq], A_{SF} - activity of radionuclide in synovial fluid [MBq]. S_{int} and S_{SF} were simulated for various thickness of sub-intima and sizes of synovial joint. Results are presented in Tables 3 and 4. Using S_{int} and S_{SF} it is possible to calculate required activity of radionuclide for the case, when a part of total activity is absorbed in intima. If D_{int} [Gy] - therapeutic dose for the target tissue, administered activity A_0 [MBq] of radionuclide can be calculated according to Eq.5 as follows:

$$A_0 = \frac{D_{int}}{K \cdot (S_{int}\delta_{int} + S_{SF}(1-\delta_{int}))} \quad (8)$$

where S_{int} и S_{SF} – S-factors for activity accumulated only in intima or only in synovial fluid (Eq.8), δ_{int} – share of total activity captured by layer of lining cell, K [h] – exposition coefficient (Eq.6). .

3. Model validation

For validation of mathematical model the physical phantom of the joint was manufactured (Figure 2) from tissue-equivalent material (ABS-plastics). Tissue-equivalence of material was confirmed by measuring Hounsfield number for ABS-plastics on Philips Precedence 16 SPECT/CT system.

Central volume of the phantom imitating the synovial capsule was filled with water solution containing 40 MBq of ^{188}Re activity measured by dose-calibrator "RADMEDRIG-04P".

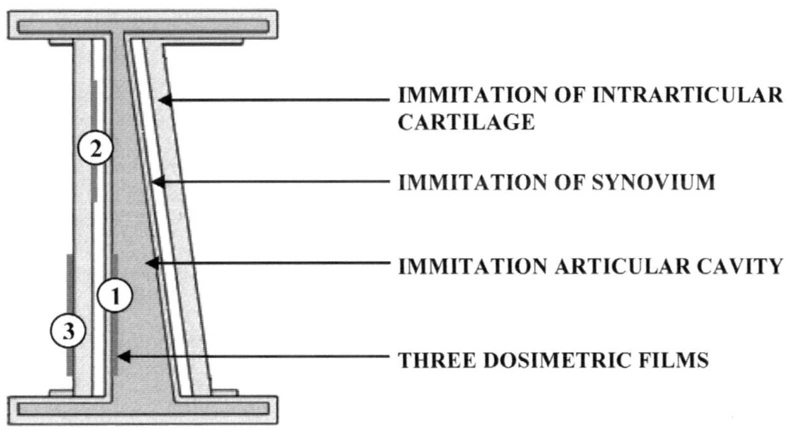

Figure 2. Working drawing of synovial joint phantom and film dosimeters positioning.

The dose at the surface of phantom after 24-hour exposition were measured by film dosimeters GafChromic EBT [10]. Dynamic range of the dosimetric film is from 0.5 to 16 Gy. Three dosimetric films were arranged so that not to overlap each other to measure dose at various distances from the source distributed in articular cavity (positions 1, 2 and 3).

Table 2. Experimentally measured and simulated absorbed doses at different detector positions in phantom with ^{188}Re-source distributed in articular cavity.

Detector position	Absorbed dose, Gy	
	Film dosimeter GafChromic, Gy	GATE simulation, Gy
1	3,5±0,5	3,8
2	2,0±0,3	2.2
3	0,5±0,2	0,7

Table 2 contains experimentally measured absorbed dose values in dosimetric films placed at different distances from the phantom cavity filled with radioactive solution (pos. 1, 2, 3 on Figure2), and GATE simulated absorbed doses. The simulated values are in good agreement with experiment.

4. Results and discussions

Calculated S-factors for two different distributions of ^{188}R and for the various sizes of joints, intraarticular gap and layer of lining cells are shown in Table 3 (a), 3 (b) and 4.

Table 3 (a). S-factor for ^{188}Re activity distributed only in intima, $S_{int}\left[\frac{Gy}{MBq\cdot h}\right]$.

Intraarticular gap, mm	Thickness intima, mm	Joint area, cm^2							
		13	28	50	79	113	154	201	254
1	0.35	0,102	0,046	0,026	0,017	0,012	0,009	0,007	0,007
	0.45	0,094	0,043	0,024	0,015	0,011	0,008	0,006	0,006
	0.55	0,088	0,040	0,022	0,014	0,010	0,007	0,006	0,006
3	0.35	0,088	0,040	0,022	0,014	0,010	0,007	0,006	0,006
	0.45	0,081	0,037	0,021	0,013	0,009	0,007	0,005	0,005
	0.55	0,077	0,034	0,019	0,012	0,009	0,006	0,005	0,005

Table 3 (b). S-factor for ^{188}Re activity distributed only in intima, $S_{int}\left[\frac{Gy}{MBq\cdot h}\right]$.

6	0.35	0,084	0,038	0,021	0,014	0,010	0,007	0,005	0,005
	0.45	0,078	0,035	0,020	0,013	0,009	0,007	0,005	0,005
	0.55	0,073	0,033	0,019	0,012	0,008	0,006	0,005	0,005

Table 4. S-factor for ^{188}Re activity distributed only in synovial fluid, $S_{SF}\left[\frac{Gy}{MBq\cdot h}\right]$.

Intraarticular gap, mm	Thickness intima, mm	Joint area, cm^2							
		13	28	50	79	113	154	201	254
1	0.35	0,095	0,043	0,024	0,016	0,011	0,008	0,006	0,006
	0.45	0,090	0,041	0,023	0,015	0,010	0,008	0,006	0,006
	0.55	0,086	0,039	0,022	0,014	0,010	0,007	0,006	0,006
3	0.35	0,050	0,023	0,013	0,006	0,006	0,004	0,003	0,003
	0.45	0,048	0,022	0,012	0,006	0,006	0,004	0,003	0,003
	0.55	0,046	0,021	0,012	0,005	0,005	0,004	0,003	0,003
6	0.35	0,027	0,012	0,007	0,005	0,003	0,002	0,002	0,002
	0.45	0,026	0,012	0,007	0,004	0,003	0,002	0,002	0,002
	0.55	0,025	0,011	0,006	0,004	0,003	0,002	0,002	0,002

Figure 3A presents dependence of S-factor (δ_{int} =0.5) on intraarticular gap for different intercondylar distances R: 40, 60, 80 and 100 mm. Increasing the intraarticular gap from 1 to 6 mm leads to S-factor reduce by 1.2 times at R = 40 mm, and 1.3 times at R = 100mm. The width of the intraarticular gap and intercondylar distance can be estimated using computer tomography.

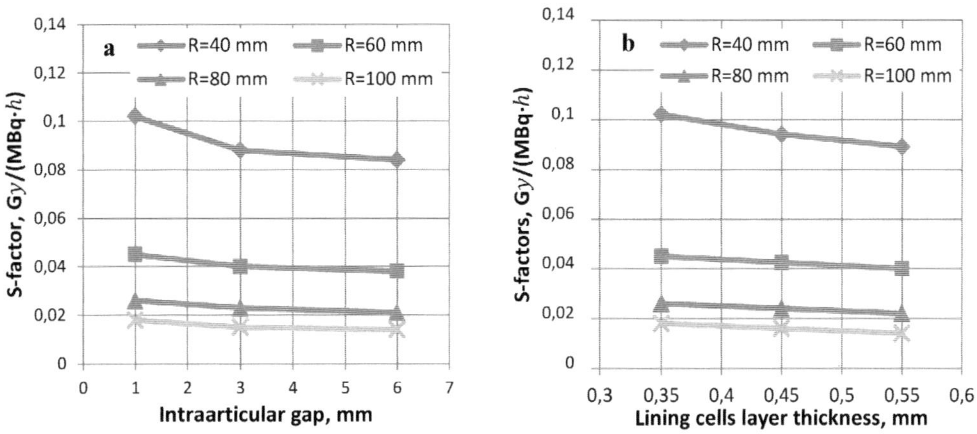

Figure 3. S-factor dependences on intraarticular gap width (a) and on thickness of the lining cells layer (b) for the number of intercondylar distances R: 40, 60, 80 and 100 mm.

Figure 3B presents dependence of S-factor ($\delta_{Int} = 0.5$) on thickness of the lining cells layer of synovium for different intercondylar distances R: 40, 60, 80 and 100 mm. It can be seen that increase of the lining cells layer thickness from 0.35 to 0.55 mm leads to S-factor reduction by 1.2 times at R = 40 mm, and 1.3 times at R = 100 mm. Lining layer thickness could be measured by ultrasound scanner or by MRI.

5. Dosimetric protocol

The next steps of dosimetric planning may be used for clinical applications of RSO:

1. CT scanning for evaluation of intercondylar distance and width of intraarticular gap;
2. US scanning for evaluation of intima and effusion thickness;
3. Determination of S-factor using Table 3 and 4.
4. Leakage control (WB scintigraphy);
5. Evaluation of radionuclide distribution (SPECT-CT).
6. Determination of immobilization time;
7. Calculation of administered activity using formula 3 and 4.

6. Conclusion

Study of the S-factor dependence on geometrical parameters of synovial joint carried out using GATE simulation shows that dose planning procedure for RSO must take into account the individual characteristics of each patient. Increase of intercondylar distance, intraarticular gap width and thickness of lining cells layer requires to increase administered activity of the radiopharmaceutical.

The proposed approach for dosimetry planning of RSO meets the modern requirements of clinical practice. Such approach allows calculation of required activity based on individual anatomical structure of synovial joint. It is possible to calculate S-factors for any parameters of synovial joint on the base of Monte-Carlo simulation and present them in a convenient form for clinical use. Necessary parameters can be determined by complex of diagnostic procedures.

Presented method has high flexibility and reproducibility and provides an opportunity for physicians to conduct extensive retrospective analysis. thus such approach will lead to improvement of RSO effectiveness.

Acknowledgements
This article was prepared with the support of National Research Nuclear University MEPhI Academic Excellence Project and Competitiveness Program of National Research Nuclear University MEPhI.

References

[1] Deutsch, E., Brodack, J.W., Deutsch. K.F. Radiation synovectomyrevisted. European Journal of Nuclear Medicine 199320: 1113-1 127

[2] ChrapkoB, ZwolakR, NocunA, GolebiewskaR, MajdanM. Radiationsynovectomy with 90Y colloid in the therapy of recurrent knee joint effusions in patients with inflammatory joint diseases. RheumatolInt 2007;27:729–34.

[3] Zuderman L, Liepe K, Zöphel K, et al. Radiosynoviorthesis (PCO): influencing factors and therapy monitoring. Ann Nucl Med. 2008;22(9):735–41. DOI: 10.1007/s12149-008-0167-7. Epub 2008 Nov 28.

[4] Liepe K, Zaknun JJ, Padhy A, et al. Radiosynovectomy usingyttrium-90, phosphorus-32 or rhenium-188 radiocolloids vs. corticoid instillation for rheumatoid arthritis of the knee. Ann NuclMed. 2011;25(5):317–23. DOI: 10.1007/s12149-011-0467-1.Epub 2011 Feb 19.

[5] B. R. B. Walters, I. Kawrakow, and D. W. O. Rogers, "History by historystatistical estimators in the beam code system,"Med. Phys.29, 2745–2752(2002).

[6] LANL,MCNPX 2.6.0 Users's Guide, Technical Report No. LA-CP-071473, 2008.

[7] G. Battistoni, S. Muraro, P. Sala, F. Cerutti, A. Ferrari, S. Roesler,A. Fasso, and J. Ranft, "The FLUKA code: Description and benchmarking," inHadronic Shower Simulations Workshop, edited by M. Albrowand R. Raja (AIP Conference, Fermilab, Batavia, IL, 2006), Vol. 896, pp.31–49.

[8] A. Ferrari, P. Sala, A. Fasso, and J. Ranft,FLUKA: A Multi-Particle Trans-port Code, Technical Report No. INFN/TC 05/11, SLAC-R-773, CERN-2005-10, 2005.

[9] International Commission on Radiation Units and Measurements. ICRU-44: Tissue Substitutes in Radiation Dosimetry and Measurements, 1989.

[10] // GafChromic: http://gafchromic.ru/.[2015]

Efficiency of the confocal method of laser endomicroscopy in complex diagnoses of diseases of common bile duct

S G Anaskin[1], D N Panchenkov[2], V B Chertyuk[3], D V Sazonov[4], F G Zabozlayev[4], O V Danilevskaya[4], N V Mokshina[4], I D Korniletsky[1]

[1] National Research Nuclear University MEPhI, Mockow, Russia
[2] Department of endoscopic surgery of SEI HPE MSMSU them. A. I. Evdokimov, Mockow, Russia
[3] Institute of Emergency children's surgery and traumatology, Mockow, Russia
[4] Federal scientific clinical center of FMBA of Russia, Mockow, Russia

Main author e-mail address: asg72@list.ru

Abstract. One of the more frequent manifestations of diseases of the bile ducts are its' strictures or stenoses that could be of either malignant or benign nature. Current methods of diagnosing this pathology include computer tomography (CT) scan, magnetic resonance cholangiopancreatography (MRCP), endoscopic ultrasound (EUS) and endoscopic retrograde cholangiopancreatography (ERCP). However, these methods are not always informative, which makes this a current and topical problem.
A fundamentally new method that broadens the capabilities of ERCP when diagnosing diseases of the bile duct accompanied by the development of strictures or stenoses is probe-based confocal laser endomicroscopy (pCLE). The method is based on the principle of confocal fluorescence microscopy.
The most elaborate complications arise with the presence of the pre-existing pancreatobiliary pathology: pseudotumoral chronic pancreatitis, acute cholangitis, etc. Early stage cholangiocarcinoma diagnosis can be difficult (and not always possible) even with the help of modern research methods. For the timely diagnostic it is advantageous to conduct pCLE and targeted biopsy of the zone with most manifested changes.
In all instances, the first use of the pCLE method for diagnostic purposes allowed us to clarify and correctly verify the diagnosis.
When concerning the diseases of the bile duct, the modern stage of pCLE development can be of critical importance when other methods are not effective.

1. Introduction

One of the most prevalent manifestations of the bile ducts disease are their strictures or stenoses, which may have both malignant and benign nature and may be accompanied by the syndrome of obstructive jaundice.

The most frequent reasons of formation of strictures or stenoses of bilious channels are the following malignant diseases: cholangiocarcinoma, pancreatic cancer, cancer of a gall bladder. Occur among benign diseases: syndrome of Mirizzi, primary sclerosing cholangitis, autoimmune pancreatitis, iatrogenic intraoperative damages of bilious channels, strictures of a biliodigestivny anastomosis [1].

A.C.Burnet et al., have analyzed these literatures over the last 10 years, chose 17 large publications including the description of 1556 cases of various diseases bilious ducts with the diagnosed stricture where the endoscopic retrograde cholangiopancreatography and a brush biopsy were a final diagnostic method [2]. Results of this research show rather low level of sensitivity – only 41%, and the prognosis of a negative outcome made 58% that allows to speak about need of further search of additional methods of intravital bilious ducts diseases diagnostics which are followed by development of strictures or stenoses.

Malignant tumors of bilious ducts are the most dangerous and complex in timeous diagnostics. Cholangiocarcinoma (carcinoma of the bile ducts) is a relatively rare malignant tumor originating from mutated epithelial cells of the bile ducts, which in its histological structure refers to adenocarcinoma. In Europe and North America the incidence of cholangiocarcinoma is 1-2 cases per 100 000 population per year, but its level have been increasing for recent decades [3]. The tumor affects mainly extrahepatic bile ducts: 60-80% in the liver gates (Klatskin's tumor) and 10-30% in the distal duct. Generally, cholangiocarcinoma develops on the background of primary sclerosing cholangitis (PSC), accompanied by ulcerative colitis It is believed that the risk of cholangiocarcinoma increases with the following congenital cystic diseases: liver fibrosis, cystic enlargement of the intrahepatic ducts (Caroli syndrome), cysts of the common bile duct, cystic liver, microanatomy (complexes of von Meyenberg), as well as biliary cirrhosis [1]. At the same time, there is no specific factor in the anamnesis at most patients with a cholangiocarcinoma.

Some authors mark the fact that the risk of developing carcinoma of the extrahepatic bile ducts in 10 years or more after cholecystectomy significantly reduces, that may indicate the relation between the tumor and biliary stones [3, 4].

Timeous diagnostics of cholangiocarcinoma, even when it is possible to do a radical operation, represents certain difficulties and depends on the capabilities of available diagnostic equipment and the qualifications of diagnosticians. That fact that the early stage of a disease lack of any symptomatology and discovery of it is an fortuity, makes diagnostics more difficult.

In cases when the patient already has an accompanying pathology of pancreatobiliary area, early diagnostics of a cholangiocarcinoma is greatly hampered in consequence of the existing morpho-functional changes of this area, and also by focus of the doctor on the pathology established earlier.

Diagnostics at patients with a cholangiocarcinoma with suffering chronic pseudo-tumorous pancreatitis, a cholangitis, cicatricial strictures of extrahepatic bilious ducts is especially difficult, because in many cases one of components of complex treatment is endoscopic prosthetics or stenting of bilious ducts. The stent which is stays in a duct makes the capture of biopsy material for cyto- or histological research impossible without its preliminary removal. Besides, long-term presence of a stent in a bile duct necessarily causes visual and morphological changes of the epithelium, which complicates timely diagnosis of cholangiocarcinoma [2].

A number of authors believe that a long course of chronic indurative pancreatitis or sclerosing cholangitis in excess of 10 years, may be accompanied by the development of pancreatic cancer or cholangiocarcinoma [1, 4]. Surgical treatment of cholangiocarcinoma is ineffective, mainly due to the low accessibility of tumor and its late diagnosis. Most patients have recurrence to medical help at the stage when the tumor is inoperable, because in the early stages disease runs oligosymptomatic. In this case, the main methods of treatment are adjuvant chemotherapy and radiation therapy, palliative care. The argument in favor of endovascular and endoscopic interventions for inoperable tumors is the elimination of jaundice and itching from dying patients.

The methods and ways of diagnostics of this pathology including the spiral computer tomography (SCT), a magnetic resonance cholangiopancreatography (MRHPG), endoscopic ultrasonic research (EUZI), an endoscopic retrograde cholangiopancreatography (ERHPG) which are available now, but aren't always informative, leaving a problem of early cholangiocarcinoma diagnostics actual and didn't solved until the end. In this regard, the growing demand for new technologies with high reliability to detect pancreatic cancer or cholangiocarcinoma in patients with chronic pseudo-tumor-like pancreatitis, cicatricial stenosis of the common bile duct and obstructive jaundice who had undergone endoprosthesis of the bile ducts.

A fundamentally new method which allows to expand the possibilities of diagnostic ERCP in diseases of the bile ducts, accompanied by the development of strictures or stenosis, is probe confocal laser endomicroscopy (Clem) (probe-based Confocal Laser Endomicroscopy – pCLE) [5, 6, 7]. This study is conducted using the device of Cellvizio®, developed by Mauna Kea Technologies (Paris, France).

The method is based on the principle of confocal fluorescent microscopy. At the fiberoptic probe is a transfer of laser radiation at a wavelength of 488 nm from a workstation. Falling on the biological tissue, a portion of the light is absorbed and laser-induced fluorescence effect causes the illumination of the tissue, which is identified by confocal microscope and processed by computer, allowing to obtain dynamic monochrome image with a frequency 11 Hz and with a resolution of 600 mkn. (figure 1).

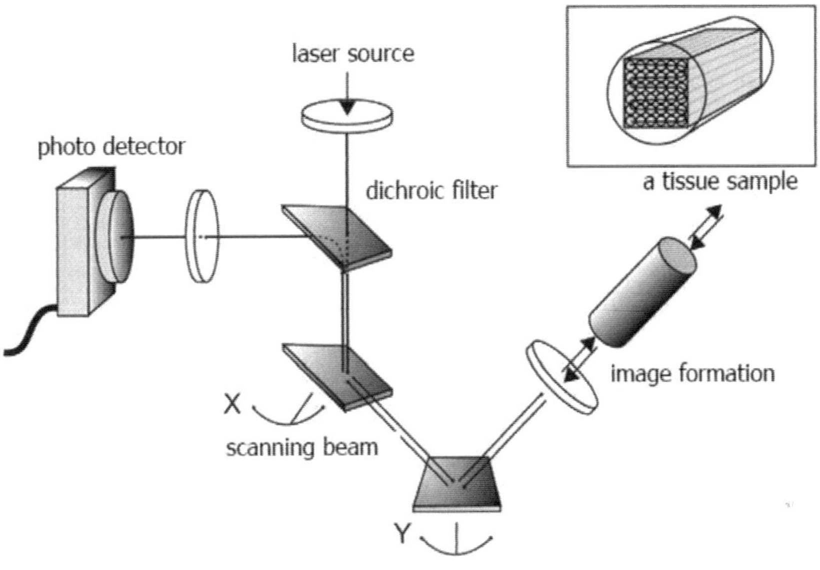

Figure 1. Scheme of Probe-based Confocal Laser Endomicroscopy (pCLE).

Only certain tissues in living organisms possess autofluorescence. The main endogenous luminescing biological tissues include the following groups of substances: flavins, proteins and porphyrins. Each of the luminescing has a characteristic spectra for absorption and emission. Mostly they are concentrated in tissues rich in collagen and elastin that does not require the introduction of additional fluoresceins [8]. Epithelial cells of the gastrointestinal tract, as well as tumors do not have sufficient volumes of these structures, in this regard, the induction of fluorescence effect using parenteral administration of sodium fluorescein (Novartis) [9-14].

For investigation of the bile duct apply a probe of Holangioflex which resolving power reaches 3,5 mkn, and the optical field makes 600 mkn, with a depth of research of 0-50 mkn. After performance of ERHPG, the minisand is carried out on system of delivery to an interest zone. For this purpose use a catheter of "OASIS Wilson-Cook" with X-ray positive tags or its analogs. Just before the beginning of scanning enter sodium fluorescein in a dose from 0,5 to 1,0 ml.

Considering the low frequency (11 Hz) of change of a personnel, it is important to get a picture without blur effect, i.e., the moving speed of the probe relative to the tissue should be minimal. The study is a preliminary evaluation of data, with further detail when viewing recorded movies. Video analytics using the classification of microendoscopic images, which lay in the basic for the evaluation of the shape, color, size, visible structures and estimation of dynamic components [15-17]. The following elements are important for an assessment of the endomicroscopic picture received at research of bilious ducts. Thin and light strips, mainly meet at bening whereas increase in their width testifies in favor of a neoplasia. Inflammatory changes are shown by numerous white strips, dark granular structures with the repeating deepenings, a thickening of strips of a reticular network.

As examples, showing the high diagnostic capability of the Clem method in the early diagnosis of cholangiocarcinoma or exception, against the background of other diseases of the pancreatobiliary zone, describe our own clinical observations.

2. Clinical observation

2.1. Clinical observation №1. Patient G., born in 1960, was observed in the Federal scientific clinical center of FMBA of Russia since 2005 courses of inpatient treatment in the departments of gastroenterology and surgery, with a clinical diagnosis of chronic pseudo-tumor-like pancreatitis.

It is known from the anamnesis that at the beginning of 2005 the patient rubed through pancreatonecrosis. In July, 2005 in connection with the stenosis of a duodenum which developed after the postponed pancreatonecrosis operation is executed: a laparotomy, a gastroenterostomy on a long loop with an interintestinal anastomosis according to Brown. On June 30, 2006 concerning mechanical jaundice ERHPG is made, symptoms of pseudo-tumorous pancreatitis, a prelum of an intrapancreatic part of the general bilious duct inflammatory infiltrate of a head of a pancreas, expansion of the general bilious and pancreatic ducts come to light. Produced biliodigestive endoprosthesis effect, the jaundice resolved.

During 2007 – 2010 years to the patient planned X-ray endoscopic interventions on replacement of stents, nazo-biliary drainage for the purpose of a decompression of biliary system were carried out. It should be noted that during the periods between hospitalization the patient did not keep to a diet, abused alcohol, did not accept the appointed drugs and refused possible radical operation (a gastropankreatoduodenalny resection). In October, 2010 there was a recurrence of mechanical jaundice in this connection, in a hospital on a sore point operation was executed: the laparotomy, external drainage of the general bilious duct, jaundice is stopped.

In 2010 - 2011 passed numerous hospitalization concerning chronic pseudo-tumorous pancreatitis, an acute purulent cholangitis, complex conservative therapy with an effect was carried out. In March, 2011 at control inspection considering categorical refusal of the patient of radical operation, short service life of plastic endoprosthesis in connection with their constant obturation biliary sludge and big losses of bile on the external drainage sick the nitinolous self-expanding stent of uncovered type with simultaneous removal of an external drainage from the general bilious duct (figure 2) was established. In May, 2012 in connection with recurrence of the mechanical jaundice caused by stent obturation a hyperplastic mucous membrane of the general bilious duct, to the patient it is executed transdermal and transhepatic external - internal drainage of the general bilious duct (figures 3, 4). When performing a fistulography and EGDS with survey of a big duodenal nipple there were suspicions on possible tumoral defeat of a distal part of the general bilious duct. The brush biopsy was executed, however convincing data were not received for oncological process.

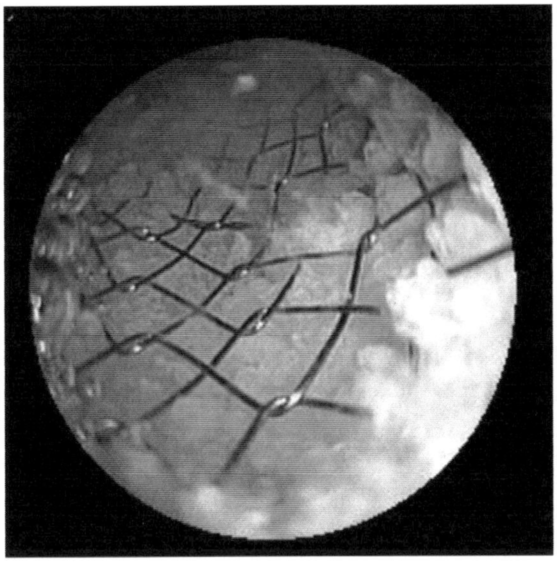

Figure 2. Nitinol self-expanding stent type uncovered in the lumen of the common bile duct.

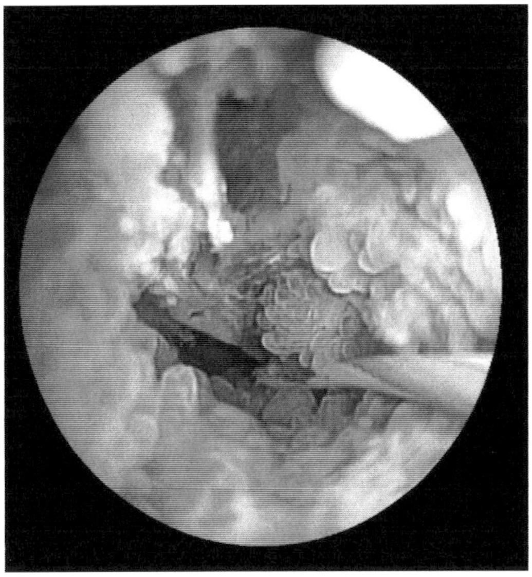

Figure 3. Stent obstruction hyperplastic mucosa of the common bile duct (the lumen of the duct is celebrated metallic conductor).

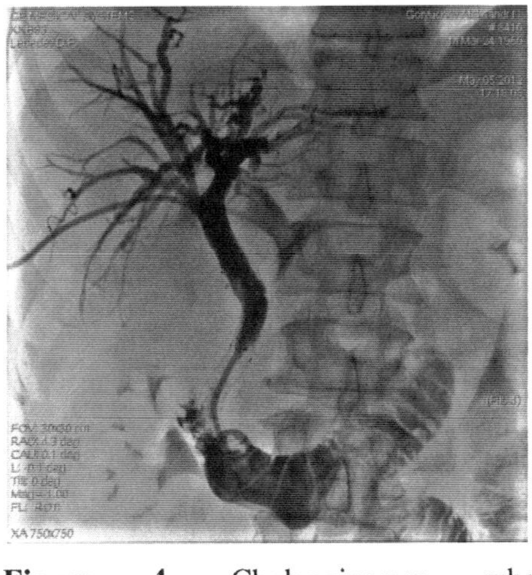

Figure 4. Cholangiogram when transcutaneous-transhepatic external-internal drainage of common bile duct.

For the purpose of clarification and possible verification of the diagnosis for the first time in Federal research and clinical centre of FMBA of Russia was used new method – Clem, which was mastered and implemented endoscopic service Center in 2011 with the aim of differential diagnostics of lung diseases.

Via the created duct of a gepaticocholangiostomy the flexible EB-530T endoscope is carried out. Via the tool channel of the endoscope by means of a miniprobe of "Cholangioflex" it is executed CLEM (figure 5) at which the hyperplasia of fabrics in the field of standing of a stent, and also signs of focus of a cholangiocarcinoma is revealed that it is confirmed with results of planned histological research after the executed biopsy (figures 6,7). The patient is advised by the oncologist, adjuvant chemoradiotherapy is appointed.

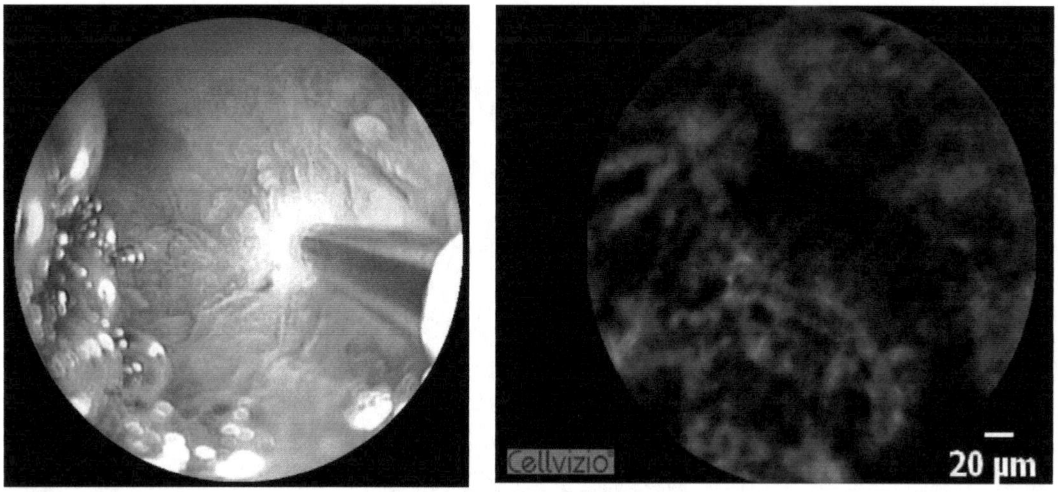

Figure 5. pCLE: revealed hyperplasia of the tissues in the region of the stent state with signs of focal cholangiocarcinoma.

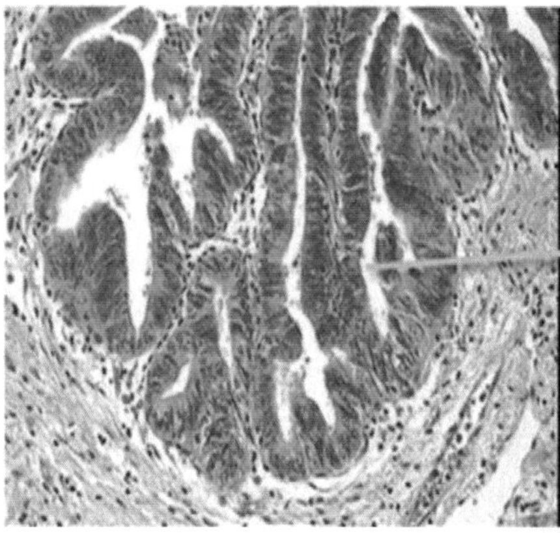

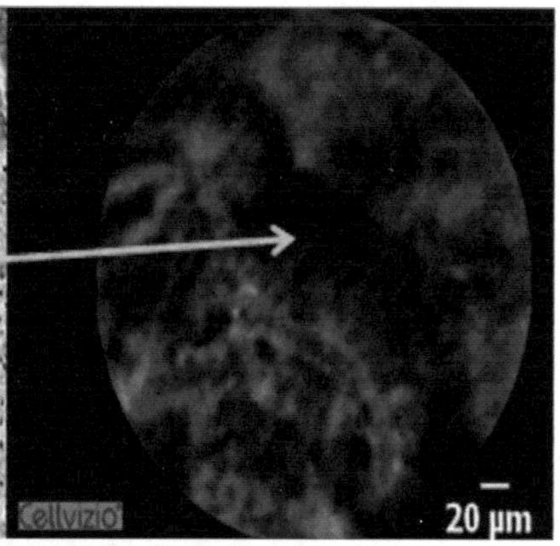

Figure 6. Microreport mucosa of the common bile duct, painting the hematoxylin-eosin staining (X200 zoom).

Figure 7. Microendoscopic dynamic monochrome image with the focus of cholangiocarcinoma.

2.1. Clinical observation №2. Patient H., born in 1956, was admitted in the surgery Department, Federal scientific clinical center of FMBA of Russia 29.01.2014 y. with referral diagnosis of post-cholecystectomy syndrome, stricture of the common bile duct, obstructive jaundice, functioning choledochotomy from . 24.12.2013.

Considers himself ill since August 2013, when he first began to note pain in the right upper quadrant and epigastric, change in urine color, later came the yellowness of the skin. Was admitted to the hospital at the place of residence, the examination diagnosed with cholelithiasis, chronic calculous cholecystitis, stricture of the common bile duct. Routinely laparotomy, cholecystectomy, drainage of the common bile duct by Pikovsky. The postoperative period was uneventful, after removal of the drainage on the 9th day, patient was discharged in satisfactory condition. In November 2013 he had a relapse of jaundice, in connection with which the patient was performed percutaneous external-internal drainage of the bile ducts. Attempted ERCP at that time failed. Sent to Federal research and clinical centre for diagnosis and definition of further tactics of treatment.

At admission clinical laboratory tests within normal limits. Ultrasound of internal organs of the abdominal cavity from 30.01.2014: the condition after cholecystectomy. Diffuse changes of the liver and pancreas. Intrahepatic bile ducts not enlarged, increased echogenicity, visualized a single haplotye include "comet tail" – pneumobilia, in the projection of the right equity and the common bile duct – drainage. The diameter of the common hepatic duct 1.3 cm, walls thickened, low echogenicity, with irregular contours that do not allow the exclusion reprocess. Determined by a single reactive modified l/nodes in the course hepatoduodenal ligament. At the time of inspection free fluid in the abdominal cavity and the pelvis is not defined.

At gastroduodenoscopy, ERCP from 07.02.2014 g. revealed a long stricture of the common hepatic duct (figure 8), produced crustaceana replacement percutaneous drainage of the bile duct. Conclusive for malignancy, compression from the outside is not received. The result of cytological examination number: 2013-19/14 (brush biopsy): clusters of cells are proliferating epithelium of the duct with a tendency to the formation of large papillopathy structures.

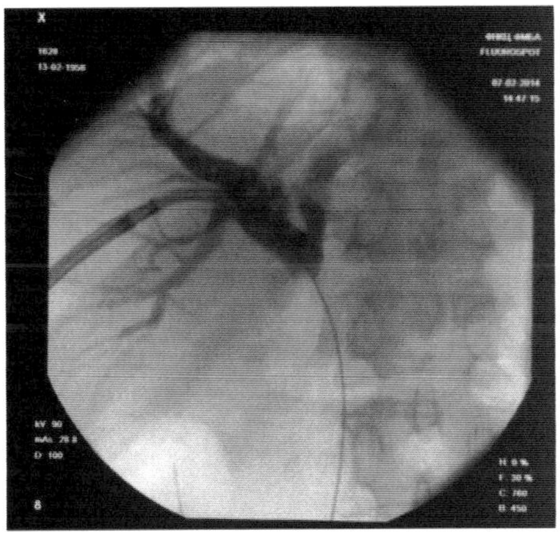

Figure 8. Contrast X-ray gram of the long stricture of the common hepatic duct.

10.02.2014 was completed radial endosonography ULTRASONIC transducer, confocal laser endomicroscopy (figure 9). Confirmed cholangitis, reactive hyperplasia of the mucosa of the common bile duct, signs of neoplasia in the area of the stricture histological examination not revealed. In connection with the exception of cancer the patient was discharged from the hospital 11.02.2014. recommendation for readmission after 2 week in Federal Research and Clinical Centre for endoscopic stenting of the common bile duct.

2.3. Clinical observation №3. Patient Z., born in 1937, was hospitalized in the surgery Department, Federal scientific clinical center of FMBA of Russia 09.04.2013, for examination with complaints on a periodic aching pain in the epigastrium, heartburn, belching air, lack of appetite, weight loss of 7 kg over the last month, change the color of the skin 2 weeks before admission.

At admission the clinical picture of obstructive jaundice. Biochemical analysis of blood from 7.04.2013 g.: total bilirubin. 87,7 μmol/l, AST 405 U/l, AST 405 U/l, alkaline phosphatase 745 U/l, glucose 12.2 mmol/l Other changes in clinical laboratory tests no.

Ultrasound of internal organs of the abdominal cavity from 09.04.2013 g: the common bile duct is dilated at the level of the gate of the liver to 1.41 cm, in the region of the distal common hepatic duct is determined volumetric education up to 3.2 cm, decreased echogenicity, heterogeneous structure. Intrahepatic bile ducts are dilated to the level of subsegmental ducts. The gallbladder is enlarged, the lumen thick bile.

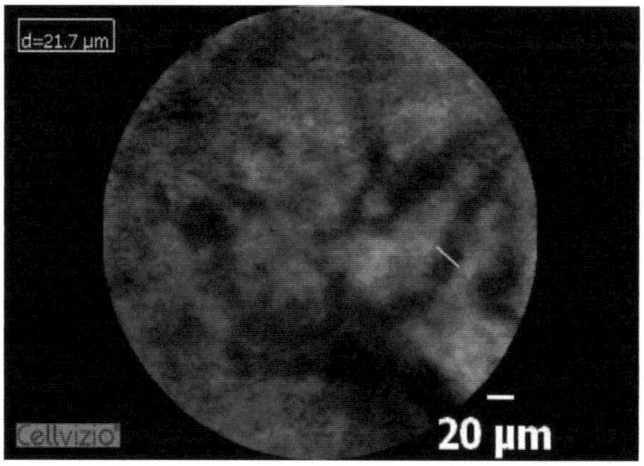

Figure 9. Dynamic monochrome microendoscopic the image with a picture of cholangitis and reactive hyperplasia of the mucosa of the common bile duct.

10.04.2013 was completed videogastroduodenoscopy, ERCP identified a biliary hypertension, signs of tumors of the common hepatic duct, taken by brush biopsy. With the aim of biliary decompression produced naso-biliary drainage of the common hepatic duct (figure 10). The result of cytological examination number:7317-20/13: expressed proliferation of epithelium of the ducts with formation of branching and papillary structures.

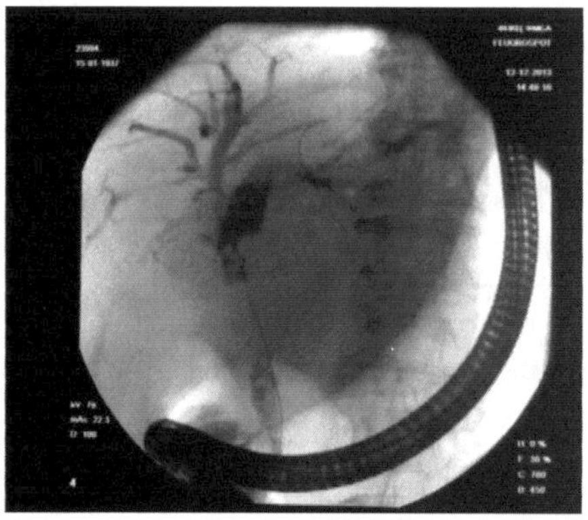

Figure 10. Cholangiogram made through nasobiliar drainage installed in the common hepatic duct. Identified during stucture the common bile duct.

When abdominal CT with contrast enhancement (12.04.2013,) and microcholangiography (16.04.2013) identify changes to best meet the neoplastic process of the common bile duct, signs of biliary hypertension. In order to clarify the diagnosis 26.04.2013 G. performed confocal laser endomicroscopy (figure 11). Revealed the formation of the common bile duct, solid structure, tight and rigid relative to the surrounding tissues during instrumental palpation, stenosing the lumen of the duct. Conclusion: visual and microscopic picture of the adenomatous structure with signs of malignization, most likely adenocarcinoma. The planned histological study confirmed the diagnosis of cholangiocarcinoma. Radical surgery because of severe concomitant cardiac pathology is not recognized by the portable produced endoscopic stenting of the common bile duct self-expanding nitinolous stent uncovered type. The patient was seen by oncologist assigned to adjuvant chemoradiotherapy.

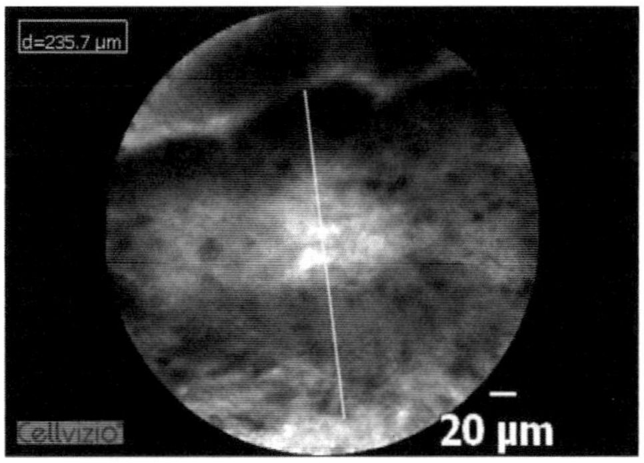

Figure 11. Dynamic monochrome microendoscopic image: painting adenomatous structure with signs of malignancy.

3. Discussion

Most difficult diagnosis of malignant neoplasms of the bile duct on the background of already existing diseases of the pancreatoduodenal zone, for example, chronic pancreatitis. Hyperplastic (pseudo tumor) chronic pancreatitis occurs in 4-6% of patients with chronic pancreatitis. Complications of disease of this form happens in 30% of patients and develop approximately 7-15 years [1]. Sclerotic processes in the head of the pancreas lead to the clinical picture, resembling a compression of the ducts of the biliary tract and pancreatic duct (Wirsung duct). The violation of outflow of bile in such cases causes mechanical jaundice. In alcoholic and hereditary chronic pancreatitis there is a reaction of precipitation of calcium and proteins inside the pancreatic ducts. Normal precipitation is prevented PSP-proteins that are secreted arenaline cells. The level of this protein in patients with chronic pancreatitis significantly reduced, thereby increasing calcium in pancreatic juice, its microcrystallization. In the lumen of the duct are formed calcification, plugging of ducts, there is a pressure increase in them. Ultimately, developing fibrosis.

In recent years, endoprosthesis of bile duct is widely used for the treatment of chronic pancreatitis complicated by development of stenosis of the common bile duct, obstructive jaundice. For the first time endoprosthetics of common bile duct produced by Soehendra and Reynders-Frederix in 1979 [4]. In the hands of an experienced endoscopist, this method did not seem so difficult, and its low injury rate and effectiveness make it attractive in clinical practice. Currently uses a variety of stents - nitinol self-expanding stents and plastic stents. The main disadvantage of the latter is high probability of obstruction of the prosthesis after 2-6 months after the setting as a result of deposition in it of salts of bile acids and bacterial films. To maintain their free clearance require regular remedial endoscopic lavage of the lumen of the prosthesis, which in the end does not increase their duration, but only allows you to save for a short time, their drainage function [2].

Self-expanding metal stents maintain patency longer than plastic, about 6 months or more. The use of metal stents is associated with a risk of germination hyperplastic tissue epithelial duct or a tumor that requires subsequent recanalization. Over time, standing uncovered biliary stent may be complicated by hyperplasia of the cuboidal epithelium of the common bile duct, leading to obstruction of the stent. To eliminate this effect use of covered nitinol stents, in which between cells sealed in a plastic membrane that prevents its germination. On this background, regardless of the type of stent cholangiocarcinoma diagnosis is delayed because of changes to the zones, technical difficulties in performing biopsies and orientation of doctors to another, previously established pathology.

The suspected cholangiocarcinoma occurs when unexplained etiology of biliary obstruction. Laboratory studies reflect only the degree of cholestasis. Diagnosis is based on ultrasound or CT. If these methods do not allow to verify the diagnosis, you must hold MRCP or ERCP with percutaneous transhepatic by cholangiographies. In some cases, ERCP not only diagnose the tumor, but also allows you to perform a tissue biopsy brush that provides a histological diagnosis without fine-needle biopsy under ultrasound or CT scan. For diagnosis verification it is necessary to perform a biopsy with subsequent histological examination of the obtained material [1].

Applied in our clinical observations, the Clem method allows not only to clarify but also to verify the diagnosis when other modern diagnostic methods proved to be uninformative.

To date, there are only a few foreign works devoted to the method of Clem in the pathology of hepatopancreatoduodenal region. Three multicenter studies conducted in Europe and the United States demonstrated a high specificity 73-88% and sensitivity 83-97% intraductal the Clem for strictures of the bile ducts of unknown etiology [18]. Some foreign authors consider that at the present stage of development of the Clem diagnosis in diseases of the bile ducts may be critical for the inefficiency of other methods [5, 6].

4. Conclusion

Diagnosis of cholangiocarcinoma at an early stage of the disease even with the help of modern methods of research are complex and not always possible. The greatest difficulties arise in the development of tumors on the background of already existing pancreatobiliary pathology: chronic pseudo-tumor-like pancreatitis, suppurative cholangitis, etc. For timely diagnosis appropriate to the conduct of the Clem and target biopsy of the area with the most pronounced changes. This clinical case shows the need for further controlled studies to determine specificity and sensitivity of the method, the Clem case of various pathologies of the bile and pancreatic ducts.

Acknowledgments

This work was supported by Competitiveness Program of National Research Nuclear University MEPhI.

References

[1] Galperin E. I., Tyuzheva T. G. 2011 Lectures on gepatopankreatobiliarny surgery with CD. (M.:vidar) 536 pages.

[2] Chan C.H., Telford J.J. 2012 Endoscopic management of benign biliary strictures//*Gastrointest. Endosc. Clin. N. Am.* Vol.22(3). – R.511-537.

[3] Nordback I.H., Pitt H.A., Coleman J. et al. 1994 Unresectable hilar cholangiocarcinoma: percutaneous versus operative palliation//*Surgery* Vol.115. p.597-603.

[4] Soehendra N., Reynders-Frederix V. 1979 Palliative gallengangdrainage//*Dtsch. Med. Wochenschr.* Vol.104. p.206-209.

[5] Caillol F., Filoche B., Gaidhane M., Kahaleh M. 2013 Refined probe-based confocal laser endomicroscopy classification for biliary strictures: Paris Classification//*Dig. Dis. Sci.* Vol.58(6) River of 1784-1789.

[6] Giovannini M., Bories E., Monges G., Pesenti C., Caillol F., Delpero J.R. 2011 Results of a phase I-II study on intraductal confocal microscopy (IDCM) in patients with common bile duct (CBD) stenosis//*Surg. Endosc.*Vol.25(7) – River of 2247-2253.

[7] Neumann H., Kiesslich R., Wallace M.B., Neurath M.F. 2010 Confocal laser endomicroscopy: technical advances and clinical applications//*Gastroenterology* Vol.139(2). – p.388-392.

[8] Goujon D., Zellweger M., Radu A., Grosjean P., Weber B.C., Van den Bergh H., Monnier P., Wagnières G. 2003 In vivo autofluorescence imaging of early cancers in the human tracheobronchial tree with a spectrally optimized system//*J. Biomed. Opt.* Vol.8(1). – p.17-25.

[9] Shahid M.W., Crook J.E., Meining A., Perchant A., Buchner A., Gomez V., Wallace M.B. 2011 Exploring the optimal fluorescein dose in probe-based confocal laser endomicroscopy for colonic imaging//*J. Interv. Gastroenterol.* Vol.1(4). p.166-171.

[10] Miehlke S., Morgner A., Aust D., Baretton G., Madisch A. 2011 Probe-based confocal laser endomicroscopy in double balloon enteroscopy//*Z. Gastroenterol.* Vol.49(12) p.1529-1534.

[11] Newton R.C., Kemp S.V., Yang G.Z., Elson D.S., Darzi A., Shah P.L. 2012 Imaging parenchymal lung diseases with confocal endomicroscopy//*Respir. Med.* Vol.106(1) p.127-137.

[12] Wu K., Liu J.J., Adams W., Sonn G.A., Mach K.E., Pan Y., Beck A.H., Jensen K.C., Liao J.C. 2011 Dynamic real-time microscopy of the urinary tract using confocal laser endomicroscopy//*Urology*.Vol.78(1). p.225-231.

[13] Van den Broek F.J., Van Es J.A., Van Eeden S., Stokkers P.C., Ponsioen C.Y., Reitsma J.B., Fockens P., Dekker E. 2011 Pilot study of probe-based confocal laser endomicroscopy during colonoscopic surveillance of patients with longstanding ulcerative colitis//*Endoscopy*. Vol.43(2). p.116-122.

[14] Becker V., Van den Broek F.J., Buchner A.M., Dekker E., Wallace M.B., Von Delius S., Schneider A., Schmid R.M., Meining A. 2011 Optimal fluorescein dose for intravenous application in miniprobe-based confocal laser scanning microscopy in pigs//*J. Biophotonics*. Vol.4(1-2). p.108-113.

[15] Wallace M., Lauwers G.Y., Chen Y., et al. 2011 Miami classification for probe-based confocal laser endomicroscopy//*Endoscopy*.Vol.43(10) p.882-891,

[16] Meining A., Shah R.J., Slivka A., Pleskow D., Chuttani R., Stevens P.D., Becker V., Chen Y.K. 2012 Classification of probe-based confocal laser endomicroscopy findings in pancreaticobiliary strictures//*Endoscopy*. Vol.44(3) p..251-257.

[17] Meining A., Chen Y.K., Pleskow D., Stevens P., Shah R.J., Chuttani R., Michalek J., Slivka A. 2011 Direct visualization of indeterminate pancreaticobiliary strictures with probe-based confocal laser endomicroscopy: multicenter experience//*Gastrointest. Endosc*. Vol.74(5). p.961-968.

[18] Davids P.H., Goren A.K., Rauws E.A. et al. 1992 Randomized trial of self-expanding metal stents versus polyethylene stents for malignant bile duct obstruction//*Lancet*. Vol.340. – River of 1488-1492.

Nuclear 3D organization and radiosensitivity

Y A Eidelman[1], S V Slanina[1], A V Aleshchenko[1], O V Sen'ko[2], A D Kononkova[3], S G Andreev[1,4]

[1] Institute of Biochemical Physics, Russian Academy of Sciences, Kosygin str. 4, 119334, Moscow, Russia
[2] Dorodnicyn Computing Centre, Russian Academy of Sciences, Vavilov st. 40, 119333 Moscow, Russia
[3]National University Higher School of Economic, Kochnovsky pr. 3, 125319, Moscow, Russia
[4]National Research Nuclear University MEPhI, Kashirskoye shosse 31, 115409, Moscow, Russia

E-mail: andreev_sg@mail.ru

Abstract. Current mechanisms of radiation-induced chromosomal aberration (CA) formation suggest misrepair of chromosomal lesions being in spatial proximity. In this case CAs have to depend on pattern of chromosomal contacts and on chromosome spatial organization in a cell nucleus. We were interested in whether variation of nucleus 3D organization results in difference of radiation induced CA formation frequency. Experimental data available do not provide information sufficient for definite conclusions. To have more deep insight in this issue we developed the biophysical modeling technique taking into account different levels of chromosome/nuclear organization and radiation damage of DNA and chromosomes. Computer experiments on gamma irradiation were carried out for two types of cells with different 3D organization of nuclei, preferentially peripheral and internal. CA frequencies were found to depend on spatial positioning of chromosomes within a nucleus which determines a pattern of interchromosomal contacts. For individual chromosomes this effect can be more pronounced than for genome averaged. Since significant part of aberrations, for example dicentrics, results in cell death, the proposed technique is capable of evaluating radiosensitivity of cells, both normal and cancer, with the incorporation of 3D genome information. This predictive technology allows to reduce uncertainties of prognosis of biological effects of radiation compared to phenomenological methods and may have variety of biomedical applications, in particular, in cancer radiation therapy.

1. Introduction

The term radiosensitivity is used widely as a parameter of an effect of interest. In radiobiology and in radiation therapy radiosensitivity often refers to cell death. For example, the dose of *e*-fold decrease in cell survival is used as a radosensitivity parameter. In this study we use this term to characterize the sensitivity of chromosome to radiation damage which are experimentally observed as chromosomal aberrations (CAs). CAs are divided into inter- and intrachromsomal, simple and complex, *etc.*; the examples and definitions are presented in figure 1.

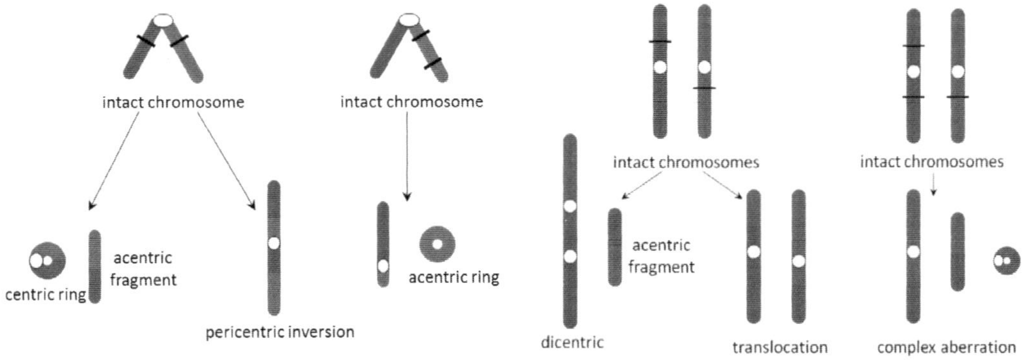

Figure 1. Pathways of induction of intra- and interchromosomal CAs. Chromosomes are schematically represented as colored sticks. Different chromosomes and/or chromosome arms are depicted in different colors. Centromeres are depicted as yellow circles.

According to the current view on radiation-induced CA formation mechanisms, aberrations are formed as a result of misrepair of chromosomal lesions being in spatial proximity [1]. Some CAs (for example, dicentrics and centric rings, see figure 1) lead to cell death, hence they can serve as a cytogenetic marker of lethal effects of radiation and as a quantitative characteristics of radiosensitivity [2, 3].

On the other hand, some CAs involving particular loci correlate with several types of malignancies [2]. The inversion between genes RET and PTC1 in chromosome 10 is associated with thyroid cancer; the translocation between genes AF4 (chromosome 4) and MLL (chromosome 11) is a predictor of acute lymphoblast leukemia; the translocation between genes ABL (chromosome 9) and BCR (chromosome 22) is a prognostic factor of chronic myeloid leukemia.

To enable CA induction, at least two lesions should be formed in chromosomal loci either being in contact at the time of irradiation ("contact first" pathway) or coming in contact shortly after irradiation ("breakage first" pathway) [1]. In both cases CAs depend heavily on chromosomal contact probability and hence on spatial organization of chromosomes in a nucleus and on contact pattern within chromosomes.

The present work is aimed at the question whether differences in nucleus 3D organization result in differences in radiation induced CA frequencies. As a marker of nucleus 3D organization, radial distribution of chromosomes in an interphase nucleus is introduced. We present the modeling approach for calculating yields of radiation induced CAs of various types taking spatial organization of a nucleus into account. This polymer physics based technique models structure of 46 human chromosomes, dynamics of megabase subunits of all chromosomes, as well as DNA double-strand break induction, repair and misrepair. CA frequencies are shown to depend on radial distribution of chromosomes. For individual chromosomes this effect can be more pronounced than for genome averaged. The approach presented demonstrates the principal opportunity of predicting radiosensitivity of both normal and malignant cells that may be beneficial for applications in tumor radiotherapy.

2. Methods

Multiscale modeling of 3D organization and dynamics of all 46 chromosomes in a human cell nucleus is performed by extension of methods developed previously [4, 5]. Each chromosome is represented as a coarse-grained chain of semi-rigid elements corresponding to megabase sized chromosomal subunits, or domains [5, 6]. Interphase chromosome structures are established in the course of mitotic chromosome decondensation. Subunits from the same chromosome interact through a Lennard-Jones potential, subunits from different chromosomes through a truncated Lennard-Jones potential with only the excluded volume component. A Metropolis method with the dynamic variant of Monte Carlo

(MC) technique is used, at each MC step positions of elements are randomly changed [4, 5]. G1 state of the nucleus is reached when gyration radii of all chromosomes cease to change. In order to make correct comparisons between our predictions and FISH data on radial distribution of chromosomes in human lymphocyte nuclei [7], we design the following scheme of modeling of chromosome location in a nucleus. The spherical nucleus with 6 μm diameter is divided into 5 concentric layers of equal volume. Chromosome center of mass is placed in one of the four layers (except of the one adjacent to the nuclear shell) with probabilities p_1, p_2, p_3 and p_4. Model 1: probabilities p_i are chosen so that distribution of domains of a given chromosome in five layers agrees with the experimental distribution of FISH signal [7]. Radial distribution of chromosomal domains correlates with that for centers of mass but they are not the same: center of mass can be in one layer while part of domains in other layers, especially for large chromosomes. Model 2: probability p_1 equals p_4 for the same chromosome in model 1, p_2 equals p_3 in model 1, *etc.* Thus the chromosomes which are preferentially peripheral in model 1 (i.e. high p_4 and low p_1) are preferentially central in model 2 (i.e. high p_1 and low p_4) and *vice versa*.

Modeling of chromosomal aberration formation at the contacts of damaged subunits is performed as follows. DNA double-strand breaks (DSBs) induced by γ-rays are distributed uniformly (on average) over the genome; they are eliminated in the course of repair by non-homologous end joining pathway; rates of lesion contact formation and decay are determined on the basis of calculated dynamics of all chromosomal subunits [4, 5]; if two damaged subunits come in contact before they are eliminated, they interact (with probability *p*, free parameter) resulting in exchange aberration formation. Identification of CAs as simple or complex is done by classification used in [8] for mFISH-detected aberrations. The term "apparently simple" used hereafter means all CAs classified as simple by mFISH technique. They include real simple aberrations as well as complex aberrations misidentified as simple (e.g. if one of exchanges comprising the complex CA is an inversion not detectable by mFISH).

3. Results

Two types of cells with different spatial organization of nucleus were studied. In this work the 3D organization of nuclei is characterized by the radial distribution of chromosomes. In cells of the first type (nucleus model 1) radial distribution of chromosomes in a nucleus is modeled so that it is able to reproduce the experimental distribution measured by the FISH technique in human lymphocytes [7], see Methods. In model 1 most of big chromosomes are distributed mainly at nucleus periphery and most of small chromosomes occupy the middle of the nucleus. In cells of the second type (nucleus model 2) we introduce another, hypothetical distribution of chromosomes.

To elucidate the impact of spatial distribution of chromosomes in nuclei on formation of radiation induced chromosomal aberrations we determine how structural parameters of chromosomes, chromosomal contacts and chromosomal aberrations are shaped by spatial distribution of chromosomes in nuclear space. We build radial profiles of chromosomes in nuclei by exploring model 1, figure 2. Next, we generate another nucleus model in which, unlike nucleus of type 1, radial distribution of chromosomes is more centered, figure 3. For both nucleus types we calculate the following structural characteristics: radial distribution of chromosomal domain density, domain pairwise contact radial distribution, total number and number of intra- versus inter-chromosomal contacts, contact distribution of genomic separation between contacting subunits.

One can observe from figure 4 b that chromatin is distributed in model 1 preferentially at nuclear periphery. 72% of chromatin is located at distance more than 2 μм from the center, i.e. further than 2/3 of nucleus radius. It corresponds to relatively homogeneous chromatin density, figure 4 a, red curve. In model 2 chromatin is located generally closer to the nucleus center: only 54% of chromatin is further than 2/3 of nucleus radius (figure 4 b) and chromatin density is dramatically higher at low distances than at high distances (figure 4 a). We conclude that cell type dependence of chromatin density radial distribution is shaped by distribution of chromosomes in 3D space of cell nucleus.

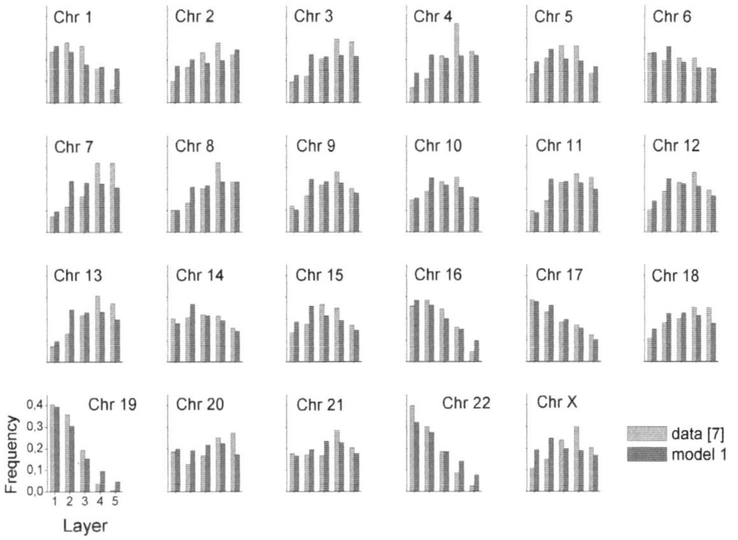

Figure 2. Radial distribution of chromatin in 23 chromosome pairs in cells of first type, model 1. Simulated distribution of chromosomes in cells of first type (model 1) are compared with FISH data [7] for human lymphocyte cells. A nucleus is divided into five spherical layers of equal volume, layer 1 is central, layer 5 is adjacent to the nuclear shell.

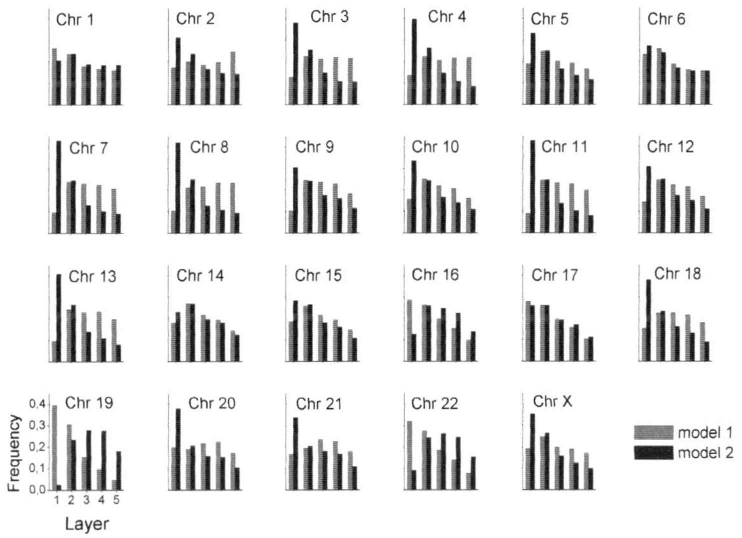

Figure 3. Radial distribution of chromatin in 23 chromosome pairs in cells of second type, model 2. Data on chromosome distributions generated by model 1 are presented for comparison.

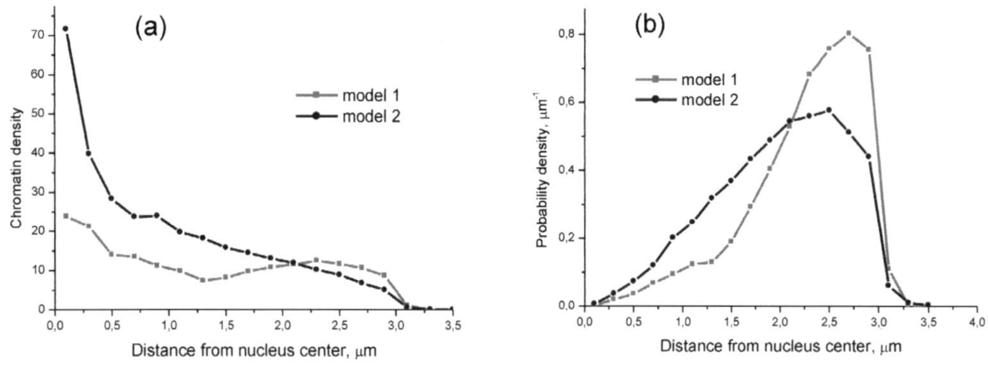

Figure 4. Radial distribution of chromatin in nuclei of two types. (a) - radial dependence of density of chromosomal subunits, or megabase domains in a nucleus. (b) - probability density of location of chromosome subunits in a spherical layer at distance r, r+δr from nucleus center.

Difference in spatial arrangements of chromosomes for nuclei of two types results in differences in radial distributions of total, intra and inter-chromosomal contacts (figures 5, 6). Two models differ significantly in total number of interchromosomal contacts (37200 vs 47900) whereas total number of intrachromosomal contacts is almost unchanged (18200 vs 19200).

Distributions of intrachmosomal contacts over genomic distance between contacting subunits, n(s), are well correlated for nuclei types 1 and 2, figure 7. Integral of n(s) equals the total number of intrachromosomal non-neighbors contacts. Comparing two models, increase of total contacts by 2.6% exists in model 2 with respect to model 1 for genome averaged n(s) distribution.

To predict frequency of chromosomal aberrations formed at contacts of damaged chromosomal loci it is necessary to assess the probability of contact-to-exchange formation, i.e. the probability that a contact of two damaged sites results in an exchange aberration. This parameter was determined by modeling simulation and description of experimental dose response curve for simple dicentrics induced by low LET γ-radiation in human lymphocytes [8]. These data were obtained by labeling of all chromosomes with multiple color fluorescence probes, or multicolor FISH (mFISH), figure 8.

The term complex (exchange) means a sum over all type of exchanges where number of damaged chromosome subunits is more than two. Scoring of aberrations in our *in silico* experiments is

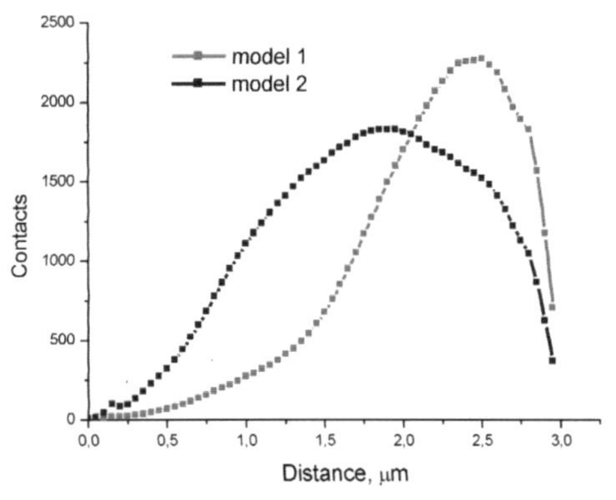

Figure 5. Radial distribution of total (intra- and interchromosomal) contacts for two nucleus types. Model 1 – 55400 contacts per nucleus, model 2 – 67100 contacts.

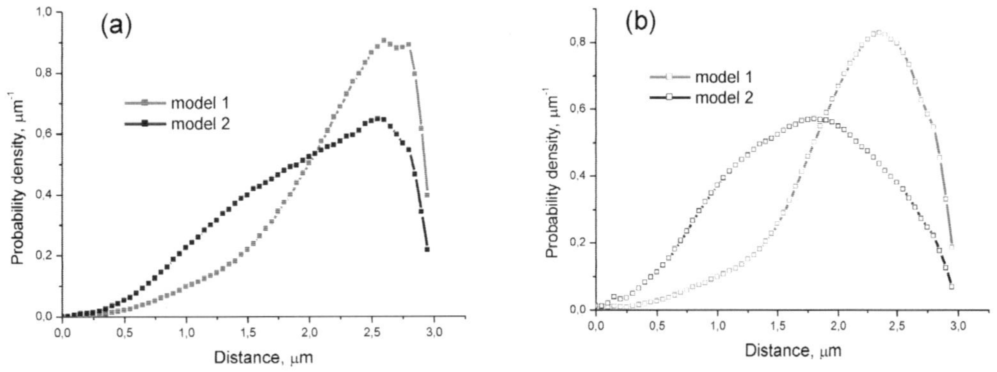

Figure 6. Intra- and interchromosomal contacts radial distributions for two nucleus models. Contacts between neighbors are excluded. (a) – intrachromosomal. Model 1 – 18200 contacts per nucleus, model 2 – 19200 contacts. (b) – interchromosomal. Model 1 - 37200 contacts per nucleus, model 2 – 47900 contacts.

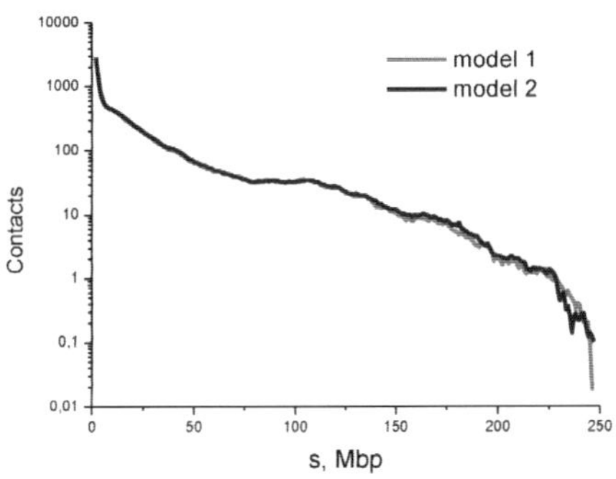

Figure 7. Distribution of intrachromosomal contacts over genomic distance between contacting subunits for two nucleus models, n(s). Distribution is genome averaged, i.e. averaged over all chromosomes. Only non-neighbor contacts are considered.

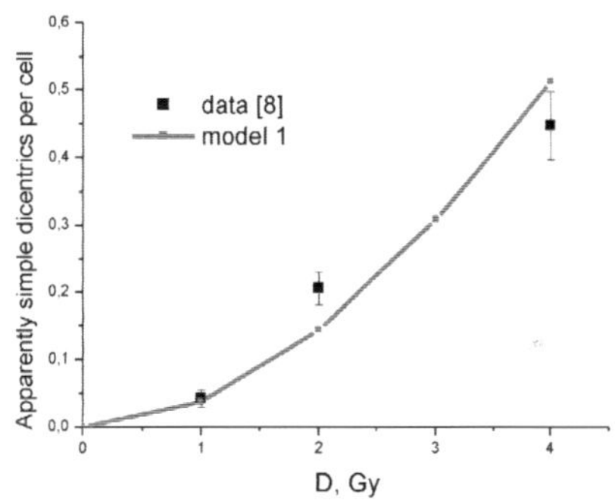

Figure 8. Dose response for γ-ray induced apparently simple dicentrics, observed by mFISH experiment on human lymphocytes [8] and by computer experiments with multilevel modeling technique developed here. Simulation for model 1, contact-to-exchange probability p=0.013.

performed in concert with the procedure applied in experimental mFISH scoring [8]. One can observe from figure 9 that aberrations yield, both simple and complex, is increased for nuclei of type 2 with respect to type 1. This conclusion is in agreement with previous observations (figures 5, 6) about increased yield of contacts per nucleus type 2.

Figure 10 demonstrates the results of *in silico* experiments to determine the aberration frequencies, both simple and complex, for nuclei with different spatial organization. These data suggest that in nuclei with more centric chromatin (model 2) there are more aberrations than in nuclei with more peripheral chromosome distribution (model 1). The frequency difference depends on type of aberrations, and in general complex aberrations are more sensitive to chromosome distribution than simple aberrations (the same is seen for the dose curves, see figure 9). Therefore, nucleus model 2

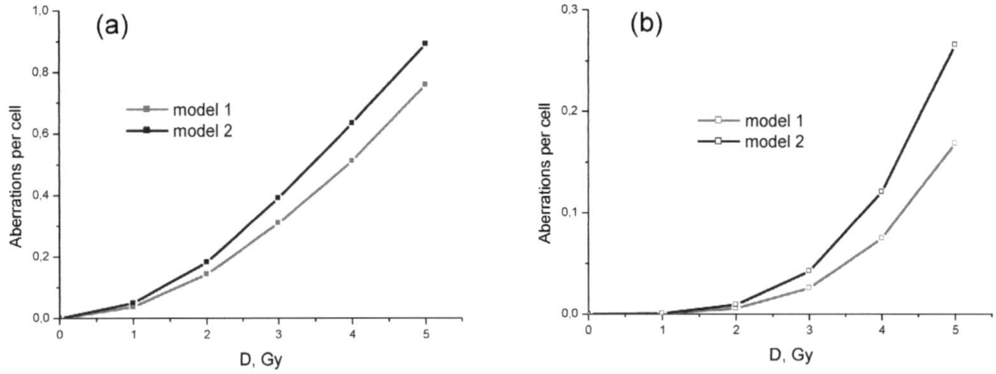

Figure 9. Dose response curves for γ-ray induced chromosomal aberrations predicted for nuclei differing in spatial organization. p=0.013. (a) - apparently simple dicentrics; (b) - visible complexes.

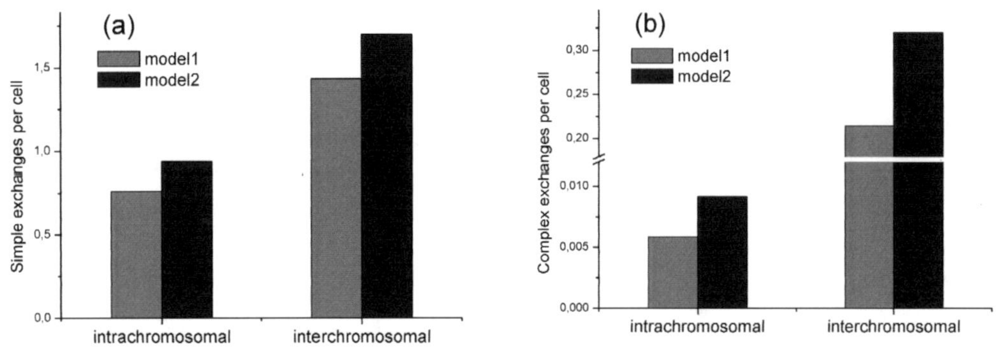

Figure 10. Predicted frequencies of simple and complex aberrations induced by γ-rays at dose 5 Gy in two cell types with different spatial nuclear organization (model 1, 2). (a) – simple exchange aberrations, intra- and interchromosomal. (b) – complex aberrations, intra- and interchromosomal.

where contacted domains are located preferentially in the center of nucleus, than in model 1 (figures 5, 6) is more radiosensitive in accord with the criterion of complex aberration formation.

Next, we study chromosomal aberration frequencies for specific chromosomes. We choose pairs of chromosome 4 and 11, as well as 9 and 22, because these chromosomes contain important proto-oncogenes. Their specific rearrangements, due to exchange aberrations of translocation type, may result in human malignancy with high incidence frequency. As nononcogenic reference point we selected the pair of chromosomes 4 and 7. In model 1 these chromosomes are amongst the most peripheral [7], whereas in model 2 they are located near the nucleus center. This predicts that contacts between them have to be rare for model 1 and frequent for nucleus model 2.

Assessment of radiation induced aberrations between pairs of specific chromosomes for two models reveals complex relationship between nuclear structure and exchange aberrations (figure 11). Exchange yield may be either independent (figure 11 a) or dependent (figure 11 b, c) on nuclear organization type. The highest difference is 4.9 times for exchanges between chromosomes 4 and 7.

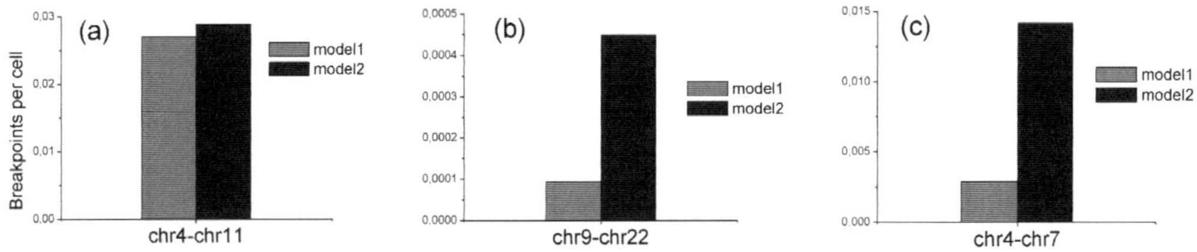

Figure 11. Frequencies of exchanges (simple plus complex) between specific chromosomes. (a) – chromosomes 4 and 11, (b) – chromosomes 9 and 22, (c) – chromosomes 4 and 7. γ-rays, D=5 Gy.

The results in figure 11 suggest that different cell types may have very different radiosensitivity with respect to specific chromosome rearrangements while having similar radiosensitivity with respect to total aberrations yield.

4. Discussion

The general aim of this work is to assess complex relationships between nuclear organization and radiation induced chromosomal aberrations. Experimental information is highly obscure in this field. The multilevel mechanistic modeling approach presented here demonstrates a wealth of opportunities to have deeper insight in this problem.

The new information about contacts obtained in *in silico* experiments reveals that the shape of genome-averaged distribution of genomic separation between contacting subunits within a chromosome n(s) is almost independent of type of nucleus organization considered. The total number of contacts, excluding pairs of neighboring domains, differs between models slightly, by 2.6%. These estimates are in accord with the data in figure 6 where the total number of intrachromosomal contacts (integral on r) differs by 4% between models. On the other hand, interchromosomal contact differences between nucleus models 1 and 2 are much higher, about 30%. We propose the following explanation of this finding. Larger chromosomes give larger contribution to the yield of chromosome aberrations [4] and "surface" mechanism of exchange aberrations formation (i.e. interchromsomal exchanges are formed at the contacts of damaged subunits on the surface of chromosomal territories) operates for at least simple exchanges [4]. Larger chromosomes have larger surface area. In model 1 large chromosomes are positioned mainly near the nuclear membrane and each chromosome contacts few others. In model 2 large chromosomes are preferentially close to the center of the nucleus and they contact many other chromosomes. Small chromosomes in model 2 are mainly placed at the periphery and have few contacts relative to model 1. Since smaller chromosomes have low overall impact on total number of interchromosomal contacts, these findings reveal that interchromosomal contacts have to be more abundant in nucleus model 2 with more centric positions of larger chromosomes. However, we still cannot exclude the possible impact of another factor influencing contact differences, as to limited number of nuclei simulated (50) for each type of model, although the statistics of radiation damage simulations by Monte Carlo is 50000 per dose point per model, or 500000 in total. This question requires further investigation.

To envision three-dimensional view on nuclear organization and 3D pattern of chromosomal contacts we generated *in silico* the snapshot of nucleus configurations of two types, corresponding to models 1 and 2, figure 12. Multiple contacts of loci on chromosomes 9 and 22 (figure 12 b) may result in interchromosomal exchange aberrations between these chromosomes in model 2 whereas rare or no contacts (figure 12 a) do not result in exchange aberration in model 1. Since these chromosomes contain oncogenes ABL and BCR, the high accurate prediction technologies of this kind of rearrangements are requested and it will have important scientific and bio-medical significance.

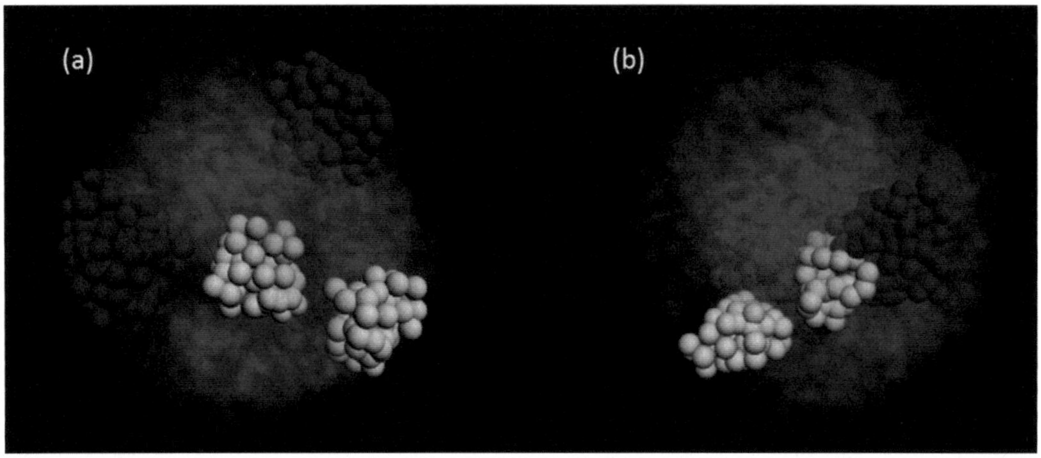

Figure 12. Visualization of contacts for two computer models of nuclear 3D organization. (a) – model 1. (b) – model 2. Colored balls represent 1 Mbp chromosomal domains. Two chromosomes 9 (141 Mb) and 22 (50 Mb) are in different colors, chromosome 9 in dark blue, chromosome 22 in cyan. Transparent light gray balls are domains from all other chromosomes.

We show that different spatial organization of a nucleus may underlie different frequencies of intra- and interchromosomal contacts which affect the frequency of γ-ray induced chromosomal aberrations. The idea that the problem of elucidation of chromosome aberration mechanisms primarily comes down to a problem of chromosomal contacts was proposed in [9] on the basis of polymer modeling of human chromosomes. This proposal is supported by chromosome conformation capture technique (Hi-C) measurements of intrachromosomal contacts [10]. The results of the present work further confirm the conclusion in [9] and uncover the complex aspects of the contacts vs chromosome aberrations problem. Further studies are required to quantify chromosome structure and chromosome-chromosome interactions in order to explore complex relationships between nuclear 3D organization and chromosomal aberrations.

Talking about perspectives, one can mention that the proposed *in silico* approach allows assessment of radiosensitivity of normal and cancer cells by incorporating differences in their nuclear 3D organization, as well as future generalization to ionizing radiation of different quality, i.e. ions of different linear energy transfer (LET). Our approach is capable to take into account variations of individual radiosensitivity which may be related, in particular, to variations of 3D genome structure. Spatial organization of a nucleus is proposed to serve as a prognostic marker of sensitivity to radiation induced cancer and as a tool of distinguishing sensitive human subpopulations. This may be of great significance in molecular epidemiology and risk prediction studies, including risks from cosmic particles [11]. These ideas are in line with the modern tendency of using radiobiological and genomic information in radiation therapy. Current studies are aimed at exploring a role and visualization of spatial distribution of complex DNA double strand breaks in human tumor therapy by carbon ions [12]. Another important task concerning our aims of research is a physical and biological determinants of radiotherapy toxicity under individual plans of irradiation [13]. Combining *in vivo* and *in silico* data on genomics, nuclear 3D structure and radiosensitivity one can hope to identify patients with reduced toxicity or improved tumor cure. Chromosomal exchange aberrations like translocations may serve as a test to identify patients with increased radiosensitivity. Detection of chromosomal translocation in blood lymphocytes was proposed to be considered as a perspective predictive assay for detection of radiosensitive individuals [14]. Our multilevel modeling approach enables additional studies of matters mentioned above in order to predict outcomes of future experiments and allows to acquire new information in the fields where conducting experiments on patients is impossible.

In conclusion, the work presented demonstrates that 3D organization of a nucleus may affect frequencies of radiation induced chromosomal aberrations of different complexity through spatial distribution of contacts and chromosome topology. The predictive mechanistic approach allows to assess radiosensitivity of normal vs cancer cells with lower uncertainties than purely phenomenological methods. The results and *in silico* technologies presented may promote a search of new directions for individualization of radiotherapy planning though it is obviously only beginning of the road.

Acknowledgements

This work was supported by grant 14-01-00825 from Russian Foundation for Basic Research and Competitiveness Program of National Research Nuclear University MEPhI.

References

[1] Bryant P E Origin of chromosome aberrations: mechanisms 2007 in: *Chromosomal alterations: methods, results and importance in human health* ed G Obe and Vijayalaxmi (Berlin Heidelberg: Springer-Verlag) pp 177–99
[2] Durante M and Loeffler J S Charged particles in radiation oncology 2010 *Nat. Rev. Clin. Oncol.* **7** 37–43
[3] Carante M P and Ballarini F Calculating Variations in Biological Effectiveness for a 62 MeV Proton Beam 2016 *Front Oncol.* **6** 76
[4] Eidelman Yu A, Ritter S, Nasonova E, Lee R, Talyzina T A and Andreev S G Prediction of dose response for radiation induced exchange aberrations taking cell cycle delays into account 2006 *Radiat. Prot. Dosim.* **122** 185–7
[5] Eidelman Y A, Slanina S V, Salnikov I V and Andreev SG Mechanistic modelling allows to assess pathways of DNA lesion interactions underlying chromosome aberration formation 2012 *Rus. J. Genet.* **48** 1247–56
[6] Eidelman Y and Andreev S G Biophysical study of the globular organisation of interphase chromosomes 2002 *Radiat. Prot. Dosim.* **99** 217–8
[7] Boyle S, Gilchrist S, Bridger J M, Mahy N L, Ellis J A and Bickmore W A The spatial organization of human chromosomes within the nuclei of normal and emerin-mutant cells 2001 *Hum. Mol. Genet.* **10** 211–9
[8] Loucas B D and Cornforth M N Complex chromosome exchanges induced by gamma rays in human lymphocytes: an mFISH study 2001 *Radiat. Res.* **155** 660–71
[9] Andreev S G and Edel'man Iu A Globular model of interphase chromosome and intrachromosomal exchange aberrations 1999 *Radiats. Biol. Radioecol.* **39** 10–20
[10] Zhang Y, McCord R P, Ho Y J, Lajoie B R, Hildebrand D G, Simon A C, Becker M S, Alt F W and Dekker J Spatial organization of the mouse genome and its role in recurrent chromosomal translocations 2012 *Cell* **148** 908–21
[11] Norbury J W *et al.* Galactic cosmic ray simulation at the NASA Space Radiation Laboratory 2016 *Life Sci. Space Res. (Ams.)* **8** 38–51
[12] Oike T *et al.* Visualization of complex DNA double-strand breaks in a tumor treated with carbon ion radiotherapy 2016 *Sci. Rep.* **6** 22275
[13] Scaife J E, Barnett G C, Noble D J, Jena R, Thomas S J, West C M and Burnet N G Exploiting biological and physical determinants of radiotherapy toxicity to individualize treatment 2015 *Br. J. Radiol.* **88** 20150172
[14] Huber R, Braselmann H, Geinitz H, Jaehnert I, Baumgartner A, Thamm R, Figel M, Molls M and Zitzelsberger H Chromosomal radiosensitivity and acute radiation side effects after radiotherapy in tumour patients - a follow-up study 2011 *Radiat. Oncol.* **6** 32

IOP Conf. Series: Journal of Physics: Conf. Series **784** (2017) 012010 doi:10.1088/1742-6596/784/1/012010

Assessing the impact of lyophilization process in production of implants based on the bacterial cellulose using Raman spectroscopy method

E V Timchenko, P E Timchenko, E V Pisareva, M Yu Vlasov, V V Revin, N A Klenova and A A Asadova

Samara National Research University, 443086, Samara, Moskovskoye shosse 34, Russia.

E-mail: laser-optics.timchenko@mail.ru

Abstract. In this article we present the research results of lyophilization process influence on the composition of hybrid materials based on the bacterial cellulose (BC) using Raman spectroscopy method. As an object of research was used BC, as well as hybrids based on it, comprising the various combinations of hydroxyapatite (HAP) and collagen. Our studies showed that during the lyophilization process changes the ratio of the individual components. It was found that for samples hybrid based on BC with addition of HAP occurs increase of PO_4^{3-} peak intensity in the region 956 cm^{-1} with decreasing width, which indicates a change in the degree of HAP crystallinity.

Introduction

Bacterial cellulose (BC) is a unique biopolymer, based on which you can solve many problems of medicine. It also used in the food, pulp and paper, electronics, biotechnology and chemical industries. BC, unlike plant-derived cellulose, synthesized by microorganisms in pure form without lignin impurities and other related components [1]. With essentially the same chemical structure as the plant cellulose, BC shows two very important qualities - subtle porosity and mechanical strength. BC is produced by the acetic acid bacteria and it is biodegradable and biocompatible.

According to researches [2,3], faster tissue regeneration and wound closure occurs using biomaterials based on BC compared to bandages. In another experiment [4] a primary effect of BC bandaging material was examined on laboratory rats. It is proved that the presence of BC may promote wound healing due to accumulation of the extracellular matrix.

Using biocompatible biologically active compounds is a method of improving pure BC and allows to get closer to "more ideal" dressing material having improved biocompatibility, antimicrobial properties, and also capable of retaining water [5].

Additional treatment at creation of implants based on BC can be a process of lyophilization (freeze-drying), which can increase the shelf life of biological materials, simplify storage conditions. However, the influence of lyophilization on the structure of implants based on BC are not known.

To investigate the effect of lyophilization process on the component composition of based on BC implants may be used physical methods of control.

In the article [6] Raman spectroscopy method was used to estimate structural differences of celluloses of different origin: produced in the presence of pectin and xyloglucan, as well as commercial cellulose and cellulose extracted from apples. It found that different Raman bands may be used for the successful and rapid determination of the crystallinity degree.

In the article [2] shows the main wavenumbers corresponding to the components of the bacterial cellulose. Thus, a peak at 1095 cm^{-1} is a sealing ring and stretching vibrations C-O-C of glycosidic bond stretching. In the region 1270-1500 cm^{-1} bands C-C-H, O-C-H, C-O-H, H-C-H and CH_2 are related to stretching vibrations of cellulose.

The purpose of research was to study the influence of lyophilization process on the composition of hybrid materials based on BC using Raman spectroscopy.

1. Materials and methods of research

In this study bacterial cellulose (BC) has been investigated, as well as hybrids, based on BC, including the different combinations of HAP and collagen: BC and HAP, collagen and BC, BC and HAP with collagen.

To obtain the BC was used the strain Gluconacetobacter sucrofermentas B-11267 [11]. The obtained membrane was washed with distilled water, dried in an oven at 80°C to constant weight. HAP was prepared according to the original method [8].

To prepare composites were added to the solution: 1% HAP (HAP + the BC); 0.5% of collagen (collagen + the BC), 1% HAP and 0.5% collagen (the BC + HAP + collagen). A part of composites further lyophilized using standard method [9].

The spectral characteristics of the materials based on BC were studied using an experimental stand, including the high-resolution digital spectrometer Andor Shamrock sr-303i with built-in cooling camera DV420A-OE, a fiber optic probe for Raman spectroscopy RPB785, combined with laser module LuxxMaster LML-785.0RB-04 (up to 500 mW, wavelength 785 nm) [12]. Separation of Raman spectrum from the background of autofluorescence carried out by means polynomial approximation of a fluorescent component and subtracting it from the recorded spectra.

2. Results and discussion

As a result of studies were obtained spectra of lyophilization and illyophilization types of the bacterial cellulose.

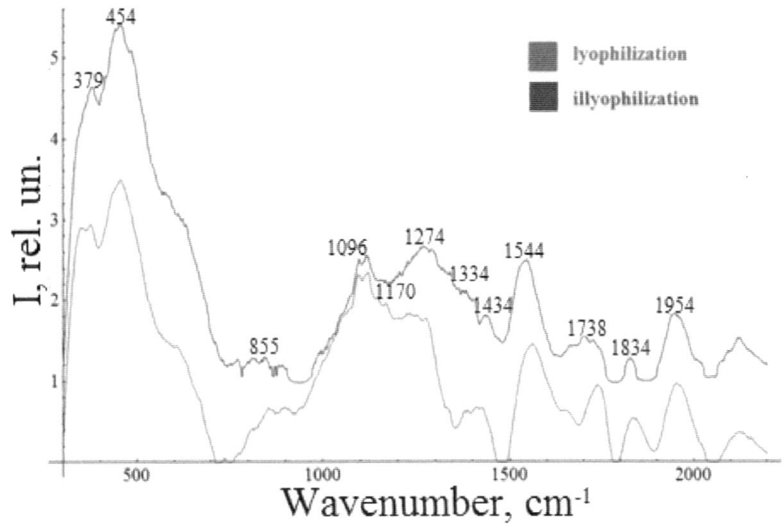

Figure 1. The normalized by the average value, Raman spectra of lyophilization and illyophilization samples BC

Figure 1 shows that in illyophilization samples in the wavenumber range 1180-1270 cm^{-1} dominated the curves H-C-C, H-C-O, H-C-H, C-O-H of cellulose compared to with lyophilization samples. These changes are due to the fact that the sample has been dried during the lyophilization. Is also seen that the lyophilization process does not change the main peak 1096 cm^{-1}, which corresponds to the cellulose.

The following figure 2 shows the spectral analysis of various studied hybrids based on BC.

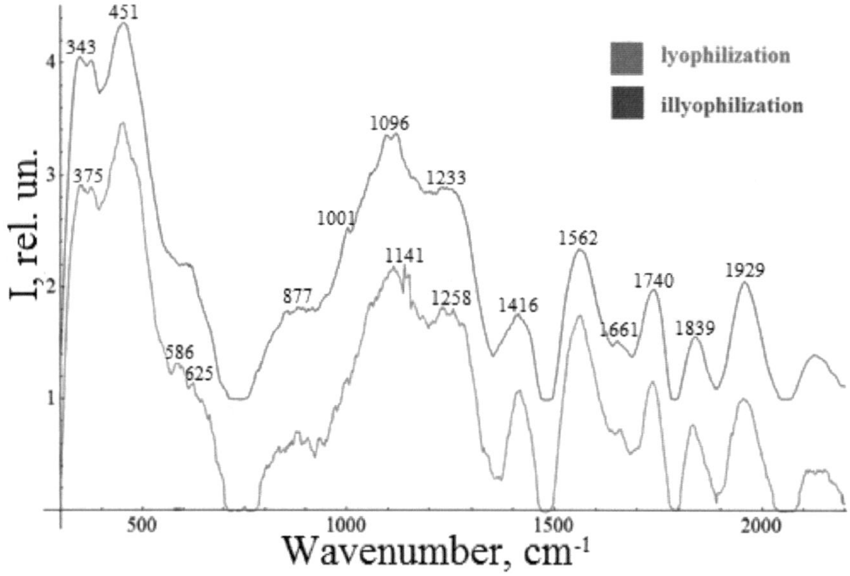

(a)

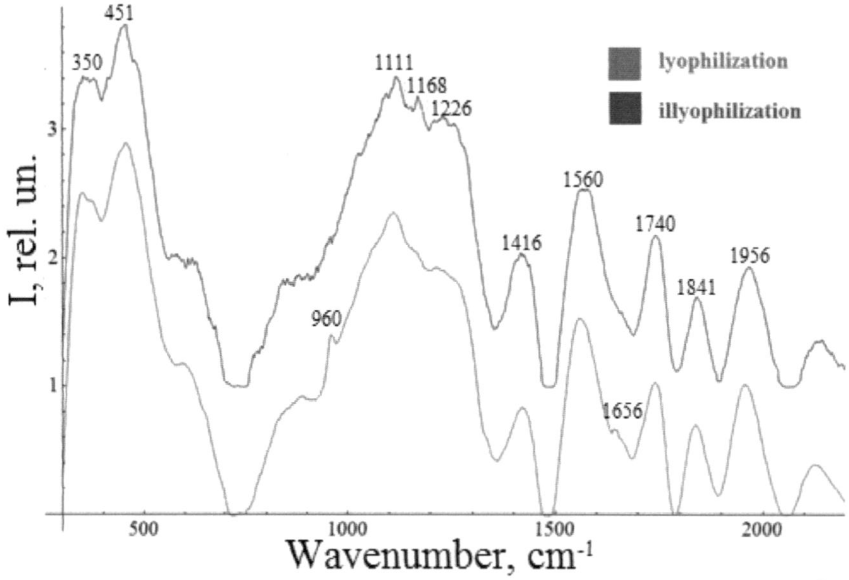

(b)

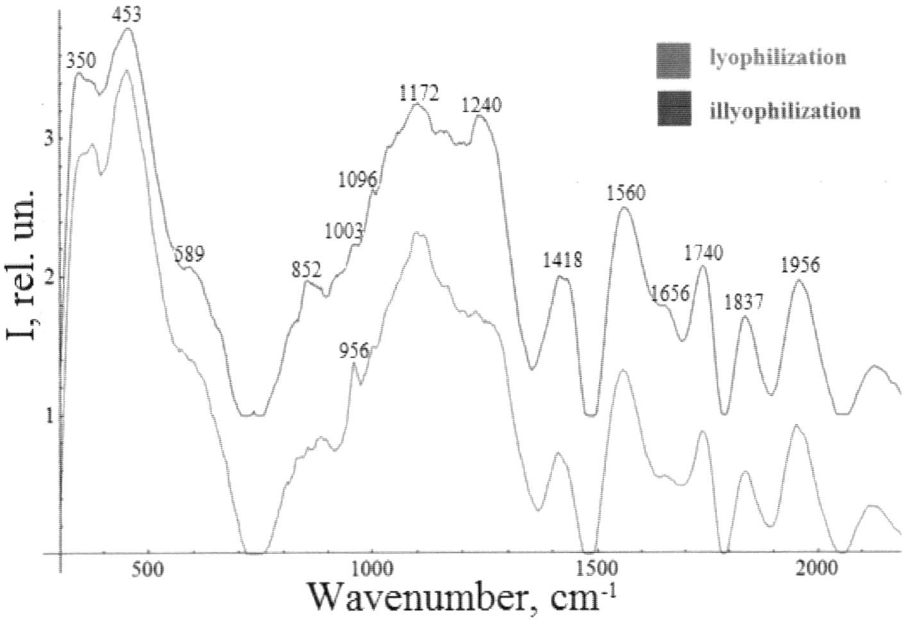

(c)

Figure 2. Normalized to the average value of the Raman spectra of lyophilization and illyophilization hybrid based on BC with the addition of (a) collagen, (b) HAP, (c) HAP and collagen

Analysis of figure 2a shows that in the region 1230-1270 cm^{-1}, corresponding to vibrations of Amide III, there is no significant changes during the lyophilization process, which indicates that the lyophilization process has no damaging effect on the collagen matrix in the composition of the BC.

From Figure 2b, it is seen that after lyophilization of samples with HAP, observed the increasing of PO_4^{3-} peak intensity in the region 956 cm^{-1} with decreasing width. In the works of authors [7, 10] have shown that the line of symmetric stretching vibration $PO_4^{3-}(\nu_1)$ has a high sensitivity to the mineral encirclement: the frequency and form of this line depends on the local surroundings and change as a result of substitution of anionic groups and changes in the degree of crystallinity. The substituted carbonate-apatite B-type phosphate ν_1 line appears in the range of 955–959 cm^{-1}. In a crystalline unsubstituted hydroxyapatite line ν_1 shifted to 962–965 cm^{-1}. Finally, at a frequency line 945–950 cm^{-1} indicates the presence of a disordered lattice phosphate apatite. In general, the phosphate band ν_1 is a superposition of three components; it usually has an asymmetric shape due to the disordered vibrations contribution of phosphate and unsubstituted hydroxyapatite. It is obvious that process of lyophilization effects on the ratio of these components, apparently resulting to decrease in the proportion of highly crystalline apatite.

Conclusions

Experimentally found features of Raman spectra for various composite implants based on BC.

It was found that during the lyophilization process changes the ratio of the individual components. Thus there is an amplification of cellulose line in the normalized spectrum, indicating at reducing the proportion of the individual components. Moreover, for the hybrid based on BC samples supplemented with HAP revealed an increase in the PO_4^{3-} peak intensity near 956 cm^{-1} with decreasing width, which indicates a change in the degree of HAP crystallinity.

Acknowledgments
This work was supported by the Ministry of Education and Science of the Russian Federation.

References

[1] Belgacem, M. N. Monomers, Polymers and Composites from Renewable Resources / M. N Belgacem, A. Gandini. – Amsterdam : Elsevier,2008. – 553 p.

[2] Fu, L. P. Zhou, S. Zhang, G. Yang Evaluation of bacterial nanocellulose-based uniform wound dressing for large area skin transplantation /L. Fu, P. Zhou, S. Zhang, G. Yang //Mater Sci Eng C. - 2013. - №33.- p. 2995–3000

[3] Fu, L. Skin tissue repair materials from bacterial cellulose by a multilayer fermentation method / L. Fu ,Y. Zhang, C. Li, Z.H. Wu, Q. Zhuo, X. Huang// J Mater Chem. – 2012.- №22. - p. 12349–12357

[4] Park, S.U. The possibility of microbial cellulose for dressing and scaffold materials/S.U. Park, B.K. Lee, M.S. Kim, K.K. Park, W.J. Sung, H.Y. Kim//Int Wound J. - 2014. - №11, p. 35–43

[5] Nakayama A., High mechanical strength double-network hydrogel with bacterial cellulose A. Nakayama, A. Kakugo, J.P. Gong, Y. Osada, M. Takai, T. Erata Adv Funct Mater. – 2004. - №14. - p. 1124–1128

[6] Szymańska-Chargot , M. Sensing the Structural Differences in Cellulose from Apple and Bacterial Cell Wall Materials by Raman and FT-IR Spectroscopy/ M. Szymańska-Chargot , J. Cybulska, Zdunek// Sensors. - 2011. - №11(6). – p.5543-5560

[7] Tarnowski, C.P. Mineralization of Developing Mouse Calvaria as Revealed by Raman Microspectroscopy/ C.P. Tarnowski, M.A. Ignelzi, M.D. Morris// Journal of Bone and Mineral Research. - 2002. - № 6. p. 1118–1126.

[8] Bio-implants to restore the structure and volume of bone tissue, Russian patent № 2372892, appl. №2008124263/14, publ. 20.11.2009 Bull. № 32, 5 p.

[9] Blankov, B.I. Application lyophilization Microbiology / B.I. Blankov, D.L. Klebanov. - M.: Medicine, 1961. - 263 p.

[10] Kiseleva, D.V. The use of micro-Raman spectroscopy to study the structural features of biogenic apatite / D.V. Kiseleva // YEARBOOK-2009, Proc. of the IGG UB RAS, 2010. - № 157, P. 332–335

[11] Revin, V.V. Получение бактериальной целлюлозы и нанокомпозиционных материалов/ V.V. Revin, E.V. Liyaskina, N.A. Pestov.- Saransk: publishing Mordovian University, 2014. 128 p.

[12] Timchenko, E.V. Application of Raman spectroscopy to assess the demineralization of bone grafts in their preparation /E.V. Timchenko, P.E. Timchenko, L.T. Volova, M.N. Milyakova, N.A. Maksimenko, L.A. Taskina // Journal of Optical. - 2015. - issue 3. – T.82 - P. 30-36.

IOP Conf. Series: Journal of Physics: Conf. Series **784** (2017) 012011 doi:10.1088/1742-6596/784/1/012011

Electrophysiological predictors of sudden cardiac death on physical exercise test in young athletes

L A Balykova[1], A A Kotlyarov[2], S A Ivyanskiy[1], A A Shirokova[1], K A Miheeva[1], L M Makarov[3]

[1] N.P.Ogarev Mordovia State University, Saransk, Russia
[2] Obninsk Institute for Nuclear Power Engineering Moscow, Russia
[3] Center for Syncope and Cardiac Arrhythmia in Children and Adolescents of Federal Medico-biological Agency Clinical Hospital Moscow, Russia.

E-mail: larisabakykova@yandex.ru

Abstract. The problem of sudden death of young athletes continues to be actual. Among its reasons, primary electric myocardium diseases along with organic heart troubles (cardiomyopathies, cordites, anomalies of coronary arteries) take an important place. The most frequent variant of channelopathesis long QT syndrome (LQTS). Both inherited and acquired LQTS may be the reason of sudden cardiac death during physical activity and have to be revealed prior to sports admission. LQTS diagnostics in young athletes become problematic due to secondary exercise-related QT prolongation. Physical load test may reveal myocardium electric instability and enhance LQTS diagnostics accuracy without genetic testing. The aim was to study electrophysiological parameters of myocardium repolarization and reveal the signs of electrical instability as predictors of the life-threatening arrhythmias in young athletes during physical exercise test. In conclusion, electrophysiological myocardium parameters during physical exercise test noted to be markers of electrical myocardial instability and in combination with the other Schwartz criteria, was evidenced the inherited or acquired LQTS. QTc prolongation in athletes at the peak of exercise as well as in early recovery period were noted to be additional predictor life-threatening arrhythmias and sudden cardiac death in young athletes

1. Introdution

Professional athletes are a special risk group for sudden cardiac death SCD [1]. Undiagnosed myocardium diseases (myocarditis, cardiomyopathies), coronary artery anomalies, channelopathies, heart defects and other considered cardiac reasons can be the causes of fatal events in sport activities [2,3]. Nevertheless quite often the cause of athletes' death remains unresolved. Therefore, thorough medical examination of children to reveal cardiovascular diseases prior to sport admission is so important. However, the volume of this examination so far is under discussion [4-5]. American preparticipation screening protocol means careful evaluation of anamnesis, family history and carrying out a physical examination of athletes, whereas European ones, which proved their efficiency for SCD prevention, recommends in addition carrying out standard 12-leads ECG [6,7].

According to the order of the Ministry of Health №314H from 01.03.2016 Russian preparticipation screening includes not only physical examination and resting ECS, but an obligatory echocardiography and exercise test [8]. However up till now accurate aims of exercise testing in young athletes have not yet been established. Traditionally in sports medicine an exercise test was used to assess physical working capacity and to reveal latent coronary failure [9.10]. But now, there are a lot of evidences concerning stress test importance for diagnostics of inherited arrhythmias, especially long QT syndrome [11, 12]. Different variants of LQTS respond differently to stress. In some cases this helps to make an accurate diagnosis without genetic testing [12, 13].

Different variants of LQTS respond differently to stress. In some cases this helps to make an accurate diagnosis without genetic testing [12, 13].

On the other hand, some of young athletes have secondary QT prolongation (associated with athletic training and other problems), which requires in-depth examination to rule out the primary LQTS [14].

The aim of our study was to establish the norm values of electrophysiological parameters (characterized myocardium repolarization) on exercise stress and to develop simple efficient technologyf or noninvasive long QT diagnostics and SCD prevention in athletes.

2. Methods

100 young athletes 11-16 yrs old (including 56 boys) - sport school attendants and 100 untrained healthy children of a similar age and gender were enrolled in our study. All the children underwent complex examination, including exercise bicycle test by Bruise protocol with synchronic ECG recording. Intervals RR and QT were measured manually. Corrected QT interval was calculate by Bazett and Fridericia formulas [15, 16]. Dispersions of absolute and corrected QT interval as well as QT delta (variability) during test were calculated.

3. Results

Potentially dangerous changes - inverted T–waves, long QTc (exceeding 460-470 ms) and short QTc (less than 320ms) were not revealed at resting ECG in any of the athletes. Nonsignificant QTc prolongation (440-460 ms) and shortening were recorded in 2 athletes. Our data corresponds to Basavarajaiah S. findings that QTc prolongation at rest occurs only in 0.4% of adult athletes and requires careful monitoring, temporary deconditioning and additional examination (including exercise test) to rule out primary LQTS, particularly if QTc exceeds 500 ms [17].

In our study mean QTc in the athletes at rest was compared to that in untrained, but absolute QT at rest and at the 1–st stage of exercise was higher in the athletes, due to bradycardia and myocardial hypertrophy.

Dynamics of myocardium repolarization in the young athletes during exercise–test was similar to that in healthy untrained adolescents (18). Gradual, smooth QT interval shortening in both groups were recorded, which correlated with gradual intensification of exercise. But at peak load QT in the athletes was considerably lower than in untrained due to hyperadaptation of repolarization processes to the heart rate grows (19). During recovery period QT duration in athletes returned to the original values more promptly than in untrained.

QTc interval in both groups increased ad maximum during low-intensity load despite the heart rate growth, but did not exceed 450-460 ms for boys and girls respectively(Figure 1). Minimal QTc values in both groups were recorded at peak load. But the athletes showed a more marked QTc shortening (less than 390 ms) compared to untrained- so-called “QTc hyperadaptation”. During recovery period QTc returned to the original level in most of the athletes similar to the control group.

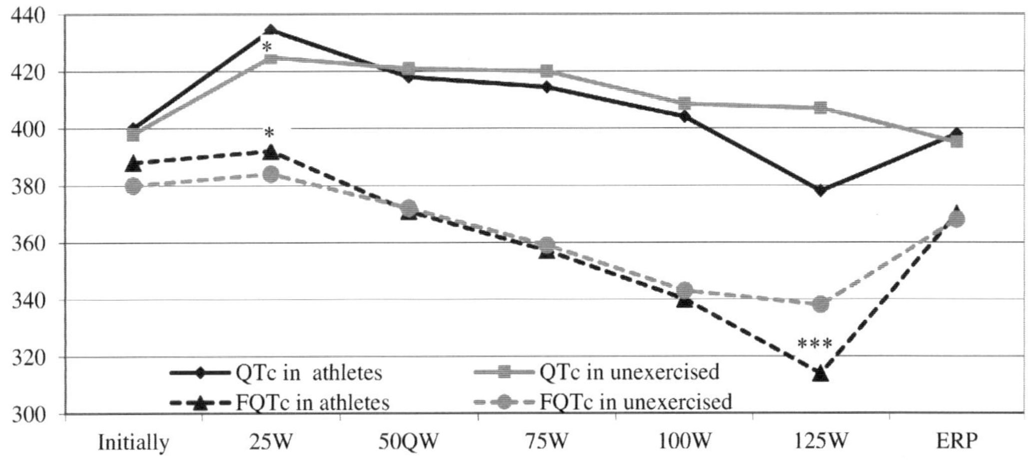

Figure 1. Dynamics of the QTc interval during exercise load test. ERP – early recovery period, * - athletesvs untrained p<0,05; *** - athletes vs untrained p<0,005/

As established by Doctor I.A. Gorbunova, QTc prolongation in athletes exceeding 450 ms in early recovery period suggests myocardium pathology and requires additional e:***:nation to rule out primary and secondary LQT. According to Peter J. Schwarts latest recommendations QTc duration over 480 ms during early recovery in adults is an additional diagnostic criteria of LQT syndrome (20). But Sy R.W. and coauthors (whose findings underlie this thesis) stated that LQTS can be diagnosed with high probability if QTc already exceeds 450 ms (21).

As Dong-Cheng showed QT adaptation to RR in early recovery period is impaired both in primary and secondary QT prolongation [22].17 children in our group had no adequate QTc short ending at peak load or early recovery period, in 3 out of them rhythm disturbances during test were recorded. Thorough medical examination showed myocarditis in 1 athlete, congenital and acquired LQTS in 2 athletes, signs of myocardial athletic remodeling, medication-induced damage or/and overtraining - in 14.

As an example, we can present a case history and exercise-test protocol of a 15-year old girl who has been in gymnastics for 9 years. Her training regime has been 8-9 hours per week. Her family history is negative, but she had one presyncopy not connected with exercise. 12-lead ECG showed bradycardia, normal QTc. EcoCG indices also were normal for athletes. Graded exercise test resulted in significant QTc prolongation (up to 500 ms per 75 Wt), which persisted up to 8 minutes of recovery. Measurement data were similar in automatic and manual estimation. 24 hours monitoring revealed bradycardia, T-wave alteration, QTc prolongation over 460 ms for more than 80% of the assessment time. Potential LQTS was diagnosed by the sum of 3.5 scores by Swartz's criteria.

Dynamics of the QTc interval, estimated by different formulas was similar in both groups, but BazettQTc values were higher than the Fridericia ones. We have determined the norm values of an absolute and corrected (by Bazett and Fridericia formulas) QT interval in young athletes during exercise test.

Dispersions of absolute and corrected QT interval during moderate exercise (50-100W) in athletes exceeded the corresponding indices in untrained children, while at peak load they were considerably lower. Like in a control group QT dispersion in athletes increased significantly during early recovery, up to the highest value in the test (did notexceeded35-40ms). Although QT dispersion alone cannot be a strong predictor of SCD, its prognostic significance is considerably growing in the presence of structural myocardial pathology.

IOP Conf. Series: Journal of Physics: Conf. Series **784** (2017) 012011 doi:10.1088/1742-6596/784/1/012011

Table 1. Dynamics of the QTc interval, estimated by different formulas.

	QT, ms	QTcBazett, ms	QTcFridericia, ms
Rest	373 (334-401)	400 (375-448)	390 (357-402)
1 st. (25W)	340 (310-366)	423 (400-460)	390 (369-410)
2 st. (50 W)	295 (270-308)	418 (396-449)	373 (344-395)
3 st. (75 W)	265 (247-289)	413 (394-445)	356 (330-380)
4 st. (100W)	240 (228-254)	404 (389-419)	335 (325-355)
5 st. (125 W)	228 (220-235)	392 (374-405)	325 (312-337)
6st. (150W)	220 (212-228)	380 (360-390)	315 (305-328)
Recovery	320 (300-340)	390 (361-450)	378 (330-400)

Table 2. Dynamics of myocardium repolarization dispersion (50, 5 and 95 ‰) in young athletes during loading test.

	QTd, ms		*dQTc*, ms		*FdQTc*, ms	
	Untrained	Athletes	Untrained	Athletes	Untrained	Athletes
Initially	22 (20-30)	28 (20-30)	(18-30)	(20-37)	(17-31)	(18-29)
25 W	18 (10-30)	22 (14-30)	18 (14-32)	(15-31)	(19,5-26)	(14-26)
50 W	14 (10-20)	15 (10-21)	(15-31)	(13-38)	(16-26)	(14-28)
75 W	12 (10-18)	13 (10-20)	23 (12-26)	19 (16-31)	(11-21)	(12-27)
100 W	8 (5-10)	7 (5-10)	15 (8-17)	12,5 (14-17)	(7-14)	(7-14,5)
125 W	7 (5-10)	6 (5-7,5)	(8-16)	(7-14)	(7-13)	(7-14)
Early recovery period	26 (19-35)	20 (15-30)	(16-40)	(13-37)	(18-35)	(16-36)

Another significant sign of electrical instability of the myocardium is QT variability that is the difference between maximal and minimal QTc during the exercise test. As Wong showed in healthy untrained children this index was less than40 ms, while in patients with LQTS it exceeds 60-70 ms [23]. In our study permissible QT variability in athletes was 50-60ms.More significant QT variability was noted in 21% of athletes, 17 of whom had cardiac remodeling and overtraining syndrome, 3 - structural cardiovascular diseases and 1 - probably LQTS.

4. Conclusions

In young athletes, as well as in the untrained maximal QTc (by Bazett) have been recorded at the initial stages (not exceeding 460 ms) and minimal - at the peak of exercise.

QTc prolongation over 390 ms at peak load and over 450 ms at min 4 of recovery, as well as QTc variability more than 50 ms during exercise test in young athletes can be considered as a signs of cardiac pathology. They require an extensive medical examination to rule out primary and secondary LQT.

Maximal QT and QTc interval dispersions have recorded at the initial stage of exercise and during recovery, they did not exceed 35 and 40 ms, respectively.

Assessment of myocardial repolarization dynamics during exercise in young athletes can be a significant diagnostic method for revealing some conditions associated with SCD.

References

[1] Priori SG, Blomström-Lundqvist C, Mazzanti A, Blom N, Borggrefe M, Camm J, Elliott PM, Fitzsimons D, Hatala R, Hindricks G, Kirchhof P, Kjeldsen K, Kuck KH, Hernandez-Madrid A, Nikolaou N, Norekvål TM, Spaulding C, Van Veldhuisen DJ. 2015 ESC Guidelines for the management of patients with ventricular arrhythmias and the prevention of sudden cardiac death: The Task Force for the Management of Patients with Ventricular Arrhythmias and the Prevention of Sudden Cardiac Death of the European Society of Cardiology (ESC). Endorsed by: Association for European Paediatric and Congenital Cardiology (AEPC). Eur Heart J. 2015 Nov 1;36(41):2793-867.

[2] Harmon KG, Drezner JA, Wilson MG, Sharma S. Incidence of sudden cardiac death in athletes: a state-of-the-art review. Heart 2014;100:1227–1234.

[3] Schmied C, Borjesson M. Sudden cardiac death in athletes. J Intern Med 2014;275:93–103.

[4] Assanelli D1, Levaggi R, Carré F, Sharma S, Deligiannis A, Mellwig KP, Tahmi M, Vinetti G, Aliverti P Cost-effectiveness of pre-participation screening of athletes with ECG in Europe and Algeria. Intern Emerg Med. 2015 Mar;10(2):143-50.

[5] Menafoglio A, Di Valentino M, Porretta AP, Foglia P, Segatto JM, Siragusa P, Pezzoli R, Maggi M, Romano GA, Moschovitis G, Gallino A. Cardiovascular evaluation of middle-aged individuals engaged in high-intensity sport activities: implications for workload, yield and economic costs. Br J Sports Med 2015 Jun;49(11):757-61.

[6] Maron B.J., Thompson P.D., Ackerman M.J., Balady G., Berger S., Cohen D., Dimeff R., Douglas P.S., Glover D.W., Hutter A.M. J.R, Krauss M.D., Maron M.S., Mitten M.J., Roberts W.O., Puffer J.C. Recommendations and considerations related to preparticipation screening for cardiovascular abnormalities in competitive athletes: 2007 update: A scientific statement from the American heart association council on nutrition, physical activity, and metabolism: endorsed by the American College of Cardiology Foundation // Circulation. – 2007.-№115.- P.1643-1655.

[7] Corrado D., Pelliccia A., Bjørnstad H., Vanhees L., Biffi A., Borjesson M., Panhuyzen-Goedkoop N., Deligiannis A., Solberg E., Dugmore D., Mellwig K.P., Assanelli D., Delise P., van-Buuren F., Anastasakis A., Heidbuchel H., Hoffmann E., Fagard R., Priori S.G., Basso C., Arbustini E., Blomstrom-Lundqvist C., McKenna W.J., Thiene G.. Cardiovascular pre-participation screening of young competitive athletes for prevention of sudden death: proposal for a common European protocol Consensus Statement of the Study Group of Sport Cardiology of the Working Group of Cardiac Rehabilitation and Exercise Physiology and the Working Group of Myocardial and Pericardial Diseases of the European Society of Cardiology // Eur. Heart. – 2005.-№26.-P.516–524.

[8] The order of the Russian Ministry of Health from 01.03.2016 N 134N "About the Order of the organization of delivery of health care to the persons playing physical culture and sport ..." The Official Internet portal of legal information http://www.pravo.gov.ru

[9] Chou R, Arora B, Dana T, Fu R, Walker M, Humphrey L Screening asymptomatic adults with

resting or exercise electrocardiography: a review of the evidence for the U.S. Preventive Services Task Force. Ann Intern Med. 2011 Sep 20;155(6):375-85

[10] Sarma S, Levine BD. Beyond the Bruce Protocol: Advanced Exercise Testing for the Sports Cardiologist. CardiolClin. 2016 Nov;34(4):603-608.

[11] Dillenburg R.F. and Hamiltion R.M. Is Exercise Testing Useful in Identifying Congenital Long QT Syndrom? Am Heart J. 2002 Oct. 140(4): 678-83.

[12] Takenaka K, Ai T, Shimizu W, et al. Exercise stress test amplifies genotype-phenotype correlation in the LQT1 and LQT2 forms of the long-QT syndrome. Circulation. 2003;107:838–844.

[13] Wong J.A., Gula L.J., Klein G.J., Yee R., Skanes A.C., KrahnA.D. Utility of Treadmill Testing in Identification and Genotype Prediction in Long-QT SyndromeCircArrhythmElectrophysiol2010;3;120-125;

[14] Corrado D, Pelliccia A, Heidbuchel H, et al. Recommendations for interpretation of 12-lead electrocardiogram in the athlete. Eur Heart J 2010; 31(2): 243-59.

[15] Bazett H. Analysis of the time relations of electrocardiograms. Heart 1920;7:353-70.

[16] Fridericia L. Die systolendauer im Elektrokardiogramm bei normalen Menschen und bei Herzkranken. Act Med. Scand. 1920;53: 469-472

[17] Basavarajaiah S, Wilson M, Whyte G, Shah A, Behr E, Sharma S
Prevalence and significance of an isolated long QT interval in elite athletes. EurHeart J. 2007 Dec;28(23):2944-9.

[18] Makarov L.M., Balykova L.A., Komolyatova V.N., Gorbunova I.A., Changes of an interval of QT in the course of assay with the dosed exercise stress at healthy teenagers of 11-15 years//the Cardiology - 2012 No. 9 of Page 15-21.

[19] Balykova L.A., Makarov L.M., Gorbunova I.A., Markelova I.A. QT Interval Adaptation in Young Athletes During Physical Load Test Abstract Congress ISHNE Moscow 2010.

[20] Schwartz P.J., CrottiL..QTc Behavior During Exercise and Genetic Testing for the Long-QT Syndrome // Circulation -2011-№124- P.2181-2184.

[21] Sy R.W., van der Werf C., Chatta I.S., Chockalingam P., Adler A., Healey J.S., Perrin M., Gollob M.H., Skanes A.C., Yee R., GulaL.J, Leong–Sit P., Viskin S., Klein G.J., Wilde A.A., Krahn A.D. Derivation and validation of a simple exercise-based algorithm for prediction of genetic testing in relatives of LQTS probands // Circulation -2011-№124- P.2187-2194.

[22] Dong-sheng Gao, Wei-yi Fang, Christine Chiu-Man, Joel Kirsh, Gil Gross, Robert M HamiltonQT hysteresis in long-QT syndrome children with exercise testing. Chin. Med. J.2007; Feb;120(3):179-82.

[23] Wong JA, Gula LJ, Klein GJ, Yee R, Skanes AC, Krahn AD. Utility of Treadmill Testing in Identification and Genotype Prediction in Long-QT Syndrome. CircArrhythmElectrophysiol. 2010;3:120-125.

Use of semiconductor nanocrystals to encode microbeads for multiplexed analysis of biological samples

Mikhail A Berestovoy[1#], Regina S Bilan[1#], Victor Krivenkov[1], Igor Nabiev[1,2] and Alyona Sukhanova[1,2*]

[1]Laboratory of Nano-Bioengineering, National Research Nuclear University MEPhI (Moscow Engineering Physics Institute), Moscow, Russian Federation,
[2]Laboratoire de Recherche en Nanosciences, LRN - EA4682, Université de Reims Champagne-Ardenne, Reims, France.
[#]These authors contributed equally to the study.

E-mail: alyona.sukhanova@univ-reims.fr

Abstract. Microbeads encoded with semiconductor quantum dots (QDs) are suitable tools for multiplexed analyses of various biological markers using flow cytometry. We have prepared a panel of microbeads encoded with QDs of different colors emitting with different luminescence intensities using the layer-by-layer deposition technique, which consists in layering of alternately charged polyelectrolytes and negatively charged QDs onto the surface of microbeads. This method allows QDs to be separated with one or several polymer layers in order to prevent Förster resonance energy transfer (FRET) and the resultant quenching of QD fluorescence in multicolor microbeads.

1. Introduction

In recent years, interest in optically encoded microbeads has been growing due to their potential applications in medical diagnostics [1]. Microbeads can be encoded with either organic fluorophores or semiconductor quantum dots (QDs) [2–6]. The use of microbeads encoded with organic fluorophores with different fluorescence maxima requires complex equipment with multiple excitation lasers for decoding, which complicates the analysis and increases the cost of instrumentation. QDs constitute a new class of fluorophores that have a number of advantages over routinely used fluorescent organic dyes, such as a high brightness, resistance to photobleaching, a wide excitation spectrum, and a narrow emission spectrum (Figure 1) [7, 8]. The unique optical characteristics of QDs make it possible to use them in multiplexed analyses, as well as in multicolor imaging [9], because QDs of different colors can be simultaneously excited using a single radiation source, whereas their fluorescence peaks can be effectively separated and detected in different channels of a flow cytometer. Therefore, multiplexed analysis by means of flow cytometry using QD-encoded microbeads is easy to perform and does not require sophisticated equipment.

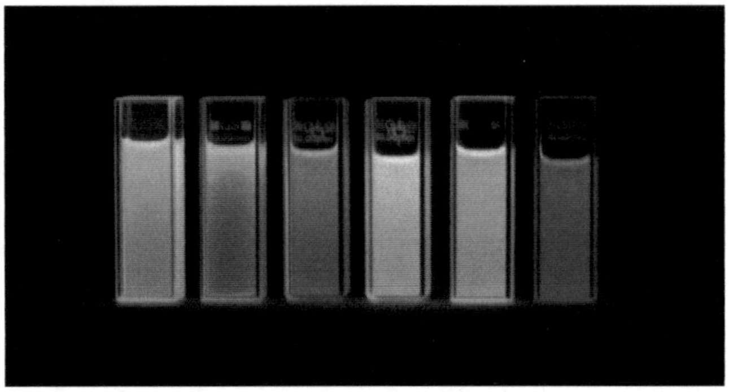

Figure 1. Fluorescent semiconductor nanocrystals of different sizes simultaneously excited using a single light source (http://lnbe.mephi.ru/ru).

A number of approaches have been designed for engineering QD-encoded microparticles: embedding of QDs into the microbeads during the manufacturing [3], incorporation of QDs into pores of swollen microbeads, [2, 4], doping of mesoporous microbeads with QDs [5], the concentration-controlled flow-focusing technique [6], and layer-by-layer deposition [10]. By using several colors of QDs and several intensity levels, it is theoretically possible to create thousands or even millions of individual optical codes [2].

In this study, we have prepared a panel of microbeads encoded with QDs of different colors at different luminescence intensity levels using the layer-by-layer deposition technique, which consists in layering of alternately charged polyelectrolytes and negatively charged QDs onto the surface of microbeads. We established the number of QDs that was adsorbed onto the microbeads and investigated some physicochemical aspects of electrostatic adsorption of water-soluble negatively charged QDs onto a positively charged microbead surface. Fluorescence lifetime measurements were used for investigation of FRET between QDs in the solution and within QD-encoded microbeads. The technique of optical encoding developed in this study makes it possible to fabricate various multicolor microbead populations for multiplexed flow-cytometry analysis.

2. Materials and Methods

2.1 Preparation and characterization of water-soluble QDs

CdSe/ZnS core/shell QDs [7] with fluorescence maxima at 508, 590, and 640 nm and CdSe/ZnS/CdS/ZnS multishell QDs [9] were synthesized and kindly provided by Dr. Pavel Samokhvalov (Laboratory of Nano-Bioengineering, Moscow Engineering Physics Institute, Moscow, Russia). Originally water-insoluble QDs coated with trioctylphosphine oxide (TOPO) were solubilized by replacing TOPO with D,L-cysteine [11] and subsequently replacing the cysteine with derivatives of polyethylene glycol containing thiol and carboxyl terminal groups ($SH\text{-}(PEG)_{12}\text{-}COOH$, Thermo Fisher Scientific, USA) as described earlier [12]. The QD concentrations were determined from their extinction spectra at the positions of the first excitonic maxima [13]. The concentration of multishell QDs was estimated from the experimental values of molar extinction coefficient and molecular weight (100,000 g/(mol·cm) and 92,600 g/mol, respectively) (data not shown). The hydrodynamic sizes of water-soluble QDs were determined by dynamic light scattering measurements using a Zetasizer NanoZS analyzer (Malvern, UK).

2.2 Preparation of QD-encoded microbeads

Optical encoding of microbeads with QDs was performed using the adapted procedure of layer-by-layer deposition of alternately charged polymer layers and water-soluble QDs described in our previous studies (Figure 2) [14, 15]. According to the adapted protocol, $3 \cdot 10^7$ carboxylated melamine resin microbeads with a diameter of 4.08 μm (Microparticles GmbH, Berlin, Germany) were suspended in 0.5 ml of ultrapure water and mixed with 0.5 ml of 2 mg/ml poly(allylamine hydrochloride) (PAH, Mw = 15,000 Da, Sigma-Aldrich, USA) dissolved in 0.5 M NaCl. The suspension was sonicated and incubated while shaking for 20 min; then, the microbeads were washed from excess polymer by triple centrifugation (2000*g* for 5 min) and resuspension in ultrapure water. For deposition of the next polymer layer, the microbeads were resuspended in 0.5 ml of ultrapure water and mixed with 0.5 ml of 2 mg/ml poly(sodium 4-styrensulfonate) (PSS, Mw = 70,000 Da, Sigma-Aldrich, USA) dissolved in 0.5 M NaCl. As with the first layer, the suspension was sonicated and incubated while shaking for 20 min, and the microbeads were washed by triple centrifugation and resuspension in ultrapure water. The procedure of polymer deposition and washing was repeated until five alternately charged polymer layers were formed on the microbead surface, the outer layer consisting of positively charged PAH. Then, the microbeads were resuspended in 100 μl of ultrapure water, and various amounts (from 0.4 to 146 μg) of negatively charged water-soluble QDs were added while stirring the suspension. The mixture was incubated while shaking for 20 min, after which the microbeads were washed by three cycles of centrifugation and resuspension. For preparation of single-color microbeads, the next stage was the deposition of five layers of alternately charged polymers, as described above. In the case of two- or three-color microbeads, the deposition of QDs of different colors was alternated with deposition of positively charged PAH layers, after which five alternately charged polymer layers were deposited. The resultant encoded microbeads were stored at 4°C in the dark.

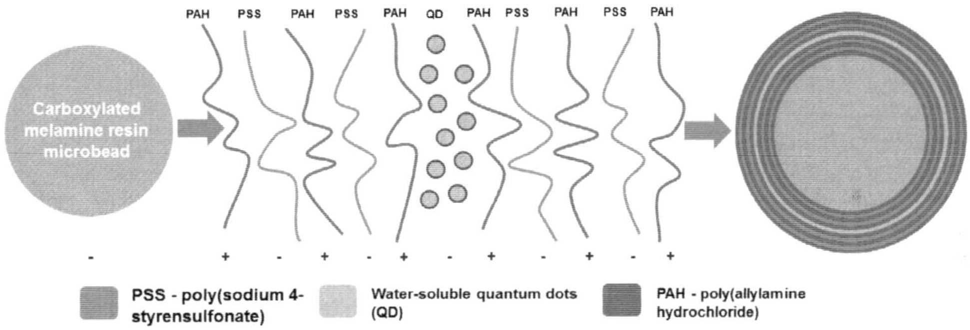

Figure 2. The method of layer-by-layer deposition [10] was used to obtain a population of microbeads encoded with QDs with different fluorescence peaks and fluorescence intensities: the surface of microbeads was successively coated with layers of differently charged polyelectrolytes and negatively charged QDs.

2.3 Analysis of photoluminescence properties

The fluorescence spectra of all samples were measured using an Infinite 200 PRO multimode reader (TECAN, Switzerland). The fluorescence decay times of QDs and QD-encoded microbeads were measured by means of a specially designed experimental setup. All the samples were excited with the use of a Tsunami pulse laser (Spectra Physics) (wavelength, 395 nm; pulse duration, 300 fs; pulse repetition rate, 8 MHz; pulse energy, 0.5 nJ). The photoluminescent signal from a sample was collected with an optical condenser and supplied to a monochromator, which was adjusted to the photoluminescence maxima of QDs. The signal from the detector (an avalanche photodiode) was analyzed using a PicoHarp time-correlated single photon counting (TCSPC) system (PicoQuant) with a set resolution of 256 ps.

2.4 Flow cytometry

QD-encoded microbeads were analyzed with the use of a FACSCanto flow cytometer (Becton Dickinson, USA). The three-color spectral code of the microbeads was excited with a blue laser (488 nm), and the fluorescence intensity was analyzed in three channels: FITC (530/30 nm), PE (575/25 nm), and PerCP (695/40 nm). The data were processed using the FACSDiva software (Becton Dickinson, USA).

2.5 Fluorescence microscopy

For studying the fluorescence of the QD-encoded microbeads, we used a Carl Zeiss Axio Scope A1 fluorescent microscope (Carl Zeiss, Germany). Before the experiment, QD-encoded microbead samples were mixed with a 20% glycerol solution and fixed on microscope slides. Imaging of the specimens was performed with the use of two optical filters, FITC (excitation wavelength, 495 nm; emission wavelength, 520 nm) and Alexa Fluor 568 (excitation wavelength, 568 nm; emission wavelength, 593 nm), and a Zeiss N-achroplan 40×/0.65 Ph2 lens. Image processing was performed using the ZEN2012 SP1 software (blue edition) (Carl Zeiss, Germany).

3. Results and Discussion

The layer-by-layer deposition technique for optical encoding of microbeads consists in applying layers of alternately charged polyelectrolytes and negatively charged QDs on the surface of microbeads. Polystyrene sulfonate (PSS) and polyallylamine (PAH) contain sulfo and amino groups ensuring the formation of multiple salt bonds between the polymers. For layering the polymers, preliminarily carboxylated microbeads are incubated with PAH to form a positively charged polymer layer on their surface. Then, excess PAH is removed from the solution, and negatively charged PSS is added. The layering is repeated until five polymer layers, the outer layer consisting of PAH, are formed. Then, the microbeads are incubated in the presence of negatively charged QDs carrying carboxyl groups on their surface. The QDs are electrostatically adsorbed onto the microbeads and form their optical code. Finally, five additional polymer layers are deposited onto the microbeads to give them stability.

Populations of monocolor microbeads encoded with QDs with fluorescence maxima at 508, 590, and 640 nm were studied in detail using flow cytometry. The range of fluorescence was analyzed in different channels of the flow cytometer (Figure 3). Fluorescence of microbeads encoded with QDs was excited using a single light source (a laser with a wavelength of 488 nm). Microbeads encoded with QDs had a high level of florescence and were detected at a high resolution.

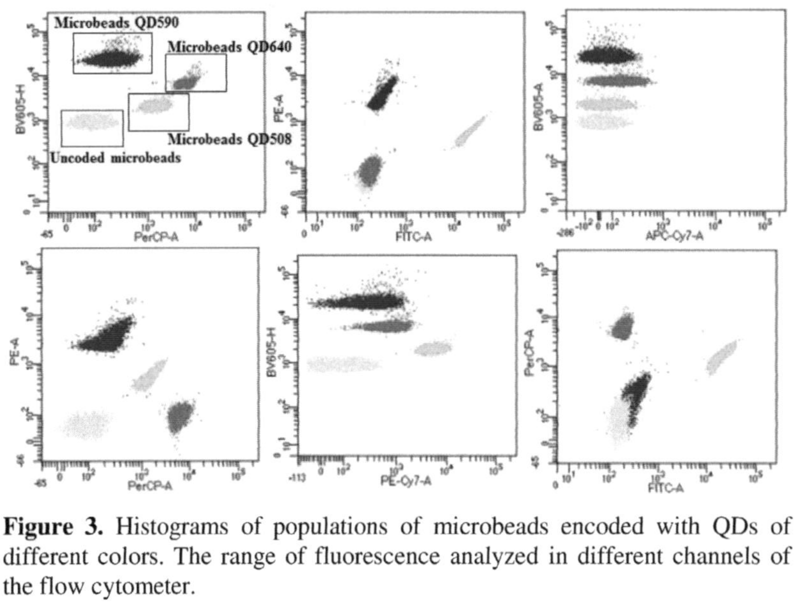

Figure 3. Histograms of populations of microbeads encoded with QDs of different colors. The range of fluorescence analyzed in different channels of the flow cytometer.

We estimated the maximum number of adsorbed QDs on the surface of microbeads. For this purpose, we compared the absorbance spectra of QDs in the stock solution and in the supernatant after incubation with microbeads. From the values of molecular weight and emission maximum of QDs, we calculated that number of QDs adsorbed on the surface of a single microbead, 4.08 μm in diameter, is $1.8–2·10^6$.

We performed fluorescence lifetime measurements of QDs in the solution and the same QDs adsorbed on microbeads. The photoluminescence decays of QDs emitting at 508 nm (QDs_{508}) in the solution and those adsorbed on microbeads are shown in Figure 4. The decay was significantly faster for QDs adsorbed on microbeads than for QDs in the solution. Specifically, adsorption onto microbeads shortened the decay time of QD_{508} fluorescence from 16.1 to 5.1 ns. This may have been accounted for by FRET between QDs [16] or specific influence of PAH, which interacts with functional groups on the QD shell surface.

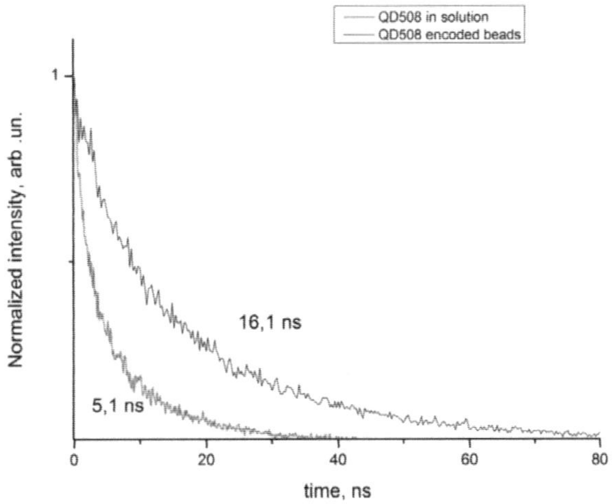

Figure 4. Fluorescence lifetime of QDs_{508} in the solution and those adsorbed on microbeads.

In multi-color QD-encoded microbeads, distance-dependent optical effects, such as FRET and photon reabsorption, may occur, affecting the resultant relative levels of the QD fluorescence intensity [6]. A way to reduce these optical effects is to spatially separate QDs of different colors, so that the distance between them is larger than the Förster energy transfer radius (5–8 nm for QDs) [17]. Furthermore, homo-FRET within a monochromatic QD population may occur due to distribution of QD sizes within the population and, hence, broadening of their fluorescence peak [16]. An overlap between the absorbance and emission spectra of the QD population leads to quenching of the blue side and enhancement of the red side of the photoluminescence peak, causing a red shift of the photoluminescence spectrum.

In order to prevent FRET between two QD populations, we spatially separated them with one or several polymer layers during the encoding process. We prepared single-color microbeads encoded with QDs_{508} (1) and two types of bicolor microbeads encoded with QDs_{508} and QDs_{590} that are spatially separated from each other by one (2) or three (3) polymer layers. The decay times of the samples were measured upon excitation at 508 nm (Table 1). All the three samples were characterized by comparable decay times of about 8 ns, which were almost independent of the number of polymer layers between two types of QDs and were considerably shorter than that in the QD_{508} solution (16 ns). If there were to be FRET, the decay times in cases (2) and (3) would be shorter than in case (1). However, reduction of decay time was not observed in these samples, which confirms the absence of

IOP Conf. Series: Journal of Physics: Conf. Series **784** (2017) 012012 doi:10.1088/1742-6596/784/1/012012

FRET. This result has allowed us to use an effective strategy of FRET prevention in preparing multi-color microbeads with a hundred of optical codes.

Table 1. The average decay time (t_{avr}) values for QDs_{508} in a solution, single-color microbeads encoded with QDs_{565}, and two types of bicolor microbeads encoded with QDs_{508} and QDs_{590} spatially separated by one and three polymer layers, respectively.

Sample	QDs_{508} in solution	Single-color QD_{508}-encoded microbeads	Bicolor microbeads encoded with QDs_{508} and QDs_{590} separated by:	
			one polymer layer	three polymer layers
t_{avr}, ns	16.0	7.35	7.86	9.5

The populations of QD-encoded microbeads have been studied in detail using fluorescence microscopy. QD-encoded microbead samples were mixed with 1% glycerol and fixed on microscope slides. Imaging of the specimens was performed with the use of two optical filters, FITC (excitation wavelength, 495 nm; emission wavelength, 520 nm) and Alexa Fluor 568 (excitation wavelength, 568 nm; emission wavelength, 593 nm), and a Zeiss N-achroplan 40×/0.65 Ph2 lens. The microbeads had a high brightness and were easily discerned under the microscope (Figure 5).

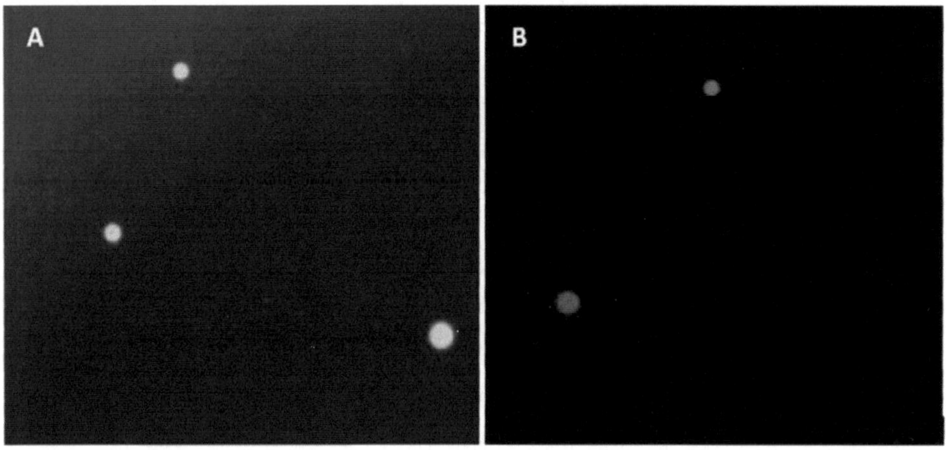

Figure 5. Typical fluorescence microscopy images of microbeads of different sizes encoded with (A) QDs_{508} and (B) QDs_{590}. All the images were taken at a magnification of 40×.

4. Conclusions

In this study, we have investigated some physicochemical aspects of electrostatic adsorption of negatively charged QDs on a positively charged microbead surface. In particular, we have estimated the maximal number of QDs that may be adsorbed in a single stage on the surface of a single microbead. We have also developed a technique of multistage QD adsorption for preparation of a panel of QD-encoded microbead populations with a wide range of fluorescence intensities. Using fluorescence lifetime measurements, we have found that the layer-by-layer deposition technique prevents FRET between QDs of different colors adsorbed at different stages.

The layer-by-layer technique for optical encoding of polymer microbeads with water-soluble QDs developed in this study makes it possible to obtain microbead populations for multiplexed analysis of biological markers using flow cytometry.

Acknowledgments
This study was supported by the Federal Targeted Program for Research and Development of the Ministry of Education and Science of the Russian Federation (grant no. 14.578.21.0054, contract no. RFMEFI57814X0054).

References
[1] Dunbar S A 2006 Clin. Chim. Acta 363 71
[2] Han M, Gao X, Su J Z and Nie S 2001 Nat. Biotechnol. 19 631
[3] Wang G, Leng Y, Dou H, Wang L, Li W, Wang X, Sun K, Shen L, Yuan X, Li J, Sun K, Han J, Xiao H and Li Y 2013 ACS Nano 7 471
[4] Cao Y C, Huang Z L, Liu T C, Wang H Q, Zhu X X, Wang Z, Zhao Y D, Liu M X and Luo Q M 2006 Anal. Biochem. 351 193
[5] Gao X and Nie S 2004 Anal. Chem. 76 2406
[6] Fournier-Bidoz S, Jennings T L, Klostranec J M, Fung W, Rhee A, Li D and Chan W C 2008 Angew. Chem. Int. Edit. 47 5577
[7] Baranov A V, Rakovich Y P, Donegan J F, Perova T S, Moore R A, Talapin D V, Rogach A L, Masumoto Y and Nabiev I 2003 Phys. Rev. B 68 165306
[8] Samokhvalov P, Linkov P, Michel J, Molinari M and Nabiev I 2014 Proc. SPIE *of* Colloidal Nanoparticles for Biomedical Applications **8955** 89550S
[9] Sukhanova A and Nabiev I 2008 Crit. Rev. Oncol. Hematol. 68 39
[10] Susha A S, Caruso F, Rogach A L, Sukhorukov G B, Kornowski A, Möhwald H, Giersig M, Eychmüller A and Weller H 2000 Colloid. Surfaces A 163 39
[11] Sukhanova A, Venteo L, Devy J, Artemyev M, Oleinikov V, Pluot M and Nabiev I 2002 Lab. Invest. 82 1259
[12] Sukhanova A, Even-Desrumeaux K, Kisserli A, Tabary T, Reveil B, Millot J M, Chames P, Baty D, Artemyev M, Oleinikov V, Pluot M, Cohen J H and Nabiev I 2012 Nanomed. Nanotechnol. 8 516
[13] Yu W W, Qu L, Guo W and Peng X 2003 Chem. Mater. 15 2854
[14] Brazhnik K, Sokolova Z, Baryshnikova M, Bilan R, Efimov A, Nabiev I and Sukhanova A 2015 Nanomed. Nanotechnol. 11 1065
[15] Sukhanova A, Susha A S, Bek A, Mayilo S, Rogach A L, Feldmann J, Oleinikov V, Reveil B, Donvito B, Cohen J H and Nabiev I 2007 Nano Lett. 7 2322
[16] Chou K F and Dennis A M 2015 Sensors-Basel 15 13288
[17] Mićić O I, Jones K M, Cahill A and Nozik A J 1998 J. Phys. Chem. B 102 9791

Semiconductor quantum dot toxicity in a mouse in vivo model

Svetlana V Bozrova[1], Maria A Baryshnikova[1,2], Igor Nabiev[1,3] and Alyona Sukhanova[1,3]

[1] Laboratory of Nano-Bioengineering, National Research Nuclear University MEPhI (Moscow Engineering Physics Institute), 115522 Moscow, Russia
[2] Blokhin Russian Cancer Research Center, Russian Academy of Medical Sciences, 115478 Moscow, Russia
[3] Laboratoire de Recherche en Nanosciences, EA4682-LRN, Université de Reims Champagne-Ardenne, 51100 Reims, France

E-mail: alyona.sukhanova@univ-reims.fr

Abstract. Quantum dots (QDs) are increasingly widely used in clinical medicine. Their most promising potential applications are cancer diagnosis, including *in vivo* tumour imaging and targeted drug delivery. In this connection, the main questions are whether or not QDs are toxic for humans and, if they are, what concentration is relatively harmless. We have carried out *in vivo* experiments with CdSe/ZnS fluorescent semiconductor core/shell QDs, which are currently the most widely used in research.

1. Introduction

QDs are highly fluorescent inorganic semiconductor crystals with a diameter from 2 to10 nm [1]. They may consist of a semiconductor (CdSe, CdS, CdTe, InP, InAs, or PbSe) core alone or have a core/shell structure (usually, with a ZnS shell), the shell protecting the QD from oxygenation and enhancing the photoluminescence quantum yield [2]. Regarding optical properties, semiconductor QDs are characterized by an exceptionally bright photoluminescence and rock-solid photo- and chemostability. They have broad quasi-continuous absorption and narrow, sharp emission spectra with an approximately Gaussian shape and large (>100 nm) Stocks shifts [3]. The high brightness of fluorescent QDs is a result of high molar adsorption coefficients (several times higher than those of fluorescent dyes and proteins) combined with a high quantum yield [4]. Owing to the broad absorption spectra, different populations of QDs can be excited at the same wavelength, which can be very far from their respective emission bands, depending on the QD core size and composition. The size of QDs can be varied in the process of their synthesis, and the QD fluorescence colour directly depends on their diameters.

In recent years, QDs have been increasingly widely used in biology for cell targeting, imaging, and drug delivery due to their unique optical and physicochemical properties [5]. They are likely to replace the commonly used organic dyes because of their considerable advantages over the organic fluorophores [6]. The use of QDs in laboratory practice is also extending. Their most promising potential applications are in cancer diagnosis, including *in vivo* tumour imaging.

Cancer remains one of the leading causes of death in the world [7]. One of unsolved problems is differentiation of tumour-affected tissue from healthy tissue during surgery. In addition, noninvasive determination of metastases is highly demanded. These problems can be solved by *in vivo* tumour imaging. Many researchers address the issue of selecting the best method for *in vivo* tumour imaging using different substances, such as fluorescent proteins [8–13]. Advanced optical characteristics of QDs as compared to organic dyes call for detailed investigation of different aspects of QD use for *in vivo* optical imaging of tumours. In this connection, the behaviour of QDs in living organisms is of special interest. Certainly, the main questions are whether or not QDs are toxic for humans and, if they are, what concentration of QDs is relatively harmless.

2. Results and discussion

We have carried out *in vivo* experiments with fluorescent semiconductor CdSe/ZnS core/shell QDs, which are currently the most widely used in research. QDs coated with trioctylphosphine oxide (TOPO) were synthesized, solubilized, and modified with electrically neutral derivatives of polyethylene glycol (PEG) containing both thiol and carboxyl groups to make them soluble in water and aqueous buffer solutions and protect them from clustering [14].

Several stages of purification after coating QDs with PEG were performed so as QDs not to bear any toxic admixtures, such as unbound PEG or other substances. The stability of QDs was accurately measured during one month before the experiments in three different solutions, phosphate buffer, RPMI medium, and mouse blood serum, at different temperatures: room temperature and animal body temperature (37°C). Since QDs were stable under all these conditions, we could be sure that they would not aggregate or decompose shortly after they enter a living body. In addition, the physical characteristics of QDs, including their sizes and emission and absorption spectra (Figures 1, 2), were measured before the *in vivo* experiment.

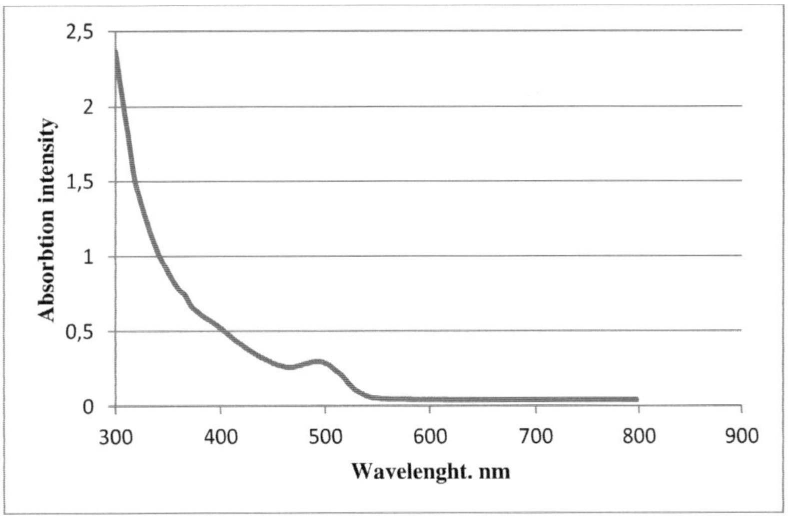

Figure 1. The absorption spectrum of CdSe/ZnS core/shell quantum dots.

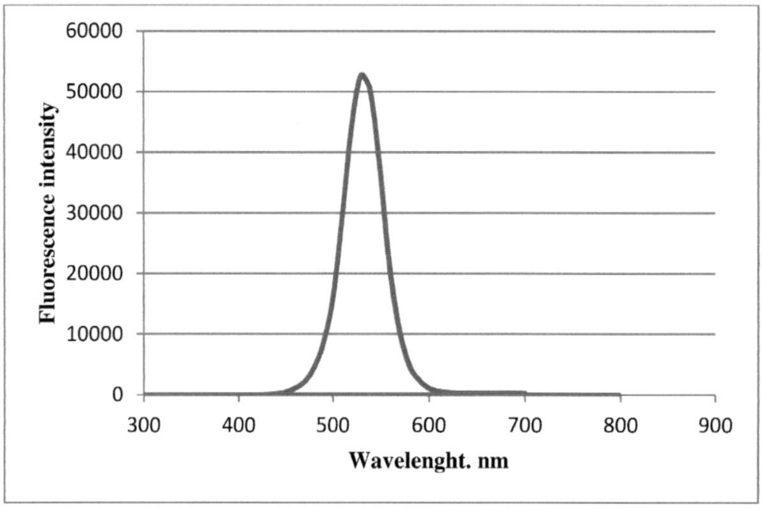

Figure 2. The emission spectrum of CdSe/ZnS core/shell quantum dots.

Seventy DMA/B6 hybrid mice were used to estimate the QD toxicity *in vivo*. A QD solution was injected intravenously. A pure solvent and PEG solution were used as negative controls. The mice were examined for about one month after QD injection. The survival rate was estimated throughout the study. Only in the group with the highest concentration of QDs did animals start to die during this period (survival rate, 75%).

During the experiment, we visually accessed the state of health of every animal and measured its weight (Figure 3). No significant weight loss was observed during four weeks.

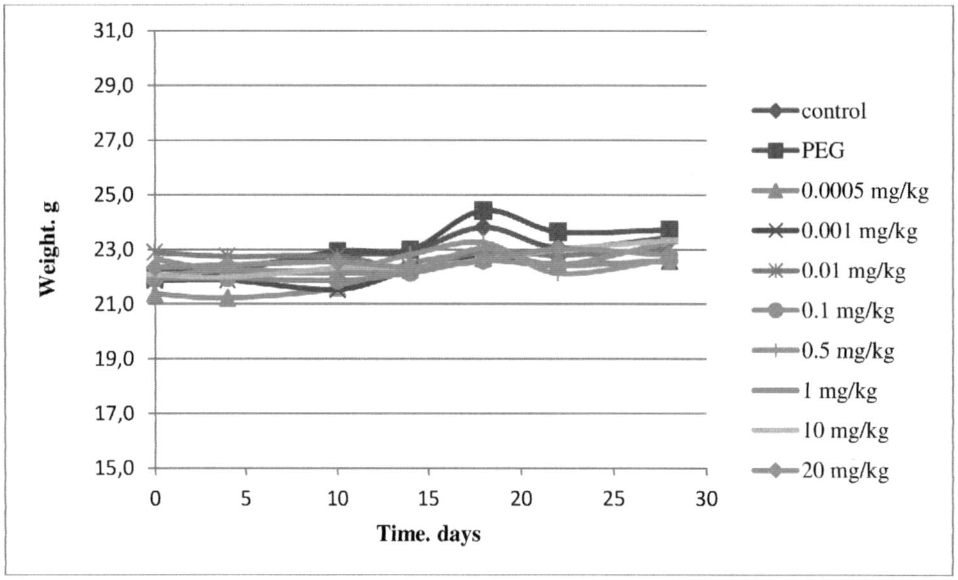

Figure 3. Animal weight assessment during the experiment.

Special attention was paid to the state of the animals' hair and eyes. It was found that mice tolerated QDs well even at a concentration as high as 10 mg/kg. The mouse hair, teeth, and eyes looked like those of healthy animals (Figure 4).

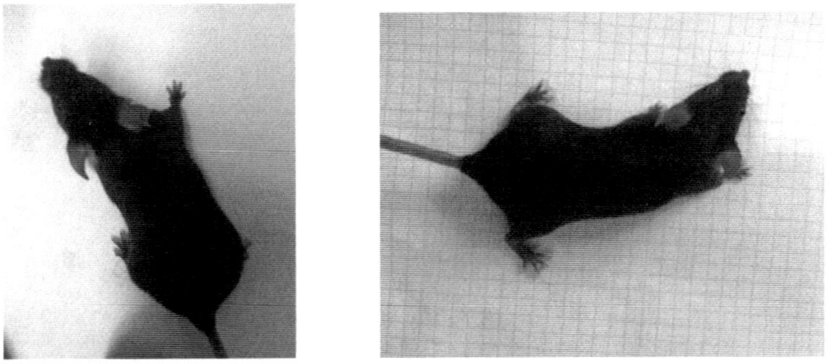

Figure 4. A mouse before the experiment (left) and a mouse three weeks after the injection of 10 mg/kg of QDs (right).

One month after the injection, the internal organs were investigated. We used paraffin sectioning and haematoxylin–eosin staining to assess the state of the liver and kidneys every week during four weeks. Their morphology was the same as in healthy control mice (Figure 5).

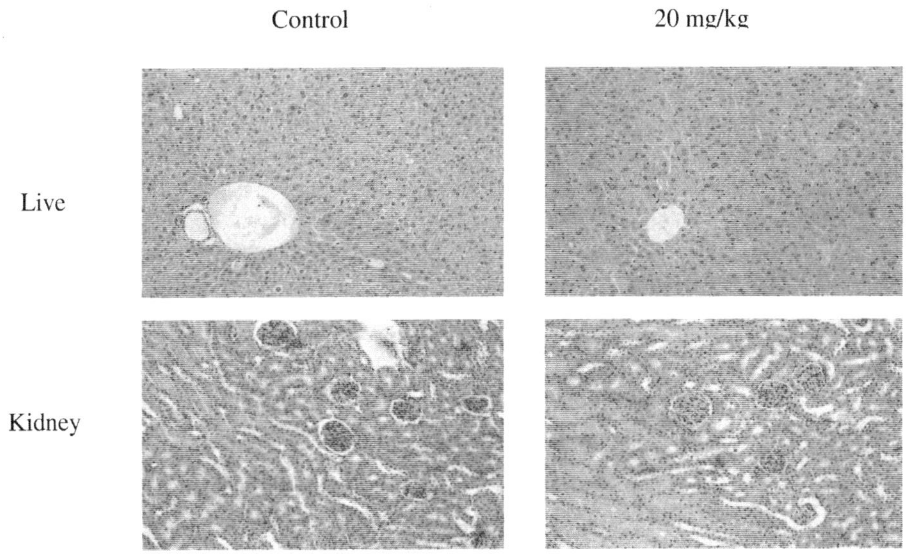

Figure 5. Liver and kidney haematoxylin–eosin staining at the fourth week of the experiment.

3. Conclusions

These results allow us to conclude that CdSe/ZnS QDs have no considerable toxic effect in the mouse *in vivo* model at concentrations up to 10 mg/kg, because there were no changes in organ morphology, physical parameters, and visually assessed general condition of mice and no weight variation during the four weeks of the experiment.

However, the QDs were found to be toxic at the highest concentration studied, 20 mg/kg body weight, with two-thirds of the animals dying shortly after the experiment was started. Further investigations will be performed to compare the single doses used in this experiment and the minimum therapeutic dose, so that we be able to use the mouse model for testing applications of QDs and their conjugates with biomolecules for tumours detection.

Acknowledgments
This study was supported by the Federal Target Program for Research and Development of the Ministry of Education and Science of the Russian Federation (grant no. 14.584.21.0012, contract no. RFMEFI58415X0012).

References

[1] Medintz I L, Uyeda H T, Goldman E R and Mattoussi H 2005 Nat. Mater. 4 435
[2] Hines M A and Guyot-Sionnest P 1996 J. Phys. Chem. 100 468
[3] Zhou D, Piper J D, Abell C, Klenerman D, Kang D-J and Ying L 2005 Chem. Commun. (Camb). 4807
[4] Leatherdale C A, Woo W-K, Mikulec F V. and Bawendi M G 2002 J. Phys. Chem. B 106 7619
[5] Smith A M, Ruan G, Rhyner M N and Nie S 2006 Ann. Biomed. Eng. 34 3
[6] Sukhanova A, Devy J, Venteo L, Kaplan H, Artemyev M, Oleinikov V, Klinov D, Pluot M, Cohen J H M and Nabiev I 2004 Anal. Biochem. 324 60
[7] Siegel R, Naishadham D and Jemal A 2013 CA. Cancer J. Clin. 63 11
[8] Condeelis J and Weissleder R 2010 Cold Spring Harb. Perspect. Biol. 2 a003848
[9] Gao X, Cui Y, Levenson R M, Chung L W K and Nie S 2004 Nat. Biotechnol. 22 969
[10] Erogbogbo F, Yong K-T, Roy I, Hu R, Law W-C, Zhao W, Ding H, Wu F, Kumar R, Swihart M T and Prasad P N 2011 ACS Nano 5 413
[11] Hu R, Yong K-T, Roy I, Ding H, Law W-C, Cai H, Zhang X, Vathy L A, Bergey E J and Prasad P N 2010 Nanotechnology 21 145105
[12] Elias D R, Thorek D L J, Chen A K, Czupryna J and Tsourkas A 2008 Cancer Biomark. 4 287
[13] Hoffman R M 2002 Cell Death Differ. 9 786
[14] Murray C B, Norris D J and Bawendi M G 1993 J. Am. Chem. Soc. 115 8706

Quenching of quantum dots luminescence under light irradiation and its influence on the biological application

K V Vokhmintcev[1], C Guhrenz[2], N Gaponik[2], I Nabiev[1,3], P S Samokhvalov[1]

[1] Laboratory of Nano-Bioengineering, National Research Nuclear University MEPhI (Moscow Engineering Physics Institute), 115409 Moscow, Russian Federation
[2] Physikalische Chemie, TU Dresden, 01062 Dresden, Germany
[3] Laboratoire de Recherche en Nanosciences, LRN-EA4682, Université de Reims Champagne-Ardenne, 51100 Reims, France

E-mail: Vokirill@gmail.com

Abstract. Applications of quantum dots (QDs) can be limited by their stability under intensive irradiation. We have studied the stability of core/shell QD photoluminescence under the irradiation typical for the laboratory or industrial environment. Our results show that QD photoluminescence stability is determined by the structure and thickness of their shell; the latter factor has the strongest effect on the QD resistance to irradiation. It has been shown that the observed photodegradation of QDs is basically caused by the transfer of excited charge carriers outside the QD. An increase in the carrier tunneling length has been shown to be an efficient way to prevent quenching of QD photoluminescence. Storage of chloroform QDs solution under the light should be avoided due to the decrease of the biological applicability.

1. Introduction

Size-tunability of the absorption and fluorescence properties of colloidal quantum dots (QDs) makes them an attractive material for modern science and technology. Size-controllable optical properties of QDs allow their applications in multiplexed biolabeling, solar cells, light emitting diodes, catalysis, etc. QDs may gradually replace organic dyes from these fields because of their higher linear absorption cross-sections, photoluminescence (PL) quantum yields (QY), and photostabilities. However, in contrast to organic dyes, QDs are known to be susceptible to environmental factors such as oxygen; in addition, intense laser irradiation is known to alter their PL QY. Therefore, understanding of the mechanisms and factors causing instability of QD optical properties is one of the most important issues for the development of new generations of QDs and QD-based devices.

In this study, we have investigated the influence of long-term visible-light irradiation on the PL QY of QDs in order to model their storage and/or operation under common laboratory or industrial conditions. Two types of light sources with the same light power output (15 mW/cm^2) were used for QD irradiation: a 40 W incandescent lamp and an array of light-emitting diodes with a radiation wavelength of 405 nm. Three types of core/shell QDs with different shell compositions and thicknesses, CdSe/ZnS(3ML), CdSe/ZnS(1ML)/CdS(1ML)/ZnS(1ML), and CdSe/CdS(7ML)/ZnS(1ML), all with initial PL QYs exceeding 70%, were used to estimate the effects of the shell thickness and confinement potential on the stability of their PL.

2. Materials and methods

2.1. Materials

Cadmium oxide (powder, 99.5%), zinc oxide (powder, 99.99% trace metals basis), 1-octadecene (ODE, technical grade, 90%), oleylamine (OLA, technical grade, 70%), selenium (powder, 100 mesh, 99.5%), thiourea (ACS reagent, ≥99.0%) tri-n-octylphosphine (TOP, technical grade, 97%), tri-n-octylphosphine oxide (TOPO, reagent grade, 99%), 2-ethylhexanoic acid (2-EHA, 99%), tri-n-octylamine (TOA, 98%), triethylene glycol dimethyl ether (TEGDME,ReagentPlus, 99%), cysteine hydrochloride hydrate (98%), sodium hydroxide and anhydrous solvents (chloroform, hexane, acetone, methyl acetate and 2-propanol) were purchased from Sigma-Aldrich; n-hexadecylphosphonic acid (97%) was purchased from PlasmaChem GmbH. All chemicals were used as received without any purification.

2.2. Synthesys of the QDs

CdSe cores were synthesized via the standard procedure reported in [1]. 1 mmol of cadmium oxide, 2 mmol of 2-ethylhexanoic acid, and 8 ml of ODE were placed in three-neck round bottom flask. The flask was heated to 220°C under argon atmosphere to complete dissolution of CdO. Then the flask was cooled to room temperature and 0.5mmol of n-hexadecylphosphonic acid, 2 ml of try-n-octylamine was added to flask. Then the reaction mixture was heated to 120 oC and evacuated in order to remove residual water. Selenium precursor was prepared by the dissolution of 2 mmol of Se in 2 ml of TOP and 0.5 ml of ODE under the argon flow. The selenium precursor was quickly injected in to the cadmium precursor solution at 240 ^{o}C. Reaction solution was kept under the argon atmosphere and continuous stirring. The growth of the CdSe cores was controlled via the UV-vis spectroscopy and the reaction was stopped by the cooling on reaching predetermined size. CdSe cores were separated from other reaction media by the precipitation with 2-poropanol and then were purified by the cycles of precipitation/dissolution with using of acetone and hexane.

QDs cored were covered by CdS and ZnS in different proportions. Solutions of cadmium 2-ethylhexanat, zinc 2-ethylhexanat and thiourea in triethylene glycol dimethyl ether were used as cadmium, zinc and sulphur precursor respectively. Amount of precursors which are necessary for cores covering were calculated using a spherical nanoparticle model according to the SILAR technique [2]. Solution of cores was evacuated and heated up to 180°C under argon atmosphere and continuous stirring, then the precursors were added dropwise, after the end of precursor addition the reaction media was maintained under the same conditions for 20 minutes in order to reach the end of shell formation reaction. Then QDs were purified by the same method as cores with using of 2-propanol, hexane and acetone.

2.3. QDs solution irradiation

Obtained QDs were dissolved in chloroform as a most common solvent, optical density at 400 nm for all solutions were equal to 0,1. Each solution was divided into two parts and each part was irradiated by UV or visible light under the continuous stirring. Incandescent lamp (40W) was used as the source of the visible light, massive of the UV LEDs were used as the source of the UV light. The light flow was the same for the both light source and equal to 15 mW/cm^{2}. Solutions were irradiated for 3 hours and in order to evaluate the dynamic of optical properties of the obtained solution the photoluminescence quantum yield was measured (PLQY). PL QY was measured relative fluorescein (QY = 95 %) for CdSe/ZnS(3ML), relative rhodamine 6G (QY = 95%) for CdSe/ZnS/CdS/ZnS and relative rhodamine 102 (QY = 96 %) for CdSe/CdS(7ML)/ZnS. Choice of the comparison dye is due to the region of the QD's luminescence. PLQY was measured every 10 minutes during the first hour, every 20 minutes during the second hour and every 30 minutes during the third hour, also PLQY was measured after 2.5, 5 minutes of the light irradiation.

2.4. Solubilization of the irradiated QDs

In order to investigate the influence of the different type of the light irradiation on the size of the agglomerates of the solubilized QDs, CdSe/CdS/ZnS QDs irradiated for 0.5, 1, 2 and 3 hours were solubilized with using of cysteine by the following procedure. QDs were purified by the precipitation with using of methyl acetate and collect by the centrifugation. Ethanol solution of DL-cysteine was added dropwise to the chloroform solution of purified QDs and stirred for 5 minutes. In order to remove cysteine not bounded with QD's surface, solution was twice purified by the centrifugation, supernatant decantation and addition of pure methanol. Then QDs were dissolved in 0.1 M water solution of NaOH and sonicated for 15 minutes. Then the size of the solubilized QDs was measured by the dynamic light scattering.

3. Results and discussion

Irradiations of QDs with an incandescent lamp and a LED array have revealed two parallel counteracting processes: an increase in and quenching of the PL. The kinetic curves of these processes are shown in figure 1, the values of the relative PLQY are presented in table 1.

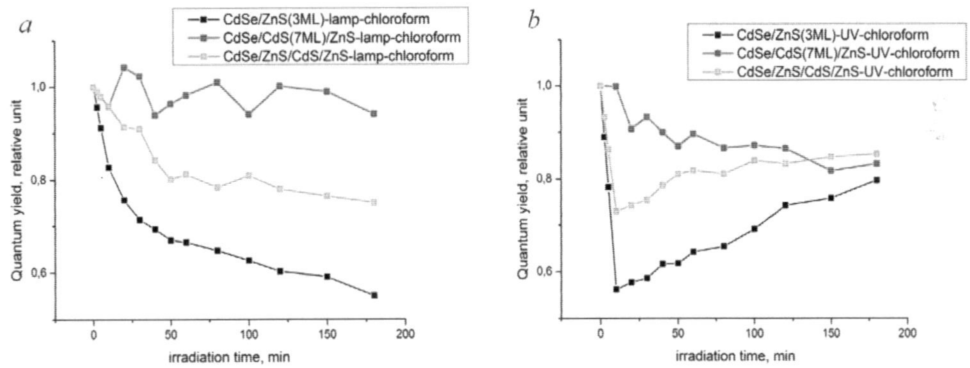

Figure 1. Dynamic of the PLQY for CdSe/ZnS/CdS/ZnS irradiated with visible (*a*) and UV (*b*) light.

Table 1. Relative photoluminescence quantum yield of QDs irradiated with UV or visible light for 3 hours

QDs type	UV irradiation	Visible irradiation
	Relative photoluminescence quantum yield,%	
CdSe/ZnS(3ML)	0.79(0.56)*	0.55
CdSe/ZnS/CdS/ZnS	0.85(0.73)*	0.75
CdSe/CdS(7ML)/ZnS	0.83	0.94

*- the minimum QD's PL QY value after the end of the photodarkening

In the case of CdSe/ZnS(3ML) QDs, irradiation with an incandescent lamp caused a slow PL decline to 55% of the initial PLQY value, while irradiation with LEDs caused a rapid PL quenching to 56% followed by slow restoration of luminescence to 79% of the initial level.

Multishell CdSe/ZnS/CdS/ZnS QDs exhibited a similar kinetic of PL QY decrease, but the degree of PL quenching was much lower. Thus, irradiation with an incandescent lamp caused a 25% drop of

PL, while irradiation with a LED array resulted in a 27% loss of PL QY followed by restoration to 85% of the initial PL QY during the remaining period of irradiation.

Thick-shell CdSe/CdS(7ML)/ZnS QDs exhibited a totally different response to irradiation with both light sources, without considerable PL quenching under visible light irradiation and without PL restoration under UV light irradiation. These QDs were found to be almost completely stable under long-term irradiation; under irradiation from either source. The reference samples representing the same solutions of QDs incubated in the dark did not exhibit detectable variation of PL QY, thus proving that the observed effects originated from light irradiation rather than from other possible environmental factors.

Transfer of the excited charge carriers from the core of QDs to the surface ligands or the surrounding molecules is known to be an efficient pathway of QD PL quenching [3]. We suppose that the difference in charge transfer rates is the reason of the observed differences in the PL quenching kinetics of the QDs studied. This difference can be related to the variation in (i) the shell thickness or (ii) the confinement potentials determined by the structure and composition of the shells. Specifically, a 3ML-thick ZnS shell ensures a high potential barrier of moderate length for electrons and holes to tunnel through and to escape the QD core. The mutishell structure provides a slightly higher potential barrier than a 3ML-thick ZnS shell does, due to the strong quantum confinement effect in monolayer-thick shell layers [4], while having the same barrier length. Finally, thick-shell CdSe/CdS(7ML)/ZnS QDs have the lowest energy of potential barrier created by CdS, but the barrier length is 2.5 times larger than that in other QDs. From the obtained results, one can conclude that the barrier length has a much stronger effect on the stability of QD colloidal solutions to irradiation, and a sufficiently thick shell could completely protect QDs from photoinduced degradation.

The difference in PL quenching kinetics under irradiation from different light sources can be attributed to different spectral overlaps of the light source emissions and QD absorption spectra. The entire incident photon flux from the blue LED array falls within the absorption range of all types of QDs, while a major part of the incandescent lamp irradiation cannot be absorbed by QDs. However, we can conclude that, even in the latter case, significant quenching of QD PL can occur under long-term irradiation.

Formation of the wide gap material shell on surface of the cores is one of the most effective methods for increase the QDs properties [5] due to the passivation of the surface trap states and preventing of charge transfer in QD or out. In the paper [6] authors collective report about the PbSe QDs treatment by the chlorine, which cause the increase of the QD's PLQY due to the formation of the $PbCl_x$ layer on QD's surface. In our case the restoration of PLQY of CdSe/ZnS(3ML) and CdSe/ZnS/CdS/ZnS irradiated by UV light. It is well known that QDs is very promising photocatalyst and cause the decomposition of chloroform to the formation of hydrogen chloride and carbon dioxide [7]. Interaction of hydrogen chloride with QDs cause the formation of $ZnCl_x$ layer on its surface, which is dielectric and form giant confinement potential, that cause the restoration of the PLQY. The absence of PLQY in case of CdSe/CdS(7ML)/ZnS QDs is caused by shell width, $ZnCl_x$ formation don't cause the changes in charge recombination kinetic.

The influence of the longterm storage of chloroform QDs solution under different type light irradiation on their biological application was investigated on the example of CdSe/ZnS/CdS/ZnS QDs. The first step on way to the QDs biological application is QDs transfer from hydrophobic phase to water. The transfer was carried out by the exchange of hydrophobic ligands to cysteine on multishell QDs irradiated by UV or visible light for 0.5, 1,2 and 3 hours. Then the size of the solubilized QDs was measured with using of dynamic light scattering, the mean size of the solubilized QDs are presented in the table 2.

Obtained results show, that irradiation causes the agglomeration of the QDs during the solubilization and UV light irradiation cause the significantly greater increase compare to visible light. The mean size of the particles in not irradiated solution is 4.1 nm, which correspond to the QD size, mean sizes of the particles irradiated by UV light for 2 hours and visible light for 3 hours are 11.7 and 7.5 respectively, that close to the sizes of the trimmer and dimmer agglomerates. Significant greater effect of UV irradiation on the described process compare to visible irradiation allows to propose that the observed effect is connected with the chlorination of the QDs surface during the irradiation of the chloroform solution. Cysteine binds with the QDs surface by the formation of S-S and Zn-S bond between QD surface and –SH group. Substitution of sulfur atoms by chlorine atoms cause the decrease of the amount of the surface site for bond formation and hence decrease the concentration of the ligand molecules on the QD's surface, that cause the decrease of the steric stabilization level that cause the agglomerate formation. Difference between levels of agglomeration for QDs irradiated by UV and visible light is due to the fact that UV light irradiation causes more intensive QD's surface chlorination that cause low level of stabilization and bigger agglomerate formation. The size of the particles is the one of the most parameter for biological application as it determine the ability to penetrate throw the different kinds of membranes in biological system, hence it necessary to avoid the long term storage of QDs in chloroform solution under the different type of light irradiation or propose the another type of the solvent in which photodegradation process of QDs are absence, it will be the subject of the further studies.

Table 2. Mean size of the solubilized QDs.

Irradiation time, h.	Size of UV irradiated QDs, nm	Size of vis. irradiated QD, nm
0	4.18	4.18
0.5	7.5	4.9
1	10.7	-
2	11.7	-
3	-	7.5

4. Conclusions

The results obtained in this work could be important for application-driven design of novel QDs. The high stability of PL QY is an advantage in engineering of photostable fluorescent biolabels or QD-LEDs; on the other hand, QDs that are prone to fast photoinduced PL quenching could be efficient tags in high-resolution stimulated emission depletion (STED) optical microscopy.The main conclusions for this study is (i) Shell structure and solvent have an unambiguous effect on the stability of colloidal QDs under intense irradiation, (ii) transfer of the photoexcited charge carriers is the reason of QDs luminescence quenching, (iii) formation $ZnCl_x$ passivation monolayer on the QD's surface can be the reason of observed photobrightenning in chloroform, (iv) Shells width, rather than confinement potential created by it, is a key factor that determines photostability of QDs, (v) Storage of QDs in the form of chloroform colloidal solution should be avoided if any surface ligand exchange is necessary in the future – eg. solubilization using bifunctional thiols for biology applications.

5. Acknowledgments

This study was supported by the RFBR grant № 16-34-60253 mol_a_dk.

References

[1] Krivenkov V A, Samokhvalov P S, Linkov P A, Solovyeva D O, Kotkovskii G E, Chistyakov A A and Nabiev I 2014 Surface ligands affect photoinduced modulation of the quantum dots optical performance ed D L Andrews, J-M Nunzi and A Ostendorf p 91263N

[2] Li J J, Wang Y A, Guo W, Keay J C, Mishima T D, Johnson M B and Peng X 2003 Large-Scale Synthesis of Nearly Monodisperse CdSe/CdS Core/Shell Nanocrystals Using Air-Stable Reagents via Successive Ion Layer Adsorption and Reaction J. Am. Chem. Soc. 125 12567–75

[3] Vokhmintcev K V., Samokhvalov P S and Nabiev I 2016 Charge transfer and separation in photoexcited quantum dot-based systems Nano Today 11 189–211

[4] Samokhvalov P, Linkov P, Michel J, Molinari M and Nabiev I 2014 Photoluminescence quantum yield of CdSe-ZnS/CdS/ZnS core-multishell quantum dots approaches 100% due to enhancement of charge carrier confinement ed W J Parak, M Osinski and K I Yamamoto p 89550S

[5] Dabbousi B O, Rodriguez-Viejo J, Mikulec F V., Heine J R, Mattoussi H, Ober R, Jensen K F and Bawendi M G 1997 (CdSe)ZnS Core−Shell Quantum Dots: Synthesis and Characterization of a Size Series of Highly Luminescent Nanocrystallites J. Phys. Chem. B 101 9463–75

[6] Bae W K, Joo J, Padilha L A, Won J, Lee D C, Lin Q, Koh W, Luo H, Klimov V I and Pietryga J M 2012 Highly Effective Surface Passivation of PbSe Quantum Dots through Reaction with Molecular Chlorine J. Am. Chem. Soc. 134 20160–8

[7] MartÍn C A, Baltanás M A and Cassano A E 1996 Photocatalytic Reactors. 3. Kinetics of the Decomposition of Chloroform Including Absorbed Radiation Effects Environ. Sci. Technol. 30 2355–64

IOP Conf. Series: Journal of Physics: Conf. Series **784** (2017) 012015 doi:10.1088/1742-6596/784/1/012015

Long-Term Effects of Stem Cells on Total-Body Irradiated Mice

M V Vyalkina, I B Alchinova, E N Yakovenko, Yu S Medvedeva, I N Saburina and M Yu Karganov

Institute of General Pathology and Pathophysiology, Baltiyskaya str., 8, Moscow, 125315, Russian Federation

E-mail: nedzumy@bk.ru

Abstract. C57Bl/6 mice were exposed to γ-radiation in a sublethal dose of 7.5 Gy. In 3 hours injection 10^6/mouse of bone marrow multipotent mesenchymal stromal cells stem cells intravenously to experimental group was done. Methods used: body weight measurement, open field behavior, subfraction composition of blood serum (laser correlation spectroscopy, LCS), histological examination of the spleen, liver, and pancreas, count of T and B cells, white blood formula. After 1.5 and 3 months the general trend towards intermediate position of the parameters observed in the experimental between those in intact and irradiated controls attests to partial protective/restorative effects of the injected cells.

1. Introduction

Bone marrow multipotent mesenchymal stromal cells (BM MMSC) are the second major stem cell population in bone marrow. They are used for restorative organ therapy, because they can be relatively easy isolated from a small volume of bone marrow aspirate and expanded in a necessary amount in a short period of time. In vitro BM MMSC are able to differentiate into cells of different types of connective tissues – bones, cartilage, tendons, skeletal muscles, adipose tissue, as well as supporting stroma. The accumulated preliminary data showed multiple mechanisms and pathways of BM MMSCs contribution to injury reparation. Systemic administration of BM MMSC is accompanied with integration of injected cells into parenchyma of recipient's organs, especially in the regions of lung, liver, kidney, muscle and pancreas injury.

Current concepts on radiation-induced insults are based on the assumption that an effective treatment modality should be administered immediately, within a few hours after radiation exposure to protect and prevent the death of the critically irradiated cells. Cell therapies to reconstitute failing organs, aside from the hematopietic bone marrow (BM) stem cells transplantation, by indefinitely replacing the affected cells in most cases are not highly practical. Multipotent mesenchymal stromal cells from different sources, mainly from the BM, cord blood and adipose tissues were proposed for regenerative medicine [1].

2. Experimental Design, Instruments and Methods

2.1. BM MMSC preparation

To obtain the primary human BM MMSC culture, the mononuclear cell fraction was isolated from bone marrow of healthy donors' iliac crest by centrifugation in Ficoll gradient. The obtained cell suspensions were centrifuged for 7 minutes at 1000 rot/min, the pellets were resuspended in growth medium (α-MEM, 17 % fetal calf serum (FCS), L-gluthamine (2mM), 1% penicillin-streptomycin). Cells were seeded on Petri dishes in high density (1×10^5 cells/cm^2). In 24 hours floating cells were removed with growth medium, the adhered culture was washed with Hank's solution, then the fresh growth medium was added, and cells were further incubated for 4-8 days. When the primary culture reached 70% confluency, cells were washed with Versene solution, incubated in 0.25% trypsin solution (37°C, 3-5 min), discarded from plastic and centrifuged (7 min, 1000 rot/min). The cells in the pellet were resuspended in growth medium to inactivate trypsin, centrifuged one more time and seeded on Petri dishes at low density (100 cells/cm^2) in full growth medium (DMEM/F12 (1:1), L-gluthamine (2mM), 1% penicillin-streptomycin, 10 ng/ml basic fibroblast growth factor (bFGF), 1:100 insulin-transferrin-selenite, 10% FCS).

During culturing process, BM MMSC culture became less heterogenic and after the second passage was mainly (more than 90%) represented with small cells with the average diameter of 6-9 μm.

The obtained BM MMSC culture expressed such surface markers as CD73 (95%), CD90 (79%), CD105 (98%) and did not express CD14 (1,1%), CD34 (1,31%), CD 45 (0,6%), CD49b (1%). The resulting marker profile is an important criteria of purity and quality of BM MMSC culture with the prevalence of precursor cells, which possess high clonogenic, proliferative and differentiation potential. These obtained and characterized BM MMSC cultures were further used in experimental work.

2.2. *Experimental groups*

The study was performed on C57Bl/6 mice weighing 25-27 g. The mice were exposed to γ-radiation in a sublethal dose of 7.5 Gy at 5.4 R/min radiation intensity. Groups: intact control (n=11), irradiated control, irradiation (n=19), experiment, irradiation + injection 10^6/mouse of the studied stem cells intravenously (n=41), green protein, irradiation + injection of the studied cells labeled with green protein (n=3). Tests were performed 3 and 6 weeks and 3 months after irradiation/injection.

2.3. *Methods*

2.3.1. Body weight. High radiation doses cause bowel and liver damages during the first weeks after irradiation, which impairs digestion and leads to changes in body weight. In our previous experiments we have found interstrain differences in changes of body weight after irradiation, which may be caused by the development of metabolic syndrome [2]. The animals were weighed on a Mettler Toledo scales after 1-h food deprivation.

2.3.2. Open field behavior. The effect of the potent stress factor on the behavior of rodents was studied using the open field test. Vertical and horizontal activities were measured over 3 min. The testing was performed in a square chamber (45×45 cm) at 40 lux illumination. In this test, behavioral phenomena caused by the conflict between fear motivation and exploratory activity are studied.

2.3.3. Subfraction composition of blood serum - Laser correlation spectroscopy. The method of laser correlation spectroscopy (LCS) is based on the analysis of the spectrum of quasielastic light scatter during coherent monochromatic laser irradiation of microparticles in biological fluids [3]. The distribution histograms in the serum provide qualitative information on the mean particle sizes and their relative content. Strict correspondence of certain fragments of the spectrum to biological nature of serum components can be determined after additional studies.

Blood was taken routinely: 100 μl of whole blood is transferred to plastic tubes with 100 μl physiological saline. The samples were left at room temperature for 0.5-2 h and then centrifuged at 5000 rpm for 15 min. The supernatant was collected in plastic tubes.

2.3.4. Histological examination of the spleen, liver, and pancreas. After the end of the experiment, the animals were sacrificed by anesthetic overdose and the organs (liver and intestine) were taken for histological examination for evaluation of the damaging effects of irradiation. The organs were fixed in 4% neutral formalin. The slides were prepared routinely.

For simplification of data processing, three injury severity degrees were specified depending on the severity of morphological changes [4].

Cytometry and white blood formula were performed routinely.

3. Results and Discussion

After 3 weeks: no differences by body weight were revealed between the irradiated controls and experimental mice. By the parameters of horizontal and vertical locomotor activities and leukocyte and T lymphocyte counts, the experimental mice were between the intact control and irradiated mice. No differences by B cell count were revealed between the irradiated controls and experimental group (figure 1).

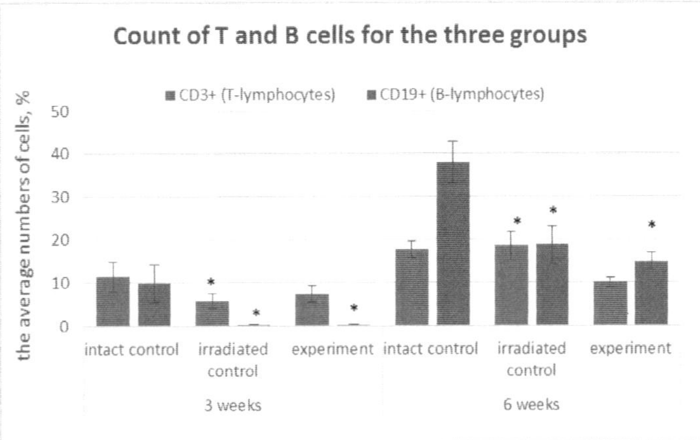

Figure 1. Count of T and B cells for intact control, irradiated control and experimental groups.

*- P_U < 0.05 in comparison with intact control group.

In the subfraction spectrum of blood serum, a tendency towards accumulation of small particles is noted, which is a typical effect of radiation. It can be hypothesized that the increase in the contribution of small particles into light scatter during the 3rd week is related to tissue destruction. Similar processes are observed in groups of irradiated mice, but injection of cells probably levels out the individual differences (figure 2).

Histological analysis revealed that severe abnormalities of the spleen, liver, and pancreas were considerably less incident in the experimental group in comparison with the control groups (intact and irradiated). In the blood, an increase in neutrophil count (stab and segmented) and a decrease in lymphocyte count were noted both in the experimental and irradiated control group in comparison with intact control.

After 6 weeks: the mean body weight increased in all groups; the experimental group occupies an intermediate position between the intact control and irradiated mice by this parameter. In the blood, neutrophil (stab and segmented) and lymphocyte blood counts in the experimental group occupy an intermediate position between the corresponding values in the intact control and irradiated mice. Leukocyte count and count of T and B cells in the experimental group were lower than in the control groups (intact and irradiated) (figure 1). Exhaustion of the pool of mature B and T cells is observed 3 weeks after irradiation. Precursors of these cells in the bone marrow are destructed by radiation. By the 6th week, the release of young lymphocytes into the blood is noted. It can be hypothesized that injection of the studied cells in the experimental group delays T cell maturation. LCS showed that the contribution of large particles into light scatter increased; these shifts are opposite to typical changes induced by irradiation (accumulation of small particles due to cell destruction and release of nucleoproteins and other components into the medium) (figure 2). In the experimental group, the histological picture of the spleen was considerably worsened with predominance of the 4th degree alterations. Round fluorescent cells were detected in cryosections of the bone marrow from mice receiving green protein-labeled cells.

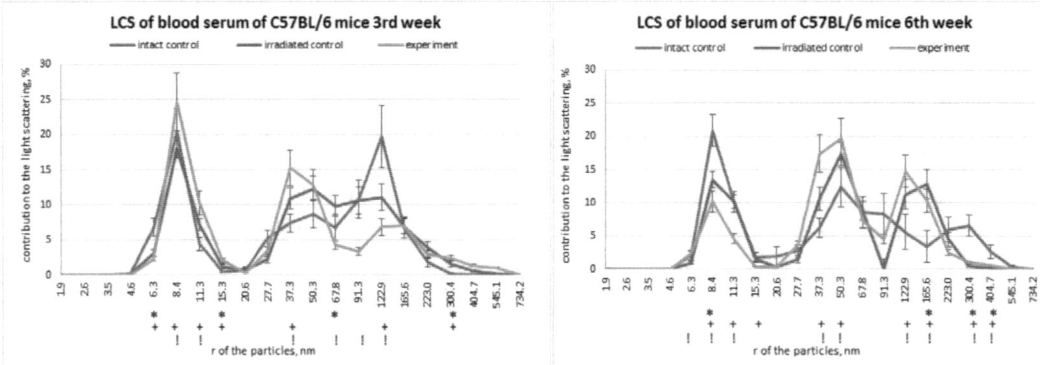

Figure 2. Changes in LC spectra of mice blood serum. Particle size (nm) is along the X axis; contribution of particles with different hydrodynamic radius to light scattering in % is along the Y axis. * - $P_U < 0,05$ – intact/irradiated groups; + - $P_U < 0,05$ – intact/experimental groups; ! - $P_U < 0,05$ – irradiated/experimental.

After 3 months: lymphocyte count returned to normal. The distribution of subcellular components in blood serum remained unchanged. Severe disturbances are less incident in the liver and spleen, but more incident in the pancreas, although the morphological picture of the pancreas in the experimental group is better than in the group of irradiated control. Round fluorescent cells were detected in cryosections of the bone marrow and liver of mice receiving green protein-labeled cells.

4. Conclusion

Detection of labeled cells in the bone marrow and liver suggests that the observed effects are caused by the experimental influence. The general trend towards intermediate position of the parameters observed in the experimental group between those in intact and irradiated controls attests to partial protective/restorative effects of the injected cells. This is also confirmed by amelioration of the histological picture of the liver and pancreas after 3 weeks. Deterioration of the histological picture of the spleen by the 6th week can be determined by the reaction to lymphocyte release after a period of radiation-induced suppression. The absence of animal deaths at the final stage of the experiment (from the 6th week to the 3th month), normalization of lymphocyte count, stability of the subfraction composition of the blood, lower incidence of severe liver and spleen damages attest to long-term nature of the observed effects. The obtained results agree with data of other authors about the prospects of using of stem cells after irradiation in high doses [5, 6, 7].

References

[1] Gaberman E, Pinzur L, Levdansky L, Tsirlin M, Netzer N, Aberman Z and Gorodetsky R 2013 Mitigation of Lethal Radiation Syndrome in Mice by Intramuscular Injection of 3D Cultured Adherent Human Placental Stromal Cells *PLoS ONE* **8 (6)** e66549 doi:10.1371/journal.pone.0066549

[2] Medvedeva Yu, Arkhipova E and Alchinova I 2015 The effects of gamma-irradiation in sub-lethal doses in mice with different radiosensitivity *American Journal of Life Sciences* **3 (1-2)** 13-17 doi: 10.11648/j.ajls.s.20150303.13

[3] Karganov M, Alchinova I, Arkhipova E and Skalny A V 2012 *Laser Correlation Spectroscopy: Nutritional, Ecological and Toxic Aspects,* Biophysics ed. A N Misra (InTech) pp 1-16 ISBN 978-953-51-0376-9, doi: 10.5772/35254

[4] Alchinova I, Arkhipova E, Medvedeva Yu, Cherepov A, Antipov A, Lysenko N, Noskin L and Karganov M 2015 The complex of tests for the quantitative evaluation of the effects of radiation on laboratory animals *American Journal of Life Sciences* **3 (1-2)** 5-12 doi: 10.11648/j.ajls.s.20150303.12

[5] Lange C et al. 2011 Radiation Rescue: Mesenchymal Stromal Cells Protect from Lethal Irradiation *PLoS ONE* **6 (1)** e14486 doi:10.1371/journal.pone.0014486

[6] Chang P, Qu Y, Liu Y, Cui S, Zhu D, Wang H and Jin X 2013 Multi-therapeutic effects of human adipose-derived mesenchymal stem cells on radiation-induced intestinal injury *Cell Death and Disease* **4** e685 doi:10.1038/cddis.2013.178

[7] Yang X, Balakrishnan I, Torok-Storb B and Pillai M M 2012 Marrow Stromal Cell Infusion Rescues Hematopoiesis in Lethally Irradiated Mice despite Rapid Clearance after Infusion *Advances in Hematology* **2012** Article ID 142530, doi:10.1155/2012/142530

IOP Conf. Series: Journal of Physics: Conf. Series **784** (2017) 012016 doi:10.1088/1742-6596/784/1/012016

Quantification and imaging of HER2 protein using nanocrystals conjugated with single-domain antibodies

S Glukhov[1], M Berestovoy[1], P Chames[2], D Baty[2], I Nabiev[1,3] and A Sukhanova[1,3]

[1]Laboratory of Nano-Bioengineering, National Research Nuclear University MEPhI (Moscow Engineering Physics Institute), Moscow, Russian Federation
[2]INSERM U1068 and CNRS UMR7258, Cancer Research Center, Marseille, France, Institut Paoli-Calmettes and Aix-Marseille, Marseille, France
[3]Laboratory of Research in Nanoscience, LRN - EA4682, University of Reims Champagne Ardenne, Reims, France

E-mail: alyona.sukhanova@univ-reims.fr

Abstract. This study dealt with quantification and imaging of human epidermal growth factor receptor 2 (HER2), an important prognostic marker for cancer diagnosis and treatment, using specific quantum-dot-based conjugates. Fluorescent inorganic nanocrystals or quantum dots (QDs) are extremely highly resistant to photobleaching and have a high emission quantum yield and a continuous range of emission spectra, from the ultraviolet to the infrared regions. Ultrasmall nanoprobes consisting of highly affine anti-HER2 single-domain antibodies (sdAbs or "nanobodies") conjugated with QDs in a strictly oriented manner have been designed. QDs with a fluorescence peak maxima at wavelengths of 562 nm, 569 nm, 570 nm or in the near-infrared region were used. Here, we present our results of ISA quantification of HER2 protein, *in situ* imaging of HER2 protein on the surface of HER2-positive SK-BR-3 cells in immunohistochemical experiments, and counting of stained with anti-HER2 conjugates HER2-positive SK-BR-3 cells in their mixture with unstained cells of the same culture in flow cytometry experiments. The data demonstrate that the anti-HER2 QD–sdAb conjugates obtained are highly specific and sensitive and could be used in numerous applications for advanced integrated diagnosis.

1. Introduction

Several types of fluorescent labels are used in biological and medical applications: heterocyclic organic dyes, fluorescent proteins, and fluorescent inorganic nanocrystals or quantum dots (QDs). The main advantages of QDs over other fluorescent dyes are an extremely high resistance to photobleaching, bright photoluminescence, and the possibility of obtaining a continuous range of emission colors from ultraviolet to infrared [1]. These characteristics make it possible to use QDs for imaging both fixed and live biological samples [2], as well as for obtaining contrast images of samples of different thicknesses in multiphoton applications [3, 4, 5]. Antibodies carrying fluorescent labels have become widely used for imaging specific protein markers, cells, and tissues in biology and medicine.

Detection of specific protein cancer markers in clinical samples is important for theranostics, a promising approach at the interface of therapeutics and diagnostics [6, 7]. Human epidermal growth factor receptor 2 (HER2) has been found to be an extremely important marker for detection of malignant tumors, assessment of the stage of their development, and medical prognosis [8]. The development of antitumor therapy where HER2 serves as a target is underway [9, 10].

Immunoglobulins of cartilaginous fishes belonging to the nurse shark antigen receptor (IgNAR) family [11, 12] and similar to G-type immunoglobulins of camelids [13] stand apart from other immunoglobulins serving as antibodies in mammals. The antigen binding surface of these antibodies consists of a single globular domain, in contrast to two domains in typical mammalian immunoglobulins. An isolated variable domain of a camelid similar to G-type immunoglobulin capable of antigen binding is termed a single-domain antibody (sdAb). The size of sdAbs (weighing as little as 13 kDa) makes them the smallest sensors or capture molecules for specific nanoprobes currently used in biomedicine [14]. The technique for obtaining recombinant sdAbs allows their primary amino acid sequence to be additionally tagged.

In this study, we estimated the performance of nanoprobes containing photoluminescent nanocrystals with different spectral characteristics and highly affine sdAbs against HER2 in detecting this cancer marker.

2. Materials and methods

2.1. Cell cultures

The SK-BR-3 and MDA-MB-231 cell cultures were kindly provided by Dr. Baryshnikova (Laboratory of Experimental Diagnosis and Biotherapy of Tumors, Blokhin Cancer Research Center, Moscow, Russia). The cultures were grown on cover slips or chamber slides to 80% confluence in RPMI 1640 supplemented with 10% bovine serum at 37°C in an atmosphere of 5% CO_2.

2.2. Antibodies

Single-domain antibodies against HER2 containing a C-terminal Cys–$(His)_6$ peptide [5, 15, 16] were kindly provided by Dr. Baty (CNRS UMR 7258, Inserm U1068, Centre de Recherche en Cancérologie de Marseille).

2.3. Quantum dots

CdSe/ZnS core/shell QDs with emission peaks maxima at 562 nm, 569 nm, and 570 nm (QDs_{562}, QDs_{569}, and QDs_{570}, respectively) dissolvable in an organic solvent were kindly provided by Dr. P. Samokhvalov (National Research Nuclear University MEPhI, Moscow, Russia). The PbS/CdS/ZnS core/core-periphery/shell QDs with an emission peak maximum in the near infrared region (1120 nm, QDs_{NIR}) were kindly provided by Dr. Sungjee Kim (Department of Chemistry Pohang University of Science & Technology Pohang, South Korea).

2.4. Transfer of quantum dots to the aqueous phase

The procedure is described in detail elsewhere [14]. In brief, 10 mg of QDs was dissolved in 1 ml of chloroform, an equal volume of methanol was added, and the mixture was stirred and centrifuged at 14,000 rpm. The supernatant was removed, and the pellet was dissolved in methanol, after which 200 μl of a 10 mg/ml solution of DL-cysteine in methanol was added dropwise while stirring to a final volume of 1 ml. The QDs were settled by centrifugation at 14,000 rpm and washed with methanol three times, centrifuging the mixture at the same speed every time. The pellet was vacuum-dried and dissolved in water while adding 1 M NaOH dropwise and stirring in a vortex. The resultant QD solution was sonicated and filtered through a filter with a mesh size of 0.22 μm. The size variation of the dissolved nanoparticles was estimated by dynamic light scattering using a ZetaSizer ZS device (Malvern). A 150 mg/ml solution of the complex polymer $HS–(CH_2)_m–EG_n–OH$ (Prochimia) was added to the aqueous suspension of 10 mg of QDs prepared for water-solubilization, and the mixture was incubated at 4°C for at least 12 h. Unbound molecules were removed by centrifugation on Amicon Ultracel-10K filters. Milli-Q water filtered through a filter with a mesh size of 0.22 μm was used for washing. The final purification of the QDs was performed in PD MiniTrap Sephadex G-25 columns (GE Heathcare). The size variation of the nanoparticles was estimated by dynamic light scattering using a ZetaSizer ZS device (Malvern).

2.5. Bioconjugation
The procedure is described in detail elsewhere [14]. In brief, a solution of the N-(p-maleimidophenyl) isocyanate crosslinker was added to 50 μl of a 2 mg/ml aqueous solution of QDs coated with modified polyethylene glycol buffered with sodium borate (pH 8.5) to a 50-fold molar excess of the crosslinker over the QDs. The reaction mixture was incubated at room temperature in the dark for 30 min while gently stirring. The reaction of QD activation was terminated by gel filtration of the preparation through Sephadex G-25 columns equilibrated with 50 mM phosphate buffer saline (PBS).

Disulfide bonds in the antibody preparation were reduced by incubation in a tris(2-carboxyethyl)phosphine solution at a molar ratio of 1:10 for 30 min; the preparation was purified in PD MiniTrap Sephadex G-25 columns. The sdAbs treated with tris(2-carboxyethyl)phosphine were mixed with the maleimid-activated QDs at a molar ratio of 10:1. The reaction was carried out at room temperature in the dark for 2 h while gently stirring. The preparation of the conjugates was purified in Superdex 200 columns equilibrated with 50 mM PBS.

2.6. Agarose gel electrophoresis
Electrophoresis of the QDs and QD–sdAb conjugates was performed in 0.5% agarose gel at a potential gradient of 1 V/cm in tris-acetic buffer saline for 30–60 min. Gel documentation was performed using a ChemiDoc MP system (Bio-Rad).

2.7. Immunosorbent assay
The adsorption of HER2 protein (10 mg/ml) was carried out in 50 mM carbonate/bicarbonate buffer saline (pH 9.4) in an immunological plate at 4°C overnight. The plate was washed with PBS three times, with tween 20 added to a volume concentration of 0.1% (PBST). The plate was blocked with a 1% casein solution in PBS (Thermofisher) at room temperature for 1.5 h on a rotator and washed with PBST three times. A 0.3 mg/ml solution of QD–sdAb conjugates was used to prepare a series of successive dilutions with PBS to ratios of 1:100, 1:1000, and 1:10,000. The conjugates diluted 1:100 in the absence of the antigen were used as a negative control. The incubation was carried out at room temperature in the dark for 1 h. After the end of the immune reaction, the plate was washed four times with PBST. The results were read at excitation and emission wavelengths of 500 and 562 nm.

2.8. Immunohistochemistry
The protocol [17] with our modifications was used. The cells were washed from the medium with D-PBS and fixed with a 4% formaldehyde solution in 1× PBS at 4°C for 10 min. Nonspecific binding sites were blocked by incubating the slides in a 1% bovine serum albumin (BSA) solution in PBS at room temperature for 40 min. Then, they were incubated in the presence of QD_{569}–sdAb solution diluted 1:50, 1:100, or 1:300 in the blocking solution of BSA in PBS for 1 h. After the immune reaction, the slides were washed with D-PBS–0.1% Tween 20 three times. Total DNA was stained with Hoechst solution (10 μg/ml) at 4°C overnight. Finally the slides were washed twice with D-PBS and once with PBS. The preparations were mounted in a 50% glycerol solution in PBS and stored at 4°C overnight before imaging.

2.9. Microscopy
Fluorescent microscopy was carried out using a Zeiss Axioscope microscope with an A-Plan 40×/0.65 lens upon fluorescence excitation with a mercury lamp. The fluorescence detection channels were 370–430 nm for Hoechst and 520–555 or 593–610 nm for QDs_{569}.

2.10. Flow cytometry

The attached SK-BR-3 and MDA-MB-231 cell cultures were grown to 80% confluence, detached from the culture vessels with 0.05% EDTA in PBS without Ca^{2+} or Mg^{2+} cations, and washed twice with an FACS buffer saline (D-PBS containing Ca^{2+}/Mg^{2+} and 0.5% BSA). 10^5 cells in 50 µl of the FACS buffer saline were incubated in the presence of monoclonal anti-HER2 antibodies conjugated with Alexa Fluor 488 in the dark at 4°C for 1 h and then washed with the FACS buffer saline. After that, stained cells were counted using a flow cytometer.

The possibility of counting HER2-expressing cells with the use of QD_{570}–sdAb conjugates was estimated as follows. The SK-BR-3 cells grown to 80% confluence were detached with 0.05% EDTA in PBS without Ca^{2+} or Mg^{2+} cations and washed twice with an FACS buffer saline (D-PBS containing Ca^{2+}/Mg^{2+} and 0.5% BSA). 10^5 cells in 50 µl of the FACS buffer saline were incubated in the presence of QD_{570}–sdAb conjugates at a concentration of 30 µg/ml in the dark at 4°C for 1.5 h and then washed with the FACS buffer saline. After that, conjugate-stained cells and unstained cells from the same culture were mixed in different proportions. The stained cells were counted using a flow cytometer, and the proportions of the experimentally detected stained cells and the introduced into the analyzed mixture stained cells were compared.

3. Results and discussion

The conjugates of sdAbs and QDs were obtained as described in the "Materials and Methods". The design of the sdAbs in combination with the selected cross-linker and surface functional groups of the QDs made it possible to conjugate all the sdAbs to the QDs in the same predetermined orientation ensuring the maximum accessibility of the antigen-binding sites of the antibodies for the antigen molecules [14]. In order to preliminarily estimate the results of conjugation between QDs and sdAbs, we compared the electrophoretic mobilities of unconjugated QDs and QD–sdAb conjugates (Figure 1).

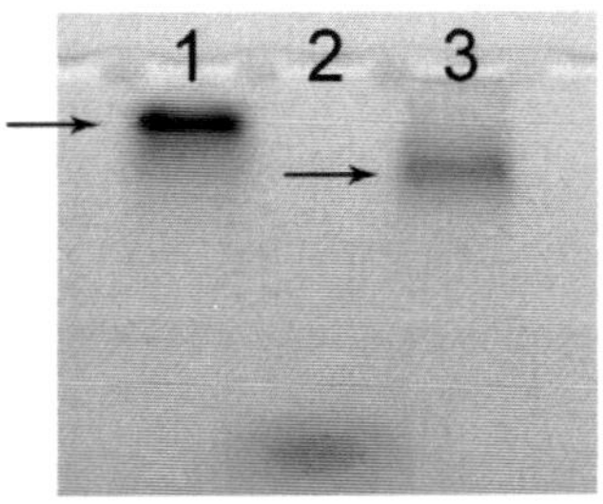

Figure 1. The electropherogram of QDs and QD–sdAb conjugates in agarose gel. Lanes: 1, QDs with an emission peak in the near-infrared region (QDs_{NIR}); 2, bromophenol blue; 3, conjugates of QDs_{NIR} with anti-HER2 sdAbs

The changes in mobility indicated successful conjugation of QDs and sdAbs. The potential applications of the conjugates were estimated in experiments with immune-sorbent assay (ISA), immunohistochemistry (IHC), and flow cytometry.

In order to estimate the possibility of using the conjugates in ISA, recombinant HER2 was adsorbed to wells of an immunological plate at a concentration of 10 mg/ml, after which, solutions of conjugates of the QDs with emission maximum at 562 nm and anti-HER2 sdAbs (QD_{562}–sdAbs) with different concentrations were added to the wells. The results of the analysis (Figure 2) have shown that the HER2 antigen can be detected using QD_{562}–sdAb conjugates at a concentration of 3 µg/ml.

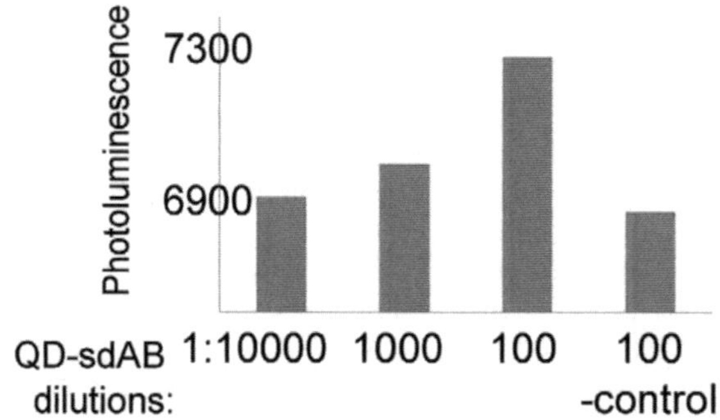

Figure 2. Analysis of the effect of the QD_{562}–sdAb conjugate concentration on the intensity of the fluorescence signal of the conjugates specifically binding HER2. The ordinate shows the fluorescence signal intensity; the abscissa shows successive dilutions of the original 0.3 mg/ml solution of QD_{562}–sdAbs. In the negative control, the antigen concentration was zero and the conjugate concentration was 3 μg/ml.

The possibility of using conjugates of anti-HER2 sdAbs and QDs with an emission peak at 569 nm (QDs_{569}) in IHC was employed in experiments with cultured SK-BR-3 cells, which are known to overexpress HER2 protein located on the cell membrane surface [18]. Cultures of MDA-MB-231 cells, which do not show high level of HER2 expression [18], served as a negative control. The HER2 expression in the cell cultures was confirmed by flow cytometry (data not shown) shortly before the IHC procedure. The fluorescent images demonstrated the staining of the cell membrane of the HER2-positive cells (Figures 3, B–D) and only background staining of the cytoplasm and nucleus in HER2-negative cells (Figures 3, F–H). Thus, the experiments showed highly specific detection of the HER2 surface antigen by means of IHC using QD_{569}–sdAb conjugates.

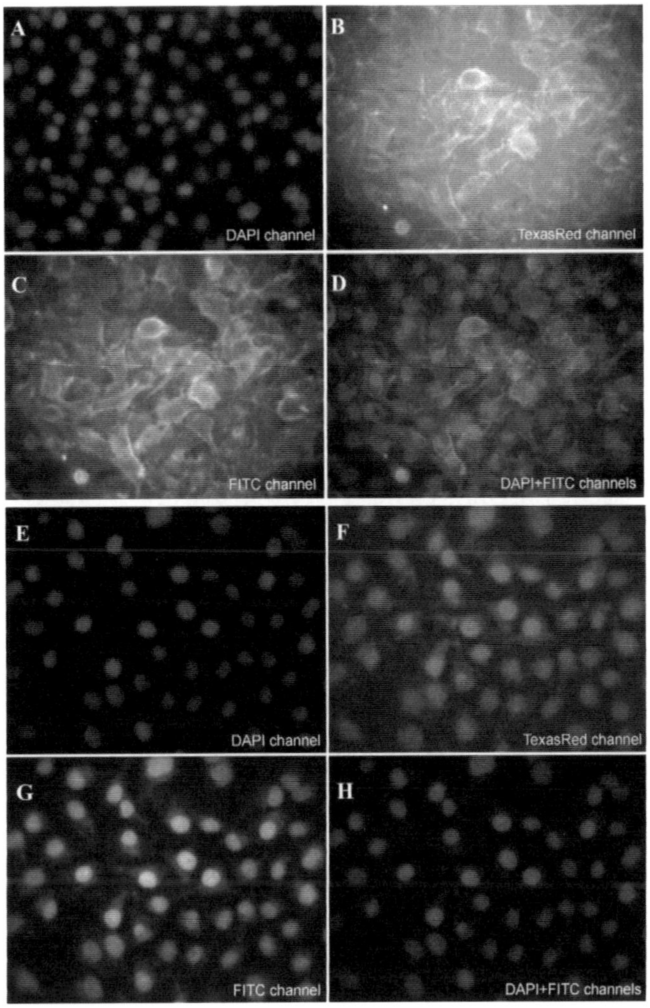

Figure 3. Microscopic images of cultures of (A–D) SK-BR-3 and (E–H) MDA-MB-231 immunostained with QD_{569} conjugated with anti-HER2 sdAbs. The fluorescent detection channels were: 370–430 nm for Hoechst (DAPI channel, panels A, D, E, H), 520–555 nm for QDs_{569} (FITC channel, panels C, D, G, H), and 593–610 nm (TexasRed channel, panels B, F). Optical magnitude, 40×.

The possibility of using conjugates of anti-HER2 sdAbs and QDs with an emission peak at 570 nm (QDs_{570}) to count of stained with anti-HER2 conjugates HER2-expressing cells in a mixture with other cells was demonstrated in flow cytometry experiments. A portion of SK-BR-3 cells was incubated with anti-HER2 conjugates as described in the "Materials and Methods" and mixed with unstained cells. Subsequent estimation of the percentage of correctly recognized cells stained with QD_{570}–sdAb conjugates has shown that the technique allows the detection of HER2-positive cells even if their proportion in the population is as small as 0.5% (Figure 4).

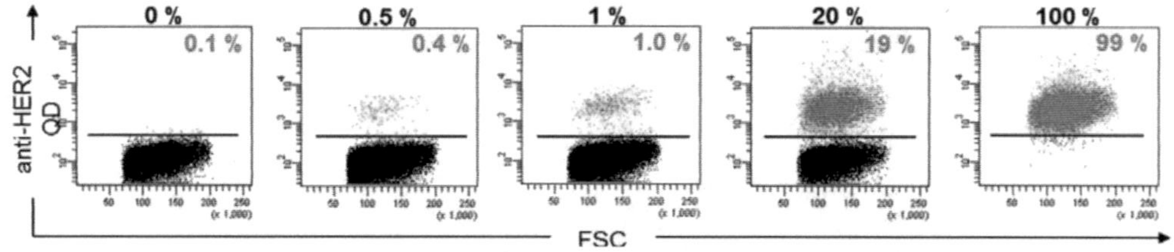

Figure 4. Results of counting the cells stained with anti-HER2 sdAbs conjugated with QDs_{570}. The black numbers above the plots show the percentages of SK-BR-3 cells stained with the conjugates that were included into the analyzed mixtures; the pink numbers at the upper right corners of the plots show the percentages of stained SK-BR-3 cells detected in the analyzed mixtures by means of flow cytometry.

4. Conclusions

The results of our study allow us to propose a method for fabrication of QD-based nanoprobes with advanced functional characteristics, including an increased sensitivity. The ultraminiature nanoprobes consisting of anti-HER2 sdAbs conjugated with QDs in a highly oriented manner developed in our studies have numerous implications for advanced integrated diagnostic procedures.

Acknowledgments

We a grateful to Dr. Pavel S. Samokhvalov (National Research Nuclear University MEPhI, Moscow, Russia) for kindly providing us with quantum dots fluorescing in the visible spectral region, Dr. Sungjee Kim (Department of Chemistry Pohang University of Science & Technology Pohang, South Korea) for kindly providing us with quantum dots fluorescing in the near-infrared region, Dr. Tina Van den Broeck, Line De Kimpe, and Dr. Frans Nauwelaers (BD Biosciences, Erembodegem, Belgium) for the assistance with flow cytometry experiments, and Dr. Baryshnikova (Laboratory of Experimental Diagnosis and Biotherapy of Tumors, Blokhin Cancer Research Center, Moscow, Russia) for the assistance with immunohistochemical experiments.

This study was supported by the Federal Target Program for Research and Developments of the Ministry of Education and Science of the Russian Federation (grant no. 14.584.21.0012, contract no. RFMEFI58415X0012).

References

[1] Sukhanova A, Devy J, Venteo L, Kaplan H, Artemyev M, Oleinikov V, Klinov D, Pluot M, Cohen J H M and Nabiev I 2004 Anal. Biochem. 324 60
[2] Bilan R, Fleury F, Nabiev I and Sukhanova A 2015 Bioconjugate Chem. 26 609
[3] Larson D R 2003 Science. 300 1434
[4] Deerinck T J 2008 Toxicol. Pathol. 36 112
[5] Hafian H, Sukhanova A, Turini M, Chames P, Baty D, Pluot M, Cohen J H M, Nabiev I and Millot J-M 2014 Nanomed. Nanotechnol. 10 1701
[6] Wang S, Zhang C, Wang G, Cheng B, Wang Y, Chen F, Chen Y, Feng M and Xiong B 2016 Theranostics 6 1877
[7] Kelkar S S and Reineke T M 2011 Bioconjugate Chem. 22 1879
[8] Harris L, Fritsche H, Mennel R, Norton L, Ravdin P, Taube S, Somerfield M R, Hayes D Fand Bast R C 2007 J. Clin. Oncol. 25 5287
[9] Khalil H S, Langdon S P, Goltsov A, Soininen T, Harrison D J, Bown J, Deeni Y Y 2016 Oncotarget, doi: 10.18632/oncotarget.12425
[10] Chotprakaikiat W, Allen A, Bui-Minh D, Harden E, Jobsri J, Cavallo F, Gleba Y, Stevenson F K, Ottensmeier C, Klimyuk V, Savelyeva N 2016 OncoImmunology 5 e1166323

[11] Roux K H, Greenberg A S, Greene L, Strelets L, Avila D, McKinney E C, Flajnik M F 1998 Proc. Natl. Acad. Sci. U.S.A. 95 11804
[12] Greenberg A S, Avila D, Hughes M, Hughes A, McKinney E C and Flajnik M F 1995 Nature 374 168
[13] Conrath K, Wernery U, Muyldermans S and Nguyen V Dev. Comp. Immunol. 27 87
[14] Sukhanova A, Even-Desrumeaux K, Chames P, Baty D, Artemyev M, Oleinikov V and Nabiev I 2012 Nat. Protoc. Exch. doi:10.1038/protex.2012.042
[15] Rakovich T Y, Mahfoud O K, Mohamed B M, Prina-Mello A, Crosbie-Staunton K, Van Den Broeck T, De Kimpe L, Sukhanova A, Baty D, Rakovich A, Maier S A, Alves F, Nauwelaers F, Nabiev I, Chames P and Volkov Y 2014 ACS Nano 8 5682
[16] Even-Desrumeaux K, Fourquet P, Secq V, Baty D and Chames P 2012 Mol. Biosyst. 8 2385
[17] Wu X, Liu H, Liu J, Haley K N, Treadway J A, Larson J P, Ge N, Peale F and Bruchez M P 2003 Nat Biotechnol. 21 41
[18] Subik K, Lee J F, Baxter L, Strzepek T, Costello D, Crowley P, Xing L, Hung M C, Bonfiglio T, Hicks D G and Tang P 2010 Breast Cancer (Auckl) 20 35

Domestic Development of Single-Photon Emission Computed Tomography (SPECT) Unit with Detector based on Silicon Photomultipliers

S Grishakov[1], O Ryzhikova[2], V Sergienko[3], A Ansheles[4] and S Novikov[5]

[1] UNIX medical company, Moscow, Russia
[2] Russian Cardiology Research and Production Complex, Moscow, Russia
[3] Russian Cardiology Research and Production Complex, Moscow, Russia
[4] Russian Cardiology Research and Production Complex, Moscow, Russia
[5] Institute of oncology, St.Peterburg, Russia

E-mail: medphys@mail.ru

Abstract. The idea of creating a single-photon emission computed tomography unit with solid-state photomultipliers is not new [1], as the problems of analog-to-digital conversion with a lot of noise and a wide range of values of intrinsic spatial resolution of the detector in a center and relevant fields of view could not be solved by means of gamma-camera detector architectures based on vacuum photomultipliers. This paper offers a new SPECT imaging solution that is free from these problems.

1. Introduction

Radionuclide diagnostics is a technique that allows evaluating structure-functional state of organs and tissues of human body by means of different diagnostic radiopharmaceuticals administered in the body to acquire imaging information about the state of cellular systems under examination.

Radiopharmaceuticals are specific chemical or biochemical compounds labeled with gamma-emitting radionuclides with a short half-life. Gamma radiation emitted from the body of a patient is recorded by a gamma camera detector and the received information is converted into a functional image of the examined organ after computer processing. Spatiotemporal pattern of the radiopharmaceutical distribution provides the information about shape, size and location of the organ, as well as about presence of lesions in it.

Radionuclide diagnostics is widely used in oncology (brain tumors, scintigraphy of malignant breast tumors, bone and liver scintigraphy), endocrinology, cardiology, urology and nephrology.

2. From prototype to industrial design

Nowadays, the most common gamma camera units are based on vacuum photomultipliers. Known user problems are defined by specifications of gamma camera detectors [2], this can be described by two values having determining competitive advantage at the market. These are the values of the spatial resolution of the detector in the center and in the relevant field of view at the level of 0.5 (FWHM) and the value of energy resolution of the unit for 99mTc. The problem of detector energy resolution values for 99mTc gets prominent coverage in literature as well as in this article. The authors focus on devices electronic component which determines detector energy resolution values. For silicon photomultiplier detector, channels number defines this value. Channel is a complex of the silicon photomultiplier (detector pixels), buffers and comparators with external control, analog-digital, time-to-digital converters and high-speed data bus.

The creation of a gamma camera detector with the field of view of 400 x 540 mm^2, based on silicon photomultiplier, requires minimum 13824 pixels. This detector architecture allows obtaining the values of the spatial resolution of the detector in the center and in the relevant field of view at the level of 0.5 (FWHM) with minimal differences. Practically, the absolute values equality could not be obtained [3] because the measured modulation transfer function is unimodal and similar to a Gaussian function [4]. The values of 2.9 mm at the level of 0.5 FWHM and of 5.3 mm at the level of 0.1 FWHM were obtained. This fact is confirmed by the advance advertising of the new unit by General Electric [5]. The unit parameters of the spatial resolution are 2.5 mm and 3.8 mm respectively.

The value of the internal energy resolution for 99mTc [6] is determined by a manufacturer during the unit acceptance tests and depends on the specified detector architecture. In our design the detector is made of large single-crystal CsI(Tl) pixels, and silicon photomultipliers are joined by optical glue. Such architecture and recovery algorithm for gamma event allow reaching internal energy resolution of 9.2% for 57Co [7]. According to authors [8], the energy resolution of the detector is inversely proportional to the square root of the energy of gamma quanta of the radioactive isotope used in the measurement. Consequently, the recalculation of the internal energy resolution for 99mTc gives the value of 8.5%.

Obviously, the cost of single-photon emission computed tomography units with solid-state pixels is higher than the cost of units with detectors based on the vacuum photomultipliers. This situation will not change until the market is filled with modern analogues. For this purpose, the price of components of such units should fall for a manufacturer. For detectors on silicon photomultipliers, this is decisive for entering the competitive market.

Alternative way, when a manufacturer introduces new technology, can be applied in the modernization of the existing technical base. GE used this option for the introduction of CZT technology. Other manufacturers will do the same. The time will show if we should look for our own way to solve the problem or we should rely upon the experience of the colleagues.

3. Conclusions

Nowadays there is a tendency towards the use of modern materials in gamma-camera detectors and SPECT. The SPECT imaging unit was developed in Russia. Its parameters can compete with world latest developments. The standard [4] underlines that its application to the detectors with the architecture of many pixels is incorrect. So we need new applications to the standards for a new generation of gamma cameras, and a direct comparison of the specifications of gamma camera based on the vacuum photomultipliers and silicon photomultipliers is also incorrect. However, the market has its own rules, and a buyer will choose a cheaper unit capable to satisfy minimal necessary requirements of a user.

References

[1] Narkevich B, Physical and technical support of nuclear medicine: current state and development prospects, , Radiation oncology and nuclear medicine, No**1**, 2012.

[2] Arsvold D, Vernik M. Emission Tomography: Fundamentals of PET and SPECT. Technosphera, Moscow, 2009.

[3] UNIX Medicl Company. Test protocol No. **01**-1 of 12 August 2015 for production prototype of gamma-camera No.**13411.1008799.13.108** The Ministry of Industry and Trade of the Russian Federation, 2015.

[4] NEMA NU 1-2012, Performance Measurements of Gamma Cameras, National Electrical Manufacturers Association, P. **8,9,** 2013.

[5] http://www3.gehealthcare.com/en/products/categories/nuclear_medicine/spect_and_spect-

ct/discovery_nm-ct_670_czt

[6] NEMA NU 1-2012, Performance Measurements of Gamma Cameras, National Electrical Manufacturers Association, pp. **11-13,** 2013.

[7] GOST R IEC 60789-2010, Characteristics and test conditions of radionuclide imaging devices. Anger type gamma cameras. Standartinform, Table 1, p. **3**, 2012.

[8] Vartanov N, Samoylov P, Applied scintillation gamma-spectroscopy, Atomizdat, Moscow, 1975.

IOP Conf. Series: Journal of Physics: Conf. Series **784** (2017) 012018 doi:10.1088/1742-6596/784/1/012018

Improvement of antigen detection efficiency with the use of two-dimensional photonic crystal as a substrate

Dmitriy Dovzhenko[1,2], Vladimir Terekhin[1], Kirill Vokhmincev[1], Alyona Sukhanova[1,3], Igor Nabiev[1,3]

[1] Laboratory of Nano-Bioengineering and [2] Department of the Physics of Micro- and Nanosystems, National Research Nuclear University MEPhI (Moscow Engineering Physics Institute), Moscow, Russian Federation
[3] Laboratoire de Recherche en Nanosciences, LRN-EA4682, Université de Reims Champagne-Ardenne, Reims, France

E-mail: dovzhenkods@gmail.com

Abstract. Multiplex detection of different antigens in human serum in order to reveal diseases at the early stage is of interest nowadays. There are a lot of biosensors, which use the fluorescent labels for specific detection of analytes. For instance, common method for detection of antigens in human serum samples is enzyme-linked immunosorbent assay (ELISA). One of the most effective ways to improve the sensitivity of this detection method is the use of a substrate that could enhance the fluorescent signal and make it easier to collect. Two-dimensional (2D) photonic crystals are very suitable structures for these purposes because of the ability to enhance the luminescent signal, control the light propagation and perform the analysis directly on its surface. In our study we have calculated optimal parameters for 2D-dimensional photonic crystal consisting of the array of silicon nano-rods, fabricated such photonic crystal on a silicon substrate using reactive ion etching and showed the possibility of its efficient application as a substrate for ELISA detection of human cancer antigens.

1. Introduction

Photonic crystals (PhC) are widely used in many scientific areas nowadays including optoelectronics, photonics [1], gas sensing [2] and biosensing [3–5]. The numbers of directions in which the PhC have refraction index periodicity determines their classification within the three principle groups: one- (1D), two- (2D) and three- (3D) dimensional PhC. Each group of PhC has its own advantages and disadvantages in applications, depending on the physical properties of PhC material and fabrication technique. For instance, 1D PhC may be easily fabricated by electrochemical etching of silicon wafers and have large surface area, what makes them suitable for fabrication of different sensors. However, by using 1D PhC it is difficult to achieve a very low mode volume and high Q-factor in order to observe strong coupling effect. 3D photonic crystals, on the other hand, have great confinement in all three dimensions, but their fabrication is much more sophisticated, and the need for periodicity in all dimensions makes it almost impossible to change refractive index inside the structure and thus vary their photonic properties. 2D PhC could be fabricated using a well-developed electron beam lithography technique from a silicon wafer, which is a conventional material for semiconductor industry. Rather high refractive index of silicon (around 4 in the visible range) leads to a great refractive index contrast in air. The presence of one "free" dimension in a 2D photonic crystal, in which the confinement of electromagnetic waves is only due to the index waveguiding, opens up an opportunity to alternate refractive index of the media inside the PhC by introduction of analyte, and hence to affect on photonic properties of the system. Moreover, 2D PhC allow to control the propagation of light in plane of periodicity, also, the very high Q-factors along with low mode volumes could be achieved. Thereby, weak coupling regime between the cavity of 2D photonic crystal and the luminophores embedded inside the PhC structure can be possible, and the luminescence inside the cavity could be enhanced [6]. These properties make 2D PhCs appear to be the best compromise for application in biosensing, for example as a substrate for improvement of luminescence detection techniques. In this study we used 2D PhC in form a of an ordered array of silicon nanorods on a silicon wafer substrate for detection of antigens in human serum samples in enzyme-linked immunosorbent assay (ELISA). We have used quantum dots (QDs) as the fluorescent markers because of their high photostability, fluorescence quantum yield almost reaching 100 % and ability to be conjugated with the detection antibodies using well developed approaches [7]. We have also made a finite difference time-domain (FDTD) calculation of the optimal parameters for 2D PhC array made of silicon nanorods, fabricated photonic crystal with desired properties on a silicon substrate using electron beam lithography and reactive ion etching (BOSCH process) and have demonstrated the possibility of its efficient application as a substrate for ELISA detection of human cancer antigens.

2. Materials and methods

2.1 Materials

Cadmium oxide (powder, 99.5%), zinc oxide (powder, 99.99% trace metals basis), 1-octadecene (ODE, technical grade, 90%), oleic acid (OA, technical grade, 90%), selenium (powder, 100 mesh, 99.5%), oleyalmine (OLA, technical grade, 70%), thiourea (ACS reagent, ≥99.0%), tri-n-octylphosphine (TOP, technical grade, 97%), tri-n-octylphosphine oxide (TOPO, reagent grade, 99%), triethylene glycol dimethyl ether (TEGDME, ReagentPlus, 99%), anhydrous solvents (chloroform, hexane, ethanol, methanol and 2-propanol) and hydrogen peroxide solution (30 wt% in H2O) were purchased from Sigma-Aldrich; n-hexadecylphosphonic acid (97%) was purchased from PlasmaChem GmbH. All chemicals were used as received without additional purification. P-type <100> Si wafers were purchased from Telecom-STV.

2.2 Calculation method

For simulation of electromagnetic wave propagation in a 2D photonic crystal and further determination of photonic structure properties we used FDTD approach. Software from MIT (MEEP) was used for calculations. Different types of structures, cavities, lattice periods and rod morphology for rod-type 2D photonic crystal have been tested in order to achieve a good mode confinement and wide enough photonic bandgap for TM polarization with the cavity eigenmode at 565 nm wavelength corresponding to the maximum of QD photoluminescence that have been used as fluorescent markers.

2.3 Photonic crystal fabrication and characterization technique

Monocrystalline silicon wafers with 380 μm thickness have been used as an initial substrate. Electron beam lithography has been used to create a Ni pattern mask for subsequent reactive ion etching with the use of BOSCH process with the following parameters: chamber pressure – 5 mTorr, etching gas – SF_6 10 sccm, passivation gas – C_4F_8 100 sccm, O_2 10 sccm, etch coil power 450 W and cathode power – 20 W. Scanning electron microscope (SEM) studies have been performed using RAITH 150 TWO in order to estimate the resulting sample morphology after the fabrication.

2.4 Surface modification of the photonic crystal substrate

Surface functionalization of the previously fabricated silicon substrates included three stages: (i) chemical oxidation of the surface of nanorod array, (ii) modification of the oxidized surface with the amine groups and (iii) modification of the exposed amine functional groups by succinic anhydride. Surface oxidation was carried out with using Piranha solution – the mixture of sulfuric acid, hydrogen peroxide and deionized water in a volume ratio of 3:1:8, respectively. Silicon substrate was placed in the cool oxidative solution with the temperature of 4° C and held there for 2 hours under gentle stirring. Excess of the oxidative solution was washed away from the surface of the silicon wafer by the deionized water. Surface modification with the amine groups was carried out with using (3-aminopropyl) triethoxysilane (APTES). 5 μl of APTES was dissolved in 15 ml of ethanol, and oxidized silicon substrate was placed into the obtained solution for 12 hours under slight agitation. In order to remove unreacted APTES, silicon substrate was placed for three times into a fresh portion of pure chloroform for 15 minutes. Then the substrates were dipped in a solution of succinic anhydride in N, N-dimethylformamide for 30 minutes under continuous stirring, and resultant photonic crystal substrates with carboxyl surface functionality were finally washed with chloroform according to the procedure used to remove unreacted APTES.

2.5 Preparation of water-soluble core-shell CdSe/ZnS quantum dots.

Core CdSe QDs used in this study were synthesized from CdO and TOPSe by injection technique, and further coated by a ZnS shell using the SILAR technique. The details of the procedure is described in details in [8]. After the synthesis, QDs were modified with DL-cysteine to render them water-soluble, and further treated using thiol-PEG- carboxylate to achieve high stability of colloidal solution and fluorescence properties.

3. Results and discussion

In our study we have investigated the possibility of improvement of antigen detection efficiency with the use of photonic crystal as a surface for immunocomplex formation. Two candidate types of 2D photonic crystals were considered, one of which represents a crystal slab with holes forming a PhC, and an array of nanorods on a substrate. In order to achieve good embedding of the liquid samples inside PhC structure and to have an ability to regenerate the surface for multiple detections, we have chosen the structure consisting of nanorods with line of nanorods missing in the center forming a defect. This type of structure is schematically shown in Figure 1.

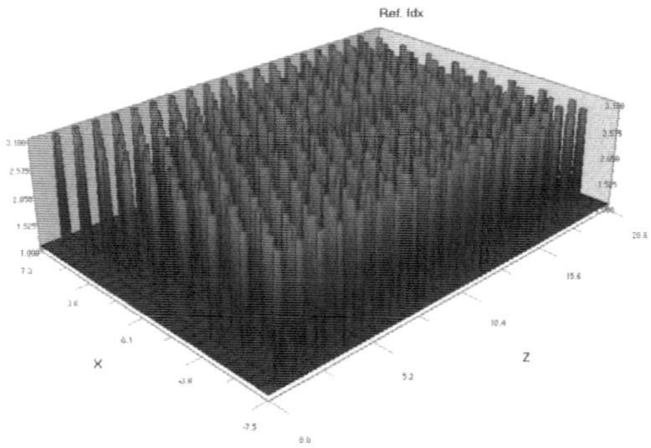

Figure 1. Schematic representation of a 2D photonic crystal structure used in this work.

The following parameters were found to be optimal to achieve photonic bandgap big enough to cover the photoluminescence spectrum of QDs by the FDTD calculation: lattice period 280 nm, diameter of individual nanorod around 100 nm and nanorod height of 600 nm. SEM image of a 2D PhC fabricated using such parameters is demonstrated in Figure 2.

Figure 2. SEM image of 2D photonic crystal surface. Tilted view from the top.

The next step was to perform a modification of the surface for further antigen immobilization using the procedure described in materials and methods section. After the surface modification and immunocomplex formation, we measured the luminescent signal of quantum dots using TECAN Infinite® 200 or TECAN Spark® 10M fluorescence readers. The fabricated arrays demonstrated a full compatibility with this standard model of fluorescent readers by and good repeatability of the obtained signal. Typical luminescence spectrum measured from QDs immobilized on the 2D PhC substrate is shown on Figure 3.

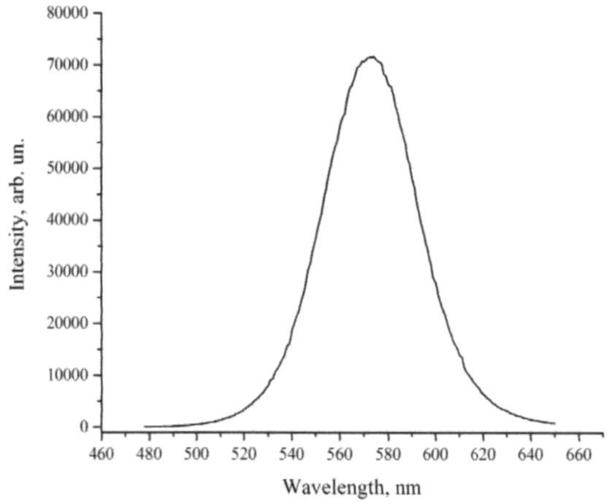

Figure 3. Luminescence spectrum of the quantum dots deposited on the 2D photonic crystal substrate.

4. Conclusions

In this study we have calculated optimal parameters for two-dimensional photonic crystal consisting of the array of silicon nanorods in order to enhance the photoluminescence of embedded QDs, fabricated photonic crystal with such properties from a silicon wafer using BOSCH process. Surface modification technique has been developed for efficient immobilization of antibodies on photonic crystal surface. Finally we have demonstrated the possibility of its application for the immunodetection of human cancer antigens using a standard fluorescent reader.

Acknowledgments
This study was supported by the Ministry of Education and Science of the Russian Federation, grant no. 14.575.21.0065 ID RFMEFI57514X0065.

References

[1] Calusine G, Politi A and Awschalom D D 2014 arXiv Prepr. arXiv1405.4539 11123 12
[2] Levitsky I 2015 Sensors 15 19968
[3] Mora M D La, Ocampo M, Doti R, Lugo J and Faubert J 2013 State Art Biosens. Gen. Asp. 141
[4] Skivesen N, Têtu A, Kristensen M, Kjems J, Frandsen L H and Borel P I 2007 Opt. Express 15 3169
[5] Zhang W, Ganesh N, Block I D and Cunningham B T 2008 Sensors Actuators B Chem. 131 279
[6] Yuan X W, Shi L, Wang Q, Chen C Q, Liu X H, Sun L X, Zhang B, Zi J and Lu W 2014 Opt. Express 22 23473
[7] Resch-Genger U, Grabolle M, Cavaliere-Jaricot S, Nitschke R and Nann T 2008 Nat. Methods 5 763
[8] Dovzhenko D S, Martynov I L, Samokhvalov P S, Eremin I S, Kotkovskii G E, Sipailo I P and Chistyakov A A 2014 Proc. SPIE 9126 91263O

Gate simulation of Compton Ar-Xe gamma-camera for radionuclide imaging in nuclear medicine

L Yu Dubov, V N Belyaev, A K Berdnikova, A I Bolozdynia, Yu A Akmalova, and Yu V Shtotsky

National Research Nuclear University «MEPHI», Moscow, Russian Federation

E-mail: LYDubov@mephi.ru

Abstract. Computer simulations of cylindrical Compton Ar-Xe gamma camera are described in the current report. Detection efficiency of cylindrical Ar-Xe Compton camera with internal diameter of 40 cm is estimated as1-3%that is 10-100 times higher than collimated Anger's camera. It is shown that cylindrical Compton camera can image Tc-99m radiotracer distribution with uniform spatial resolution of 20 mm through the whole field of view.

1. Introduction

Taking all into consideration we conclude that behavior of S-factor in different cases has a strong influence on effective therapeutic activity. Method has high flexibility and reproducibility and gives opportunity to physician for wide retrospective in-depth analysis, thus such approach will lead to improvement of RSO effectiveness.

Radionuclide imaging is a powerful tool for medical diagnostics [1]. Single photon emission computer tomography (SPECT) based on Anger scintillation gamma cameras with typical spatial resolution of about 10 mm is commonly used for imaging distribution of radioisotopes emitting single gamma quanta. However, the low efficiency of mechanically collimated gamma cameras significantly affects their performance, particularly it is not possible to reduced exposition time or to lower the activity of administered radionuclide. The absence of a collimator allows to use much more of the available radiation.

Detectors with an electronic collimation based on the Compton effect, having a high luminosity and an acceptable spatial resolution of ~ 10 mm, were considered in a number of previous studies [3,4]. In the current work parameters of a full size cylindrical gaseous Argon-Xenon Compton gamma camera are optimized through Monte Carlo simulations in Gate/Geant4 environment [2].Also, the requirements for event selection and reconstruction algorithms for 3D Compton images are discussed.

2. Compton gamma camera

The determination of the source location via the Compton effect is based on the dependence between the scattering angle Θ and the energy of the recoil electron E_1 after Compton scattering of a gamma quantum with the energy E:

$$\cos(\theta) = 1 - m_e c^2 \left(\frac{1}{E_1} - \frac{1}{E} \right)$$

Journal of Physics: Conference Series **784** (2017) 012019 doi:10.1088/1742-6596/784/1/012019

Compton camera (see Figure 1) consists of a scatter detector (SD) and an absorption detector (AD). If the energy of the gamma rays is known it is necessary to determine the energy of the recoil electron E1 in the RD for calculation of the scattering angle Θ. Compton gamma camera detects the arrival direction of the gamma ray within the events acceptance cone, which is determined by the scattering angle Θ. Event cone axis is the extension of straight line connecting the point of interaction in the SD (x1, y1, z1) and the AD (x2, y2, z2).The apex of the cone is the point of interaction (Compton scattering) in the RD.

Uncertainty in the determination of the events acceptance cone are defined by a finite energy resolution of the SD, an error in the coordinates of events and by the Doppler broadening of the scattered γ-ray energy [3].

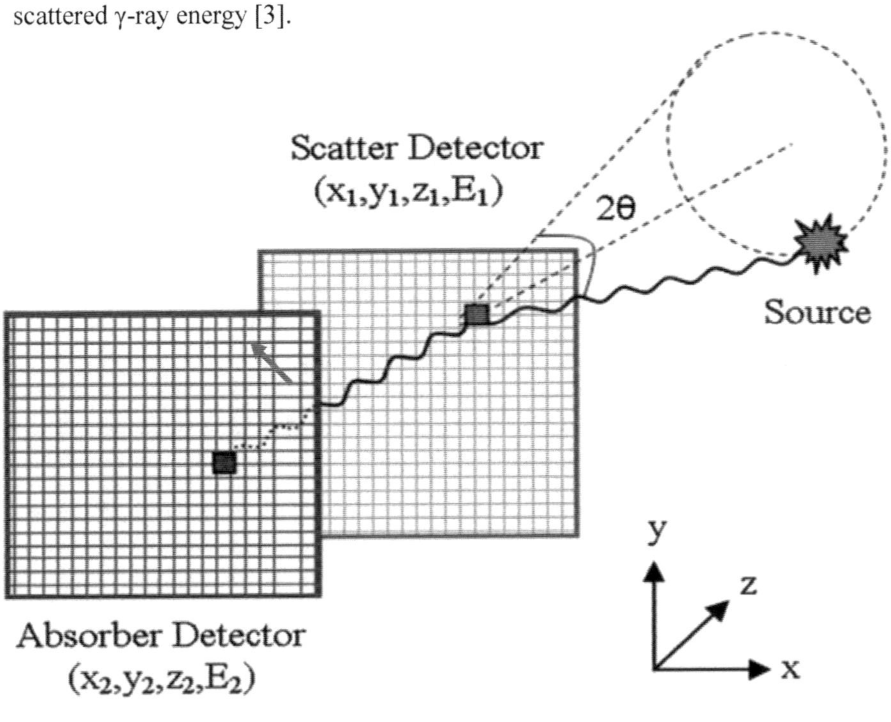

Figure 1. Compton camera operation principle.

3. Optimal parameters of gaseous detectors

3.1. General requirements for the Compton camera detectors

The scatter detector (SD) determines the coordinates of Compton interaction and the energy of Compton electron. SD must have high energy resolution and the prevailing probability of Compton scattering of the gamma rays η_C. The probability of photoelectric effect and multiple Compton scattering should be as low as possible.

Total absorption detector (AD) determines the coordinates and energy of photoelectric absorption. AD should have a high probability of the photoelectric effect η_{PE} for gamma rays. Contributions of single and multiple Compton scattering (followed by photoelectric effect) in AD should be minimal.

The detector for medical imaging should enclose the investigated object in order to achieve maximal efficiency. The optimal geometry of the Compton gamma camera for this task is a coaxial system of two cylinder detectors (SD and AD) of considerable size (about 1 m in diameter and in length). The only economically available detectors with such a size may be gaseous detectors. The aim of the computer simulation is to optimize the parameters of the gaseous scatter and absorber detectors of the Compton gamma camera to obtain the highest possible registration efficiency for gamma rays emitted by radioisotope Tc-99m (140 keV).

3.2. Selection of the optimal parameters of the gaseous detectors

Selection and optimization of parameters of gaseous detectors are carried out through Monte Carlo simulation in Gate\Geant4 environment, and the simulation results were analyzed in the MathLab environment.

The results of simulation have demonstrated that the most suitable gas for the SD is argon (η_C>80%) and for the AD is xenon (η_{PE}> 90%). Required detection efficiency can be achieved by increasing the gas pressure or the thickness of the detector.

Increasing the pressure from 10 bar to 20 bar in the SD at a constant scattering efficiency causes the grow of multiple Compton scattering. The ratio of probabilities of single (η_C) and multiple (η_{MC}) Compton scattering ($\kappa_C = \eta_C / \eta_{MC}$) reduce from 30 to 18. Multiple scattering worsen the quality of the reconstructed image. Optimal parameters for SD gaseous chamber are thickness of about 35 cm and argon pressure of 15 bar. It provides the scattering efficiency ε_C = 10 % and the single/multiple ratio κ_C = 22.

Similar calculations for the AD have shown that increasing the pressure at the same photoelectric absorption efficiency (η_{Ph}) causes growing the probability of photoelectric effect following Compton scattering (η_{C+Ph}), which also worsen image quality. The ratio of probabilities ($\kappa_{Ph} = \eta_{Ph} / \eta_{C+Ph}$) decreases from ~ 11 to ~ 9 by increasing the pressure from 10 bar to 20 bar. The obtained results demonstrate that the optimal parameters for AD gaseous chamber is a thickness of ~ 30 cm and xenon pressure of 15 bar. This ensure the effectiveness of the photoelectric absorption ε_{Ph} = 72% with the accept able contribution of multiple events κ_{Ph}= 9.5.

3.3. Accuracy of the interaction position determination

Position of Compton scattering or photo absorption events determined by the detector corresponds to the center of ionization region created by the secondary electrons. To estimate the spatial resolution of SD and AD the deviations of event position from the real point of interaction were calculated.

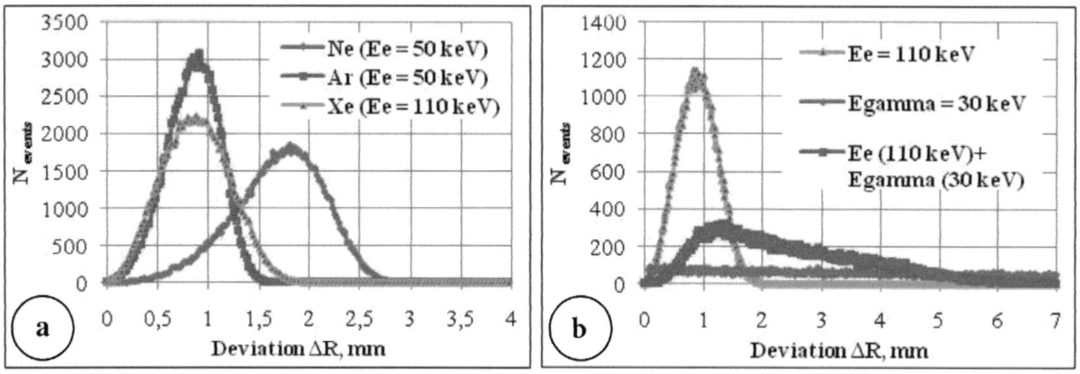

Figure 2. Distribution of deviation ΔR of determined position from the real point of interaction: a) position accuracy for Compton scattering in Ne and Ar (50 keV), for photo absorption in Xe (110 keV); b) position accuracy for photo absorption in Xe (photoelectron 110 keV, characteristic X-rays 30 keV and the total photoelectric absorption (140 keV).

For Ar-SD typical energy of secondary electron is 50 keV, while for Xe-AD – 110 keV. The simulation results are presented in the Figure 2a. Average deviation ΔR_e for Ar-SD and Xe-AD is ~ 1 mm with a mean square deviation of 0.3-0.4 mm. However, the photoelectric effect in Xe-AD is followed by a characteristic radiation with an energy of ~ 30 keV. It leads to an additional error in determination of the coordinates of the photo absorption.

The deviation of the characteristic radiation absorption from the point of X-ray emission are shown in the Figure 2b. The mean deviation of the coordinate for X-ray in Xe-AD is ΔR_γ = 8 mm. The energy

Journal of Physics: Conference Series **784** (2017) 012019 doi:10.1088/1742-6596/784/1/012019

weighted average deviation of the coordinate for "total photo absorption " is $\Delta R_{e+\gamma}$ = 2.4 mm, and the FWHM of this distribution is about 3 mm.

If the detector electronics able to distinguish between single and multiple interactions, coordinate accuracy in Xe-AD can reach 1 mm.

4. Events selection in the gaseous compton camera

For identification of the initial γ-ray direction two coincident events have to be registered: Compton scattering in SD and photo absorption in AD. But there are some other combinations of physical events which cause coincidence in SD and AD. Computer simulations of the cylindrical Compton gamma camera were carried out using Gate/Geant4 framework for the development of algorithm for the proper coincidence selection. Mathematical model of the camera consists of two enclosed cylinders - Ar-SD and Xe-AD (Figure 3).

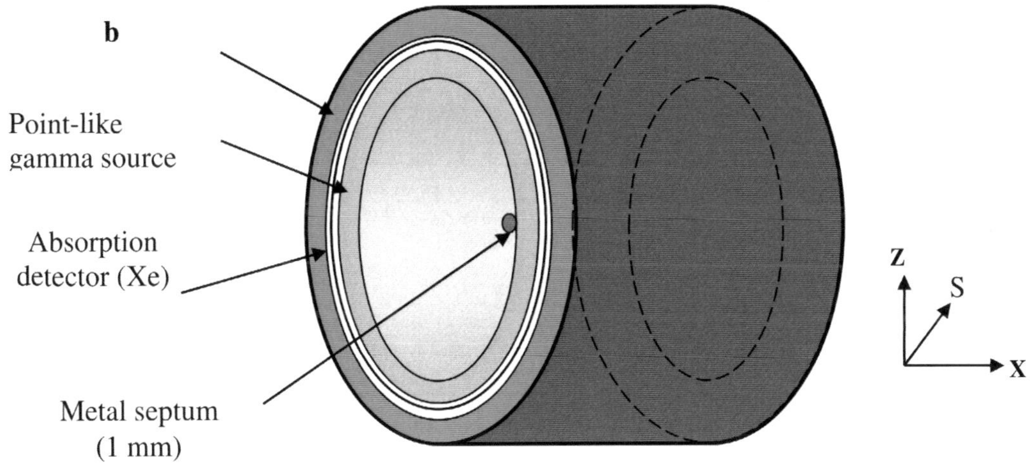

Figure 3. Model of the cylindrical Ar-Xe Compton gamma camera.

The gas pressure in both detectors is 15 bar. The inner diameter of the SD is 400 mm, outer diameter is 1100 mm. The inner diameter of the AD is 1100 mm, outer – 1700 mm. Between the detectors is a thin (1 mm) metal septum is placed. Detectors length is 500 mm. A point-like isotropic mono energetic gamma radioactive source with energy of 140 keV is placed in the center of the gamma camera. 10^6 initial γ-rays were simulated.

All the coincidences observed in Ar-Xe gamma camera were divided into eight types (see Table 1). Coincidences of the "0" type without Compton scattering occur with the photoelectric effect of the primary gamma rays in the Xe-AD and the subsequent registration of the characteristic radiation (~ 30 keV) in the Ar-SD. These coincidences do not provide any useful information and are background. Coincidences of the "1" type - "useful" coincidences, caused by Compton scattering of the primary gamma quanta in Ar-SD and following photoelectric absorption of the scattered radiation in the Xe-AD. Coincidences of the "2-7" type, resulting from the multiple Compton scattering of the primary gamma rays in the SD and the AD distort useful information.

Table 1 shows the classification of different types of coincidences and their relative contribution.

The efficiency of coincidences registration is ~ 3% when the restriction only on the total deposited energy (E_{SD}+E_{AD}) > 100 keV is used. The efficiency become 1% when selecting the energy as follows 1 < E_{SD} < 29 keV, E_{AD}> 70 keV and (E_{SD}+E_{AD}) > 100 keV. Table 1 demonstrate that such selection lead to the maximal contribution of "useful" coincidences of the type "1" is (87.9%); the contribution of coincidences of the type "0" is reduced from 15.8% to 0.2%; the contribution of coincidences of the type "4" (two Compton scatterings + Photo absorption) is practically unchanged (~ 12%).The contribution of other types does not exceed 0.5%.

It is important to take into account that about 50% of the gamma quanta emitted from the source escape through the open ends of the gamma camera without interaction.

Table 1. Classification and the contribution of the different types of coincidences («C» – Compton scattering, «Ph» – photoelectric effect)

Type	Contribution		Origin of coincidences	
	Without an energy selection in each detector	With an energy selection in each detector	1st event	2nd event
0	15,8%, (background)	0,2%, (background)	PE – photoelectron registration (110 keV) and a characteristic γ-quantum (30 keV) with different detectors (Ph)	
1	67,2%, (useful)	87,9%, (useful)	SD (one C)	AD (Ph)
2	1,5%	0%	AD (one C)	SD (Ph)
3	1,1%	0,1%	SD (one C)	AD (one C)
4	12,2%	11,5%	SD (two C) SD (one C)	AD (Ph) AD (one C+Ph)
5	0,5%	0%	AD (two C) AD (one C)	SD (Ph) SD (one C+Ph)
6	0,5%	0%	AD (one C)	SD (one C)
7	1,1%	0,3%	More than 2C	

5. Spatial Resolution of the point-like gamma sources

To estimate the spatial resolution of cylindrical Compton camera the image of two point-like gamma sources was simulated. For the image reconstruction algorithm of simple cone back projections in combination with iterative maximum likelihood estimation method (MLEM) was used [4]. Figure 4 shows a reconstructed image of two point-like gamma sources (140 keV) placed symmetrically with respect to Z axis at a distance of 80 mm from each other.

Reconstruction with a simple cone back projections method provides a spatial resolution of ~ 40 mm. A subsequent usage of the MLEM allowed to improve the spatial resolution up to ~ 20 mm. Selection of the most effective image reconstruction algorithm concerned with the design of the Compton gamma camera and largely determine its spatial resolution. It should be noted that unlike the flat Compton camera spatial resolution in the cylindrical Compton chamber is uniform throughout the field of view.

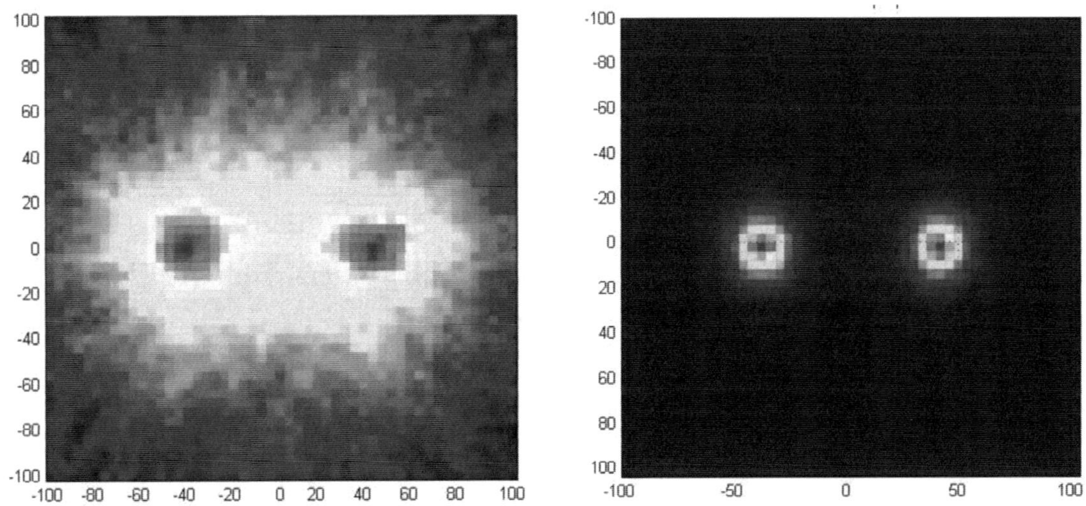

Figure 4. Reconstructed OXY plane images of two point-like isotropic gamma sources (140 keV) placed at the distance ±40 mm from Z: a) simple cone backprojection method; б) iterative maximum likelihood estimation method (after8iterations).

6. Conclusion

The following results and operating parameters of the detector were obtained via computer simulations:

- argon scatter detector with a thickness of 350 mm (at 15 bar) provides an efficiency of 140 keV gamma radiation scattering ~ 10 % and a spatial resolution ~ 1 mm;
- xenon absorption detector with a thickness of 300 mm (at 15 bar) provides an efficiency photoelectric absorption of scattered radiation~ 70 % and a spatial resolution ~ 3 mm;
- efficiency of registration of coincidences which are selected based on the energy deposit in both detectors is ~ 1 %;
- FWHM of the point-like gamma source spread function (spatial resolution) obtained with a simple backprojection reconstruction method is ~ 40 mm, and ~ 20 mm with a reconstruction based on the iterative MLEM algorithm.

Acknowledgements
This article was prepared with the support of NRNU MEPhI Academic Excellence Project.

References

[1] «Webb's physics of medical imaging» / M. A. Flower. CRC Press, 2012
[2] GAMOS 3.0.0 User's Guide, 2011, <http://fismed.ciemat.es/GAMOS/
[3] Bolozdynya A., Egorov V., Koutchenokov A.at. al.,NIM, A385, p. 225-238, 1997
[4] T. Hebert, R. Leahy, M. Singh, Opt. Soc. Amer., v. 7, N 7, p. 1305-1313, 1990

Synthetic techniques of radiopharmaceuticals production labeled with C-11 for PET in cardiology

V S Dyubkov[1,2], I V Ekaeva[2], T A Katunina[2], A S Rumyantsev[2], A V Silchenkov[2] and T V Tuflina[2]

[1] Electrophysical facilities Department, National Research Nuclear University MEPhI, Kashirskoe shosse 31, Moscow 115409, Russian Federation
[2] Bakulev Scientific Centre for Cardiovascular Surgery, Rublyovskoe shosse 135, Moscow 121552, Russian Federation

E-mail: vsdyubkov@mephi.ru

Abstract. Positron emission tomography (PET) and PET-Computerised Tomography (CT) are unique, non-invasive diagnostic techniques, in which the local, temporal and quantitative distributions of radioactive labelled substances are measured to investigate physiological processes. It is well known that PET centre of Bakulev Scientific Centre for Cardiovascular Surgery is the oldest one in Moscow. During more than fifteen years a large number of patients have received PET scans. Due to main stream of Scientific Centre, emphasis is placed on examining the heart functioning. For the diagnosis innervation of the heart muscle a number of radiopharmaceuticals are used, including PET radiopharmaceuticals such as ^{11}C-CGP 12177, ^{11}C-meta-hydroxyephedrine as well as its synthetic analogues labelled with other PET radionuclides (^{18}F, ^{68}Ga). ^{11}C-meta-hydroxyephedrine is one of the most perspective radiopharmaceutical for an investigation of cardiac receptors function due to required materials availability for a radio synthesis in Russia. The main advantage of proposed ^{11}C-meta-hydroxyephedrine synthesis technique is the use of a catalyst which allows one decrease reaction time from 5 minutes to 30 seconds. Obtained results allow one decrease reaction time of methylation and increase radiochemical and technological yields.

1. Introduction

Cardiac neurotransmission imaging with PET allows in vivo assessment of presynaptic reuptake and neurotransmitter storage as well as of regional distribution and activity of postsynaptic receptors. Thus, the biochemical processes that occur during neurotransmission can be investigated in vivo at a micro molar level using radiolabeled neurotransmitters and receptor ligands. PET of cardiac neurotransmission characterizes myocardial neuronal function in primary cardioneuropathies, in which the heart has no significant structural abnormality, and in secondary cardioneuropathies caused by the metabolic and functional changes that take place in different diseases of the heart. One of the many radiopharmaceuticals that have been designed and tested to assess cardiac neurotransmission is ^{11}C-meta-hydroxyephedrine. ^{11}C-meta-hydroxyephedrine is a false neutransmitter that has the same neuronal uptake mechanism as norepinephrine, but is not degraded by monoamine oxidase and catechol methyltransferase, the enzymes responsible for the metabolism of norepinephrine in the heart. The storage and release properties seem to differ from those of the physiologic neurotransmitters. ^{11}C-meta-hydroxyephedrine can be synthesized with chemical purity greater than 95% and specific activity from 33.300 to 74.000 GBq/mmol [1].

There is a new activity at Bakulev Scientific Center for Cardiovascular Surgery on synthetic techniques of radio pharmaceutical production labeled with ^{11}C for PET in cardiology in connection with the benefits of receptor radio pharmacology and clinical cardiology mentioned above. The staff of the PET center has accumulated a big experience in the production of PET radiopharmaceuticals. ^{11}C-meta-hydroxyephedrine is one of the most perspective radiopharmaceutical for an investigation of cardiac receptors function due to required materials availability for a radio synthesis in Russia. Pharmaceuticals based on fluorine and gallium cannot be used for adequate analysis of biochemical processes because of it is not full analog of natural metabolites. In turn, a production of the other receptor radiopharmaceuticals labeled with ^{11}C is difficult because of stringent legislation in the area of precursors. Furthermore, there are no commercially available precursors for the other concerned radiopharmaceuticals in Russia. Additionally, ^{11}C-meta-hydroxyephedrine production is rather inexpensive.

2. Materials, methods and experiment

There is a facility for PET radionuclide production at Bakulev Scientific Center, namely, ^{11}C. It is a low-energy H^- radial-sector cyclotron with internal ion source CTI RDS-111 (see figure 1), providing proton beams of up to 40 μA, with an energy of approximately 10.6 MeV after losses in the vacuum and target entrance windows [2]. The self-shielded cyclotron was commissioned in 2000. Its RF system as well as ion source (see figure 2) were upgraded in 2014 in order to increase the reliability of radionuclides production [3]. Now it can produce up to 10 GBq of radionuclide ^{11}C after 1.2 cm^3 gas target irradiation filled by N_2 with 2% of $^{16}O_2$ during 5 minutes. This activity is enough for production up to 3 GBq of radiopharmaceutical. Unfortunately, manufacturing these cyclotrons was ended in September 2016.

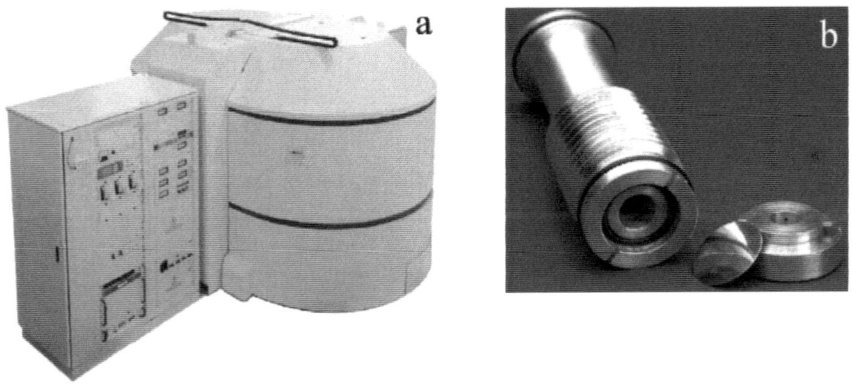

Figure 1. Cyclotron complex view (a) and its target (b).

Radiochemistry laboratory takes part in a project of a unit development for radiochemical synthesis. So, the prototype of Positom-PRO [4] synthesis unit was used for methylation.

Technique of ^{11}C-meta-hydroxyephedrine synthesis is based on using the so called "wet" production method of ^{11}C-methyliodide by means of $^{11}CO_2$ reaction with $LiAlH_4$ in diethyl ether solution (0.1 M). After that the reaction product is dissolved by 59% water solution of HI. Produced $^{11}CH_3I$ is distilled from the reaction mixture and trapped in a 0.01 M metaraminol solution under presence of acetonitrile/dimethyl sulfoxide mixture in proportion 3:1.

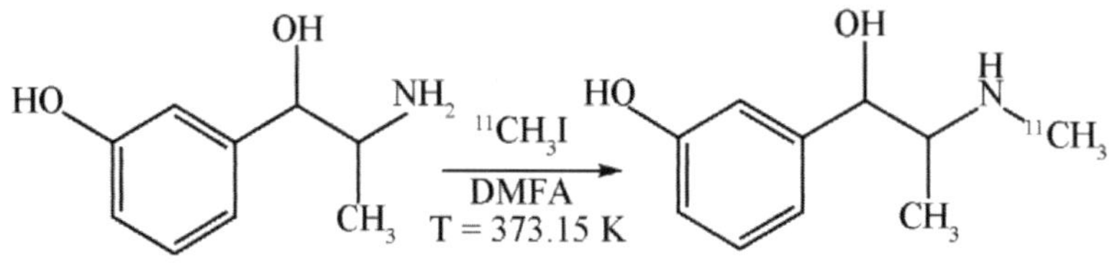

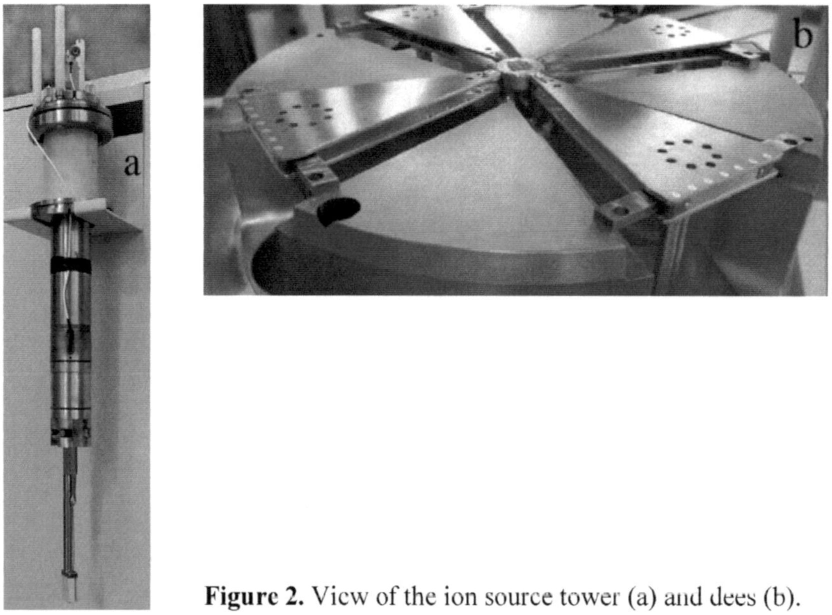

Figure 2. View of the ion source tower (a) and dees (b).

The reaction of metaraminol ^{11}C-methylation was carried out during 5 minutes under the following conditions: reactor temperature – 100°C and pressure (250 – 300) kPa. The reaction yield and composition were evaluated by radio-thin layer chromatography (TLC) on silica gel plate in a mixture of acetone/benzol in proportion 3:1. Typical chromatogram, obtained with HPLC Akvilon unit, is shown in figure 3.

The reaction of metaraminol ^{11}C-methylation was carried out in two cases. In the first case the reaction of metaraminol ^{11}C-methylation was carried out in free base form as it was done in [5]. In the second one the reaction was performed in free base form with an addition of equimolar amount of tetrabutylammonium hydroxide (TBA) [6]. Radiochemical yields for two mentioned cases are shown in figure 4.

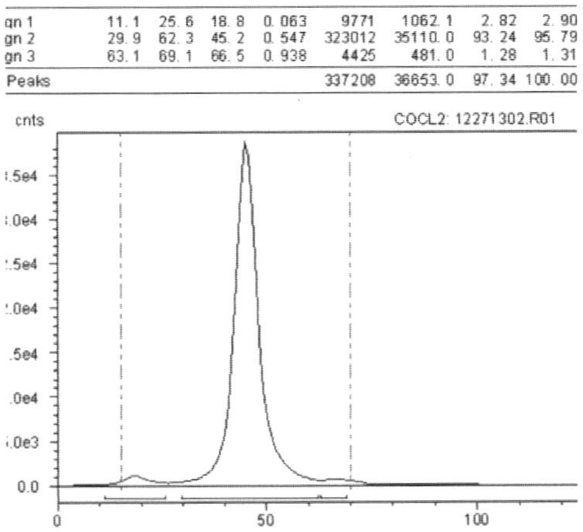

Figure 3. The analysis of compounds labelled with ^{11}C in the reaction mixture by means of radio-TLC.

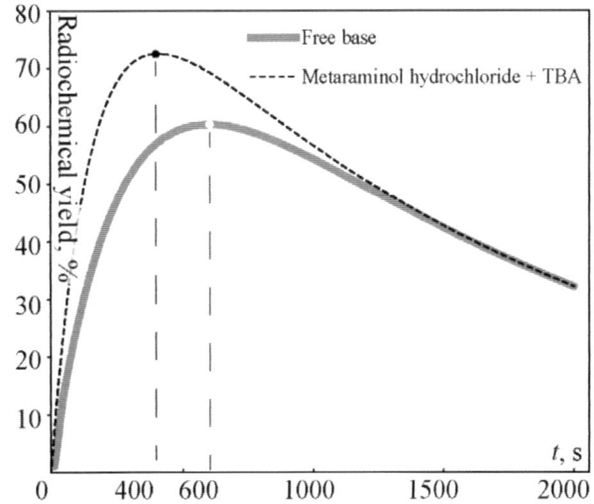

Figure 4. The radiochemical yield of ^{11}C-methylation vs reaction time.

3. Conclusion

It was obtained that the reaction catalysis of metaraminol ^{11}C-methylation by means of TBA resulted in radiochemical yield increase up to 75% while it was 60% under usual technique of ^{11}C-meta-hydroxyephedrine synthesis. It was shown that TBA implementation allows one to use metaraminol hydrochloride instead of free base metaraminol form. Developed ^{11}C-meta-hydroxyephedrine synthesis technique may help one to use PET in receptor radio pharmacology and domestic cardiology.

Acknowledgments

This work was supported by Competitiveness Program of National Research Nuclear University MEPhI.

References

[1] Carrio I 2001 J. Nucl. Med. **42**(7) 1062
[2] Korenev S A 2015 *PoS (Baldin ISHEPP XXII)* 011
[3] Dyubkov V S and Korenev S A 2015 *Proc. 6th Int. Particle Accelerator Conf. (Richmond, VA)* p 323
[4] www.positom.com
[5] www.akvilon.su
[6] Lodi F, Rizzello A and Boschi S 2008 *Computer in Cardiology* **35** p 341
[7] Boros I and Aigbirhio F I Synthesis of O-methyl-11C metomidate ([11C]-MTO) *(Radiochemical Syntheses: Radiopharmaceuticals for Positron Emission Tomography vol 1)* ed P J H Scott and B G Hockley (New Jersey: Wiley) chapter 24 pp 245–256

Simulation of proton movement in F_oF_1-ATP synthase by quantum-mechanical approach

L A Ivontsin[1,2], E V Mashkovtseva[1] and Ya R Nartsissov[1,2]

[1] Institute of Cytochemistry and Molecular Pharmacology, 24/14 6th Radialnaya str., Moscow 115404, Russia
[2] National Research Nuclear University MEPhI, 31 Kashirskoe sh., Moscow 115409, Russia

E-mail: leonid-ivoncin@yandex.ru

Abstract. Quantum-mechanical approach is applied to the description of proton transport through F_oF_1-ATP synthase which is the crucial process in ATP synthesis. Proton was described as a particle located in potential wells formed by charged centers along the half-channels. Energy spectra of bounded states were calculated using Bohr-Sommerfeld quantization, and the initial population of each quantum level was determined by Boltzman distribution. Water molecules were stochastically distributed in an inlet half-channel taking into account atomic radii. Characteristic time of proton transition between the charged centers (amino acid or water molecule) was estimated and it revealed the critical areas needed to be full with water. All possible pathways were analyzed in Monte-Carlo simulation which allows calculating of a mean time of proton transfer trough the inlet half-channel (23 ms).

1. Introduction

Adenosine triphosphate (ATP) plays a very important role in metabolism and energy exchange in all living organisms. It is a high-energy compound which appears to be a universal energy source for many physiological and biophysical processes. ATP is produced in cells from adenosine diphosphate (ADP) and inorganic phosphate (P_i) by means of protein complex F_oF_1-ATP synthase which is a unique energy transformer driven by electrochemical gradient of hydrogen/sodium ions [1]. Its catalytic reversibility is also a unrivalled complex feature which is not yet fully recognized. Structure of this enzyme is well known, but its precise working mechanism is still a point of actual scientific discussions.

Macromolecular complex F_oF_1-ATP synthase consists of two morphologically determined parts. One of them, F_1 (factor determined at first), is exposed to the aqueous phase, markedly extends from the membrane as a nearly spherical formation and it can be easily separated as a soluble fraction. The other one, F_o (oligomicin-sensitive), is embedded into the membrane and represents a highly hydrophobic protein complex [2]. Although being a highly conservative protein, it has few peculiarities depending on a specie. For example, in bacterium E.Coli (the most investigated one) F_1 part consists of nine protein subunits ($\alpha_3\beta_3\gamma\varepsilon\delta$). It carries three catalytic centers responsible for the reactions of synthesis and hydrolysis of ATP [3]. Transmembrane F_o part ($a_1b_2c_{8\text{-}15}$) includes two proton-conducting half-channels across the membrane [4]. A proton passes through the inlet half-channel to the middle of membrane, then it gets bounded to the essential amino acid residue and

IOP Conf. Series: Journal of Physics: Conf. Series **784** (2017) 012021 doi:10.1088/1742-6596/784/1/012021

moves in membrane during rotation of the rotor to the outlet half-channel which leads it to the other side of the membrane.

Key amino acid residues containing protonated groups capable of holding the protons and transmitting them to each other play the significant role in F_oF_1-ATP synthase half-channels functioning. Most essential amino acid residues were determined using site-specific mutagenesis. Two important ones belong to *a*-subunit – *a*Arg210 and to *c*-subunit – *c*Asp61 (in E.Coli) [5]. Presumably, *a*Arg210 is a positively charged residue, initiating rotation of F_oF_1-ATP rotor, but it is not directly involved to the conduction of proton through the membrane because of its location outside of the half-channels area [6]. *c*Asp61 is the final/initial point of proton movement through both half-channels and it is the residue bound to which the proton moves in membrane during the rotation process [7].

According to modern data, proton half-channels are formed by *a*- and *c*-subunits of F_o factor. First 3D-structure of the protein, obtained by means of NMR, showed that *a*-subunit folded in membrane as five vertical transmembrane helices (TMH) and thus provided access of protons to *c*Asp61 residue [8]. However, recent research of rotating ATPase structures by electron cryomicroscopy has revealed refined structure of *a*-subunit. It occurred to consist of a pair of long, horizontal, membrane-intrinsic helices in *a*-subunit next to the *c*-ring rotor [9]. Meanwhile up-to-date there is no precise understanding of TMH 3D distribution in *a*-subunit, and the position of proton half-channels is still under consideration.

There are three main approaches to description of proton transport through different media: Grotthuss mechanism, stochastic modeling, and quantum-mechanical approach. In Grotthuss mechanism, proton could hop from one molecule of water to another, in cases when formation of Zandel cation ($H_5O_2^+$) or Eigen cation ($H_9O_4^+$) is possible [10]. Stochastic modeling is based on calculation of random trajectory of a particle which is modeled by finite difference equations giving an approximate solution of the equation of motion [11]. The most comprehensive approach is a quantum mechanical consideration which takes into account the energy of electrostatic interaction of all charged particles and also purely quantum effects, inherent only to microparticles (tunneling effect and localization of bound states) [12].

In this work combined approach is used for description of proton transport through half-channels of F_oF_1-ATP synthase. Probability of proton transfer from one charged center to another one is determined in quantum-mechanical model of one-dimensional motion. The general problem of proton transport through the sequence of charged centers in the half-channel is solved using stochastic approach. This combined method could be further applied for description of energy transfer in protein channels and could lead more profound understanding of energy transforming mechanisms.

2. Methods

The main force acting on a proton is due to its electrostatic interaction with fixed charges of protein and with polarized water molecules. In the paper these charged particles, that can affect proton movement and create short-lived bound states, are called charged centers.

2.1. Structural analysis of proton half-channel

Based on 1c17 PDB structure (X-ray or NMR protein structure databank) we determined modeling areas as the region of TMH 2-5 of *a*-subunit in the inlet half-channel including all essential amino acid residues which were proposed to take part in proton transport (figure 1).

In order to take into account the interaction of a proton with polar water molecules present in the half-channel, coordinates of each water molecule should be determined (position and orientation). In the present study voids in the molecular structure of the half-channel were stochastically filled with water molecules using Van der Waals radii of atoms.

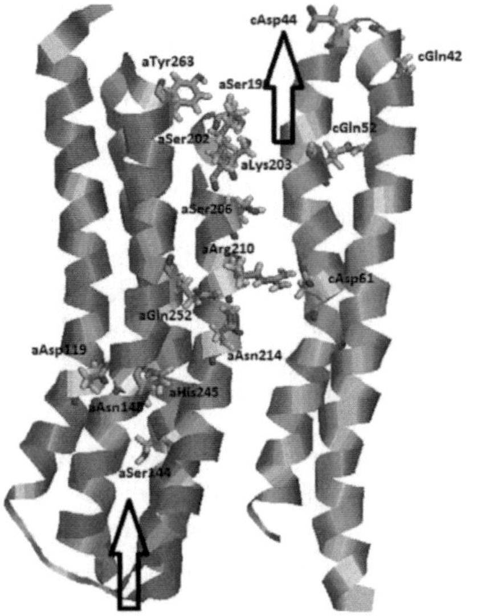

Figure 1. Half-channels structure (red – four TMHs of a-subunit, orange – the nearest of c-subunits). The arrows indicate the intended direction of proton movement. Essential amino acid residues are also labeled.

2.2. *Quantum-mechanical description of proton transport*

Potential of interaction between proton and charged centers should be determined to formulate quantum-mechanical description of proton transport mechanism. Disregarding effects of polarization of atoms and orientation of electron orbitals, sum of Coulomb potential and Lennard-Jones potential were used as interaction potential:

$$U(r)=\frac{kq_1q_2}{\tilde{\varepsilon}r}+\varepsilon\left[\left(\frac{r_{\min}}{r}\right)^{12}-2\left(\frac{r_{\min}}{r}\right)^{6}\right], \tag{1}$$

where q_1 – proton charge, q_2 – uncompensated charge of the center, $\tilde{\varepsilon}$ – permittivity, ε – binding energy between the proton and the charged center, $r_{\min}$ – bond length.

Probability estimation of the proton hop between neighbor centers was obtained by using approximation of quasi-one-dimensional proton movement along the line, connecting the centers, in the potential given by the sum of potentials of both centers.

In the case of bounded state of proton there is a discrete spectrum of possible energies which can be calculated by solving of Sturm-Liouville problem on eigenvalues of particle Hamiltonian. Analytical solution occurs only in a few cases, for example in a rectangular quantum well, so one can use more simple quasi-classical approach – a Bohr-Sommerfeld quantization. Condition for existence of allowed quantum state is equality of the following:

$$\frac{1}{\pi\hbar}\int_a^b\sqrt{2m\left(E-U(x)\right)}\mathrm{d}x=n+\frac{1}{2}, \tag{2}$$

where a and b – classical turning points, n – integer number, defining the level number with energy E in quasi-classical case.

The probability of proton occurrence at the given quantum level is determined by Boltzman distribution:

$$N=N_0\exp\left(-\frac{E}{kT}\right). \tag{3}$$

Despite indefinite proton localization in the potential well in terms of classical approach, in the case of quantum-mechanical description the penetration of particles into the next well is allowed through a sufficiently thin barrier. Tunneling probability depends on proton energy and could be defined as:

$$p = \exp\left(-\frac{2}{\hbar}\int_{x_1}^{x_2}\sqrt{2m\left(E-U(x)\right)}\mathrm{d}x\right), \tag{4}$$

where x_1 and x_2 are subject to condition $U(x_1) = U(x_2) = E$.

In quasi-classical approach flux of particle from the given well to the adjacent well is equal to product of classical frequency of particle oscillation in the well and its tunneling probability. In a model system the possible proton way can be located at the chain of charged centers so the total time of the proton transport through the channel could be determined as the sum of characteristic times of proton transfer between each pair of adjacent centers.

3. Results

3.1. The spatial arrangement of water molecules in the inlet half-channel of F_oF_1-ATP synthase

Statistical set of one million of water molecules distributions was obtained due to previously described approach. Since coordinates of each molecule were chosen randomly, there were distributions with a small number of molecules and a rather dense packing as well. Two areas with a large number of water molecules were discovered. They were located at the beginning and the end of the half-channel. In the central region of the half-channel an obtained small number of water molecules was associated with a quite dense structure of this protein area. To consider the majority number of possible proton ways through half-channels, the distribution with the maximum number of water molecules which amounted to 66, was further used in the model.

3.2. Quantum-mechanical calculation of the proton transfer between the two binding sites

In the described system oxygen atoms of water molecules (H_2O), oxygen atoms ($R\text{–}O^-$) of carboxyl and hydroxyl groups, nitrogen atoms ($R\text{–}N^-$) of the amine groups of amino acid residues can be considered as the proton binding centers. The proton can move between these centers in a random sequence. Full time of proton movement through half-channel will depend on the number of involved charged centers in the chain and the distance between them, and also on types of these centers (water or amino acid residue). Therefore, to solve the problem of proton transport it is necessary to calculate the proton transfer time between all possible pairs of the adjacent centers.

In case of sufficiently close centers location at a distance up to 5 Å, two separate wells do not appear, and a proton has common states. The characteristic time of transfer between the centers is determined by the size of well and the momentum of the proton and is about τ = 15÷30 μs. When the distance between the centers is more than 5 Å, separate wells are formed, and the proton has a localized state at each center. The characteristic time of transfer between the centers is about τ = 0.49÷3.58 ms (figure 2).

Analyzing the proton behavior in the system depending on the distance between centers the one can easily notice a sharp increase in the characteristic time of protons transport at the distance more than 5 Å when it becomes higher than the physiological timescale of 10 ms that logically leads to the restriction of the critical center distance.

All types of present in the model system distances and types of centers were analyzed and the critical distances for all combinations of center types were obtained to provide the possibility of the continuous transport of the protons.

The characteristic times of transitions between the charged centres are calculated for each considered geometrical configuration and are used to estimate the total proton transfer time via the inlet half-channel and to determine the optimal transport pathways.

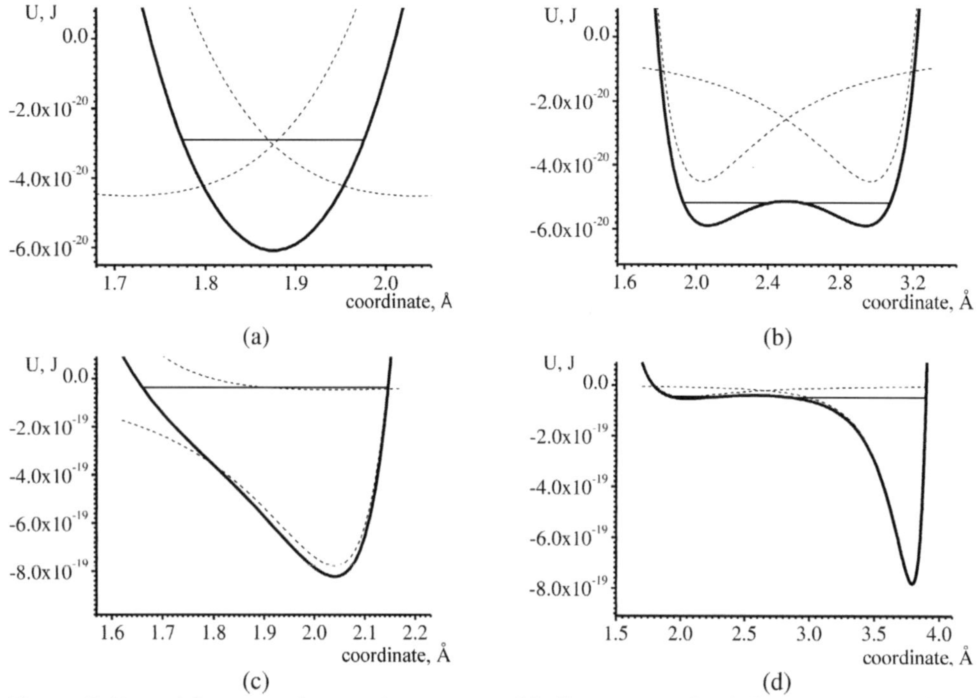

Figure 2. Potential energy of proton between two binding centers. Dash line shows potential energy of proton in field of each center given by Coulomb and Lennard-Jones potential. (a) Distance between centers ($H_2O \rightarrow H_2O$) is 3.75 Å, $\tau = 3.26 \cdot 10^{-5}$ s. (b) Distance between centers ($H_2O \rightarrow H_2O$) is 5 Å, $\tau = 4.88 \cdot 10^{-4}$ s. (c) Distance between centers ($H_2O \rightarrow R\text{–}O^-$) is 3 Å, $\tau = 1.58 \cdot 10^{-5}$ s. (d) Distance between centers ($H_2O \rightarrow R\text{–}O^-$) is 4.75 Å, $\tau = 3.58 \cdot 10^{-3}$ s.

3.3. Proton transport pathway analysis for inlet half-channel of F_oF_1-ATP synthase

All obtained pathways with different number of hops and the characteristic times of movement were carefully analyzed. Pathways with a small number of hops were unlikely and the occurrence probability increased along with a number of hops that was caused by the possibility of cyclic transitions in the applied algorithm. However, it turned out that the proton had reached the end of the half-channel in a finite time (figure 3 (a)). The average time of the proton movement through inlet half-channel was 22.83 ms. In figure 3 (b, c) there are histograms of pathway frequencies with a different number of all types of centers and a different number of water molecules in the transfer chain respectively.

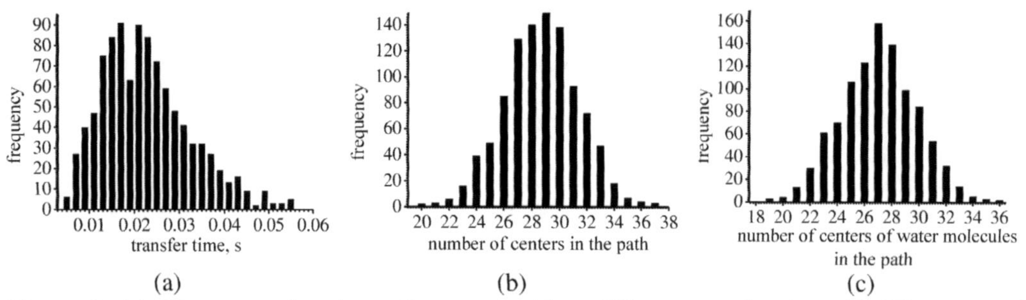

Figure 3. (a) Histogram of pathway frequency with a different transfer time. (b) Histogram of pathway frequency with a different number of centers. (c) Histogram of pathway frequency with different number of water molecules.

The most likely are the pathways with 27-29 proton binding centers. The number of water molecules located in half-channel during the protein structure assembly need not to be maximal. 19 water molecules occurred to be sufficient for proton transfer through half-channel. However, the probability of the pathways with such number of water molecules centers was not high although this type of centers was the most common. Comprehensive research revealed eight water molecules occupied in all the pathways which can be considered "essential" (figure 4) [13].

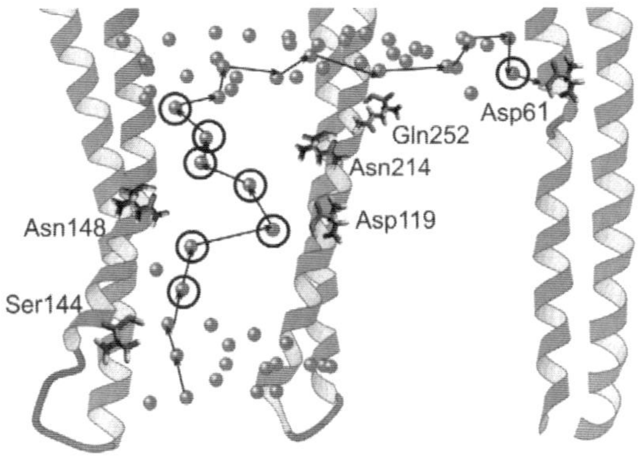

Figure 4. The example of proton pathway including only water molecules centers. Essential amino acid residues are labeled. The total transfer time is 6.84 ms, the number of binding centers is equal to 19. Water molecules critical for proton transport are highlighted by circles.

Proton transport time through inlet half-channel of F_oF_1-ATP synthase has not been yet experimentally estimated. Nevertheless, the proton transport time through the membrane could be indirectly evaluated from the enzyme turnover rate. Direct comparison of calculated in the model mean time of proton transport through inlet half-channel (22.83 ms) with estimations of published data (0.16-19.47 ms) [14-17] is challenging. Observing mismatch of defined time values can be easily accounted by an insufficient level of experimental technical possibilities and consequent low interest to the considered issue. The existing experimental values are in a wide range of time due to various approaches of their executing and a large scope of obtained characteristics. Yet mean time of proton transfer calculated in the present work is still in physiological timescale and complains with turnover rate for the whole enzyme.

4. Discussion

ATP is one of the renewable compounds in mammalian organism. The stock of ATP is not kept due to high rate of its utilization in all types of energy consuming physiological processes. Therefore, the new knowledge about the precise mechanism and regulation principles of F_oF_1-ATP synthase, a protein complex that catalyzes the synthesis of ATP, may have important scientific implications.

The main point of interest is the mechanism of electrochemical transmembrane gradient and rotational energy transformation into the chemical energy of ATP. The energy coupling can be localized in the areas of proton transfer in the transmembrane domains of the enzyme. Direct simulation and experimental study of this process are not fully realizable at the current stage of scientific technology.

Theoretical modeling of the proton transport using described combination of quantum-mechanical and stochastic approaches allows evaluating not only the time of proton movement through protein structure, but also the proton energy level and possible ways of energy transformation.

As a result of the modeling, various proton pathways were obtained. The total transfer times highly depend on the type and number of centers involved in a specific pathway. Including of hops with critical and near-critical (in the model) distances between the centers leads to a sharp increase in the total time. The low prevalence of amino acid centers in the proton pathways can be caused not only by its lower distribution but also by their localization and a deeper energy well. The exact definition of the upper boundary of the proton hopping distance is an important task which allows the determination of the maximum energy passed by the proton during its movement. Analysis of obtained set of pathways revealed critical areas inside the intramembrane part of the enzyme which are to be occupied with water molecules during the protein assembly. Any failure in this physiological process may lead to irreversible breakthrough of the proton pathway and thereafter to corruption of normal ATP synthesis.

The obtained in the model mean proton transfer time occurred to be 22.83 ms and is on the upper limit of the time range estimated using various experimental data. Any ion transfer time is obviously highly dependent on its transmembrane potential and electrochemical gradient (the quickest time of 0.16 ms [14] was measured at 100 mV across the membrane). A further consideration in the model of the proton-motive force and membrane potential will not only improve the accuracy of the calculated proton transfer time under real biological conditions, but will also be useful to analyze the mechanism of transformation and accumulation of energy in the process of the proton transfer which is an important step in understanding the entire energy conversion mechanism during the F_oF_1-ATP synthase catalytic cycle.

This study is the first attempt to apply the quantum-mechanical approach to describe the proton transport in half-channels of F_oF_1-ATP synthase. The obtained new data on the mechanism and parameters of the protons movement in the protein half-channels along with method application on other subject can also have implications in explaining the functioning of many biological structures.

References

[1] Romanovsky Yu M and Tikhonov A N 2010 Phys. Usp. 53 893–914
[2] Fillingame R H 1997 J. Exp. Biol. 200 217–24
[3] Engelbrecht S and Junge W 1997 FEBS Letters 414 485-91
[4] Altendorf K, Stalz W, Greie J and Deckers-Hebestreit G 2000 J. Exp. Biol. 203 19-28
[5] Fillingame R H and Steed P 2014 Biochim. Biophys. Acta 1837 1063–68
[6] Elston T Wang H and Oster G 1998 Nature 391 510-13
[7] Pitard B et al 1996 Eur. J. Biochem 235 779
[8] Rastogi V and Girvin M 1999 Nature 402 263-68
[9] Allegretti M et al 2015 Nature 521 237-40
[10] Agmon N 1995 Chem. Phys. Lett. 244 456-62
[11] Siva K and Elber R 2003 Proteins: Structure, Function and Genetics 50 63-80
[12] Chen D and Wei G W 2012 J. Chem. Phys 136 134109
[13] Ivontsin L A, Mashkovtseva E V and Nartsissov Ya R 2016 Biochim. Biophys. Acta – Bioenergetics 1857 56
[14] Feniouk B A, Kozlova M A, Knorre D A, Cherepanov D A, Mulkidjanian A Y and Junge W 2004 Biophys. J. 86 4094–4109
[15] Gray R E et al 1996 Academic Press, 369–89
[16] Pepe I M et al 2002 J. Photochem. Photobiol. 66 148
[17] Pitard B et al 1996 Eur. J. Biochem. 235 779

IOP Conf. Series: Journal of Physics: Conf. Series **784** (2017) 012022 doi:10.1088/1742-6596/784/1/012022

Study of the structure of 3-D composites based on carbon nanotubes in bovine serum albumin matrix by X-ray microtomography

D Ignatov[1], N Zhurbina [2], A Gerasimenko [3]

[1]Bachelor, MIET, Moscow, Russia
[2]Bachelor, MIET, Moscow, Russia
[3]Ph. D, MIET, Moscow, Russia

E-mail: androfag45@gmail.com

Abstract. 3-D composites are widely used in tissue engineering. A comprehensive analysis by X-ray microtomography was conducted to study the structure of the 3-D composites. Comprehensive analysis of the structure of the 3-D composites consisted of scanning, image reconstruction of shadow projections, two-dimensional and three-dimensional visualization of the reconstructed images and quantitative analysis of the samples. Experimental samples of composites were formed by laser vaporization of the aqueous dispersion BSA and single-walled (SWCNTs) and multi-layer (MWCNTs) carbon nanotubes. The samples have a homogeneous structure over the entire volume, the percentage of porosity of 3-D composites based on SWCNTs and MWCNTs - 16.44%, 28.31%, respectively. An average pore diameter of 3-D composites based on SWCNTs and MWCNTs - 45 µm 93 µm. 3-D composites based on carbon nanotubes in bovine serum albumin matrix can be used in tissue engineering of bone and cartilage, providing cell proliferation and blood vessel sprouting.

1. Introduction

The 3-D composites are widely used in tissue engineering. [1]. To replace lost tissues composites must have a certain porosity (pore size, pore volume, the ratio of voids to solid part of the material). The porous structure of composites provides sprouting of blood. This requires the selection of optimal porosity characteristics [2].

3-D composites were formed by laser ablation of an aqueous dispersion of carbon nanotubes in the protein matrix. Most photostability type of protein has been selected – bovine serum albumin (BSA). Carbon nanotubes (CNTs) are allotropic form of carbon filamentary structure. CNTs have a high mechanical strength, excellent thermal and electrical properties. The mechanical properties similar to the properties of protein structures.

Comprehensive analysis of the structure of the 3-D composites based on carbon nanotubes in the BSA matrix was conducted by X-ray microtomography. This method allows to obtain high-resolution visualization of structure without requiring preliminary preparation of samples.

2. Materials and methods

X-ray Microtomograph Scyscan 1174 was used to study the structure of the 3-D composites based on CNTs in BSA matrix. The voltage and power of the X-ray tube current is 20-30 kV and 400 mA, respectively. 14-bit X-ray camera with a resolution 2000 ms (projection ratio 1304x1024 pixels) was used to detect X-ray radiation. While scanning a 3-D composite is rotated around its axis (an angle of rotation 0.1 degrees ~ 3500 projections). Samples were placed on a special specimen stage of a transparent material to X-rays, printed on a 3-D printer. Scan time of one sample was 300 minutes.

After scanning composites, shadow projections were reconstructed using multi-threaded modified algorithm of Feldkamp. Image enhancement is achieved by averaging the several shadow projections in one angular position. Also used:

- misalignment compensation of the image projection to compensate for possible displacements during scanning;
- beam-hardening correction through a linear transformation;
- ring artifacts reduction to smooth the projections by masking pixel defect in the specified area.

Various palettes and settings is selected to create a model when visualization of the reconstructed images: volumetric model, the extent of absorption, the rendering mode. Maximum intensity projection used in the three-dimensional visualization.

Quantitative analysis of composites allows to estimate composites porosity. Area (region) of interest was allocated in such a way that it contains the most homogeneous area of 3-D composite. After the allocation of the zone of interest, the image binarization was performed. In these images, the white color represents areas with the brightness within the binary threshold "solid", and areas outside this interval corresponds to the black "void".

Experimental samples of composites were formed by laser vaporization of the aqueous dispersion BSA and single-walled (SWCNTs) and multi-layer (MWCNT) carbon nanotubes. Samples 1 and 2 based on SWCNTs and MWCNTs, respectively, formed by continuous semiconductor laser (Table 1). The wavelength and the laser power was 810 W and 2 nm, respectively.

Protein concentration in the aqueous dispersion of nanotubes and BSA was ~ 25% and CNTs concentration was ~ 1 g/l. It is subjected to mechanical stirring and ultrasonic dispergation to obtain a homogeneous dispersion. Further, the irradiation of dispersion was carried out until the solid phase state of the sample. When laser irradiation of an aqueous dispersion of CNTs and BSA, intense absorption of radiation occurs, followed by heating of individual CNTs and their conglomerates. Further, the local boiling of the BSA matrix around the CNTs and their conglomerates was occurred, causing denaturation of BSA. The degree of dispersion (the size of carbon nanotubes and their conglomerates) affect the size of the formed internal pores and cavities. Thus, if size of CNTs conglomerates in the initial solution standardized, the internal structure of the 3-D composite is a more homogeneous. 3-D composites based on SWCNTs (a) and MWCNTs (b) is shown in Figure 1.

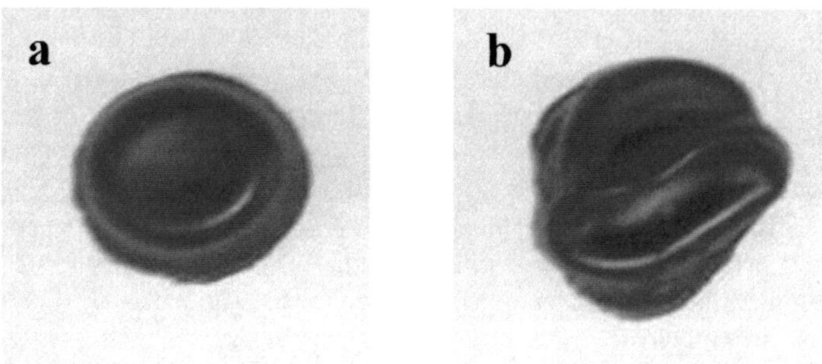

Figure 1. The 3-D composites based on SWCNTs (a) and MWCNT (b)

3. Results

Figure 2, 3 show the results of the processing of the reconstructed images of 3-D composites. Figure 2a, 3a show the three-dimensional visualization of the reconstructed images of 3-D composites based on SWCNTs and MWCNTs, respectively. Figure 2b, 3b show the reconstructed image of the 3-D composites based on SWCNTs and MWCNTs, respectively. The samples have a homogeneous structure throughout its volume. Large pores, which could prevent ingrowth of blood vessels, are not detected. Figure 2c, 3c show a two-dimensional visualization (enlarged section) of reconstructed images of 3-D composites based on SWCNTs and MWCNTs (the lighter portion, the greater the size of the object). There is a small pore system throughout the volume of the samples.

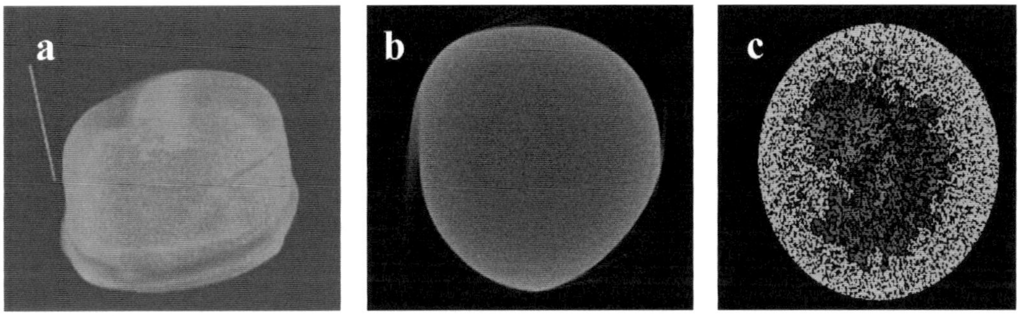

Figure 2. Two-dimensional and three-dimensional visualization of the structure of the 3-D composites based on SWCNTs. a - three-dimensional visualization, b - a reconstructed image of a 3-D composite in - a two-dimensional visualization of 3-D composite

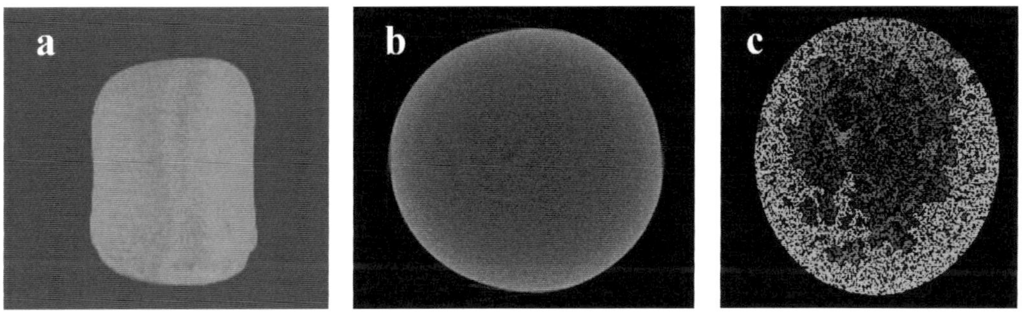

Figure 3. Two-dimensional and three-dimensional visualization of the structure of the 3-D composites based on MWCNTs. a - three-dimensional visualization, b - a reconstructed image of a 3-D composite in - a two-dimensional visualization of 3-D composite

Table 1 provides a quantitative analysis of 3-D composites based on the reconstructed images. The average pore diameter of the composite 3-D based on SWCNTs - 45 μm, on MWCNT - 93 μm. 3-D composite based on SWCNTs has 16.44% porosity, on MWCNT – 28,31 %. The proliferation and the germination of blood vessels accelerates with an average pore diameter – 1-300 μm and porosity rate – 10% - 30% [3]. The difference between the parameters of the quantitative analysis of 3-D composites based on SWCNTs and MWCNTs can be explained by the different structure of SWCNTs and MWCNTs. The average SWCNTs diameter was ~ 1.4-1.6 nm, its length was — 0.5-1.5 μm, its surface area was ~ 400 m^2/g. The MWCNTs diameter was ~ 50-60 nm, length — 3-7 μm, the wall thickness — 15-20 nm, the core space diameter — 10-20 nm, the specific surface area — 90-120 m^2/g [4].

The intense radiation absorption followed by heating of single CNTs and its conglomerates occurs during the laser radiation of the water-albumin CNTs dispersion. Next the local boiling of water-albumin matrix happens around CNTs and its conglomerates that causes BSA denaturation. The dispertion level (size of CNTs and its conglomerates) affect the size of the developed inner pores and cavities. So, if the size of CNTs conglomerates in the initial solution is normalized then the inner structure of 3-D bioconstructions become more homogenous.

Table 1. Quantitative analysis of 3-D composites based on carbon nanotubes. Sample 1 based on SWCNTs; sample 2 based on MWCNTs.

Parameters	Sample 1	Sample 2
Minimum pore diameter (μm)	13	40
Maximum pore diameter (mm)	0,14	0,31
The average pore size (μm)	45	93
Porosity (%)	16,44	28,31
The percentage of open pores (%)	0,02	0,02
The percentage of closed cells (%)	16,42	28,29

4. Conclusion

Comprehensive analysis of bulk composites by X-ray microtomography was carried out: scanning of 3-D composites, image reconstruction of the shadow projections, two-dimensional and three-dimensional visualization of the reconstructed images, quantitative analysis of the samples. It was found that 3-D composites based on SWCNTs and MWCNTs are homogeneous throughout the volume. 3-D composite based on MWCNT has greater porosity (28.31%) than 3-D composite based on SWCNTs (16.44%). The average pore size of the 3-D composites based on SWCNT and MWCNT - 45 μm and 93 μm, respectively. Such parameters: percentage porosity and average pore diameter provide cell proliferation and blood vessel sprouting. Therefore, 3-D composites can be used in tissue engineering of bone and cartilage. This work was provided by the Ministry of Education and Science of the Russian Federation (14.578.21.0221, RFMEFI57816X0221).

References

[1] Liu Y, Lim J 2013 Review: Development of clinically relevant scaffolds for vascularised bone tissue engineering Biotechnology Advances 31 688–705

[2] Dhandayuthapani B, Yoshida Y, Maekawa T, Kumar D 2011 Polymeric Scaffolds in Tissue Engineering Application: A Review, International Journal of Polymer Science 2011 1-19

[3] El-Ayoubi R, DeGrandpre C, DiRaddo R, Yousefi A, Lavigne P 2011 Design and dynamic culture of 3D-scaffolds for cartilage tissue engineering, J Biomater Appl 25 429–44

[4] Gerasimenko A Yu., Ichkitidze L P, Podgaetsky V M, M.S.Savelyev, Selishchev S V 2016 Laser nanostructuring 3-D bioconstruction based on carbon nanotubes in a water matrix of albumin Proc. of SPIE 9887 988725-1 - 988725-10

The utilization of information technology in biomedicine

E A Isaev[1,2] and P A Tarasov[1]

[1] National Research University "Higher School of Economics"
[2] Pushchino Radio Astronomy Observatory of Astro Space Center of P.N. Lebedev Physics Institute, RAS, Pushino, 142290

Abstract. Biomedicine is a branch of medicine that studies the human body, its structure and function in health and disease, pathological condition, methods of diagnosis, treatment and correction [1]. At the moment, to solve their diverse problems associated with the collection, storage, and data analysis, process modeling, biomedicine extensively uses modern technical equipment. The goal of this article - to make a brief analysis of existing technologies (big data, mobile and cloud technologies), in terms of their applicability to the needs of biomedicine.

1. Formulation of the problem

Currently, check studies in biomedicine affect the following fields: mathematical and computer modeling of the living and environmental systems; mathematical and computer modeling of the prevalence and patterns of diseases; solution of problems of medical diagnosis, prediction of outcomes of disease; evaluation of the effectiveness of medical interventions; collection, storage and operational analysis of patient data; computer genomics, proteomics; information management system, and support for biological and medical research; the creation of new drugs; cell therapy [2].

Recently it became possible in biomedicine to carry out the most complex mass of research thanks to modern advances in various fields of electrical engineering and information technology. Such example is the study of the genome of different complexity of living systems from microbes to human. The accumulated result of these experiments may contain important figures and still unknown information about the properties of the genetic code. To understand the impact of differences in the properties of the genome of an organism, it is necessary to process data of large numbers of the decipher genetic sequences. For instance, to determine the characteristics of the genome correlated with the different properties of human, such as susceptibility to diseases, it is necessary to analyze the genomes of large numbers of people living in different social environments in conjunction with environmental parameters and indicators of his medical records.

Full implementation of the kind of research up-to-date has become possible with the advent of the latest achievements in the field of information technology, in terms of handling large volumes of digital information. Frequently, the amount received in the course of experiments and observational figures is very large. For example, a genomic sequence conservation in a computer storage media require some hundreds gigabytes [3]. Therefore there is a problem not only for simple possession of these figures, but also their transfer to processing places. In the places of storage, filtering and analysis there are big problems in the processing of information of enormous size, because the traditional data mining algorithms in bioinformatics are not suitable for dealing with such volumes of information.

2. Current possibilities for transferring large amounts of data biomedicine

The only technical solution at the moment which enables to transfer this large amounts of biomedical data is a data transmission by means of laser technology, using a variety of media distribution [4]. Optical communication systems are widely used in the scientific and commercial exploitation. The electromagnetic eradiation of the optical range is used as a carrier for the information signal. Fiber optic cables are used as guide systems. Bandwidth of the fiber-optic communication lines often times the bandwidth of all other communication systems by virtue of the high carrier frequency, and wide possibilities of multiplexing [5].

Networks, currently used in commercial operations, have the speed of communication lines which is not higher than 10 Gbit/s. The only experimental commercial network linking Paris and Lyon since 2013, has a speed of 400 Gbit/s [6]. It should be noted that the speed of the specialized scientific and educational computer networks is higher than the speed of the above commercial networks supporting functions of the Internet. These networks provide access to scientific equipment and computing power only to members of the scientific community. As an example is the Pan-European research and Education Network (GÉANT). It organizes the communication between more than ten thousand institutions in Europe. After a recent upgrade its capacity increased to 500 Gbit/s [7].

It will be observed, that for the growing demands of science that speed isn't enough. Taking into account the above-mentioned rate of generation of scientific data for science, the opportunity of a significant increase of the speed of data transfer is still the priority, as well as the installation of new communication channels and new technologies that can cope with the necessary speeds. This requires improving of the infrastructure networks at all levels: upgrading the communication protocols, incorporated equipment, types of signal coding [8].

3. Big Data

The concept of "big data" includes a huge amount of information, which is characterized by the following features: a diverse membership; occurrence of a large number of sources and storage; generating considerable speed, often in real time; the availability of data in different formats (structured data, unstructured text documents, email, data, video, and audio files, etc.) [9]. Often, this concept also applies tools that are essential for working with big data. First of all, it is the various methods and techniques of analysis, applicable to large data (class methods Data Mining, mixing techniques and the integration of data, artificial neural networks, various statistical analyzes, etc.). Secondly, it is instrumental software and hardware information technologies, allowing to store and process extremely large amounts of data, including the use of massively parallel processing that scales to hundreds or thousands of nodes («shared nothing architecture»). The third component of the process are of super special hardware and software systems for large data (for example, Oracle Big Data Appliance complex) [10].

To date, grid computing for some biomedical projects using GRID-computing technology in World Community Grid infrastructure. Such projects include, for example, «Discovering Dengue Drugs – Together» [11] or «Help Conquer Cancer» (processing results of studies of proteins by X-ray crystallography to study the causes of cancer) [12] Sometimes research teams use for their work the computing power of supercomputers or clusters. For example, the power of a supercomputer "Lomonosov", set in the Lomonosov Moscow State University in 2009, used for modeling drugs (Figure 1). A computing cluster IMPB RAS made approximation of the experimental magnetic encephalography data for subsequent frequency analysis (Figure 2).

But better decisions for similar problems often use cloud computing concept. It involves the use of existing data links for remote access of researchers to computing resources and software infrastructure, including operating systems. Typically, these systems are created within the data centers, - specialized buildings for network and server equipment. The structure of the data center is required to enter the premises with special specifications, as well as qualified personnel. The main tasks of the data center are effective optimal storage and processing of data, providing users with application services, and support for the functioning of external applications [13].

Figure 1. Supercomputer "Lomonosov" [14].

Figure 2. Computing cluster IMPB Russian Academy of Sciences [15].

The using of a data center can reduce the cost of managing computing resources, technical support, increase the physical security of servers and provide a comfortable environment for the equipment and technical personnel. In addition, the data center simplifies operation of backup systems and power supply through the use of standard configurations. There is improvement and network security because servers are protected by a single, centrally-managed firewall [16].

Today the usage of cloud computing services in the scientific institutions is the only alternative to the usage of the supercomputer of any classical architecture. To make downloading one of these machines for its computational tasks extremely difficult, because, as a rule, they are used primarily for business purposes. You must also pay all the time to use it, regardless of the actual volume of calculations, and rewrite your code with programmers serving supercomputer institute. At the same time, the usage of cloud-based platform is available immediately, easily scalable, and cost much less. [13].

4. Electronic monitoring technologies

The medical electronic monitoring systems is hardware, software tools and activities which serve the purpose of monitoring the condition of the patient, as well as the collection and subsequent analysis of this information for the diagnosis of diseases, preventing their development, treatment and definition of life-threatening conditions. For statistical and analytical processing and storage of the data cloud technologies are used.

Hardware and software of electronic monitoring tools include the following elements [17]:

• Monitoring systems of various processes in the body, including the fetus during pregnancy (pressure, brain activity, sleep and waking rhythms, heart rate, etc.);

• Diagnostic systems, including rapid blood analysis system, urine, saliva, breath;
• Remoting systems with physician data store, including mobile devices and applications;
• Software solutions for health monitoring, preparation of medication schedules, planning treatment, etc.

5. Conclusion

Thus, for the storage, transfer and processing of large amounts of information in biomedicine technical solutions must be used specialized , powerful servers, disk storage, communication channels. Due to the significant increase in the volume and speed of growth of the experimental data in biology and bioinformatics, which have a complex structure and organization, as well as the need for their processing and analysis, namely the use of cloud computing technology offers great potential for the necessary computing power.

References

[1] The official site of the «Bio-Medicine» organization *URL:http://www.bio-medicine.org/medicine-definition/Biomedicine/*
[2] The official site of the University Auckland, New Zealand *URL:http://www.science.auckland.ac.nz/en/about/subjects-and-specialisations/ug-majors/biomedical-science-ug/biomedical-science-year-3.html*
[3] Tsai E, Shakbatyan R, Evans J, Rossetti P, Graham C, Sharma H, Lin C, Lebo M 2016 Bioinformatics Workflow for Clinical Whole Genome Sequencing at Partners HealthCare Personalized Medicine *J. Pers. Med.* 6(1) 12
[4] Isaev E, Tarasov P 2016 Transmission of large amounts of scientific data using laser technology *Journal of Physics: Conference Series* 740
[5] Bailey D and Wright E 2003 *Practical Fiber Optics* (Oxford: Newnes)
[6] The official website of Orange company. *URL: http://www.orange.com/en/press/Press-releases/press-releases-2013/France-Telecom-Orange-and-Alcatel-Lucent-deploy-world-s-first-live-400-Gbps-per-wavelength-optical-link*
[7] The official site of GEANT project. *URL: http://www.geant.net/Pages/default.aspx*
[8] Isaev E, Kornilov V, Tarasov P, Asur E 2013 The problem of storing and transferring large amounts of astronomical data *Proceedings of the participants of the III All-Russian scientific-practical conference "Modern continuing education and innovative development"*
[9] The official site of the company SAS (Statistical Analysis System) *URL:http://www.sas.com/en_us/insights/big-data/what-is-big-data.html*
[10] The official site of the company Oracle *URL:http://www.oracle.com/us/corporate/analystreports/esg-getting-real-bigdata-2228170.pdf*
[11] The official site of the World Community Grid Project *URL:https://www.worldcommunitygrid.org/research/dddt/overview.do*
[12] The official site of the World Community Grid Project *URL:https://www.worldcommunitygrid.org/research/hcc1/overview.do*
[13] Kornilov V, Isaev E, Isaev K 2015 Prospects for the use of data centers in the solution of problems of mathematical biology and bioinformatics *Mathematical Biology and Bioinformatics* vol 10 chapter 1 pp 60–71
[14] The official site of the Lomonosov Moscow State University *URL:http://www.msu.ru/lomonosov/img/comp1.jpg*
[15] Lahno V, Isaev E, Pugachev V, Zaitcev A, Fialko N, Rykunov S, Ustinin M 2012 The development of information and communication technologies in the Pushchino Scientific Center RAS *Mathematical Biology and Bioinformatics* vol 7 chapter 2 pp 529–544
[16] Isaev E, Dumskii D, Samodurov V, Kornilov V 2015 Information security of cloud computing *Mathematical Biology and Bioinformatics* vol 10 chapter 2 pp 567–579

[17] Adibi S 2015 Mobile Health. A Technology Road Map *Springer Series in Bio-/Neuroinformatics* vol 5

The "PHOENIX" Space Experiment: Study of Space Radiation Impact on Cells Genetic Apparatus on Board the International Space Station

M Yu Karganov[1,2], I B Alchinova[1,2], E N Yakovenko[1,2], V V Kushin[3,4], K O Inozemtsev[3,4], A Strádi[5], J Szabó[5], V A Shurshakov[4] and R V Tolochek[4]

[1] Research Institute for Space Medicine, Federal Biomedical Agency of Russia, Orekhoviy boulevard, 28, Moscow, 115682, Russian Federation
[2] The Institute of General Pathology and Pathophysiology, Baltiyskaya str., 8, Moscow, 125315, Russian Federation
[3] National Research Nuclear University MEPhI (Moscow Engineering Physics Institute), Kashirskoye shosse, 31, Moscow, 115409, Russian Federation
[4] Russian Academy of Sciences, Institute of Biomedical Problems, Khoroshevskoye shosse, 76A, Moscow, 123007, Russian Federation
[5] Hungarian Academy of Sciences, Centre for Energy Research (MTA EK), P. O. B. 49, H-1525, Budapest, Hungary

E-mail: mkarganov@mail.ru

Abstract. The preliminary results of the 1st session of Russian "PHOENIX" long-term space experiment are presented. The survival of dried human lymphocytes and mouse bone marrow cells in 199 days space flight is studied. The degree of DNA fragmentation is analysed for samples flown in different ISS compartments. It is shown that biological data correlates with the results of space radiation dose measurements.

1. Introduction

It is commonly accepted that damage of genetics apparatus of cells in space flight conditions is primarily due to the impact of mixed space radiation. Indirect effects of radiations in water-equivalent medium account for 80–90 % of the total damage in eukaryotes and prokaryotes. These effects are responsible for cell inactivation.

The Russian "PHOENIX" experiment is designed to study the effects of long-term space flight conditions on the genetic apparatus and viability of dried human lymphocytes and mouse bone marrow cells (BMC). The experiment is carried out onboard the Russian Segment of International Space Station (ISS).

2. Experimental Design, Instruments and Methods

Sealed vials with human lymphocytes and mouse BMC were packed in 12 "*Bio-ecology*" boxes arranged at 3 containers. Each box was equipped with the temperature sensor and passive space radiation dosimeter.

The containers were delivered to ISS in November 2014 and installed in *Pirs*, *Poisk* and *Zvezda* modules at panels №436, №103 and №103 respectively. Control vials with cells and travel background dosimeters were stored at the laboratory. Three boxes (one from each module) were descended to the Earth in 199 days exposure.

2.1. Methods for preparation and analysis of human lymphocytes and mouse BMC
Lymphocytes were isolated from the whole blood of healthy donors by centrifugation in a density gradient. Mouse BMC were isolated using the modified method [1]. The cell concentration was adjusted to 1×10^6 per 100 μl, this volume of the cell suspension was transferred into vials and lyophilized [2, 3].

2.1.1. DNA isolation. Lyophilized cells were rehydrated with 150 μl phosphate buffered saline and permeabilized in sucrose buffer. The samples were incubated for 10 min and then centrifuged at 3000 rpm for 15 min (Eppendorf 5702R centrifuge, Germany). The supernatant was decanted, the pellet (cell nuclei) was re-suspended in 400 μl buffer for proteinase K, and 20 μl 10 % SDS was added. After 5-min incubation, proteinase K (5 μl of stock solution with a concentration of 20 mg/ml) was added to final concentration of 250 μg/ml. The samples were incubated with proteinase K at 37° C overnight. Then, 400 μl buffered phenol was added and DNA was isolated routinely. The pellet was dried at room temperature until complete ethanol evaporation, and DNA was dissolved in TE buffer (overnight) [4]. The concentration and purity of isolated DNA were evaluated by the optical density of the obtained solution at 260 nm measured against TE buffer on the Nanodrop 1000 spectrophotometer (Thermo Scientific, USA). The samples were equalized by DNA content.

2.1.2. DNA electrophoresis in 1% agarose gel. Electrophoresis in 1% agarose gel was performed to identify and separate DNA fragments [4]. During separation in gel, the position of DNA was directly determined, as DNA bands in gel were stained with the fluorescent intercalating dye ethidium bromide in a low concentration. The mixture containing 1.0 μl 0.25 % bromphenol blue (BPB) in 50 % glycerin and 5.0 μl sample were applied into the slots of the gels. The current was set at 35 mA and dye migration front was observed. In 20 min, the current was switched off. The results were analyzed using transluminator (Biogene, Russia).

2.1.3. Evaluation of the morphofunctional parameters of cells and their components after rehydration. The cells from vials delivered from ISS and control samples remaining in the laboratory were simultaneously and uniformly rehydrated in sterile PBS. Then, aliquots of the cells suspensions were mixed with Hoechst 33258 fluorescent dye and examined under fluorescent microscope. Hoechst 33258 is the most widespread DNA-specific dye, which allows using it for nucleus contrasting in pro- and eukaryotic cells in both cultures and tissue samples as well as for detection of possible bacterial or mycoplasma contamination of the cell cultures.

2.2. Space Radiation Dosimetery
Passive dosimeters comprising CR-39™ "Tastrak" (TASL Ltd., Bristol, UK) solid state nuclear track detector (SSNTD) and LiF:Mg,Ti (93% ^{7}Li and 7% ^{6}Li) "DTG-4" (ROSATOM, Russia) thermo-luminescent detector (TLD) were applied in the experiment for the linear energy transfer (LET) spectra and dose measurements [5].

2.2.1. SSNTD treatment & track data analysis. "Tastrak" sheets were etched in 6 N NaOH solution at 70° C for 6 hours. The corresponding bulk etch rate (V_b) was 1.35±0.06 μm·h^{-1} [6]. This value agrees well with result of [7]. After chemical etching the detector's were manually scanned in 2-D and 3-D modes [6] using the Carl Zeiss® AxioScope.A1 microscope equipped with digital camera. Different geometrical parameters were measured and detector sensitivity $V = V_t / V_b$ (where V_t is the track etch rate) was calculated for every individual track.

The tracks of long-range particles with incident angles close to 90° were evaluated via the so-called "classical" method. In this case diameters of the track opening and value of removed layer thickness are applied for the sensitivity V findings. The sensitivities V were converted to LET values using the calibration function [7]. The detector's lower LET threshold turned out to be ~7.5 keV/μm (H_2O).

Oblique tracks were evaluated via the procedure firstly described in [8]. Track opening ellipse diameters and full projected length as well as the bottom sphere radius (for over-etched tracks) were applied for the sensitivity V findings. The calibration function [6] was used to obtain the corresponding LET values.

The long-range tracks were considered as primaries. Tracks observed in nuclear fragmentation events and single tracks of short-range particles (over-etched tracks) were attributed to the group of short-range secondaries.

The resulting total event LET spectra are the sum of primaries and secondaries. The track data processing, LET spectra calculations and dose rates evaluation were performed with the help of specially developed software based on MatLab®. The individual LET distributions of the secondaries and corresponding dose calculations were performed separately.

2.2.2. TLD treatment. Thermo-luminescent signal measurements were carried out using the HARSHAW-3500 TLD reader. Subsequent TLD data processing was performed with the help of "Windows Radiation Evaluation and Management System" software. The following parameters were used in the TLD treatment [9]: pre-heating at +50 °C; detector's heating range +50…+300 °C; detector's heating rate 4° C / sec; full annealing time 63 sec; integration interval +150…+250 °C.

3. Results and Discussion

3.1. Cell viability & DNA damage

Cell viability was evaluated by trypan blue exclusion test (Figure 1). The morphology of human lymphocytes and mouse BMC was analyzed after Romanovsky–Giemsa staining (Figure 2). MTT assay showed the presence of mitochondrial activity of the cells after rehydration.

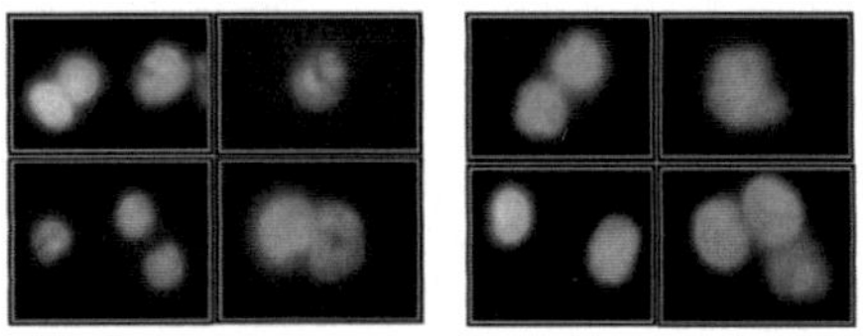

Figure 1. Hoechst 33258 staining (trypan blue exclusion test): mouse BMC (left), human lymphocytes (right). Upper left in each picture – control sample.

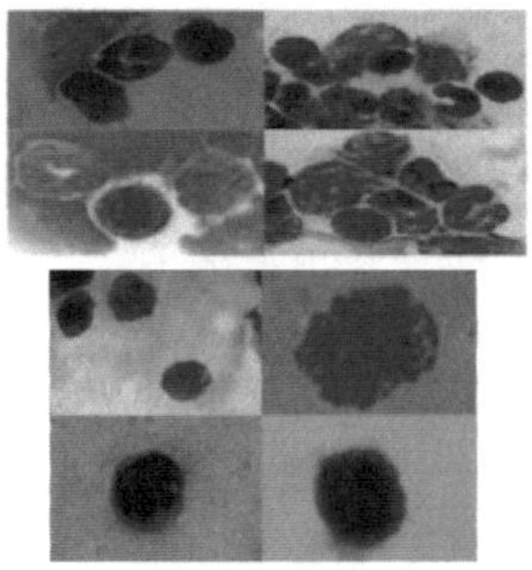

Figure 2. Romanovsky–Giemsa staining: mouse BMC (left), human lymphocytes (right). Upper left in each picture – control sample.

DNAs were isolated from the cells and analyzed by electrophoresis followed by densitometry (see Section 2.1 and [10]). Four zones in Figure 3 reveal the most pronounced differences between the biological samples from different boxes. In the material from *Poisk* and *Pirs* modules density peaks were found corresponding to 1500-2000 bp values of the DNA marker (GeneRuler 1kb DNA Ladder, Thermo Scientific). The samples from *Zvezda* and *Poisk* modules were characterized by 15000 bp peak. In all cases, the peaks of samples from the *Pirs* module are shifted to the right relatively to the

IOP Conf. Series: Journal of Physics: Conf. Series **784** (2017) 012024 doi:10.1088/1742-6596/784/1/012024

samples from the *Zvezda* and *Poisk* modules. This indicates the formation of smaller DNA fragments. Similar results were obtained for the human lymphocytes.

Double staining with fluorescent dyes ethidium bromide (necrosis) and acridine orange (visualizing live cells and specifically staining apoptotic cells) revealed cells in different physiological states. Standard methods for evaluation of the nucleus integrity revealed no apparent damage.

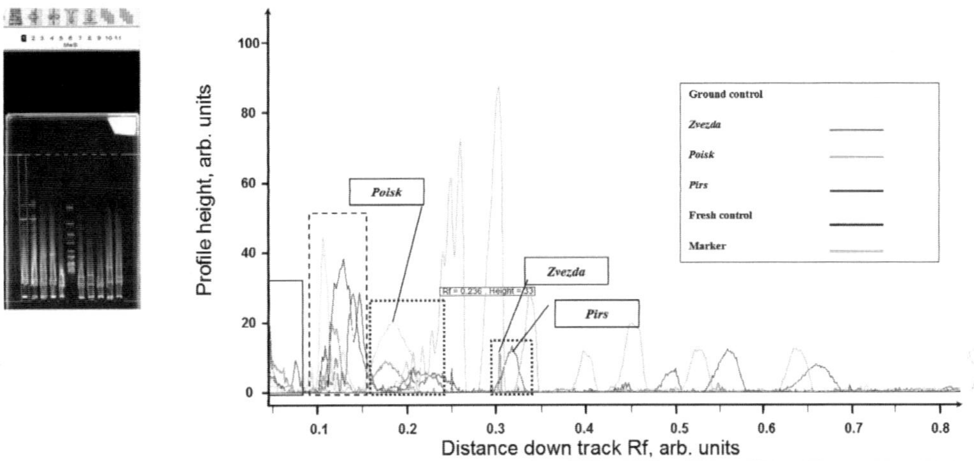

Figure 3. DNA from mouse BMC electrophoresis and densitometry data.

3.2. *Space Radiation Data*

3.2.1. LET distribution. The LET spectra measured in different locations are presented in Figure 4. The contribution from primary and secondary high-LET space radiations are presented in Figure 5 separately.

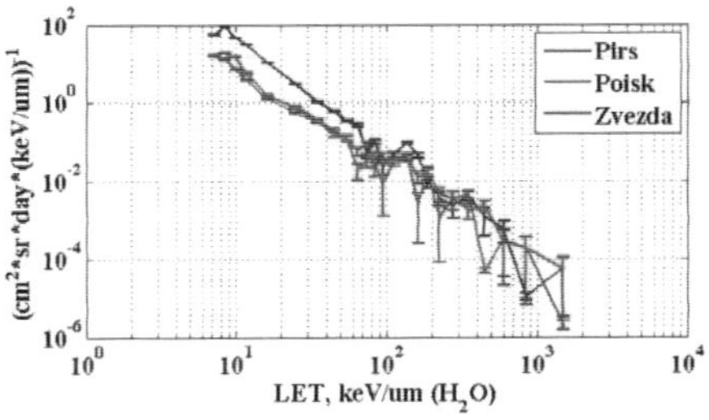

Figure 4. Total event LET spectra measured in ISS compartments.

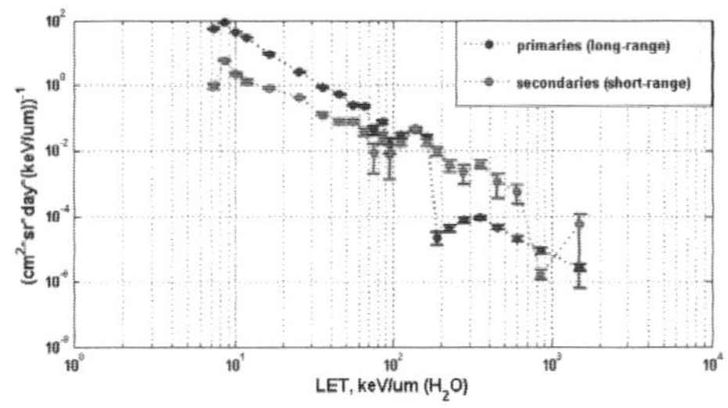

Figure 5. The LET distribution of long-range primaries and short-range secondaries produced in nuclear fragmentations (*Pirs* module).

The relatively high flux of particles with LET < 100 keV/μm (H_2O) observed in the *Pirs* module (Figure 4) is supposedly due to the thinner shielding of compartment. The measured flux of secondaries exceeds the flux of primary component at LET > 100 keV/μm in all three modules of ISS. The measured flux of secondaries is 1–2 orders of magnitude higher than flux of long-range component at LET 100–600 keV/μm. Experimental data correlates reasonably with general conclusions of [11-14].

3.2.2. Doses. The dose rates measured by means of passive detectors are summarized in Table 1.

Contribution of nuclear fragments (D^c and H^c) into the total absorbed dose D^d and dose equivalent H^d varies in 2.0–5.6 % and 19.6–36.5 % respectively. This agrees well with the result [13].

The observed effect of DNA fragmentation depends on dose rate at the compartment. The smaller fragments of DNA were produced in the *Pirs* module due to the highest dose rates from all components of space radiation. Conversely, the fragments of bigger size were produced in *Poisk* and *Zvezda* modules due to relatively small dose rates.

Table 1. Space radiation doses measured in "Bio-ecology" boxes.

ISS Compartment		*Pirs*	*Poisk*	*Zvezda*
TLD	D^a, μGy/day	693±70	492±40	271±15
SSNTD	D^b, μGy/day	115±2[e]	30±2[e]	39±2[e]
	H^b, μSv/day	988±36[e]	399±28[e]	456±33[e]
	Q^b	8.6±0.4	13.2±1.2	11.8±1.1
	D^c, μGy/day	24±2	10±2	16±2
	H^c, μSv/day	365±31	171±25	257±30
Combined	D^d, μGy/day	729±70[e]	505±40[e]	287±15[e]
	H^d, μSv/day	1602±79[e]	874±49[e]	705±36[e]
	Q^d	2.20±0.24	1.73±0.17	2.45±0.18

[a] Absorbed dose measured by TLD signal (non corrected).
[b] SSNTD dose values at LET > 10 keV/μm (H_2O) range.
[c] Dose values, attributed to nuclear fragments at LET > 10 keV/μm (H_2O) range.
[d] Total dose values convolved from TLD and SSNTD data by method [15]. The Q(LET) function proposed in ICRP pub. No. 60 was applied in the calculations.
[e] Sum of primaries and secondaries.

4. Conclusion

The preliminary results of the "PHOENIX" space experiment shows a good survival of human lymphocytes and mouse bone marrow cells nuclei after 199 days space flight, when the doses and dose equivalents varied in the range 57–145 mGy and 140–319 mSv respectively. The degree of DNA fragmentation depends on the dose rate directly. The smallest DNA fragments were detected in the samples exposed at the *Pirs* module with the maximum dose rates.

The further efforts will be focused on the analysis of data from the next sessions of the "PHOENIX" experiment.

Acknowledgments

Authors would like to express their gratitude to the Resource Center of Kurchatov complex NBIKS NCI Technologies, National research Center "Kurchatov Institute" (Moscow, Russia) for kindly support in gel documenting and image processing.

Authors are grateful to Dr. J.K. Pálfalvi (MTA EK, Budapest, Hungary) for helpful discussion.

The portion of this work was performed within the framework of the National Research Nuclear University MEPhI (Moscow Engineering Physics Institute) Academic Excellence Project (contract № 02:a03.21.0005).

The dosimetry support of the "PHOENIX" space experiment was provided under the Agreement on Scientific and Technical Cooperation between the National Research Nuclear University MEPhI (Moscow Engineering Physics Institute) and Research Institute for Space Medicine (№3-16).

References

[1] STEMCELL Technologies Inc., Video Tutorial "Isolating of MSCs from Mouse Compact Bone", access: https://www.youtube. com/watch?v=Dnvq_wMZ2fI
[2] Loi P *et al.* 2008 *PLoS ONE* **3(8)** e2978

[3] Natan D *et al.* 2009 *PLoS ONE* **4(4)** e5240
[4] Maniatis T *et al.* 1982 *Molecular Cloning: A Laboratory Manual* (Cold Spring Harbor Laboratory Press)
[5] Benton E R *et al.* 2002 *Radiat. Meas.* **35(5)** 439-455
[6] Inozemtsev K O and Kushin V V 2016 *Radiat. Meas.* **91** 44-49
[7] Szabó J and Pálfalvi J K 2012 *Nucl. Instrum. Methods in Phys. Res. A* **694** 193-198
[8] Henke R P and Benton E V 1971 *Nucl. Instrum. Methods* **97** 483-489
[9] Inozemtsev K O *et al.* 2015 *Aviakosm. Ekol. Med.* **49(2)** 16–22
[10] Freeman S E and Thompson B D 1988 *Int. J Biomed. Comput.* **22** 121-133
[11] Benton E R *et al.* 1996 *Radiat. Meas.* **26** 793-797
[12] Cacinotta F A *et al.* 1997 *Radiat. Meas.* **26** 923-934
[13] Kushin V V 2010 *Rad. Prot. Dos.* **141** 199-204
[14] Kodaira S. *et al.* 2013 *Radiat. Meas.* **49** 95-102
[15] Hajek M *et al.* 2008 *Radiat. Meas.* **43** 1231-1236

The influence of the sequence of nanoparticles injection to solution on the rate of fibrinogen-thrombin reaction

M N Kirichenko[1,2], S V Krivokhiza[1], L L Chaikov[1,2], N A Bulychev[1]

[1] Lebedev Physical Institute of the Russian Academy of Sciences, Moscow 119991, Russia
[2] National Research Nuclear University "MEPhI", Moscow 115409, Russia

Abstract. The influence of Fe_2O_3 nanoparticles on the rate of fibrinogen-thrombin reaction is studied. The nanoparticles were obtained in acoustoplasma discharge with cavitation. The sequence of nanoparticles injection appeared to change dramatically the rate and result of enzymatic reaction. In case of nanoparticles injection to fibrinogen before thrombin addition, enzymatic reaction practically stopped at the first stage. The mixing of nanoparticles with thrombin before its addition to fibrinogen leads to acceleration of gel formation in comparison with reaction without nanoparticles. We believe that Fe_2O_3 nanoparticles can modify the rate of enzymatic reaction, in one case acting as inhibitors of the reaction and as activators in other.

1. Introduction

Studies of the nanoparticles influencing on the biological systems are very important because of their wide application for different aims of modern medicine. They can be used for target and drug delivery, for visualization of metastasis, for stimulation of angio- and arteriogenesis, etc. Many papers are devoted to the investigations of nanoparticle interactions with proteins, for example, study of binding mechanism of Fe_2O_3 nanoparticles with fibrinogen by using a combination of circular dichroism, UV-vis fluorescence spectroscopic and computational methods [1]; or surface plasmon resonance study of fibrinogen and human serum albumin binding to metal oxide nanoparticles [2], etc. However, influencing of the nanoparticles on the rate of important biological reaction is not fully studied yet.

In our paper we investigated the influence of the sequence of Fe_2O_3 nanoparticles adding into solution on the rate of biological reaction of the fibrinogen cleavage by thrombin. This reaction is widely investigated by means of dynamic and static light scattering. For example, Kita et al in the paper [3] shows the time-dependence dynamic of the distributions of the scattered light intensity on particles sizes reflecting the fibrinogen-fibrin-gel transition under fibrinogen cleavage by thrombin. They determined time when the system begins to be a gel: at a gelation point the autocorrelation function $g^2(\tau)$ has no characteristic time and shows a power law behavior. Another big study of fibrin gels was conducted by Ferri et al [4] by elastic light scattering techniques and fractal analysis. They show that gels can be described as random networks of fibers of average diameter d and density ρ, entangled together to form densely packed and spatially correlated blobs of mass fractal dimension D_m and average size or crossover length ξ.

The aim of the study was to investigate dynamic of autocorrelation functions and distributions of scattered light intensity on particle sizes in fibrinogen-thrombin system without Fe_2O_3 nanoparticles and in case of different sequence of nanoparticles injection: (1) nanoparticles were added to fibrinogen solution before injection of thrombin; and (2) nanoparticles were added to thrombin solution before adding this mixture to fibrinogen.

2. Methods and materials

For our study we used a traditional DLS setup with a He-Ne laser (the wavelength of 633 nm, the power of 5 to 10 mV), an optical system ensuring spatial coherence of recorded light, a goniometer allowing measurements at various angles, a special photomultiplier, a Photocor-FCm correlator (in the «Multiple tau» regime), and a PC with Photocor-FCm and DynaLS programs. For a more detailed description of the setup and DLS method see papers [5,6].

For model fibrinogen solution preparation we used fibrinogen from human plasma (Sigma Aldrich, ≥80% of protein is clottable) diluted in Tris-HCl buffer (7.2 pH). Final concentration of fibrinogen was about 3 mg/ml. After sample preparation solution was filtered through 0.2 μm "Superpure PVDF" filter.

Solution of thrombin was prepared using thrombin from human plasma (Sigma Aldrich, lyophilized powder, ≥2,000 NIH units/mg protein) that was diluted in Tris-buffer. Final activity of thrombin after adding it to fibrinogen solution was 1.7 NIH/mg. Mean hydrodynamic radius of the thrombin in a pure solution was about 230 nm.

Fe_2O_3 nanoparticles were obtained by unique acoustoplasma method with cavitation that combines effect of the elastic oscillations of high intensity ultrasound and pulsed or steady electric fields in a liquid medium. As a result relatively narrow size distribution of the synthesized nanoparticles with specific surface properties was obtained [7, 8]. Final volume concentration of Fe_2O_3 nanoparticles in model protein solution is $C_v \approx 0.5 \cdot 10^{-7}$ ($n_p \approx 4 \cdot 10^5$ cm^{-3}). Mean hydrodynamic radius of nanoparticles aggregates in the solution after ultrasound treatment was about 280 nm.

3. Results of experiment

3.1. Dynamic of distributions of scattered light intensity on sizes in fibrinogen-thrombin system.

First series of measurements was conducted in model solution of fibrinogen with thrombin. Figure 1a shows a distribution obtained for pure fibrinogen solution. In the distribution first peak having mean radius 13 nm and an area of the peak A1=0.728 corresponds to fibrinogen monomers, second peak with mean radius 398 nm and area A2=0.049 corresponds to essential aggregate of fibrinogen, as well as third peak (the area Ai is logarithmic part of intensity, $I_i \sim A_i \cdot r_i$). Figure1 (b-f) shows temporal dynamic of distributions of scattered light intensity on particles sizes in the fibrinogen-thrombin system. Already through 20 min after thrombin addition, the distribution starts to change accordingly to enzymatic reaction of fibrinogen degradation, and after 40 min first peak size decreases from 13 nm to 2 nm as well as the area of first peak – from 0.728 to 0.028. Besides, area of second peak increases from 0.049 to 0.55.

The formation of fibrin gel is considered to have three main stages. In a first stage, cleavage of fibrinogen by thrombin at special sites leads to appearance of short peptides (radius and area decreasing of first peak in Figure1b-e) and fibrin in the solution; in a second stage – formation of protofibrills and their aggregate takes place (growth of mean radii of second and third peak in Figure1); and, finally, the whole network of fibrin fibers is forming in the solution (a set of radii appears in the distributions). These stages can overlap in time, how it can be seen in Figure 1.

In Figure 1 also presents values of amplitudes of correlation function of intensity $g^{(2)}(\tau)$ and the total intensities of scattered light (I/IL). In gel formation before sol-gel transition the total intensity of scattered light (I/IL) increases sharply, while the amplitude of correlation function decreases and the correlation function has such a shape, which gives a set of consecutive times or sizes being decomposed by exponents (Figures 1e,f). Such a picture appears already through 20-30 minutes after thrombin addition to pure fibrinogen solution (Figure 1b) and becomes more distinct in 1.5 hours. The maximal amplitude of correlation function is defined by geometry of experiment; and in case of liquid sample with Brownian particles (particle suspension) it equals to 0.35-0.4 in our geometry. One can see that already through 20 minutes after thrombin addition (Figure 1b) the amplitude is equal 0.0315. Such a decreasing of this value indicates process of gel formation in the sample. In paper [3] authors pointed out that under approaching to sol-gel transition, scattering pattern has a form corresponding to fractal gel dimension and correlation function has a power law component; and the closer sample to gelled condition, the more this component.

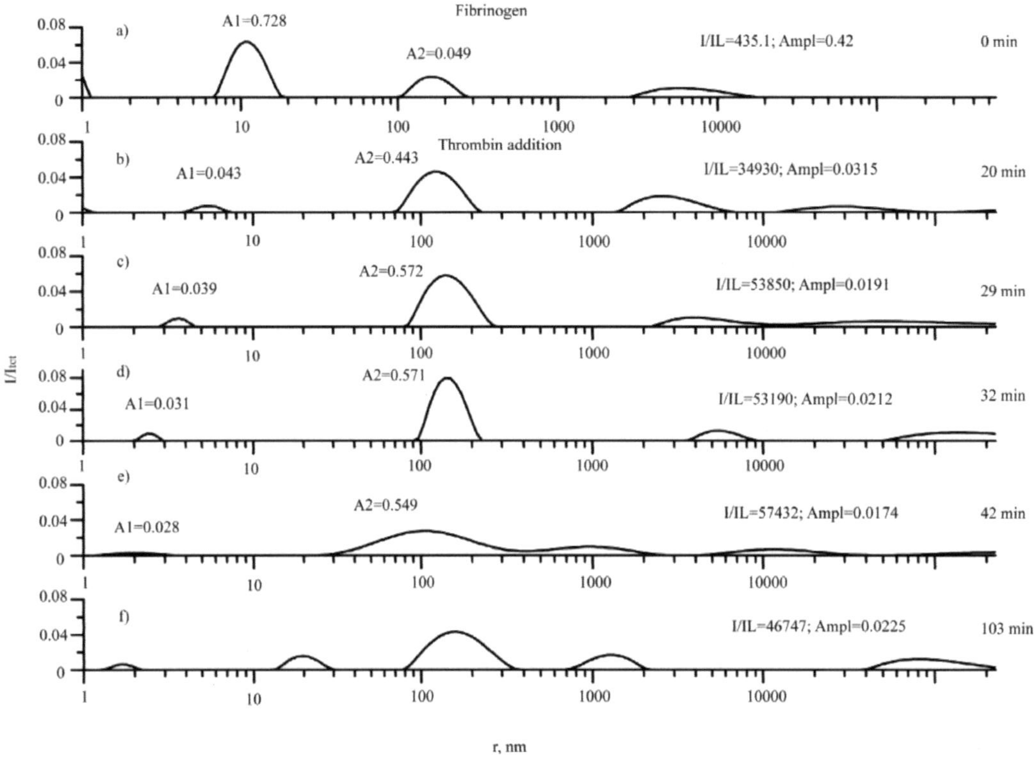

Figure 1 (a) The distribution of the particles in pure fibrinogen solution; **(b-f)** dynamic of distributions of scattered light intensity on particles sizes in model fibrinogen-thrombin system; Ai – area of the peak, I/IL – scattered light intensity, Ampl. – amplitude of the autocorrelation function.

3.2. Dynamic of particle distributions in fibrinogen-thrombin system in case of Fe_2O_3 nanoparticle mixing with fibrinogen before thrombin addition

Second series of experiments was carried out with different sequence of Fe_2O_3 nanoparticles addition to model solution. First sequence was mixing of nanoparticles with fibrinogen before thrombin injection. Figure 2a shows distribution of fibrinogen with nanoparticles. As the fibrinogen particle distribution has the aggregates with mean radius 398 nm, the addition of nanoparticles Fe_2O_3 with mean radius 280 nm does not disturb the initial distribution significantly. Further injection of thrombin to fibrinogen-nanoparticle solution leads to unexpected result. Figures 2b-e show obtained dynamic of distributions of scattered light intensity on particles sizes. The distributions do not practically change: first peak has the unchanged mean radius and area (A_1) as well as second and third peak during more than one hour after thrombin addition. The total intensity of scattered light and the amplitude of correlation function vary slightly during experiment. It means that the presence of nanoparticle in the solution prevents thrombin to cleave fibrinogen, and further gel formation is difficult. Therefore, nanoparticles can play role of the inhibitors of enzymatic reaction.

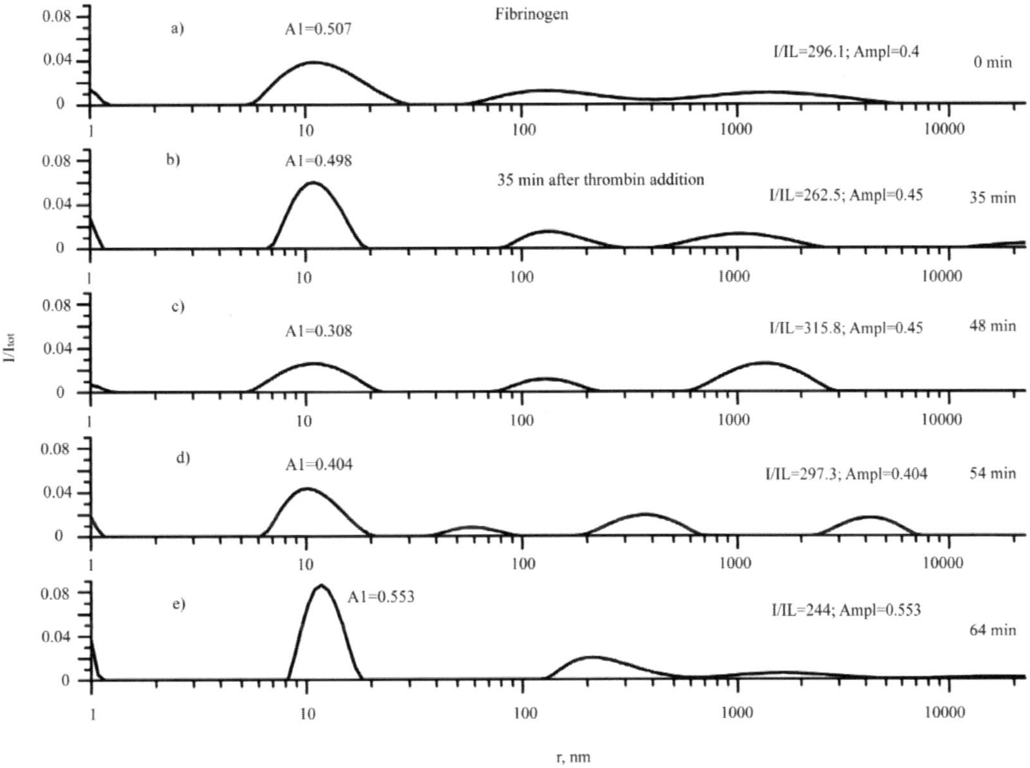

Figure 2. (a) The distribution of particle sizes in solution of fibrinogen with nanoparticles, **(b-e)** dynamic of distributions of scattered light intensity on particles sizes in model solution in case of mixing of fibrinogen with Fe_2O_3nanoparticles before thrombin addition to solution; Ai – area of the peak, I/IL – scattered light intensity, Ampl. – amplitude of the autocorrelation function.

3.3. Dynamic of particle size distributions in fibrinogen-thrombin system in case of Fe_2O_3 nanoparticle mixing with thrombin before its adding to fibrinogen solution

Second sequence of nanoparticle addition was carried out by mixing Fe_2O_3 nanoparticles with thrombin solution before its addition to fibrinogen solution (Figure 3). Figure 3b shows the distribution of scattered light intensity on sizes for solution of thrombin with nanoparticles: it has one peak with mean radius about 250 nm. Addition of thrombin-nanoparticle solution (Figure 3b) to fibrinogen solution (Figure 3a) leads to dramatically increasing of the total intensity of scattered light (from 800 a.u. to 400 000 a.u., i.e. almost in 500 times) and decreasing of amplitude of correlation function (from 0.41 to 0.042). Note that in case of enzymatic reaction without nanoparticles (Figure 1) the intensity jump was about 50 times. In case of thrombin-nanoparticle mixing, jump of the scattered light intensity is almost 500 times, which seems indicate enzyme reaction acceleration. Moreover, already through 30 min after solutions mixing distributions of scattered light intensity on radii have a set of consecutive times corresponding to autocorrelation functions with power law component (see Figures 3c, d). All these results point out fast gel formation in case of thrombin mixing with nanoparticles Fe_2O_3 before its addition to fibrinogen solution.

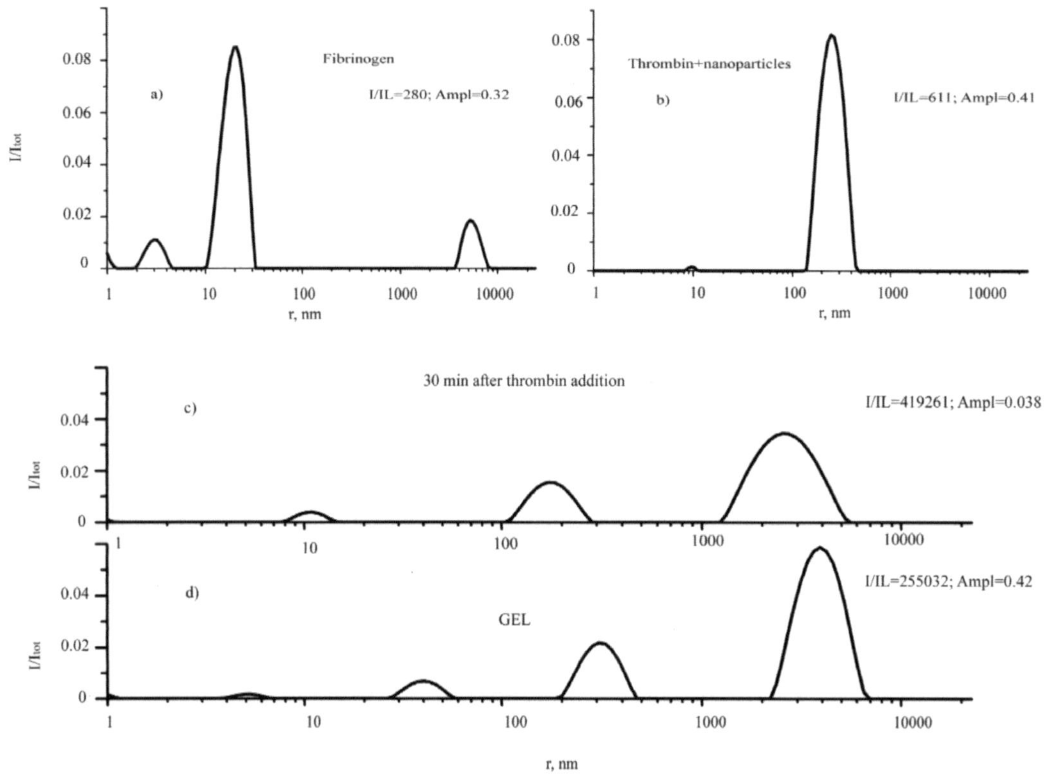

Figure 3(a) The distribution of scattered light intensity on particle sizes in pure fibrinogen solution; **(b)** the distribution for mixture of nanoparticles with thrombin, **(c,d)** dynamic of distributions of scattered light intensity on particles sizes in the solution in case of mixing of thrombin with Fe_2O_3 nanoparticles before injection to fibrinogen solution (gel formation), Ai – area of the peak, I/IL – scattered light intensity, Ampl. – amplitude of the autocorrelation function.

Conclusions

In this paper we show that different sequence of Fe_2O_3 nanoparticles addition to model solution leads to absolutely different results. In case of mixing of fibrinogen with nanoparticles before thrombin addition, enzyme reaction is practically stopped at the first stage. In case of addition of nanoparticles to thrombin before its injection to fibrinogen, the reaction was accelerated and third stage of gel formation was registered already through 30 min after reaction initiation. We believe that nanoparticles can play role of inhibitors or activators of enzymatic reaction depending on the sequence of their addition. In first sequence of addition, they could cover the fibrinogen sites binding to thrombin, thereby preventing the interaction of the protein with enzyme. In second sequence of nanoparticle injection, interaction of nanoparticles with thrombin activates the enzyme and leads increasing of the reaction rate.

Acknowledgments

This work was supported by Competitiveness Program of National Research Nuclear University MEPhI.

References

[1] Zhang H, Wu P, Zhu Z, Wang Y. 2015 Spectrochim Acta A Mol Biomol Spectrosc. 151:40-7.
[2] Canoa P et al. 2015 Biosensors and Bioelectronics 74, 376–383

[3] Kita R., Takahashi A., Kaibara M. 2002 Biomacromolecules. 3:1013–1020
[4] Ferri F et al. 2002 Physical review. E, Statistical, nonlinear, and soft matter physics. 66(1),13.
[5] Kovalenko K V, Krivokhizha S V, Masalov A V and Chaikov L L 2009 Bull. Lebedev Phys. Inst. 36(4), 65
[6] Chaikov L L, Kirichenko M N, Krivokhizha S V and Zaritskii A R 2015 J Biomed Opt, 20 (5), 057003.
[7] Bulychev N A, Kazaryan M A, Chaikov L L, Burkhanov I S, Krasovskii V I 2014 Bull. of the Lebedev Physics Ins. 41(9) 264
[8] Burkhanov IS, Chaikov LL, Bulychev N A, Kazaryan M A, Krasovskii V I 2014 Bull. of the Lebedev Physics Inst. 41(10) 297.

Synthesis and investigation of the specific activity of the DNA-doxorubicin conjugates

A V Kokorev[1], A V Zaborovskiy[2], A A Kotlyarov[1], L A Balykova[3], M A Malkina[3], I V Kargina[3], E V Gromova[3], V Yu Medvezhonkov[3], K G Gurevich[2], S A Shchukin[3] and N A Pyataev[3]

[1]National Research Nuclear University "MEPhI", Moscow, Russia
[2]Moscow State University of Medicine and Dentistry named after A. I. Yevdokimov, Moscow, Russia
[3]National Research Mordovia State University, Saransk, Russia

Main author email address: kav2972@yandex.ru

Abstract. In the present work, the method of obtaining the conjugate of the anticancer chemotherapeutic agent doxorubicin to the exogenous double-stranded DNA of the sturgeons is proposed (the source: commercial drug "Derinat"). The optimal conditions for synthesis of conjugate (pH, temperature and the mass ratio of the components), ensuring the highest degree of binding the chemotherapeutic agent to a carrier, were picked out. Clearing the conjugate from the non-encapsulated chemotherapeutic agent was being made by ultrafiltration method. The investigation of the toxicity and specific antineoplastic activity of the synthesized complex was conducted. The performance of the drug toxicity were established on the intact mice in compliance with the accepted standards. The antineoplastic activity was evaluated upon the Tumor Growth Inhibition Index and Metastasis Inhibition Index on mice with the transplanted Lewis lung carcinoma (LLC). It was demonstrated that the conjugate toxicity is approximately lower that the one of the unconjugated doxorubicin (LD 50 was equal 14.6 mg/kg and 9.9 mg/kg for the conjugate and doxorubicin, respectively). The specific antineoplastic activity was investigated in equitoxic doses of the drug. It was established that the conjugate being administered in equitoxic doses possesses a stronger antineoplastic activity, than the water-soluble drug (maximum 35% more as to the tumor volume and 51% more as to the Tumor Growth Inhibition index).

1. Introduction

Increasing the selectivity of the anticancer chemotherapeutic agents is one of the most crucial tasks of Oncopharmacology. The most popular way of solution thereof is binding the cytostatic chemotherapeutic agent to one or another vector, enabling specific binding to the receptors on the surface of the tumor cell as well as upcoming endocytosis. Our attention was attracted by the possibility to use a native exogenous DNA as such a vector. It is a fact that the exogenous DNA can be phagocytized by the animal cells [1 – 8]. It has been established that the endocytosis of DNA is of receptor-mediated nature [2 – 6]. This mechanism of the intracellular penetration of DNA is established for tumor cells too [7, 8]. These facts became a reason for using the exogenous DNA as a vector for the target-specific delivery of the anticancer drugs [10 – 12]. But in these works, not the chemotherapeutic agent alone, but its complex with the human albumin was used for the synthesis of the DNA-dox conjugate. In fact, albumin was as a linker in binding the chemotherapeutic agent to DNA. It should be noted that, in aqueous media, DNA forms strong complexes with cationic polymers only at the body temperature, and the conjugate to the native albumin having a comparable number of anionic and cationic groups, would be not stable enough [13 – 15]. As albumin can be taken up by cells, including tumor cells [18], under stress too intensively [16 – 17], it's impossible to establish the role of each vector in delivering the chemotherapeutic agent, so the necessity to use DNA was put in doubt. In this context we developed and approved the method for the synthesis of the DNA-Dox conjugate not demanding albumin using.

Work objective: the development of the method for synthesis of the conjugate of the anticancer drug doxorubicin to DNA; the investigation of the toxicity and specific antineoplastic activity of the synthesized conjugate.

2. Materials and methods

2.1. Materials.
The following reagents were used in the work:

- DNA-Na (the source: commercial drug "Derinat", Closed Joint-Stock Company "Tekhnomedservice", Russia);
- Doxorubicin hydrochloride (Dox), > 98% – "Sigma-Aldrich", USA;
- "Milli-Q" water;
- Phosphate buffer (PBS), "Sigma-Aldrich", USA;
- Hank's balanced salt solution ("BioloT" Ltd., Russia).

2.2. The synthesis of DNA- Dox conjugates.
The following was mixed aseptically: 2 ml of the DNA solution with the content of the substance of 15 mg/ml, 2 ml of the aqueous solution of doxorubicin with the drug content of 7, 5, 10 or 15 mg/ml (depending on the batch) and 2 ml of Phosphate buffer with 7.0 or 8.0 pH (depending on the batch). The batch description is given in the part "Findings". The mixture was incubated within 60 minutes at a constant temperature (4°C, 20°C and 37°C) and a continuous shaking. The obtained conjugate was cleared from the unconjugated doxorubicin by ultrafiltration method on the unique facility [19] using the cellulose membrane Q1210-55 F3 ("Orangescientific", Belgium) with the pore diameter of 12 – 14 kDa, procedure duration was 4 hours. 3 x 3 syntheses were performed, for each variable parameter (pH, temperature, reagents ratio). The degree of the drug encapsulation into the conjugate was being calculated by the formula:

$$\omega = (m_0 - m) / m_0 \cdot 100\% ,$$

where m_0 – initial quantity of doxorubicin, m – the quantity of the non-encapsulated doxorubicin. The quantity of the non-encapsulated doxorubicin was calculated due to its concentration in ultrafiltrate. The drug concentration in dialysate was being estimated spectrophotometrically on the UV-2600 Shimadsu spectrophotometer (Japan), at the wave length of 490 nm upon the preliminary plotted calibration schedule.

2.3. Experiments on animals.
All the animals were received from the "Stolbovaya" nursery and were kept under the standard vivarium conditions. Experiments on animals were performed subject to bioethical principles upon the approval by the local Ethics Committee at the Federal State-Funded Educational Institution of Higher Professional Education "Mordovia State University named after N. P. Ogaryov".

2.3.1. The investigation of the acute toxicity. The toxicity investigation was performed on 240 BALB/cWt mice. The animals were divided into 4 groups (a control and 3 experimental ones). In the control group, the animals were being administered with 0.2 ml of isotonic sodium chloride solution intravenously. In the 1st experimental group, doxorubicin was administered intravenously; in the 2nd experimental – doxorubicin conjugated to DNA; in the 3rd experimental – 1.5% DNA solution. In each experimental group, the 6 subgroups were divided (males as well as females in fives in each), in which the drugs were administered in the dosage of 2, 4, 8, 16, 24 and 32 mg/kg in equivalent of pure doxorubicin). The animals were observed within 30 days; the lethality was being registered daily. Establishing the toxicity performance was being executed by Probit Analysis. The LD 10, LD 50, and LD 100 of the substances studied as well as the 95% confidence intervals thereof were calculated.

2.3.2. The investigation of the antineoplastic activity. The investigation of the antineoplastic activity was performed on 60 C57Bl/6 mice of both sexes. The antineoplastic activity was being studied on the syngeneic tumor system from the bank of tumor strains from the Russian Cancer Research Center named after N. N. Blokhin of the Russian Academy of Medical Sciences – Lewis lung carcinoma (LLC).

The tumor tissue of LLC was transplanted in animals intramuscularly into the femur of the hind paw on the left in amount of 1×10^6 cells in the Hank's balanced salt solution. On the 22nd day upon the transplantation of the tumor cells, the animals discontinued the experiment.

The antineoplastic as well as antimetastatic activity were evaluated due to the guideline for studying the specific activity of the anticancer drugs being in force in the Russian Federation [20, 21].

The volume of the primary tumor node was being determined in the course of the trial, as well as the weight thereof – at the end of the trial. The number of the lung metastases was counted upon the fixation thereof in Carnoy's solution with using MBS-9 binocular loupe (16x magnification). The size of the tumors on the transplantation site was being established by a beam-compass and calculated their volume by the formula for ellipsoid volume:

$$V = 0{,}131 \times L \times (D1 + D2)^2,$$

where L – tumor length, $D1$ and $D2$ – two other mutually perpendicular diameters.

The drug entry was made intravenously threefold 72 hours apart, starting the treatment from the 7th day after the LLC strain transplantation. The drug dosage in the groups is shown in the Table 1. The dosage selection was made on the basis of the earlier toxicology studies (the equitoxic doses were compared). In the 1st and in the 3rd experimental group, the drug doses made a half of LD_{10}, in the 2nd and in the 4th ones – 1 LD_{10}. The DNA-dose in the 5th group was equal to the one in the group with the maximum dose of the conjugate.

Table 1. The schemes of the administered therapy.

Group	No.	Drug	Dose	
			% of LD	mg/kg
The control one	10	-	-	-
The 1st experimental	10	Water-soluble doxorubicin	0.5 of LD_{10}	2 mg/kg
The 2nd experimental	10	Water-soluble doxorubicin	LD_{10}	4 mg/kg
The 3rd experimental	10	Doxorubicin-DNA conjugate	0.5 of LD_{10}	6 mg/kg (on DOX)
The 4th experimental	10	Doxorubicin-DNA conjugate	LD_{10}	12 mg/kg (on DOX)
The 5th experimental	10	DNA	-	7.5 mg/kg (on DNA)

The antineoplastic activity was being evaluated due to the Tumor Growth Inhibition Index (TGII), which was calculated by the formula:

$$TGII = (V_k - V_0)/V_k \cdot 100,$$

where $V_к$ and V_o – average tumor volume in the control as well as in the experimental groups, respectively.

The antineoplastic activity of the drugs were estimated due to the following performances:

1. The metastasis frequency of the tumor – percentage of the animals with metastases towards the total amount of the animals in the group;
2. The mean number of the metastases per one animal in each group;
3. Metastasis Inhibition Index (MII) was being calculated by the formula:

$$MII = ((A_k \times B_k) - (A \times B)) / A_k \times B_k \times 100\%,$$

where $A_к$ and A – the frequency of lung metastasizing in mice of the control and experimental groups; $B_к$ and B – the mean number of the lung metastases per one animal in the control and experimental groups.

Statistical data manipulation was being performed with using the Student's t-test and Chi-square. The critical level of the difference significance constituted as equal 5% ($p < 0.05$).

3. The findings and the discussion thereof.

3.1. The development of the method for synthesis of DNA- Dox conjugate.

The influence of the concentration ratio of DNA-doxorubicin, pH and temperature on the efficacy of binding DNA to dox was investigated. The findings are shown in the Table 2.

Table 2. The efficacy of binding the doxorubicin by DNA at various parameters of synthesis.

pH	Mass ratio of DNA/ Dox	Temperature, °C	Conjugation rate	Dox content (mg per 1 mg of DNA)
7.0	2:1	4	74 ± 3	0.37 ± 0.02
		21	64 ± 3	0.32 ± 0.01
		37	48 ± 4	0.24 ± 0.01
	3:2	**4**	**71 ± 5**	**0.47 ± 0.02**
		21	66 ± 4	0.44 ± 0.02
		37	61 ± 6	0.41 ± 0.03
	1:1	4	The complex is instable.	
		21	42 ± 3	0.42 ± 0.01
		37	38 ± 4	0.38 ± 0.02
8.0	2:1	4	43 ± 5	0.22 ± 0.03
		21	39 ± 5	0.20 ± 0.03
		37	34 ± 6	0.17 ± 0.03
	3:2	4	44 ± 3	0.29 ± 0.01
		21	42 ± 4	0.28 ± 0.02
		37	38 ± 6	0.25 ± 0.03
	1:1	4	The complex is instable.	
		21	32 ± 6	0.32 ± 0.03
		37	29 ± 5	0.29 ± 0.02

The colloidal system was instable in acidic media, therefore the efficacy of the synthesis was investigated at the neutral pH as well as at weakly alkaline one. We can see that the neutral pH is the most optimal for the synthesis of the conjugate, whereby the degree of Dox encapsulation into the conjugate was increasing with the temperature decrease. The maximum content of the chemotherapeutic agent per a mass unit of the carrier was to be registered at the mass ratio of DNA

:Dox = 3:2. Therefore, the following conditions were chosen for the synthesis of the conjugate used in the present work: pH = 7.0, temperature = 4 °C, mass ratio DNA/ Dox = 3/2. The sequence of the synthesis operations is described above.

It is a fact that one of the main mechanisms of the antitumor effect of doxorubicin is the intercalation between the nucleotides during the DNA synthesis, what discontinues the latter. In this regard, the question arises as to whether the drug activity not to be lost during the interaction with DNA in vitro? But the findings of the earlier studies [10 – 12] and the findings obtained in the present work demonstrate that doxorubicin bound to exogenous DNA, preserves its antineoplastic activity. This phenomenon appears to be explained due to the fact that the irreversible binding of doxorubicin to the nucleotides doesn't take place under the used synthesis conditions. The links of Dox with nucleotides as part of DNA are not of covalent, but of hydrogen nature (mean bond energy = 4.9 kcal/mol) [22 – 23]. The phagocytized complexes DNA- Dox are being disintegrated by endonucleases, what is accompanied by release of the chemotherapeutic agent within the cell.

3.2. The investigation of the acute toxicity of the DNA- Dox conjugate.
The findings of the acute toxicity of doxorubicin are given in the Tables 3 and 4.

Table 3. The lethality rate of the animals at administering of various dosage forms of DOX.

Group	The substance under study	Lethality rate (abs./%) depending on the DOX dosage, mg/kg					
		2	4	8	16	24	32
1	Water-soluble doxorubicin	0	1	3	7	10	10
2	DNA- Dox conjugate	0	0	2	5	8	10
3	DNA	0	0	0	0	0	0

Table 4. The performance of the acute toxicity of the various dosage forms of DOX.

Performance	Water solution of DOX		DOX- Dox conjugate	
	Dose, mg	U_{95}*	Dose, mg	U_{95}*
LD-10	4.4	2.1 – 6.4	6.2	2.7 – 9.0
LD -50	9.9	7.1 – 13.4	14.6	10.5 – 19.9
LD -100	43.9	27.7 – 118.2	69.5	41.1 – 256.8

Note: U_{95} – 95% confidence interval.

We can see that toxic doses of the conjugate are on the average 1.5 more to compare with the water-soluble doxorubicin. The cause of the toxicity decreasing can be incomplete dissociating of the conjugate in somatic cells, a less concentration in myocard as well as inhibiting the pro-oxidant effect of doxorubicin by nucleic acid. Taking into consideration the fact that the toxicity decrease may be accompanied by the decrease of the specific antineoplastic activity of the chemotherapeutic agent, the equitoxic doses of the chemotherapeutic agent were used for further investigating.

3.3. The investigation of the antineoplastic activity.
The evaluation results for the tumor volume are presented on the Figure 1, the Tumor Growth Inhibition Index – in the Table 5.

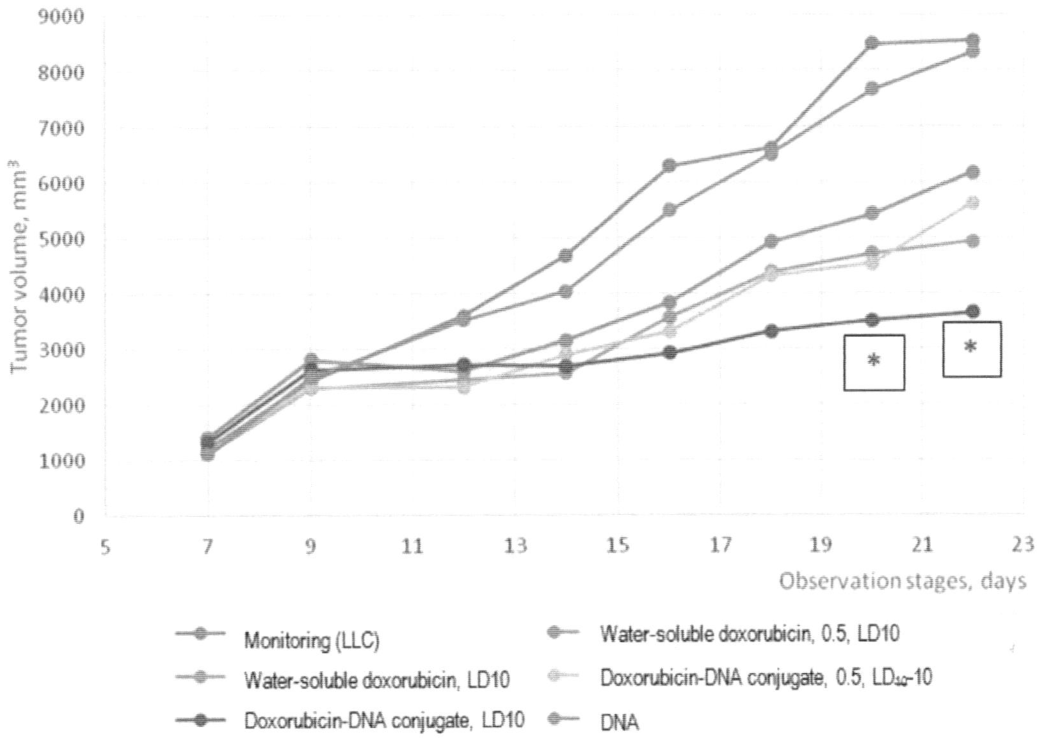

Figure 1. The changes of the tumor volume in the groups. In all the groups, except the group, being treated with DNA, the tumor volume differs significantly from the one in the control group, starting from the 14th day.

* - the significant difference of the tumor volume in the group of the water-soluble DOX from such in the group of DNA–conjugated Dox in a dose of LD 10.

Table 5. The values of the Tumor Growth Inhibition Index in the groups.

The groups of animals	TGII values at the observation stages						
	The 9th day	*The 12th day*	*The 14th day*	*The 16th day*	*The 18th day*	*The 20th day*	*The 22nd day*
Water-soluble doxorubicin, 0.5 of LD 10	-15.9	27.5	32.7	38.9	25.7	36.0	27.9
DOX-DNA conjugate, 0.5 of LD 10	4.1	35.7	38.1	47.4	34.7	46.6	34.3
Water-soluble doxorubicin, LD 10	5.0	32.1	45.4	43.2	33.7	42.4	44.1
DOX-DNA conjugate, LD 10	-8.7	24.3	42.7	53.6	50.0	64.0	67.4
DNA	-3.4	2.1	13.8	12.8	1.5	9.6	2.6

It is apparent that the inhibition of the tumor growth occurred in all the groups of the animals in the course of the chemotherapy. The statistically significant differences in the tumor volume between the control and all the experimental groups occurred on the 14th day of the observation. The significant difference as to the mentioned performance among the experimental groups was not registered earlier than on 20th day of the observation. On the 20th and 22nd days, the tumor volume in the group, being treated with the DNA-conjugated dox in a dose of LD_{10}, was 34.3% and 35.0% less than a similar performance in the group, treated with water-soluble doxorubicin ($p < 0.05$ for both points). As to mentioned dose, the difference in TGII between the group of DNA- Dox and the group of the water-soluble doxorubicin made 14.1 and 15.0 percentage points (or 50.9 and 52.8%), respectively, on the 20th and 22nd days.

The similar trend was to be registered as well for the groups of the animals, being treated with the drug in a dose of 0.5 LD 10 but these differences weren't confirmed statistically.

Table 6. The findings of the antimetastatic activity evaluation.

Groups of the animals	Tumor weight on the 22nd day, g (M ± sigma)	Frequency of metastases, %	The mean number of the surficial metastases, (M ± sigma)	Metastasis Inhibition Index
Control (with no treatment)	8.8 ± 2.9	100	94.7 ± 4.2	-
Water-soluble doxorubicin, 0.5 of LD 10	6.4 ± 2.1	100	**71.2 ± 6.9**	24.8
DOX-DNA conjugate, 0.5 of LD 10	**5.8 ± 2.4**	100	**65.3±11.2**	31.0
Water-soluble doxorubicin, LD 10	**5.1 ± 2.2**	90	**44.5±12.**	57.7
DOX-DNA conjugate, LD 10	**4.1 ± 1.8**	80	**35.6±9.8**	69.9
DNA	8.6 ± 3.1	100	90.2±8.5	4.8

Note: Significant differences from the control group are typed in bold.

In estimating the antimetastatic activity, it was established that all the studied except for the water-soluble doxorubicin in a dose of 0.5 LD 10, reduced the number of the lung metastases significantly. The number of metastases in the groups having been administered by the DNA-conjugated drug was less, than one in the groups being treated with aquatic DOX, but those difference wasn't statistically significant. The most probable mechanism of increasing the antineoplastic activity of DOX during its conjugation to DNA is a selective accumulation of the chemotherapeutic agent in the tumor tissue, caused by the endocytosis of the DNA- Dox complex.

4. Conclusion

The method of obtaining the conjugate of the antineoplastic chemotherapeutic agent doxorubicin to the exogenous double-stranded DNA is proposed. The investigation of the toxicity and specific antineoplastic activity of the synthesized complex was conducted. It was demonstrated that the conjugate toxicity is lower than the one of the unconjugated doxorubicin (LD50 was equal 14.6 mg/kg and 9.9 mg/kg for the conjugate and doxorubicin, respectively). The administration of the conjugate in equitoxic doses possesses a stronger antineoplastic activity, than the water-soluble drug (maximum 35% more as to tumor volume and 51% – as to the Tumor Growth Inhibition index).

Acknowledgements
The work is executed under the auspices of the Ministry of Education and Science of the Russian Federation as part of the State planning task of the Federal State-Funded Educational Institution of Higher Professional Education "Moscow State University named after N. P. Ogaryov", Project No. 2952.

References

[1] Rusinova, G. G., 1971. Biochemistry, 36 – 5, p. 88 – 89.
[2] Rosenkranz, A. A., Yachmenev, S. V., Jans, D. A., Serebryakova, N. V., Muravyov, V. I., Reiner, Peters and Sobolev, A. S., 1992. Receptor-Mediated Endocytosis and Nuclear Transport of a Transfecting DNA Construct. Experimental cell research, 199, p. 323 – 329.
[3] Guy, J., Drabek, D. and Antoniou, M., 1995. Delivery of DNA into mammalian cells by receptor-mediated endocytosis and gene therapy. Mol. Biotechnol., Jun. 3 (3): 237-48.
[4] Liu, J., Cao, Z. and Lu, Y., 2009. Functional nucleic acid sensors, Chem. Rev., 109, p. 1948 – 1998.
[5] Budker, V., Budker, T., Zhang, G., Subbotin, V. and Loomis, A., 2000. Hypothesis: naked plasmid DNA is taken up by cells in vivo by a receptor-mediated process. J. Gene. Med. 2, p. 76 – 88.
[6] Kensuke, M. and Masahiro, O., 2013 Endocytosis-free DNA sensing by cell surface TLR9 in neutrophils: Rapid defense with autoimmune risks. Eur. J. Immunol., 43, p. 2006–2009.
[7] Wittrup, Al. and Belting, M., 2009. Characterizing peptide-mediated DNA internalization in human cancer cells. Methods Mol. Biol. 480:101-12.
[8] Orava, E. W., Cicmil, N. and Gariépy, J., 2010. Delivering cargoes into cancer cells using DNA aptamers targeting internalized surface portals. Biochimica et Biophysica Acta., 1798 2190 – 2200.
[9] Deprez-De Campeneere, D., Baurain, R., Huybrechts, M. and Trouet, A., 1979. Comparative study in mice of the toxicity, pharmacology, and therapeutic activity of daunorubicin-DNA and doxorubicin-DNA complexes. Cancer Chemother. Pharmacol., V 2, p. 25–30.
[10] Weinberg, Yu. P., 1980. Physicochemical and biological properties of DNA-Carminomycinum complex. Medicine. No. 9, p. 18 – 22.
[11] Pyatayev, N. A., Skopin, P. I., Minayeva, O. V., Shchukin, S. A., Korovina, Ye. Yu., Zyrnyayeva, N. N., 2011. The peculiarities of the tissular distribution and antineoplastic activity of doxorubicin at administering as a conjugate of DNA to PC-1 in the rats with a transplanted carcinoma. Russian Journal of Biotherapy. V. 10 (2), p. 55 – 59.
[12] Pyatayev, N. A., Minayeva, O. V., Zyrnyayeva, N. N., Kokorev, A. V., Gurevich, K. G., Saborovskiy, A. V., Shchukin, S. A., 2014. The efficacy of the DNA-conjugated forms of doxorubicin and cisplatin at cholangiocellular carcinoma in rats. Fundamental Studies. No. 10 – 5, p. 959 – 963.
[13] Fried, M. G. and Bromberg, J. L., 1997. Factors that affect the stability of protein-DNA complexes during gel electrophoresis. Volume 18, Issue 1, p. 6 – 11.
[14] Oupický, D., Čestmír Koňák, Dash, Ph. R., Seymour, L. W. and Ulbrich, K., 1999. Effect of Albumin and Polyanion on the Structure of DNA Complexes with Polycation Containing Hydrophilic Nonionic Block. Bioconjugate Chem., 10 (5). p. 764 – 772.
[15] Fischer, Dl., Bieber, T., Brüsselbach, S., Elsässer, H. and Kissel, T., 2001. Cationized human serum albumin as a non-viral vector system for gene delivery? Characterization of complex formation with plasmid DNA and transfection efficiency. Int. J. Pharm., Aug. 28; 225 (1 – 2): 97-111.
[16] Wunder, A., Muller-Ladner, U., Stelzer, E. H., Funk, J., Neumann, E., Stehle, G., Pap, T., Sinn, H., Gay, S. and Fiehn, C., 2003. Albumin-based drug delivery as novel therapeutic approach

for rheumatoid arthritis. J. Immunol. 170: 4793 – 4801.

[17] Neumann, E., Frei, E., Funk, D., Becker, M. D., Schrenk, H. H., Muller-Ladner, U. and Fiehn, C., 2010. Native albumin for targeted drug delivery. Expert Opin. Drug Deliv., 7 : 915 – 925.

[18] Frei, E. Albumin binding ligands and albumin conjugate uptake by cancer cells. Diabetology and Metabolic Syndrome. 1: 3 – 11.

[19] Pyatayev, N. A., Minayeva, O. V., Petrov, P. S., Kokorev, A. V., Gurevich, K. G., Saborovskiy, A. V. The facility for ultrafiltration of liquids. The patent on a utility model RUS 154660, 26.11.2014.

[20] Khabriyev, R. U., 2005. Guidance on experimental (preclinical) studying of new pharmaceutical substances. Under the editorship of R. U. Khabriyev (Moscow: Medicine), p. 832.

[21] Mironov A. N., 2012. Guidance on conducting the preclinical studies of pharmaceuticals. Part I / Under the editorship of A. N. Mironov (Moscow: Griff & Co.).

[22] Agudelo, D., Bourassa, Ph., Bérubé, G. and Tajmir-Riahi, H.-A., 2014. Intercalation of antitumor drug doxorubicin and its analogue by DNA duplex: Structural features and biological implications. International journal of biological macromolecules. 66:144 – 150.

[23] Cristina Pérez-Arnaiz, Natalia Busto, José M. Leal and Begoña García, 2014. New Insights into the Mechanism of the DNA/Doxorubicin Interaction. J. Phys. Chem. B. 118 (5), p. 1288 – 1295.

Phage therapy: present and future

S G Kolesnikova[1], E N Tulyakova[1] and I Y Moiseeva[2]

[1] National Research Nuclear University MEPhI (Moscow Engineering Physics Institute), 115409, Moscow, Russian Federation
[2] Penza State University, Penza, 440026, Russian Federation

E-mail: sgkolesnikova@mail.ru

Keywords: antibiotic resistance, antimicrobial therapy, bacteriophages, infectious diseases, phage therapy, phage drugs.

Abstract. In recent years, bacteriophages are known to have become an effective alternative to antibiotic drugs. The article describes the current and potential applications of bacteriophages and phage endolysins. Also of interest is the devastating effect of phages on biofilms. The development of phage resistance is touched upon as well. Furthermore, the authors discuss the issue of laying down the rules of rational phage therapy.

Perhaps we are seeing the beginning of a new era in the principles of treatment of infectious diseases. An important problem in medicine today is the increasing antibiotic resistance of microorganisms. One solution to this problem is to use bacteriophages.

As an alternative to antibiotics, bacteriophages were included into «The Comprehensive Program for Development of Biotechnology in the Russian Federation through 2020» adopted by the Russian government on April 24, 2012.

In numerous studies bacteriophages are considered as promising anti-infective preparations.

But the question arises: is it safe to use phages, will it not lead to any adverse health consequences?

1. Medical applications

It is known that bacteriophages are found in the places where the relevant target bacteria are accumulated, e.g., in wastewater, soil, etc. Thus, it would be fair to say that if there is a bacterium, there should be a bacteriophage specific to it. This is an extremely useful feature as it makes it possible to provide highly specific treatment of infectious diseases. This justifies the numerous studies on the use of phage-based drugs in various fields of medicine.

The State Register of Medical Remedies (as of October 4, 2016) includes the following bacteriophage preparations: Dysenteric bacteriophage, Salmonella bacteriophage (of groups A, B, C, D, E), Pyobacteriophage, Klebsiella pneumoniae bacteriophage, Klebsiella bacteriophage, Typhoid bacteriophage, E. coli-Proteus bacteriophage, Intesti-bacteriophage, Streptococcal bacteriophage, E. coli bacteriophage, Staphylococcal bacteriophage, Proteus bacteriophage, and Pseudomonas aeruginosa phage.

All these are Russian-made products. The drugs are manufactured by the FSUE «NPO Microgen». Today, it is the only enterprise in the Russian Federation producing bacteriophage preparations as therapeutic agents.

Bacteriophages are specified in the list of medications approved to treat diseases, conditions and monitoring for treatment in the health care standards in the presence of the transplanted lung [1], campylobacteriosis in children [2], shigellosis [3], pertussis [4], etc.

Bacteriophages have acquired particular importance in treating gastrointestinal diseases, including acute intestinal infection, which has been confirmed in several studies. They can be used to correct dysbiosis in children [5]. There is evidence for possible use of bacteriophages in epidemic foci and emergencies as a measure of prevention of shigellosis, salmonellosis, typhoid fever, etc. [6].

Bacteriophages are used in treating ENT disorders. It was found that bacteriophages shorten the duration of treatment during exacerbations of chronic laryngitis. [7].

Bacteriophages are effectively used in surgery. A study performed on the wound infection prevention in a group of patients who had their lower limb amputated has shown that administration of phage therapy reduces the risk of postoperative complications [8]. In a comprehensive treatment of chronic osteomyelitis, the use of bacteriophages has also shown high efficiency [9].

Phage preparations can be applied in dentistry. For example, a study using the drug « Sextaphag » has proven its efficiency in treating gingivitis and parodontium [10]. Bacteriophages have shown a bactericidal effect on E. faecalis even when a root canal biofilm is formed [11].

It is also possible to use bacteriophage preparations in gynecology for the treatment of lower genital tract infections [12].

There is evidence for successful use of phage therapy in urology [13]. Positive results were observed in treating different urinary tract diseases in children [14].

It seems promising to use bacteriophages in pulmonology. Positive results have been reported in respect to pneumonia caused by Pseudomonas aeruginosa [15, 16].

1.1 Biofilms and bacteriophages

A particularly interesting area of phage therapy is the possibility of using bacteriophages to control biofilms. Biofilms pose a major problem in modern medicine. They are formed in the organs, tissues, and implanted equipment. Biofilm formation leads to the occurrence of certain chronic infections. But the most important thing is that antibiotic resistance significantly increases in a conglomerate of bacteria.

Phages produce three types of enzymes which have a devastating effect on biofilms: 1) exerting a hydrolytic effect on the biofilm extracellular polymeric substances; 2) acting on the bacterial cell wall; 3) disrupting the bacterial capsule [17].

The film-forming staphylococcal strains were shown to have high susceptibility (more than 50%) to bacteriophage preparations [18]. Another study proved the efficiency of bacteriophages in the prevention of P. aeruginosa biofilm formation and destruction of those already formed [19].

These findings support the theory of efficient use of bacteriophage preparations against biofilms.

1.2 Phage-associated lysins (PAL)

An important problem in applying phage therapy is to prevent the development of phage resistance. For this purpose, other possible applications of phage preparations are currently being developed. In addition to the possibility of using whole virus bacteriophages, research is being conducted on the use of phage-associated lysins (PAL). The PALs are endolysins, or murein hydrolases, i.e., enzymes which promote cleaving the bacterial cell wall peptidoglycan. Their action occurs only in the presence of the second factor, i.e., choline membrane protein which forms pores in the plasma membrane and provides endolysin with an access to bacterial cells [20].

The efficiency of SPZ7 endolysin against gram-negative bacteria was shown in [21]. It seems promising to use PMG1 endolysin isolated from Pseudomonas aeruginosa [22].

PAL-based drugs do have their advantages: firstly, they exclude the presence in the preparation of foreign genetic material which patients may disfavor or try to avoid; secondly, these substances may be obtained synthetically, thus eliminating the multistage process of culturing bacteriophages.

2. Bacteriophages in biomedical nanotechnologies

In studying bacteriophages, it was found that their size ranges from 20 to 200 nm. Thus, most bacteriophages may be qualified as «nanoparticles». These nanoobjects are created by nature; therefore, their distinctive features are complete naturalness and high specificity. For example, it is possible to use bacteriophages as vectors for delivering genes into the cells in order to change their functions [23]. It is particularly important to apply phage peptide libraries for obtaining tissue- and tumor-specific peptides which can be used as independent therapeutic agents or to deliver drugs to specific target organs [24]. In addition, temperate phages can be used to obtain recombinant bacterial strains. A bacteriophage is integrated into the bacterial genome as a prophage, and the microorganism receives additional, new genetic information. This contributes to bacterial synthesis of various substances, such as enzymes, hormones, regulatory proteins, drugs, etc.

3. Other uses of bacteriophages

By now, interest in bacteriophages has increased so much that there are attempts to use them in various spheres of human activity.

A Russian company, Microcosm SPC Ltd., manufactures such products containing phages as «Phagogin» (gel for intimate hygiene), «Otophag» (gel for treating ENT infections), etc. But they do not have a registration number and are not included in the State Register of Medical Remedies.

Bacteriophages can be used in veterinary medicine. They have proven effective in the treatment of colibacillosis in pigs [25]. In another study [26], it has been proven that they reduce the risk of the presence of Salmonella in chicken products.

Bacteriophages are successfully used as dietary supplements for the decontamination of food and for the prevention of a number of infections [27]. They enter into the composition of probiotic dietary supplement «Foodphage» which has proven effective as a drug for correcting dysbacteriosis [28].

Phages are also used in the cosmetic industry. In particular, JSC Faberlic has started introducing bacteriophages into personal and intimate hygiene means. MIRRA Company (Russia), which produces drugs to treat acne and inflammatory skin changes, is also involved in this activity.

4. Phage resistance

Bacteria and bacteriophages have evolved continuously and interdependently, and since these microorganisms are closely interrelated, the changes in them have been aimed at acquiring properties giving the edge over the competitors. On the part of bacteriophages, these properties imply better ways to infect and destroy the bacterial cell. Bacteria, in turn, have developed mechanisms for phage resistance.

Bacteria are able to protect themselves almost at any stage of the infection process. The first mechanism involves blockade of the surface receptors, synthesis of the extracellular matrix and competitive inhibitors to prevent phages from being adsorbed onto the bacterial cell surface. The second mechanism uses phage injection blocking. The third mechanism is triggered after bacteriophage injecting into the bacterial cell genome. Restriction endonucleases recognize and cleave alien DNA. This is known as the restriction-modification (R-M) system. One more way to prevent the development of an infection is an abortive infection when the host bacterium dies and the virus does not spread any further in a population of susceptible strains [20].

A comparative study of the susceptibility of bacterial strains to antibiotics and phages isolated from patients has shown that about 50% are bacteriophage-susceptible. This figure is higher than the susceptibility of these strains to antibiotics but also indicates the presence of phage resistance [29].

The possibility of applying bacteriophages in medical practice depends on the conditions in which phage resistance develops. But, at the moment, this problem is not given due consideration.

This issue became a matter of interest to specialists of the Medical Faculty at the INPE NRNU MEPhI. Extensive studies are underway at the Departments of Microbiology, Virology and Immunology in order to identify phage-resistant pathogens of pyoinflammatory diseases:

Staphylococcus aureus и Streptococcus pyogenes. However, it is too early to speak about any results at this stage of research.

5. Principles of rational phage therapy

Attempts are made of a wide, large-scale introduction of bacteriophages into practice. However, in the past, we had the imprudence to connive at wide, irrational, uncontrolled use of bacteriophages which resulted in the development of antibiotic resistance in a number of microorganisms. Phage therapy should initially be based on definite standards, i.e., principles of rational use. This problem has been repeatedly addressed in various research works. Based on the acquired knowledge, we can already identify the following main points:

1) Only drugs officially recommended for phage therapy and commercially produced by licensed manufacturers should be used;
2) A drug can be used provided that there are reliable data that it contains only virulent phages but not temperate ones;
3) Administration of bacteriophages can be justified only if antibiotics are found ineffective or cannot be applied;
4) Before administering phage drugs, the phage susceptibility of the target microorganism should be determined;
5) It is necessary to confirm high lytic phage activity against the specific pathogen;
6) Bacteriophages can be used both as monotherapy and in combination with antibiotics;
7) Phages should be applied strictly according to the instruction, following the recommendations on the method of administration and treatment period; the recommended phage titers should be maintained;
8) With regard to phage drugs, it is important to observe the storage precautions; prolonged storage decreases the phage titer and, as a consequence, its activity;
9) Bacteriophages can be prescribed as preventive drugs in epidemic areas in emergencies but not before the phage susceptibility of microorganisms has been determined.

6. Conclusion

Studying the effectiveness of bacteriophages leads as often as not to positive results. Currently, there are all the necessary knowledge, skills and equipment for performing accurate research in this area. Phage drugs are certainly not a panacea for infectious diseases they but represent a promising alternative to antibiotics. Perhaps, we shall have yet to learn about some features of these microorganisms. For example, previously there were no adequate data on the bacterial resistance to phages, which has become a matter of concern nowadays. It is necessary to competently treat these drugs, limiting their potential widespread use. Using bacteriophages requires clear guidelines and rules approved by the WHO.

Acknowledgments

This work was supported by Competitiveness Program of National Research Nuclear University MEPhI.

References

[1] The order of the Russian Ministry of Health of 24.12.2012 N 1381n «About the approval of the standard of primary medical and sanitary help in the presence of the transplanted lung». Access from the ConsultantPlus legal reference system (reference date: 09.10.2016)

[2] The order of the Russian Ministry of Health of 24.12.2012 N 1406n «About the approval of the standard of specialized medical care to children at campylobacteriosis of moderate severity». Access from the ConsultantPlus legal reference system (reference date: 09.10.2016)

[3] The order of the Russian Ministry of Health of 24.12.2012 N 1468n «About the approval of the standard of primary medical and sanitary help to children at the shigellosis (dysentery) of

easy severity». Access from the ConsultantPlus legal reference system (reference date: 09.10.2016)

[4] The order of the Russian Ministry of Health of 20.12.2012 N 1130n «About the approval of the standard of specialized medical care to children at whooping cough of heavy severity». Access from the ConsultantPlus legal reference system (reference date: 09.10.2016)

[5] Bekhtereva M K and Ivanova V V 2014 Place of bacteriophages in treating infectious diseases of the gastrointestinal tract *Pediatrics* **2** 24–29

[6] Bondarenko V M 2013 New horizons of bacteriophage therapy *Bull. Orenburg Sci. Cent. UB RAS* **4** 10

[7] Zaripova T N, Mukhina V I and Chuykova K I 2013 Rationale for the use of bacteriophages in treating patients with acute exacerbation of chronic laryngitis (persons of vocal and speech professions) *Otolaryngology* **14** 1008–14

[8] Fominykh E M, Nizovoy A V, Islamov R N and Samoylov O A 2010 Polyvalent bacteriophage in the prevention of postoperative infections after lower limb amputations *Kuban Res. Med. Gaz.* **7(121)** 150–154

[9] Shevalayev G A and Yefemov I A 2013 Combimed application of antibiotic, ozone and phage therapy in treating patients with chronic osteomyelitis *News of Higher Educational Institutions. The Volga Region* **2(26)** 104–114

[10] Gileva O S, Bondarenko E A, Gibadullina N V, Kobus A B and Khokhrin D V 2011 Topical phage therapy in the comprehensive treatment of gingivitis and parodontitis *Perm Med. J.* **2** 87–93

[11] Khalifa L., Shlezinger M., Beyth S., Houri-Haddad Y., Coppenhagen-Glazer S., Beyth N. and Hazan R. 2016 Phage therapy against Enterococcus faecalis in dental root canals *J. Oral Microbiol.* **8** Available at: <http://www.journaloforalmicrobiology.net/index.php/jom/article/view/32383> Date accessed: 09.10.2016

[12] Belopolskaya Kh A, Sidorova I S, Shakhgireyeva L S and Belopolskiy A A 2014 Possibilities of phage therapy with respect to gynecological infections *Difficult Patient* **9** 6–9

[13] Sybesma W, Zbinden R, Chanishvili N, Kutateladze M, Chkhotua A, Udzhmadzhuridze A, Mehnert U and Kessler T M 2016 Bacteriophages as potential treatment for urinary tract *Front. Microbiol.* **7** 1–9

[14] Zorkin S N and Shakhnovsliy D S 2013 Possibilities of bacteriophage therapy in treating patients with complicated urinary tract infections *Pediatric Pharmacol.* **4** 132–138

[15] Debarbieux L., Leduc D., Maura D., Morello E., Criscuolo A., Grossi O., Balloy V. and Touqui L. 2010 Bacteriophages can treat and prevent Pseudomonas aeruginosa lung infections *J. Infect. Dis.* **201** 1096–1104

[16] Kozlova Yu N, Repin V Ye and Mayborodin I V 2013 Specific bacteriophage for treating surgical infections caused by Pseudomonas aeruginosa in experiment *J. Exp. Clin. Surg.* **4** 425–431

[17] Drukker V V and Gorshkova V V 2012 Bacteriophages and their functioning in biofilms *Bull. Irkutsk State Univ. Ser. Biology, Ecology* **3** 8–16

[18] Voronkova O S, Sirokvasha O A, Shevchenko T M, Vinnikov A I and Kudryavtseva V Ye 2014 Susceptibility of film-forming staphylococcal strains isolated from the respiratory tract to therapeutic bacteriophage preparations *Zaporozhye Med. J.* **6 (87)** 105–108

[19] Gabisonia T G, Loladze M Zh, Nadiradze M M, Chakhunishvili N K, Alibegashvili M G, Tamarashvili N G and Pushkina V A 2016 Influence of bacteriophages on biofilms formed by Pseudomotias aeruginosa strains *Appl. Biochem. Microbiol.* **3** 312–317

[20] Elbreki M, Ross R P, Hill C, O'Mahony J, McAuliffe O and Coffey A 2014 Bacteriophages and Their Derivatives as Biotherapeutic Agents in Disease Prevention and Treatment *Hindawi Publishing Corporation Journal of Viruses* 1–20

[21] Klyachko N L, Legotskiy S A, Levashov P A, Popova V M, Belogurova N G, Timasheva A V,

Dyatlov I A and Levashov A V 2010 SPZ7 bacteriophage endolysin: Influence of the effectors on the enzyme catalytic activity in the lysis of gram-negative microorganisms *Bull. Mosc. State Univ. Ser. 2: Chemistry* **3** 222–226

[22] Chertkov O V, Chuprov-Netochin R N, Legotskiy S V, Sykilinda N N, Shneider M M, Ivanova M A, Pletneva E A, Shaburova O V, Burkaltseva M B, Kostryukova E S, Lazarev V N, Klyachko N L and Miroshnikov K A 2011 Properties of the peptidoglycan-degrading enzyme of the Pseudomonas aeruginosa φPMG1 bacteriophage *Russ. J. Bioorg. Chem.* **6** 807–814

[23] Larocca D, Baird A, Johnson W 1998 Receptor-meditated gene delivery using bacteriophage vectors Patent No6448083

[24] Nemudraya A A, Richter V A and Kuligina E V 2016 Phage peptide libraries as a source of targeted ligands *Acta Naturae* **8** **1(28)** 52–63

[25] Skoblikov N E and Zimin A A 2013 Experience in application of non-transducing bacteriophages for prophylaxy and therapy of intestinal colibacteriosis of pigs *Proc. Int. Sci. Conf.* «Bacteriophages: Theoretical and Practical Aspects of Their Application in Medicine, Veterinary and Food» p 16

[26] Pleshakova V I and Stepanov D N 2013 Salmonella bacteriophage treatment of chickens with the features of the technology of poultry breeding *Proc. Int. Sci. Conf.* «Bacteriophages: Theoretical and Practical Aspects of Their Application in Medicine, Veterinary and Food» p 17

[27] Aleshkin A V, Karaulov A V, Svetoch E A, Volozhantsev N V, Aleshkin V A, Afanasyev S S, Rubalskiy O V, Vasilyev D A, Zolotukhin S N, Afanasyev M S, Lakhtin V M and Rubalskiy M O 2013 Bacteriophages as probiotic tools to regulate microbiocenosis and decontaminate food products, animals and plants with microorganisms *Immunopathology, Allergology, Infectology* **3** 80–89

[28] Kiselyova I A 2015 Specialized bacteriophage cocktail-based product for dietary prophylactic nutrition: design, production technology, safety and efficiency assessment: *Ext. Abstr. of Cand. Sci. (Biology) Diss.* Moscow p 27

[29] Gabrielyan N I, Gorskaya Ye M, Spirina T S, Prudnikova S A and Romashkina L Yu 2011 Study of the antibiotic and phage susceptibility of nosocomial microbial strains isolated from transplant clinics patients *Journal of Transplantology and Artif. Organs* **3** 26–32

Prospects for the use of animal cell cultures in screening of pharmaceutical substances

S G Kolesnikova[1] and I Y Moiseeva[2]

[1] National Research Nuclear University MEPhI (Moscow Engineering Physics Institute), 115409, Moscow, Russian Federation
[2] Penza State University, Penza, 440026, Russian Federation

E-mail: sgkolesnikova@mail.ru

Keywords: cell cultures, in vitro studies, cytotoxicity, pharmacological cytotoxicity, testing cultures, drug substances

Abstract. Currently, there is a tendency to reduce the use of animals in conducting safety tests of chemical substances. Therefore, *in vitro* methods are a good alternative or adjunct to *in vivo* safety tests. This is especially important at the stage of pre-clinical drug trial. In 2004, the international standard for the principles of good laboratory practice (GLP) [1] was adopted which regulates chemicals trials in cell cultures. However, in Russia, until recently, this issue has been neglected. Research works have been scarce. In 2013, the standard for GLP principles and compliance monitoring was adopted in Russia [2]. The feasibility of using animal cell cultures as drug testing system has been proved by the experimental base and is now being introduced into practice [3].

1. Introduction

It should be understood that the safety of drug therapy is one of the most important tasks of modern health care. Adverse drug reactions rank No. 4–6 place among the causes of mortality in advanced countries. And of course the system for testing drugs before they are introduced into clinical practice is constantly being improved.

Any substance with biological activity intended to be subsequently used in touch with living systems, as a rule, is tested for its proper toxicity at the first stage. Certainly, the classical methods for evaluating the toxicity of biologically active substances, including pharmaceutical substances, still dominate in experiments. However, methods alternative to the classical experimental animal tests, namely models using animal cell cultures are increasingly found in toxicological studies [4, 5, 6]. Despite the extensive development and acceptance of alternative methods in the world, they have not aroused widespread in our country. In addition to solving the ethical problems associated with the mass use and death of experimental animals, these methods can significantly reduce the cost and time of preliminary studies of new chemical drugs mainly at the stage of pre-clinical trials. Another indubitable advantage of in vitro models is the ability to work with test substances directly on human or animal cell cultures which makes the obtained data more adequate when they are projected on a macroorganism. Furthermore, the use of cell cultures makes it possible to establish the nature of biological activity and the mechanism of realizing the effects of tested compounds directly at the cellular level and consider complex synergistic and/or multidirectional effects of test compound mixtures [7, 8]. Test cultures of animal cells derived from different tissues and organs exhibit different sensitivities to chemicals; and in vitro tests can thereby detect negative unpredictable reactions of individual tissues' possible hypersensitivity in a macroorganism [9]. Test systems based on cell cultures are handy and can be used to solve a variety of tasks in the field of toxicology and cell biology, namely, to study a number of pharmaceuticals (potential or already in use) of practical interest.

Currently, introduced into practice and prescribed by the General Monograph (OFS.1.7.2.0002.15) are biological test methods for interferon preparations using cell cultures. This applies to substances and dosage forms of human interferon of all types of natural and genetic engineering origin [3].

Improving the pre-clinical trial stage of pharmaceutical substances is also particularly relevant in the treatment of heart arrhythmias where side effects are up to 30% of cases. Most of the traditional antiarrhythmics are known to have a cardiotoxic effect [10, 11, 12, 13, 14]. In this context, it is necessary to expand the arsenal of pre-clinical drug testing methods for an objective evaluation of their safety. It is essential to intensify the search for new drugs nonspecifically exhibiting antiarrhythmic activity, while remaining as safe as possible, for example, of the metabolic action type (dimephosphon, mexidol, emoxypine, cytochrome c) [15, 16, 17].

2. Authors' research

2.1. Materials and methods

The in vitro studies were performed on passaged MDBK cell cultures (calf kidney cell culture) and KST (calf heart vessel cell culture). Standard culture media and solutions were used in the experiment. Culturing was performed on a steady-state monolayer basis at a temperature of 37^0C based on the Eagle's, Eagle's MEM and Eagle's DMEM culture media (figure 1). 5–10 % of bovine serum and antibiotics (penicillin and streptomycin up to100 μg/ml) were added to the culture medium. The cell lines were maintained by periodic passaging. Isolate passaging was performed as the monolayer was formed, usually after 4–5 days of culture. The cells were removed without a centrifuge, using Versene solution (0.02%) and trypsin solution (0.25%) at the ratio of 9 : 1. The seed concentration was within 70 000–100 000 c/ml. Etacisin and mexidol were tested at concentrations equivalent to therapeutic ones (0.5; 1.0 μg/ml and 5.0; 10.0 μg/ml, respectively). The combined use showed the concentration value of 2.5 μg/ml for mexidol and 1.0 μg/ml for etacisin.

2.2. Results

It was established that etacisin (1.0 μg/ml) caused a developmental delay of both cultures after the drug was added with the cell suspension. When etacisin was added on the formed monolayer, changes were observed corresponding to a pronounced cytopathic effect. Gradually, the cells were losing turgor and adhesive properties due to which the monolayer was quickly degraded (figure 2). There were multiple cell anomalies. There were cases of multinucleate MDBK and KST cell cultures, the formation of blade nuclei; the drug contributed to the formation of micronuclei (figure 3). The MDBK culture was found to be the most sensitive to the antiarrhythmic. The number of cells with two or more nuclei exceeded the benchmark by more than three times. Pyknoses of nuclei were observed. There appeared pathological mitosis forms; their number was five times larger than the normal range. Prevalent among them were pluripolar mitoses, mitoses with agglomeration, delay and fragmentation of highly condensed chromosomes (figure 4). The cell cytoplasm acquired the foamy form due to the large number of vacuoles (figure 5). Etacisin in a dose of 20 μg/ml caused a mitotic arrest at the prophase stage with the chromosome being significantly altered and did not match in shape and appearance the prophase chromosomes in the control. 3-hydroxymethyl-ethylpyridine succinate (mexidol) had no negative cytomorphological effects in both cell lines, nor did it affect the ratio of multinucleate cells as compared to the control.

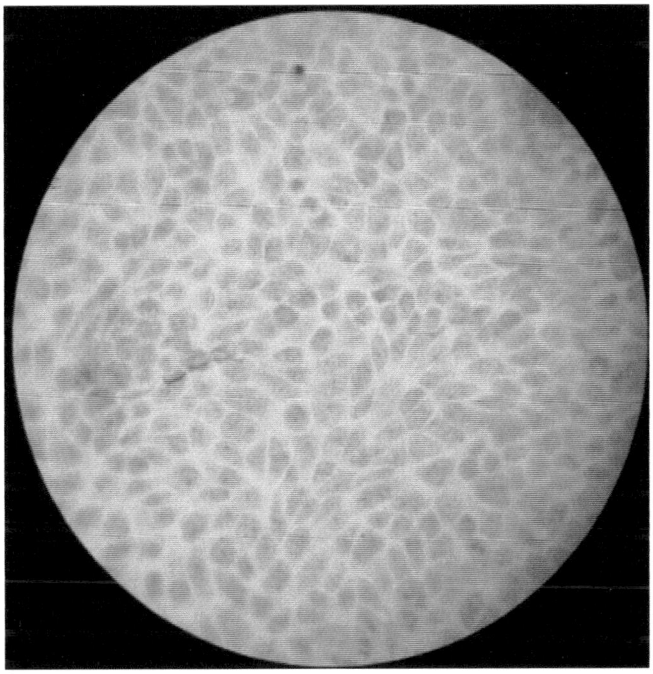

Figure 1. MDBK cell line. Normal monolayer.

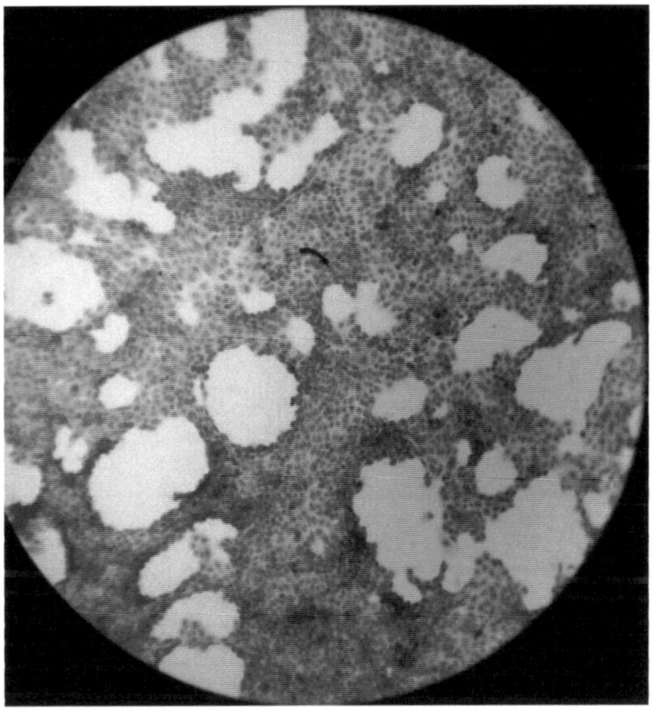

Figure 2. MDBK cell line monolayer degeneration.

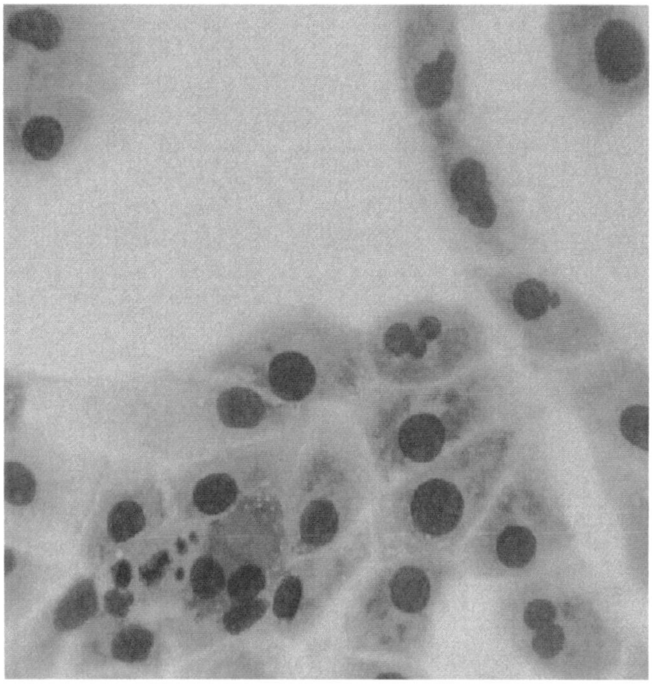

Figure 3. MDBK cell line. Pyknosis and nuclear fragmentation.

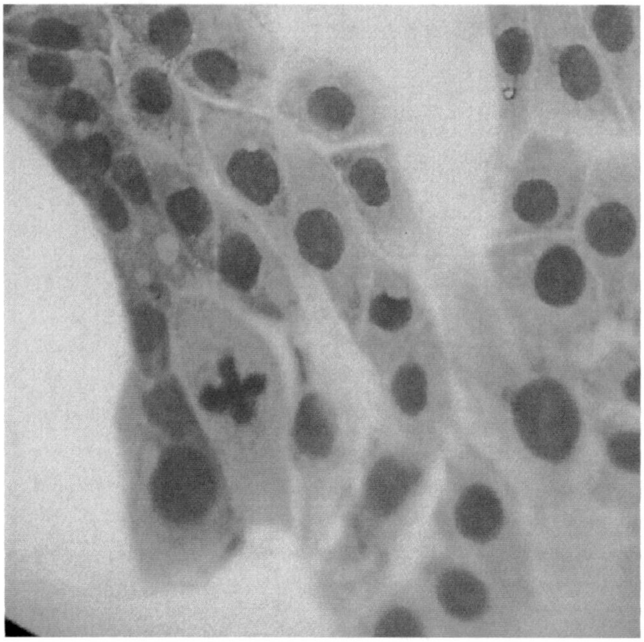

Figure 4. Pathological multiipolar mitosis.

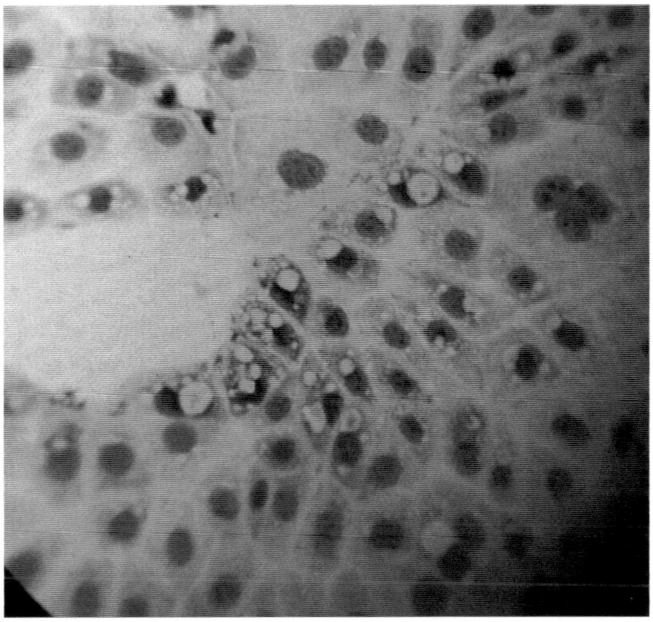

Figure 5. MDBK cell line. Multinucleate cells and large cytoplasm vacuolization (the "foaming" effect).

The karyological control showed that the modal number of chromosomes affected by etacisin corresponded to the normal values both for MDBK and KST cultures and was 48–50 (46–60 in the control); however, the drug significantly extended the variability range in the number of chromosomes within 35–85. The metaphase plates were significantly modified. Often, there appeared highly condensed chromosomes in the form of globules with unexpressed shoulders. The modal class of chromosomes affected by 3-hydroxymethyl-ethylpyridine succinate was 48–50; the variability range in the number of chromosomes was within 40–60, which also corresponded to the norm.

It was also shown that 3-hydroxymethyl-ethylpyridine succinate (mexidol) is able to partly neutralize the negative effects of etacisin when they are used together.

3. Conclusion

The results of the studies are indicative of high etacisin cytotoxicity in a dose of 1.0 μg/ml, which is equivalent to a therapeutic dose, as well as the ability of mexidol (2.5 μg/ml) to weaken the severity of the adverse effects of the antiarrhythmic agent etacisin. These studies confirm the retrospectively negative cardiotoxic effects of etacisin shown in clinical practice. The introduction of in vitro test-systems on a permanent regulated basis into the pre-clinical drug testing stage would make it possible to early detect their possible cytotoxic effects and the molecular-cellular "targets".

Acknowledgments

This work was supported by Competitiveness Program of National Research Nuclear University MEPhI.

References

[1] OECD series on Principles of Good Laboratory Practice and Compliance Monitoring - No 14 - Advisory document of the working group on Good Laboratory Practice. The application of the Principles of GLP to in vitro studies (2004)
[2] GOST 31891-2012 Principles of Good Laboratory Practice (GLP). Application of GLP

principles to in vitro studies. Russia (2013)

[3] The biological test methods for interferon preparations using cell cultures (General Monograph) OFS.1.7.2.0002.15. (reference date: 10.29.2015)

[4] Anders M 2000 S-conjugate-dependent toxicity; alternatives to animal studies. *Altern. Animal Testing & Experiment. AATEX* **1(7)** 30–36

[5] Eropkin M Yu 2004 The cell cultures as a model system in the bio chemical-toxicological studies. *Doctor Biol. Sci. Diss.* St.Petersburg 354 p

[6] Chiba K, Makino J, Ohuchi J. et al. 1999 Interlaboratory validation of the in vitro eye irritation tests for cosmetic ingredients. Evaluation of cytotoxicity test on HeLa cells. *Toxicol. In Vitro.* **1(13)** 189–198

[7] Dyakonov L P, Galnbek T V, Kulikova I L et al 2000 Animal cell culture: (Methods and Application in biotechnology) Moscow : Sputnik+ Co 398 p

[8] Zhukova O S, Abdrakhmanov I K, Gerasimova G K, Dyakonov L P 2006 Cytotoxic effect of catalytic system containing phtalocyanine and naphtalocyanine complexes and ascorbic acid on normal and tumor cells *Abstracts 46th ETCS International Meeting* "In vitro cytotoxicity mechanisms". Verona p 46

[9] Galnbek T V, Akinshina G T 2013 Modern trends and development of cell biotechnology in veterinary medicine. *Proc. of the All-Russian Research Institute of Experimental Veterinary Medicine n. a. Ya R Kovalenko* **77** 219–227

[10] Alpert J and Francis G 1994 Treatment of myocardial infarction. Trans. from Engl. Moscow: Praktika Publ. 255 p

[11] Onopriev V V, Uvarov A V, Reznikov A Y 1994 Antiarithmic effects of richjocaine *European students conference at the Charité for students and young doctors*, Berlin, Germany p 7

[12] Khankoeva A I 1998 Cardioprotector and cardiotoxic effects of antiarrhythmic compounds: ways of pharmacological correction, molecular mechanisms of action. *Extended abstract of M. D. diss.*: Volgograd 47 p

[13] Arutyunov G P and Rozanov A V 2003 Place of thrombolytic therapy in the treatment of cardiogenic shock *Heart* **1(2)** 18–19

[14] Galenko-Yaroshevskiy P A, Skibitskiy V V, Boddii V B et al 1997 Cardiotoxic effects of antiarrhythmic drug ritmidazol and their correction with sufan, befol and their combinations *Bulletin of Experimental Biology and Medicine* **12(124)** 640–644

[15] Balykova L A, Balashov V P, Shkolnikova M A and Kemaeva N N 1999 Effectiveness of dimephosphon in treating heart rhythm disturbances in children *Russian Bulletin of Perinatolgy and Pediatrics* **2** 125–128

[16] Gurevich M A 2002 Pathogenesis and treatment of heart failure, myocardial infarction. *Clin. Med.* **4(74)** 15–20

[17] Popov D A 2004 Pharmacological efficacy of antioxidant (emoxypine, dimephosphon), antiarrhythmic (trimecain) and cardiac (strophanthin) drugs in experimental anaphylaxis. *Extended abstract of M. D. diss.* Staraya Kupavna 20 p

Laser correlation spectroscopy in the diagnosis of tumor diseases of the female reproductive system (preliminary results)

A A Korneeva[1], M N Sekerskaya[2], K I Zhordaniya[2], V S Sapezhinskiy[2], N V Golubtsova[2], A E Barmashov[2], S A Gonchukov[1] and A V Ivanov[2, 3]

[1] National Research Nuclear University "MEPhI"
[2] N N Blokhin RCRC, Moscow
[3] State Research and Clinical Center for Laser Medicine of FMBA Russia, Moscow

E-mail: ivavi@yandex.ru

Abstract. The study of blood serum of cancer patients by laser correlation spectroscopy to determine the possibility of differentiation of benign and malignant tumors of the female reproductive system. We analyzed the data and assessed the applicability of the method mentioned above target.

1. Introduction

Early diagnostics of malignant neoplasms is one of the most important problems of an oncology. The method of the laser correlation spectroscopy (LCS) is applied to detection of pathologies in a human body, allowing to obtain information on subfractional structure of extracellular vesicles in biological liquids. Extracellular vesicles perform an important role in information transfer between cells in an organism. They are involved in various physiological processes – both normal, and pathological. It is possible to determine absence or existence of any disturbances in a human body by structure and the size of vesicles [1, 2]. One of method of obtaining information on extracellular vesicles is the LKS method based on definition of spectral characteristics of the laser radiation disseminated when passing in disperse medium [3]. The program of data processing allows to visualize in the form of the histogram a percentage contribution to light scattering of separate fractions of biological liquid with hydrodynamic radiuses in the range from 1 nanometer to 104 nanometer. The method of a laser correlation spectroscopy has high sensitivity concerning detection of pathologies, including oncologic. In this regard it draws to itself (himself) more and more attention as a way of diagnostics of various diseases [4-8]. An opportunity to differentiate good-quality and malignant neoplasms is of great interest.

In work [9] it is experimentally shown that the histograms received by the LCS method from samples of a blood plasma of patients with benign tumors differ from the same histograms of patients with a breast cancer. The purpose of the real work is the research of a possibility of use of a method of a laser correlation spectroscopy of blood serum for differentiation of benign and malignant tumors of female reproductive system.

IOP Conf. Series: Journal of Physics: Conf. Series **784** (2017) 012029 doi:10.1088/1742-6596/784/1/012029

2. Materials and methods

20 women aged from 30 to 70 years were examined with the LSK method: 9 patients had benign tumors and 11 malignant ones. All patients were examined in the RCRC N.N. Blokhin department of gynecology and not been treated prior to the survey. Blood sampling, the subsequent preparation of blood serum and a further research were carried out in one day. Samples of a blood of 8 ml centrifuged 10 minutes at 3000 rpm. The received Serum was transferred to sterile plastic test tubes like Eppendorf of 1 ml. For carrying out measurement used 10% Serum solution (0.2 ml of Serum, 1,8 ml of water for injections). Solution was in a test tube at the room temperature of 20 °C of 10 minutes then the test tube was placed in a measuring cell of a spectrometer.

Measurements were taken on experimental installation in which the semiconductor VCSEL with a wavelength of 855 nanometers and a power output of 1 mW is a light source. The principle of work of LCS is based on definition of spectral characteristics of the monochromatic light scattered in disperse medium. Ranges of intensity' fluctuations of diffused light processed mathematically the ProLSDRS program. As a result of processing received the histogram of particle distribution of blood serum by the sizes.

On each of samples of blood serum four measurements are taken. By means of the method described in work [10] two groups of particles were taped: to 100 nanometers and more than 200 nanometers. For each sample the average size of group of particles and weight in a percentage ratio bound to concentration are calculated. The obtained data averaged on all samples of Serum of patients with benign tumors and separately – on samples with malignant neoplasms. Differences as a part of blood serum of two groups of patients were determined by these parameters.

3. Results and discussion

For each sample of blood serum histograms of particle distribution by the sizes which characteristic type in case of benign and malignant tumors is given in fig. 1 were received.

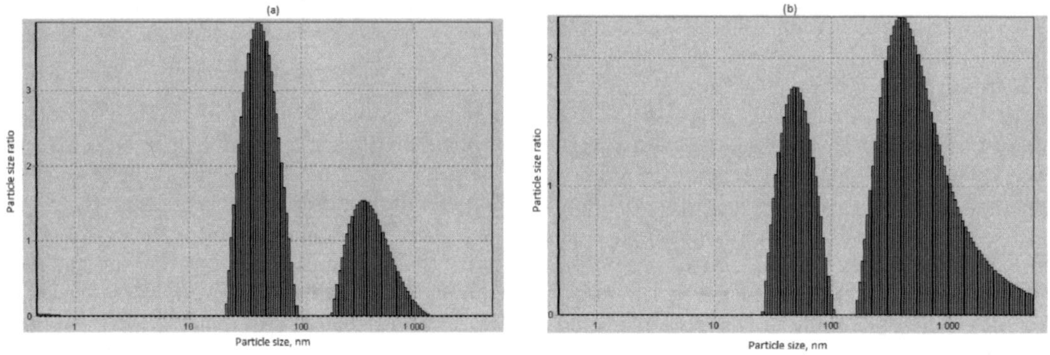

Figure 1. Histograms of particle size distribution in the case of (a) benign and (b) malignant tumors.

The analysis of histograms of blood serum at patients with benign tumors was carried out. In histograms of all patients both on weight, and on intensity of dispersion of light particles with hydrodynamic radiuses of 10 - 50 nanometers over coarse particles > 200 nanometers prevail (table. 1). Existence of sharp dependence of intensity of light scattered on the size of suspended particles (in proportion to the sixth degree of the characteristic size) allows to measure reliably distribution of insignificant number of coarse particles against the background of fine. Therefore, in spite of the fact that concentration of fine particles considerably exceeds concentration large, average values of intensity of diffused light differ by one and a half times.

The similar analysis of histograms of blood serum of patients with malignant neoplasms showed that in dispersion of light coarse particles prevail > 200 nanometers over particles of the small size of 20-70 nanometers (table 2). Histograms on samples of blood serum of patients to malignant tumors are similar among themselves in particle distribution by the sizes. Only in one sample of a ratio between a contribution of two groups of particles to dispersion and their weight are similar to histograms of samples of blood serum of patients with benign tumors that is explained by specific features of the patient. Perhaps, such difference is bound to the fact that the tumor of this patient of low degree of malignancy, that is characterized by slow advance.

Table 1. The contribution of the two groups of serum particles in the scattering of light (%) and weight (%), averaged over all histograms patients with benign tumors.

	10 – 50 nm		200 nm	
Particle contribution	The weight,%	Scattering,%	The weight,%	Scattering,%
The average contribution	66,9	49,4	1,3	30,3
The minimum contribution	40,3	22,9	0,1	12,5
The maximum contribution	93,6	75,9	2,5	48,1

Table 2. The contribution of the two groups of serum particles in the scattering of light (%) and weight (%), averaged over all histograms patients with malignant tumors.

	20-70 nm		200 nm	
Particle contribution	The weight,%	Scattering,%	The weight,%	Scattering,%
The average contribution	62,1	28,2	18,9	71,5
The minimum contribution	36,7	6,1	5,1	50,3
The maximum contribution	80,7	61,7	51,8	93,9

The group of particles of the small size in Serum ranges with malignant tumors is shifted to the area of larger sizes (20-70 nanometers) concerning the same group of particles in case of good-quality neoplasms (10-50 nanometers). Higher concentration of coarse particles which average value according to histograms of blood serum makes 19%, in comparison with average concentration of 1,3% of similar group of particles at benign tumors is observed. Most likely, such difference is bound to a high metabolic rate of clones of malignant cells and, respectively, allocation of larger number of large vesicles, unlike clones of good-quality cells.

High concentration of coarse particles with malignant tumors brings in blood serum of patients to is enlarged a contribution of these particles to light dispersion. In ranges it is shown in the form of larger intensity of diffused light from group of particles > 200 nanometers in comparison with intensity of diffused light from particles of the small size. Blood serum ranges with benign tumors yield opposite result, and excess of intensity of diffused light from fine particles concerning intensity of diffused light from large.

The research of blood serum of patients showed that criteria of differentiation of good-quality and malignant neoplasms are: shift of group of particles of small hydrodynamic radiuses to the area of the larger sizes at malignant tumors and rising of concentration of coarse particles > 200 nanometers, and also change the relation of a contribution to intensity of dispersion of light between groups of particles: to 100 nanometers and more than 200 nanometers.

The LKS method allows to distinguish benign and malignant tumors of female reproductive system with high precision of 95%. The presented results are preliminary as selection of examinees isn't numerous, but results are very indicative. A set of the database proceeds.

Acknowledgments

This work was supported by Competitiveness Program of National Research Nuclear University MEPhI.

References

[1] Komarov G D, Kuchma V R and Noskin L A, 2001, Polysystemic sanogenetic monitoring. M. MIPKRO, 343 p.

[2] Tatischeff I, 2015, Cell-derived Extracellular Vesicles Open New Perspectives for Cancer Research. Cancer Research Frontiers. Apr; 1(2): 208-224. doi: 10.17980/2015.208.

[3] Cummins H Z and Pike E R (eds.), 1974, Photon Correlation and Light Beating Spectroscopy, New York: Plenum Press.

[4] Lisova N A, 2001, A new approach to the laboratory diagnosis of disorders of homeostasis: the diagnostic capabilities of laser correlation spectroscopy in Pediatric Nephrology, Nephrology and Dialysis, V.3, №1, 32-39

[5] Akleev A V, Pashkov I A, Klopov N V, Lebedev A D, Silina A G, Pavlov M Yu and Noskin L A, 1995; Patterns of Homeostatic Alterations during Delayed Aftereffects of Radiation: Detection using Laser Correlation Spectroscop, Bulletin of Experimental Biology and Medicine, 5: 556-560.

[6] Petrova G P, Petrusevich Yu M, Boiko A V, Fedorova K V, Khlapov V P, Dombrowski I V, Ivanov A V and Papish EA 2006, Light scattering method in the clinical diagnosis of cancer. Medical Physics, №4, p.61-64

[7] Rumyantseva E I, Kovalev I E, Kovaleva O I and Karganov M Yu, 2005, Laser Correlation Spectroscopy of Macromolecular Complexes in Blood Serum Is an Efficient Method for Detecting Insulin Overdose and Correction of the Insulin Therapy for Diabetes Mellitus in Children, Doklady Biochemistry and Biophysics, vol. 402, pp. 210–213

[8] Merlich K I, Geshelin S A, Noskin L A, Nisevich I I, Omel'chenko V S, Silina A G, Sitnik A G, Brygar' V A, Tsepkolenko V A and Myzeichuk Yu I, 1993, Laser Correlation Spectroscopy in an Investigation of the Subfractional Composition of Plasma from Patients with Gastric Bleeding, Craniocerebral Injury, and Intoxication, Byulleten' Eksperimental'noi Biologii i Meditsiny, vol. 116, No 8, pp. 220-222.

[9] Merlich K I, Geshelin S A, Noskin L A, Varbanets V F, Shcherbakov A M, Omel'chenko V S, Silina A G, Nisevich I I and Lebedev A D, 1993, Subfractional Composition of Blood Plasma in Benign Tumors and Breast Cancer Studied by Laser Correlation Spectroscopy, Byulleten' Eksperimental'noi Biologii i Meditsiny, vol. 116, No 8, pp. 193-195.

[10] Kiritchenko M N, 2015, Trends sizes and concentrations of proteins and their complexes in blood plasma in vitro according to light scattering. Kand. diss, The P.N. Lebedev Physical Institute LPI

IOP Conf. Series: Journal of Physics: Conf. Series **784** (2017) 012030 doi:10.1088/1742-6596/784/1/012030

An analytical light distribution model in the optical system of a scintillation detector

Sergey Kuznetsov[1,2], **E.V. Skachkov**[2], **V.N. Belyaev**[1]

[1] National Research Nuclear University "MEPhI", Russia, Moscow, Kashirskoe shosse, 31
[2] PLC "NIITFA", Russia, Moscow, Varshavskoe shosse, 46

E-mail: kuzzz99@gmail.com

Abstract. The article describes an analytical light distribution model in the optical system of a scintillation detector. The model can be useful for scintillation detector development since it allows to make quick calculations with different parameters. Comparison of the analytical model and Geant4 calculation results has been done. The comparison of the analytical model calculation results and experimental measurements have been done. Both comparisons show model validity and a capability to be used in the research.

1. Introduction

One of the most important characteristic of the modern gamma-scanning equipment such as PET or SPECT is a spatial resolution which directly depends on detector design. Sizes, geometry, reflection surfaces of scintillation crystal, sizes and material of lightguide and light-sensitive elements location are parameters the right choice of which determines the main characteristics of the whole detector.

It is necessary to vary and select each of these parameters. Using mathematical modeling and then experiment verification is optimal for this problem. A widely used mathematical modeling method is Monte-Carlo method needs a lot of calculation power or a long time for calculation.

Because of this analytical light distribution model in optical system of scintillation detector has been developed (figure 1). The model allows to calculate a number of photons coming from point isotropic light source (scintillation) to square area (silicon photomultiplier as a light detector is considered). Also the model takes into account optical effects (refractions at interfaces, multiply reflections from crystal sides, light absorption in materials), allows to set scintillation crystall size and properties, number and properties of other materials (lightguide, optical glue etc.).

Also model can be used for gamma photoabsorbtion point estimation using, for example, Maximum likelihood estimation method [1].

Such problems have been solved in works [2], [3], but the developed models don't take into account some optical effects. The work [2] hasn't been taken interfaces refraction into account. There is no reflection from side faces contribution in work [3].

2. Assumptions in the model

The analytical model developed has next assumptions:

- Light source is isotropic.
- Light source is point.

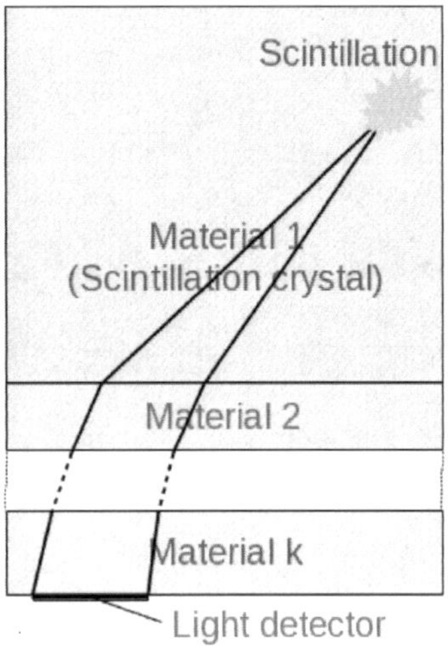

Figure 1. The model scheme.

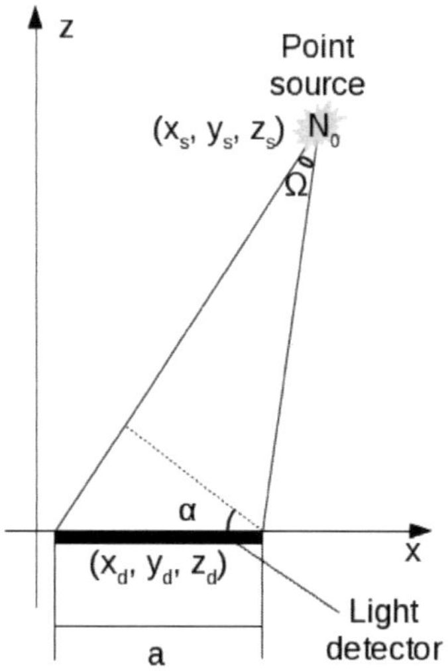

Figure 2. Number of optical photons falling on square area.

- Absolute absorption on the light detector.
- All materials are absolute homogeneous.
- The detector has a rectangle form. Other forms of crystal and lightguides are not considered.
- Multiple reflections only from the upper and side faces of the first material (scintillation crystal) are considered. For other materials only interfaces reflections are considered.

3. A number of optical photons falling on the square area

The base of the model is a function describes a number of optical photons falling on the square area. This function is derived from geometric optics. A point isotropic source which radiate N_0 optical photons and located at the point (x_0, y_0, z_0) is considered (figure 2).

The square area has a linear size a and coordinates (x_d, y_d, z_d). The infinitesimal element of this area with square dS_d is seen from the source at the elementary solid angle.

A number of optical photons falling on infinitesimal element of square area are defined by solid angle $d\Omega$ at which square area dS_d is seen:

$$dN_{point} = \frac{d\Omega \cdot N_0}{4\pi}; \qquad (1)$$

An elementary solid angle can be expressed as follows:

$$d\Omega = \frac{dS_d \cdot \cos\alpha}{R^2} = \frac{z_s \cdot dx_d \cdot dy_d}{\left((z_s - z_d)^2 + (x_s - x_d)^2 + (y_s - y_d)^2\right)^{3/2}}; \qquad (2)$$

A full number of optical photons falling on the detector is whole area integral:

$$N_{point} == \frac{N_0}{4\pi} \cdot \int\limits_{y_d-a/2}^{y_d+a/2} \int\limits_{x_d-a/2}^{x_d+a/2} \frac{z_s \cdot dx_d \cdot dy_d}{\left(z_s - z_d)^2 + (x_s - x_d)^2 + (y_s - y_d)^2\right)^{3/2}}; \tag{3}$$

The integration result:

$$N_{point} = \frac{N_0}{4\pi} \cdot \Big(\arctan\Big(f_1(x_d + a/2, y_d + a/2)\Big) - \arctan\Big(f_1(x_d - a/2, y_d + a/2)\Big) - \\ - \arctan\Big(f_1(x_d + a/2, y_d - a/2)\Big) + + \arctan\Big(f_1(x_d - a/2, y_d - a/2)\Big)\Big); \tag{4}$$

where

$$f_1(x,y) = \frac{(x - x_s) \cdot (y - y_s)}{(z_d - z_s)\sqrt{(z_s - z_d)^2 + (x_s - x_d)^2 + (y_s - y_d)^2}} \tag{5}$$

Thus, knowing the light source (scintillation) coordinates, the square area (light detector) size and coordinates, the flash number of photons, it is possible to calculate how much optical photons fall on the square area using (3). It is assumed that the light source and the detector are inside homogeneous material and its borders don't affect on the light distribution.

4. Interface reflection considering

The scintillation detector is often consist of not only crystal and light detector. There is an optical glue or grease, lightguide (in gamma-cameras Anger type) between the crystall and light detector. So optical effects on the interfaces (refraction and reflection) should be considered in the analitical model.

The reflection on interface between i and $i+1$ material is considered by equation:

$$K_i = \frac{1}{2}\Big[\Big(\frac{\sin(\psi_i - \phi_i)}{\sin(\psi_i + \phi_i)}\Big)^2 + \Big(\frac{\tan(\psi_i - \phi_i)}{\tan(\psi_i + \phi_i)}\Big)^2\Big]; \tag{6}$$

where

- K_i – refletion coefficient on interface between i and $i+1$ material;
- ϕ_i – angle of incidence on interface between i and $i+1$ material;
- psi_i – angle of refraction on interface between i and $i+1$ material.

The reflection on all the interfaces is considered above.

5. Refraction considering

Refraction is considered by more a difficult way. The area which is a light detector image can be distinguished on the first interface (figure 3). The light from source passed through the area falls on the light detector (if the light absorption in the material and reflection on the interfaces are neglected).

If the position and size of the light detector image on the first interface is determined, the refraction on the all other interfaces can be considered. So, it is needed to calculate trajectories (points where ray crosses interfaces) of four rays from the light source to the light detector corners.

To obtain a trajectory of one ray from the light source to some known point of the light detector equations system is composed. Each equation is a refraction law for corresponding interface (figure 4):

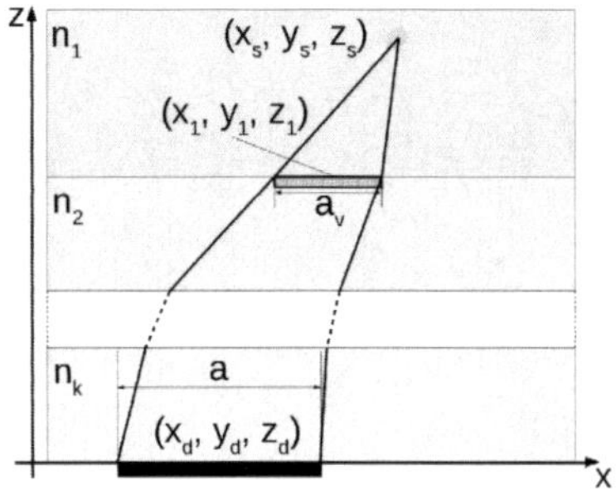

Figure 3. Interface refraction considering.

$$\begin{cases} n_1 \cdot \sin\theta_1 = n_2 \cdot \sin\theta_2; \\ \dots \\ n_i \cdot \sin\theta_i = n_{i+1} \cdot \sin\theta_{i+1}; \\ \dots \\ n_{k-1} \cdot \sin\theta_{k-1} = n_k \cdot \sin\theta_k; \end{cases} \tag{7}$$

Or expressing angles sinus by coordinates:

$$\begin{cases} n_1 \cdot \frac{\sqrt{(x_1-x_s)^2+(y_1-y_s)^2}}{\sqrt{(x_1-x_s)^2+(y_1-y_s)^2+(z_1-z_s)^2}} = n_2 \cdot \frac{\sqrt{(x_2-x_1)^2+(y_2-y_1)^2}}{\sqrt{(x_2-x_1)^2+(y_2-y_1)^2+(z_2-z_1)^2}}; \\ \dots \\ n_i \cdot \frac{\sqrt{(x_{i-1}-x_i)^2+(y_{i-1}-y_i)^2}}{\sqrt{(x_{i-1}-x_i)^2+(y_{i-1}-y_i)^2+(z_{i-1}-z_i)^2}} = n_{i+1} \cdot \frac{\sqrt{(x_i-x_{i+1})^2+(y_i-y_{i+1})^2}}{\sqrt{(x_i-x_{i+1})^2+(y_i-y_{i+1})^2+(z_i-z_{i+1})^2}}; \\ \dots \\ n_{k-1} \cdot \frac{\sqrt{(x_{k-2}-x_{k-1})^2+(y_{k-2}-y_{k-1})^2}}{\sqrt{(x_{k-2}-x_{k-1})^2+(y_{k-2}-y_{k-1})^2+(z_{k-2}-z_{k-1})^2}} = n_k \cdot \frac{\sqrt{(x_{k-1}-x_k)^2+(y_{k-1}-y_k)^2}}{\sqrt{(x_{k-1}-x_k)^2+(y_{k-1}-y_k)^2+(z_{k-1}-z_k)^2}}; \end{cases} \tag{8}$$

In this equation coordinates of the ray light and the interfaces intersections points x_i, y_i $(i = 1 : (k-1))$ are unknown. The point with index k is a known point on the light detector. Coordinates z_i are known because of they are interfaces coordinates, but width of each material is set in the model geometry definition. Thus there are $2(k-1)$ unknown and $k-1$ equations in system (8). The system obtained can't be solved.

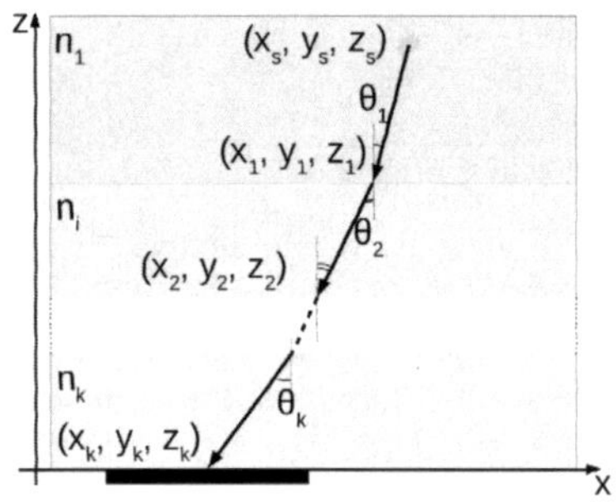

Figure 4. Ray light passing through several different optical materials.

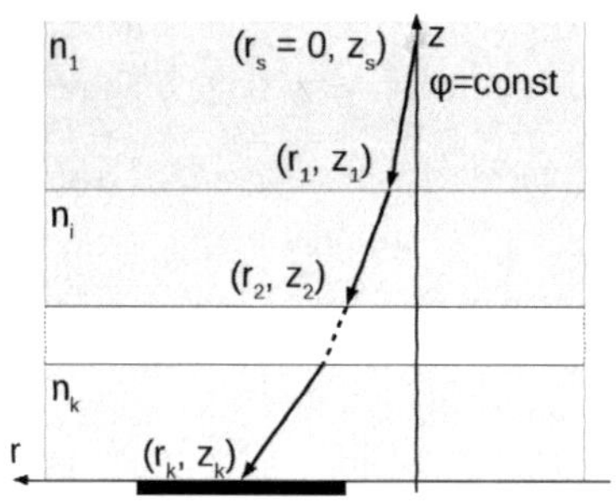

Figure 5. Transition to polar coordinates.

To determine a trajectory number of unknown in the system (8) is needed to be reduced. Transition to polar coordinates has been done for this. And it is useful to choose a new coordinate system in the way when r coordinate of source equals zero. Plane contains a ray trajectory is considered (figure 5). Coordinate ϕ is constant at the plane.

The system of equations (8) in polar coordinates(r, φ, z) has a form:

$$\begin{cases} n_1 \cdot \frac{r_1}{\sqrt{r_1^2+(z_s-z_1)^2}} = n_2 \cdot \frac{r_2-r_1}{\sqrt{(r_2-r_1)^2+(z_2-z_1)^2}} \\ \ldots \\ n_i \cdot \frac{r_i-r_{i-1}}{\sqrt{(r_i-r_{i-1})^2+(z_i-z_{i-1})^2}} = n_{i+1} \cdot \frac{r_{i+1}-r_i}{\sqrt{(r_{i+1}-r_i)^2+(z_{i+1}-z_i)^2}} \\ \ldots \\ n_{k-1} \cdot \frac{r_{k-1}-r_{k-2}}{\sqrt{(r_{k-1}-r_{k-2})^2+(z_{k-1}-z_{k-2})^2}} = n_k \cdot \frac{r_k-r_{k-1}}{\sqrt{(r_k-r_{k-1})^2+(z_k-z_{k-1})^2}} \end{cases} \tag{9}$$

Thus r_i $(i = 1 : (k-1))$ are unknown. A number of unknowns and number of equations are the same now and the system (9) can be solved using a numerical method (for example using Levenberg-Marquardt algorithm). The result of the system (9) solving is points which can be used for the trajectory construction. Also the obtained points are used for determining path length of the light ray inside of each optical environment. It is necessary for considering the light absorption in materials.

As a result, the algorithm of refraction considering looks like:

(i) Consider 4 rays from source to detector corners and for each:
  (a) Transit into the polar coordinate system.
  (b) Find the ray trajectory by solving equations system (9).
  (c) Return back to the rectangular coordinate system.

(ii) Find the center and size of light detector image on the first interface.

(iii) Using the equation for number of optical photons on the square area (3) determine how much photons fall on the detector image. The same number of optical photons (if materials absorption and interfaces reflection are neglected).

6. Considering light absorption in materials

Light absorption in materials is considered by the attenuation law:

$$N_{att.i} = N_i \cdot e^{-\frac{L_i}{\lambda_i}}; \tag{10}$$

where N_i – number of photons which falls on the material; $N_{att.i}$ – number of photons which falls out of the material; L_i – path length in the material; λ_i – the material attenuation length; i – the material number.

The infinitely number of light rays pass through the square area image on each interface. So the assumption that each ray path length is not very different from path length of the ray falling to the center of the light detector is considered. Therefore the model doesn't preciously considers refraction when linear size of detector's optical system is great in comparison with its height.

7. Horizontal multireflections at the side facets of a scintillation crystal considering

Light reflection from the side faces is considered using virtual sources method. Virtual images of source which are mirror reflections from the side facets of scintillation crystal are created. A number of photons irradiated from the virtual sources is calculated from the reflection coefficient of side facets. As mentioned above it is assumed that the reflection coefficient of the side facets is independent from the incidence angle. One reflection is considered in this way.

Virtual sources are reflected from side facets of scitillation crystal to consider multiple reflections (figure 6).

On the figure marked:

- by green – real source position;
- by blue – real souces derived from from 3 multireflelections;
- by red – crystall facets.

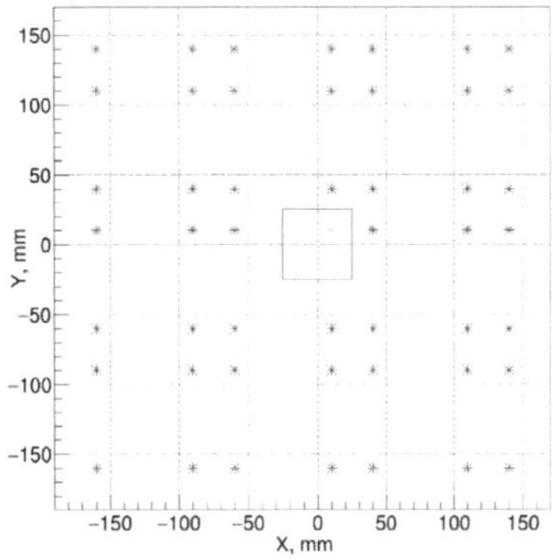

Figure 6. Light reflection from the side facets of the scintillation crystal considering.

A number of optical photons falling on the light detector is calculated using the equation:

$$N_{sq} = \sum_{v=0}^{M} \sum_{h=0}^{M} \left(N_{point}(x_h, y_v, z) \cdot K_{refl.side}^{v+h} \right); \quad (11)$$

where

$$x_h = \begin{cases} x_s, h = 0; \\ 2x_r - x_s, h = 1; \\ 2x_l - x_s, h = 2; \\ 2x_r - x_{h-3}, h \neq 0, 1, h\text{– odd}; \\ 2x_l - x_{h-1}, h \neq 2, h\text{– even}. \end{cases} \quad (12)$$

- N_{sq} – number of optical photons falling on light detector from all sourcesat one plate (without vertical reflections);
- x_r – crystal right facet coordinate;
- x_l – crystal left facet coordinate;
- x_s – real source coordinate;
- M - number of multireflections taken into account.

Values of y_v are similar to (12).

8. Vertical multireflections considering

The plane contains a real light source and virtual sources is obtained while considering horizontal multiple reflections. It is necessary to reflect the obtained plane of sources respecting to the upper and lower crystal facet to a consider vertical (along z axis) and mixed (i.e. from side and upper facet in different variations) multiple reflections (figure 7).

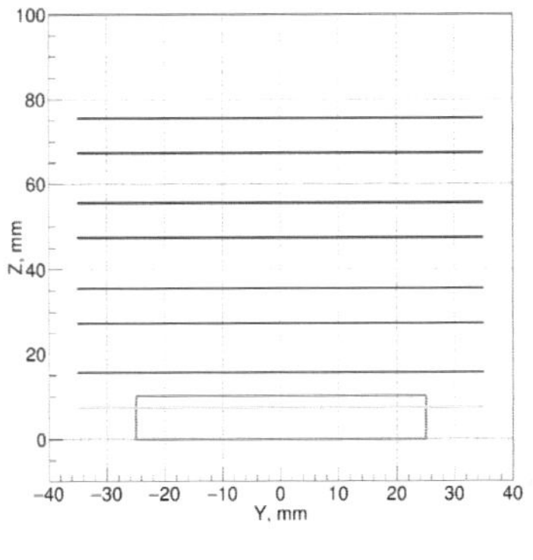

Figure 7. Reflections from the upper and lower crystal facet considering.

It can be described by the equation:

$$N_{total} = \sum_{q=0}^{T} \left[N_{sq}(z + 2qh_{cr}) \cdot K_{refl.up}^{q} + + N_{sq}(2h_{all} + 2qh_{cr} - z) \cdot K_{refl.up}^{(q+1)} \right]; \quad (13)$$

where

- N_{total} – total number of photons which are falling on light detector;
- h_{cr} – crystal height;
- h_{all} – total height of all detector optical system (assuming $z = 0$ is coordinate of lower facet of lower material);
- $K_{refl.up}$ – upper facet reflection coefficient.

Thus several plates are obtained each of which consists of point sources irradiating different number of optical photons. The contribution of each point source can be estimated using a function (3) which describes how many optical photons from the point source fall on the square area.

9. Comparing with Geant4

To verify the analytical model developed comparison of modeling results obtained from Monte-Carlo method using Geant4 [4] and calculation results using the analytical model was done.

The scintillation detector was considered as follows (figure 8): crystal CsI(Tl) (refraction index 1.79) with size 50x50x10 mm, glass with the same length and wodth but 1 mm height under the crystal (refraction index 1.52), optical grease 0.1 mm height (refraction index 1.51), silicon photomultiplier (SiPM) inside epoxy (refraction index 1.54) with 3x3 mm sensitive area. Three different cases of the reflection surfaces were modeled:

- All facets have total light absorption;
- All facets have mirror reflection with 0.95 reflection index;
- Side facets have total light absorption and upper facet has mirror reflection with 0.95 reflection index.

The crystal facet joined to glass always had no covering (reflecting or absorbing).

The point isotropic light source was placed on the different positions along X axis at fixed x, y coordinates ($z = 0$ was corresponded to light detector plane). The distribution of 10 000 000 photons were modeled using Geant4. Several positions with different z coordinate of source were considered.

Dependences of photons number on the detector versus distance between the light source and light detector at $z = 10.3$ mm coordinate of source are shown at figure 9, 10. Curves form of dependency are

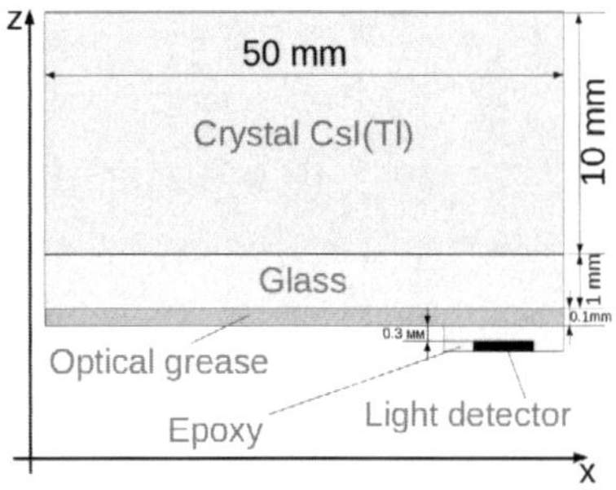

Figure 8. Scheme of scintillation detector modeling.

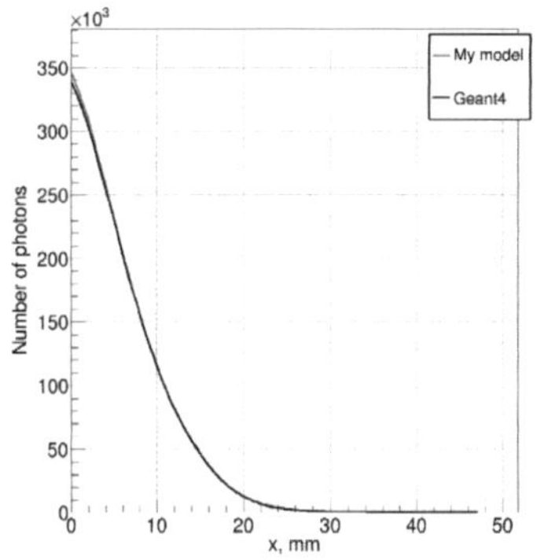

Figure 9. A number of optical photons falling in the light detector versus distance to source. All the crystal facets has mirror reflection with 0.95 index.

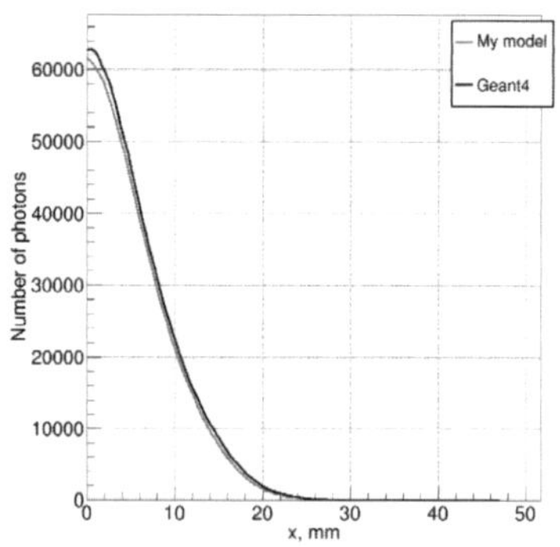

Figure 10. A number of optical photons falling in the light detector versus distance to source. All the crystal facets has total absorption.

Table 1. The maximal difference between Geant4 modeling results (statistics used: 10 000 000 optical photons) and the analytical model developed relatively to Geant4.

z source coordinate, mm	10.3	9.3	8.3	6.3
All the facets have total absorption	6.2%	5.7%	5.5%	11.8%
The upper facet has mirror reflection	4.5%	7.2%	7.2%	9.0%
All the facets have mirror reflection	2.7%	2.8%	1.9%	3.2%

the same for Geant4 and the analytical model. Maximum differences between Geant4 results and the analytical model results are shown at table 1.

The analytical model concurs with Geant4 numerically, The case when all reflecting surfaces have mirror reflectance is most precisely. Computer with CPU Intel Core i7 4770 was used for calculations. The analytical model algorithm was programmed on C++. Geant4 model used 5 thread and the analytical model - only one thread. Besides, the time needed for the calculations for the analytical model was ~ 0.07 seconds. Time needed for the calculations for the same detector using Geant4 was $\sim$ 70 seconds. I.e. speed increase about 1000 times is obtained even without optimization with negligible accuracy losing.

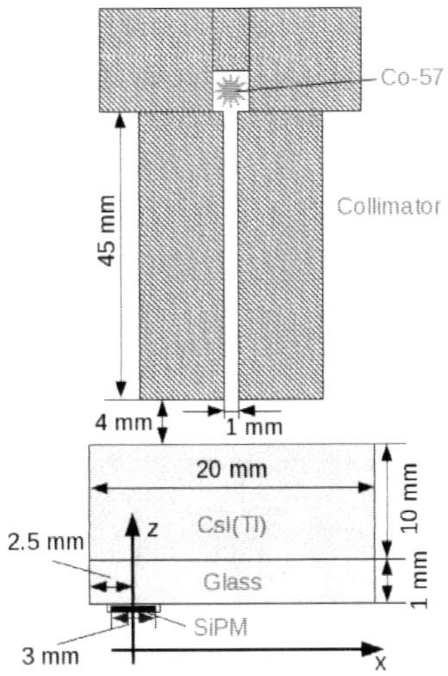

Figure 11. Experiment scheme.

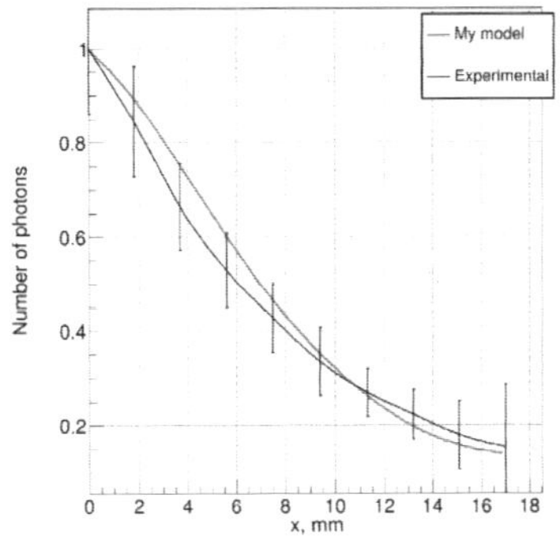

Figure 12. Comparison of experimental and the analytical model results obtained.

10. Comparing to the experiment

The comparison of the analytical model calculations and experimental measurements was done. The experiment scheme is shown at figure 11.

Gamma source Co-57 was placed in the collimator with 1 mm hole diameter and 45 mm height. A scintillation detector was placed at 4 mm distance from collimator. The scintillation detector was consisted of: crystal CsI(Tl) with 20x20x10 mm size. glass with 1 mm height and silicon photomultiplier Sensl MicroFC-30035 [5]. Metal dusting drifted on the upper (nearest to gamma source) facet was provided mirror reflection. The side facets have black covering but in spite of this its reflected light was less than the upper facet. The sensitive area of SiPM was covered by epoxy. The optical grease was placed between the glass and epoxy.

The collimator with source were moved along X axis with fixed coordinate y. Mean SiPM signal amplitude were calculated at each source position.

The analytical model describes optical photons propagation in the optical system of scintillation detector. A photoabsorption point of each gamma was unknown in the experiment, so mean value (based on exponential law of gamma attenuation) of z coordinate 9.3 mm were used. Geometry of the scintillation detector were set according to the real detector in the experiment. Other parameters were chosen as follows:

- Crystal refraction index $n_{crystal} = 1.79$;
- Glass refraction index $n_{glass} = 1.52$;
- Optical grease refraction index $n_{opt.grease} = 1.51$;
- Epoxy refraction index $n_{epoxy} = 1.54$;
- Upper facet reflection index $K_{refl.up} = 0.9$;
- Side facets reflection index $K_{refl.side} = 0.5$;

Comparison of the results obtained is shown at figure 12.

Amplitude signal versus source position derived from the analytical model concurs with experimental within error.

11. Conclusion

The analytical model of the light distribution in the scintillation detector optical system has been developed. The model considers optical effects which allow to set a different scintillation detector configuration. The model correlates with Geant4 and allows to reduce the calculation time, Besides, the model developed agrees with the experiment within error.

In the future it is possible to improve model's accuracy, to add diffuse reflection facets covering.

Acknowledgments

This work was supported by Competitiveness Program of National Research Nuclear University MEPhI.

References

[1] H. H. Barrett, C. J. H William 2009 Maximum-likelihood methods for processing signals from gamma-ray detectors *IEEE Transactions on Nuclear Science*, **56** 72535

[2] Zhi Li, G. Vandersteen, P. Bruyndonckx 2009 3D nonlinear least squares position estimation in a monolithic scintillator block, *IEEE Nuclear science symposium conference record*

[3] M. Galasso, C. Borrazzo, A. Fabbri 2015 A scintillation light radial distribution model for monolithic crystal gamma cameras *Nuclear instruments and methods in physics research A*, **786** 40-6.

[4] Geant4 official web page, https://geant4.web.cern.ch/geant4/

[5] C-Series Datasheet, Sensl, http://sensl.com/downloads/ds/DS-MicroCseries.pdf

Metal-carbon nanoclusters for SERS

A Kucherik, S Kutrovskaya, A Osipov, I Skryabin and S Arakelian

Stoletovs Vladimir State University, 87 Gorky st, Vladimir 600000, Russia

E-mail: kucherik@vlsu.ru

Abstract Metal–carbon materials exhibiting surface-enhanced Raman scattering have been synthesized by laser irradiation of colloidal systems consisting of carbon and noble metal nanoparticles. The dependence of the Raman scattering intensity on the material composition has been investigated. The possibility of detecting of the Raman spectrum of organic due DCM and rhodamine 6G is demonstrated.

1. Introduction

The formation of new photonic materials is a promising area in laser physics. Bimetallic complexes of noble metals with particle sizes from 1 to 100 nm in carbon matrix are widely used as materials for implementing surface enhanced Raman scattering (SERS) [1]. The application of carbon as a matrix, stabilizing metal nanoparticles and preventing their degradation in environment, is interesting and promising [2]. The metal-carbon materials based on carbyne are of particular interest. Carbyne is a carbon polymer material [3]. The distance between neighboring atoms in a carbyne chain is 0.128 nm and the distance between chains is 0.295 nm; these values are much smaller than the distances between atoms and between layers in graphite: 0.142 and 0.335 nm, respectively. Carbyne has a unique of the optical and electrical properties [4].

In this paper, we report the results of studying the formation of metal-carbyne under laser irradiation of colloidal systems, consisting of carbon and noble-metal nanoparticles. It is shown that this irradiation leads to the formation of clusters, in which metal nanoparticles are interrelated by carbon chains. The Raman spectra of these structures exhibit SERS.

2. Laser-induced carbon chains

We presented a new experimental realization by the two-stage laser-induced synthesis of long linear carbon chain – LLCC [5]. In our experiments, the proposed method is based on the laser irradiation of colloidal systems composed of nanoparticles shungite. The substance is a mixture of various carbon allotropes: natural origin of fullerenes and carbon nanostructured forms whose small lattices are connected by amorphous carbon). This a two stage technique boils down to the following: (1) a colloidal system prepared beforehand by laser ablation from target shungite in distilled water using YAG:Nd^{3+}- laser (1.06 μm) with a pulse duration of 1 ms and a pulse energy of up to 10 J; (2) next, such colloidal system is placed under Yb-fiber laser (1.06 μm) with a pulse duration of 100 ns and a pulse energy of up to 1mJ, and the process results in the fabrication of LLCC in liquid. The mechanism for that is due to the homogeneous symmetry interaction in liquid for the carbon system that stabilizes the linear structure [6].

On the other hand, when exposed to fluids, the stabilization of long linear carbon chains (carbyne allotropic form) can be achieved by laser ablation in the presence of, e.g. gold nanoparticles, which leads to the consolidation of the ends of linear chains on the surface of gold particles, and prevents them from instabilities/destruction: twisting further in to tangles [7].

The Raman spectra of both the original colloidal systems and systems obtained by us after irradiation by nanosecond laser radiation intensity 10^6 W/cm^2 (stage 2) are shown in Figure 1 (with the laser exposure time from 5 to 15 minutes).

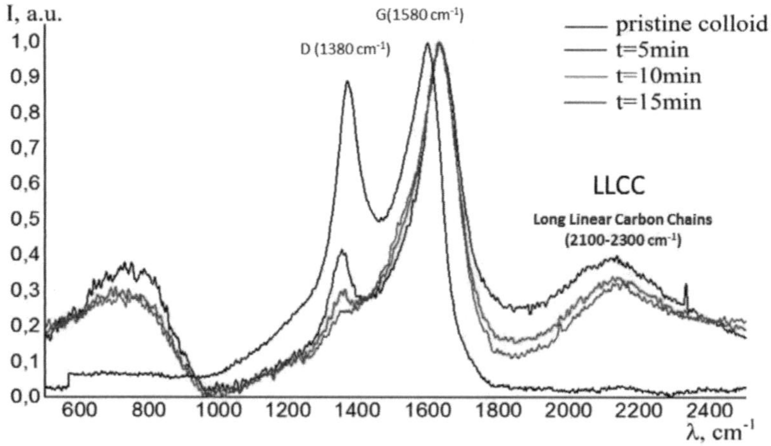

Figure 1. Raman spectra of synthesized substance with different electron orbital configurations in the experiment with shungite (for irradiated colloidal system).

We have the following specific characteristics for spectral bands: graphite G(1580 cm^{-1}) and disorder D(1380 cm^{-1}) lines for carbon substance; 2100-2300cm^{-1} – polyene chain with the carbon chain (–C≡C–)n; 1900-2200cm^{-1} – commulene connection with the carbon chain (=C=C=)n; 600-900cm^{-1} – probably, mechanical stress, deformation and distortion of linear carbon chains.

The next approach, is to stabilize the linear chain of carbon by external forces from metallic particles Au and Ag doped in the colloidal system.

3. Laser synthesis of metal-carbyne material

The carbon, gold and silver colloidal solutions were prepared by the intense mixing with changing mass ratio in water, then the ultrasonic bath was used for about 10 minutes for particle decoagulation (see Figure 2, where we show the results for different concentration in composition C:Au:Ag).

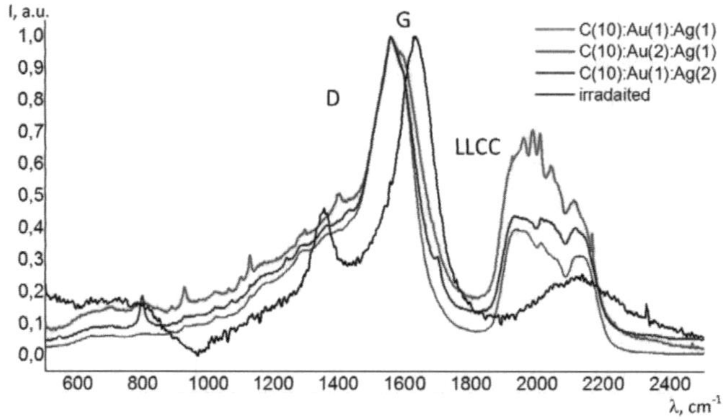

Figure 2. The Raman spectra of colloidal systems with variable composition in concentration of the mixture (C:Au:Ag) system. Laser energy W=5J; irradiation time T=5 min.

After the irradiation in similar condition [8] we have detected the peaks in the area of 1900-2200 cm^{-1}. This area corresponds to carbyne spectra [6-9]. The D-peak for all samples was reduced and forming of the LLCC-peak [7, 10] was observed. Cumulene structure (= C = C = C =) is shown by the absorption lines 1950 $cm^{-1}$and 1070 cm^{-1}. Polyyne structure of carbon atomic chains is shown by the lines of the of stretching vibrations of triple bonds C ≡ C in area of 2100-2300 cm^{-1}, and deformation oscillations at 800 cm^{-1}. The chain length change results in the shift of the line in shortwave area during the irradiation of colloidal system. The absorption line in the area of 1600 cm^{-1} can be compared with fundamental absorption line for the cumulene state of carbine (regular zigzags with straight parts consist of 4 carbon atoms). This kind of structures can be also modified under the laser radiation. According to the [11] the formation of carbon linear chains can result in the sp^3-hybridization structure formation (example: the carbine chain binds two fullerenes). In this case, the irradiation of the metal-carbon colloidal system can result in the obtaining of the more complicated C:Au:Ag systems (as in analogy [6]). The variation of spectra in area of 1900-2200 cm^{-1} corresponds to formation of the carbon chains with different length. There is a dramatic spectra modification. The change of the metal concentration results in the variation of the spectral lines.

4. Research of SERS

Thin metal-carbon films were formed on the surface of the glass substrate using a drop deposition technique [12]. Small droplets deposited on the surface of the substrate at room temperature using a glass capillary with an inner diameter of 50 μm. Evaporation drops occurred under normal conditions.

The SERS research by deposited films was performed using Senterra spectrometer (Bruker), with the pump laser wavelength of 532 nm, the power of 0.1 mW and the focal spot diameter of 2 microns.

In the first experiments we determined a limit of sensitivity of metal-carbon films. For that we deposited the solution of dye DCM with different concentration on the surface of metal-carbon films. The Raman investigations (Figure 3) are demonstrated, that limit sensitivity of our material is 10^{-6}M.

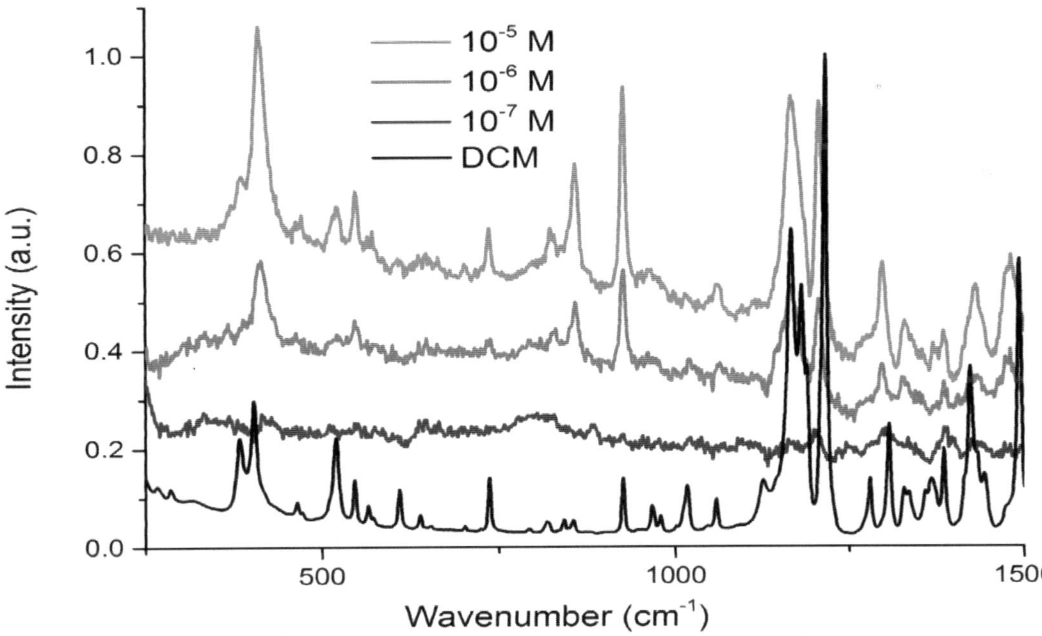

Figure 3. Raman spectra of due DCM with various concentration.

In the next experiments the standard dye Rhodamine 6G was used as a test molecule. The dye solution in ethanol (10^{-6}M) was placed on a metal-carbon structures using a micropipette. The metal-carbon surfaces are formed on an oxide glass substrates with different composition of metal nanoparticle. The Raman spectra of molecules of Rhodamine 6G on various substrates are shown on Figure 4.

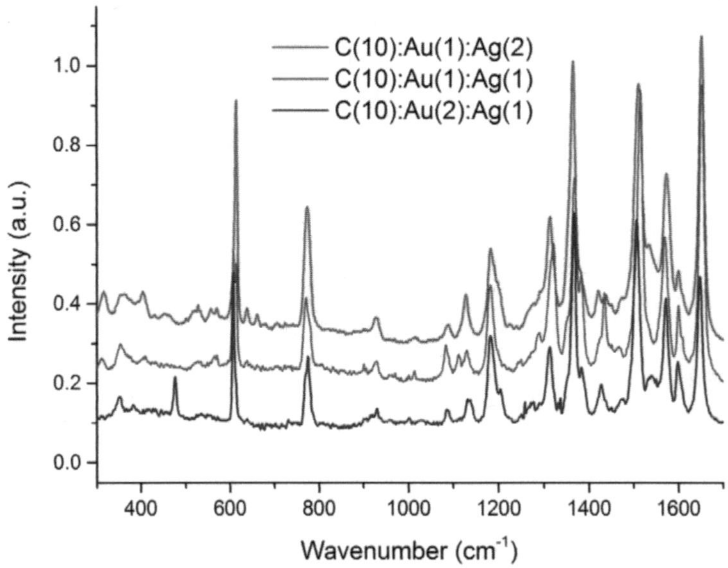

Figure 4. Raman spectra due Rhodamine 6G with variable composition of metal-carbon films.

The use of films with variable composition as a substrate of metal-carbon nanostructures under the same measurement conditions allow to detect and identify the dye on the Raman spectra with sureness in Figure 4. The changing of the gold and silver nanoparticles ratio results in varying of the amplification degree of different bands in the spectrum of the dye. The peaks of plasmon resonances for silver and gold are in the range of 410 and 540 nm, thus, various vibration modes are amplified at gold and silver particles with different gain.

5. Conclusion

The results of the experiments of laser metal-carbon cluster and complex synthesis are shown in this work. The laser with nanosecond pulse duration allows to modify the structure of carbon bonds and makes best conditions for obtaining of linear carbon chains with different hybridization. The obtained structures are planned to be used for the registration of the effect of the SERS with the possibility of the sensitivity control in different areas of the spectra because of the changing of the initial component concentration and morphology.

Raman and UV/VIS absorption spectra were measured at Center for Optical and Laser Materials Research, St. Petersburg State University.

The reported study was also supported by the Ministry of Education and Science of the Russian Federation (state project no. 2014/13), RFBR grants № 16-42-330531, №14-02-97506

References

[1] Manshina A. A., Grachova E. V., Povolotskiy A. V. et al 2015 *Sci. Rep.* **5**
[2] Zherebtsov D. A., Galimov D. M., Lashkul A. V. et al 2011 *Inorg. Mater.: Appl. Res.* **2** (5) 524
[3] Korshak V. V., Kasatochkin V. I., Sladkov A. M. et al 1961 *Dokl. Akad. Nauk SSSR* **136**, 1342
[4] Akagi K., Nishiguchi M., Shirakawa H. et al 1987 *Synth. Met.* **17** 557

[5] Kucherik, A.O., Arakelian, S.M., Garnov S.V. et al 2016 *Quantum electron*, **46**(7), pp. 627–633
[6] Pan B., Xiao J., Li J., Liu P., Wang C., Yang G. 2015 *Science Advances*. **1**(9) e1500857
[7] Shi L., Rohringer P., Suenaga K. et al. 2016 *Nature Materials* (2016) published online
[8] Kucherik A., Arakelian S., Vartanyan T. et al 2016 *Optics and Spectroscopy* **121**(2) 263–270
[9] Ravagnan L., Siviero F., Lenardi C. et al 2002 *Phys. Rev. Lett.* **89**(28) 285506
[10] Shi L., Sheng L., Yu L., An K., Ando Y,, Zhao X. 2011 *Nano Research* **4**, 759
[11] Ravagnan L., Manini N., Cinquanta E. et al 2009 *Phys. Rev. Lett.* **102** 245502
[12] Arakelyan S. M., Veiko V. P., Kutrovskaya S. V. et al 2016 J Nanopart Res **18** 155

Risk assessment for human health and terrestrial ecosystem under chronic radioactive pollution near regional radioactive waste storage

G V Lavrentyeva[1,2], M N Katkova[3], R R Shoshina[2] and B I Synzynys[2]

[1] National Research Nuclear University MEPhI, Obninsk, Russia
[2] Bauman Moscow State Technical University (Kaluga Branch), Kaluga, Russia
[3] RPA "Typhoon" of Roshydromet, Obninsk, Russia

E-mail: Lavrentyeva_G@list.ru

Abstract. An impact of the radioactive waste storage facility at the regional population was assessed under supervision of IAEA. It was made in accordance with the methodology for assessment of doses and risks to human storage using different scenarios of radionuclides releases into the environment. The following scenarios were considered: leakage of fluid, resuspension of dust, fire, flooding. Thy evaluation of radiation doses received and the risks to the human showed that the risk has been acceptable for all scenarios. An approach for an ecological risk assessment for terrestrial ecosystem is presented as five modules: selection of the ecosystem-receptor of radiation effects; determination of reference species of living organisms and their survival indices; the critical load as an absorbed dose rate is calculated from the dependence between the absorbed Sr-90 radiation dose rate and the coefficient of radioactive strontium accumulation in mollusc shells; the critical dose; risk is assessed from a part of the ecosystem territory with increased mollusc loading; uncertainties appeared at each stage of risk assessment are characterized. The risk of exposure to the repository on the ecosystem should be characterized as unacceptable.

1. Introduction

Much attention is being given to optimal human-environment relationships under the conditions of ever-increasing anthropogenic impact on the biosphere.

Thus, the developed sanitary-hygienic principles of rationing are based on an anthropocentric approach which considers a human impact of technogenic factors and is aimed at limiting pollutant intake. The anthropocentric principle of rationing was formulated by the International Commission for Radiological Protection (ICRP) in 1970 and confirmed in 1990 [1, 2]. However in 2007 ICRP [3] had stated that in developing the strategy of environmental radiation safety, it is necessary to give the evidence for protecting not only people and but also living organisms; it reflects the ecocentric principle of radiation factor rationing. Therefore many recent papers have developed the ecological approaches to rationing of radiation effects on ecosystems [4-6].

In this paper are presented the results of the risk assessment for human health and for terrestrial ecosystem under chronic radioactive pollution near regional radioactive waste storage. The territory near a regional radioactive waste storage created in Kaluga region has been chosen as a study area. This object had been put into operation in the 50-70s of the last century. Now the object is a closure.

In 1998-1999 the increased specific activity of Sr-90 had been revealed in the surveying wells and in the following this was stipulated by surface water penetration in one of the facility. At reservoir overfilling, Sr-90 and Cs-137 radionuclide escaped outside. Multi-year studies have showed that Sr-90 contributes much to radioactive contamination on the territory near the storage and in the adjacent territory [7].

2. Materials and methods

2.1. Risk assessment for human health

An impact of the radioactive waste storage facility at the regional population was assessed under [2], supervision of IAEA [8].

It was made in according with the methodology for assessment of doses and risks to the human from the effects of storage using different scenarios of radionuclides into the environment. The following scenarios were considered: leakage of fluid, resuspension of dust, fire, flooding. Calculations were carried out with the mathematical models describing selected script.

Radiation risk assessment carried out by calculation of radiation exposure for the population at different ways of radionuclides in the human body. The integral radiation risk factor of malignancy and hereditary health effects was used for the calculations. The magnitude of the risk factor for a uniform irradiation of the whole body is 0.057 Sv^{-1} [9].

2.2. Risk assessment for terrestrial ecosystem

The ecological risk assessment for terrestrial ecosystem is presented as five modules: 1) selection of the ecosystem-receptor of radiation effects; 2) determination of reference species of living organisms and their survival indices; 3) the critical load as an absorbed dose rate is calculated from the dependence between the absorbed Sr-90 radiation dose rate and the coefficient of radioactive strontium accumulation in mollusc shells; 4) the critical dose; risk is assessed from a part of the ecosystem territory with increased mollusc loading; 5) uncertainties appeared at each stage of risk assessment are characterized.

2.3. Soil sample analysis

Soil samples are collected by a special sampler with a set of Edelman hand soil drills (Eijkelkamp Firm, Netherlands). Soil sample preparation for assessing Sr-90 content had been realized according to the [10].

Soil mass of 1 kg was counter balance weighed and uniformly distributed on a stainless steel griddle in the layer of 20-30 mm in thickness. To dehumidify, the griddle with samples had been kept in a drying box at 110 ^{0}C during 5-7 hours. Dried samples were disintegrated and mixed in a mortar to the grain size of 2-3 mm. Then the mean average sample (mass 50-300 g) of air-dried soil was analytical balance weighed. The sample had been annealed in a muffle furnace at 700 – 750 ^{0}C to transform the basic mass into oxide. Strontium carrier solution (50 mg of metal) and 5 M HNO_3 of 200-300 cm3 in volume were added in a glass with annealed soil. The soil/solution mixture was boiled in a sand-bath during half an hour. A solution was separated from a solid phase by filtration. Solid soil was twice treated by boiling in 4-5 M HNO_3 during half an hour. After filtration the solid soil sediment had been washed three times with distilled water in a filter. After filtration the solutions were combined and evaporated to 100 cm^3.

A version of oxalate precipitation was used to separate strontium from other radionuclides and chemical elements and to concentrate it. Ethanedioic acid solution (95 g/l $H_4C_2O_2$) in the quantity of about 100 cm3 was added to the solution obtained after desalinization and evaporation. pH=4 in a solution was set by deacidification with an aqua ammonium solution. The solution with an oxalate precipitate was stayed for 4 hours, further a precipitate was elutriated and washed twice with hot distilled water in a filter. A precipitate and a filter were dried and then annealed at 750^{0}C in a muffle furnace during an hour. Hereafter a precipitate was put into a glass and dissolved in 3-4 M HNO_3, the solution acidity was set at 3.0 M/l (the solution volume is preferably made up to 5-10 cm^3). When the accompanying components were separated, Sr-90 was cleaned by extraction. Nitric-acid solution was put in a separating funnel of 50-100 cm^3, poured with chloroform in a 2:1 ratio of the aqueous and organic phases, shaked up and phased. After phase separation, the aqueous solution was added by 5-10 cm3 dicyclohexane-18-crown-6 in 0.2 M of a chloroform solution and extracted by shaking up during 30 s. The organic phase was washed with two portions of 2M HNO_3 solution in a 1:1 ratio; the aqueous phase was rejected. The extract (organic phase) or its aliquot part was put on a target and dried under an infrared lamp. Dry matter β-activity on the target was measured.

2.4. Plant sample analysis
Plant – great nettle (*Urtíca dióica*) - samples were collected at the same sites (Figure 1). Before analysis, fresh crude plant samples were washed from dust and soil first with tap water, then with distilled water. Then the plant mass was dried up in desiccators at 80 – 85^{0}C. A dried plant sample was milled. The optimal size of plant fragments had to be about 1-5 mm. Plant mineralization was realized by dry ashing. After digestion a sample had undergone acid treatment according to the above procedure [10].

2.5. Mollusc sample analysis
Terrestrial molluscs *Bradybaena fruticum* were collected from plants (great nettle (*Urtíca dióica*)) and soil under plants at the tested sites. At least 25 specimens were chosen in each mollusc sample. First soft tissues were removed from shells. Shells were grinded in a mortar, hereafter the direct Sr-90 activity measurements were performed. Samples had been incinerated in a muffle furnace at 450^{0}C during 6 hours. Radiochemical separation and Sr-90 content assessment were realized according to the above procedure.

2.6. Calculation of annual exposure doses for molluscs and statistical data
The annual doses of external β-exposure of mollusc tissues due to 90Sr and it daughter 90Y radionuclide contained in shells were calculated with software package ERICA Tool.

Statistical data processing was realized with R software programs [11].

3. Results and discussion

3.1. *Risk assessment for human health*
Assessment of the impact of regional radioactive waste storage facility was carried out taking into account the environmental monitoring data, including specific activities of artificial radionuclides in soil, vegetation, groundwater and surface waters [7].

It was used different scenarios of radionuclides into the environment in according with the methodology for assessment of doses and risks to the public from the effects of storage.

In the analysis of the data taken into account the recommendations of IAEA, which indicated that the radioactive waste storage facility should be designed so that the average dose to the critical group does not exceed the dose limit of 0.3 mSv/year, which corresponds to the level of risk of 10^{-5} per year [12].

The results obtained by calculation of total radiation doze from multiple paths of radionuclides in the human body for scenario «leakage of fluid» have shown the following. Total radiation dose from multiple paths of radionuclides in the human body is well below the regulated limit dose and up to Cs-137 – 1.20E-08 Sv/year and for Sr-90 – 4.66E-08 Sv/year.The main contribution to the dose for the population brings the oral route of radionuclides into the body with food.

Cs-137 is dose-forming radionuclide for the scenario "resuspension of dust" with the storage trench type. Inhalation and food intake are the main routes of public exposure. Radiation doze for the population is much less than 0.3 mSv/year: for Cs-137 is 5.8E-13 Sv/year, for Sr-90 is 1.7E-13 Sv/year.

Sr-90 is dose-forming radionuclide and the oral route is major public exposure for the scenario "flooding".

In the scenario "fire" in a trench type storage total radiation dose is determined by the radionuclide Sr-90 (75%). The inhalation and food intake are the main pathways of exposure. The total dose was 2.7E-12 Sv/year.

Radiation risk of occurrence of genotoxic effects for public health are significantly lower than the acceptable risk to the public (10^{-6}) (table 1).

Table 1. Risk assessment for all scenarios.

Scenario	Risk of Cs-137 irradiation	Risk of Sr-90 irradiation	Total risk
Leakage of fluid	6.9E-10	2.7E-09	3.39E-09
Resuspension of dust	3.3E-14	9.7E-15	4.27E-14
Flooding	2.8E-20	1.9E-16	1.90E-16
Fire	3.5E-14	1.2E-13	1.55E-13

It is worth noting that the greatest interest is the scenario of «leakage of fluid», since it is such a scenario corresponds to an accident situation in the storage. In this case the total risk from multiple ways of intake of radionuclides was for Cs-137 is 6.9E-10, and for Sr-90 is 2.7E-09. The main contribution to the overall risk for the population brings a way of intake of radionuclides with food.

3.2. *Risk assessment for terrestrial ecosystem*

Radiation and chemical monitoring of the storage territory has been carried out to determine the degree of biotope pollution. The specific activity of natural (Ra-226, Th-232, K-40) and technogenic (Sr-90, Cs-137) radionuclides in soil samples collected in the storage territory and cross-border regions has shown that artificial radionuclides make a major contribution to the radioecological situation. In addition, the specific activity of natural radionuclides is compared with the background ones, therefore their contribution to radioactive pollution is insignificant.

The concentration of heavy metals Mn, Zn, Ni, Cu, Cr, Co, Cd, Pb, Fe in the samples studied does not exceed the value of maximum allowable concentration, therefore chemical pollution is of no concern in terms of priority pollution.

Identification of reference species and indices most properly reflecting the degree and the character of pollution in a studied territory is the next stage to assess a biotope of the radioactive waste storage. The terrestrial mollusc (*Bradybaena fruticum*) has been chosen as a reference species; it turned out to be rather sensitive to radioactive pollution in the territory studied. It should be noted that freshwater and terrestrial molluscs are being profitably employed as test-objects of ecosystem Sr-90 pollution for a long time. It is shown that practically the whole Sr-90 content is concentrated in mollusc shells and its activity is governed primarily by the level of territory pollution and does not undergo systematic long-term, seasonal and taxonomical changes. It is also worthy of note that the choice of similar reference species is supported by other researchers [13-15]. Thus, it is assumed that terrestrial molluscs and their response (lowing Sr-90 accumulation in shells) should be considered as reference species for the terrestrial ecosystem polluted by radionuclides.

Critical radionuclide loads on the reference species and indices have to be determined for assessing the ecological risk. In this case by the critical load we shall basically mean such radionuclide ingress into the ecosystem which does not cause irreversible alterations in its biogechemical structure, biodiversity and productivity during a long period. A method has been applied to reveal the critical loads by plotting "dose-effect" dependencies and to analyze them [16]. To assess the storage impact on terrestrial molluscs from experimental data, the dose dependences with respect to a load gradient have been plotted. In this case the maximum permissible load is the critical point on a dose-effect curve which connects the input loads and variations in the coefficient of radioactive strontium accumulation in shells.

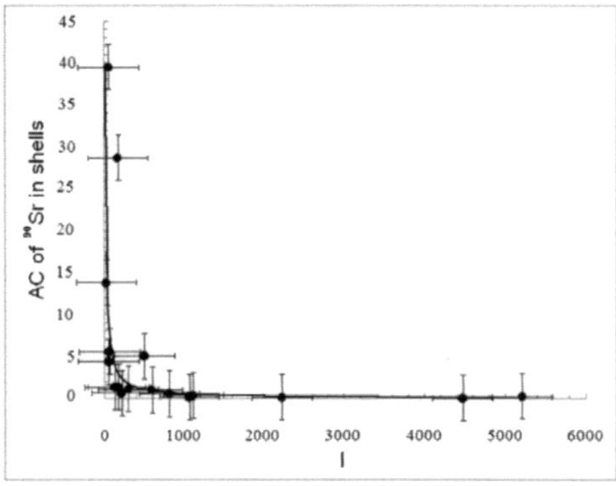

Figure 1. Accumulation coefficients of Sr-90 of mollusc shells versus radioactivity aggregation index.

The dose dependence has been analyzed to find the site with such an aggregation index value at which a 10-fold decrease in the accumulation coefficient of Sr-90 (critical load) takes place; this analysis allowed one to obtain the critical aggregation index value (18.3) corresponding to the critical coefficient value (3.99) of radionuclide accumulation (in this case, the higher the AC values, the better is for ecosystem). The excess critical value of Sr-90 accumulation coefficient causes alterations in the life-sustaining activity of molluscs, decreased activity of their nutrition and, hence, changes in the need for shell building materials and Sr-90 is proved to be one of them (Figure1).

To characterize the ecological risk from critical loads, GIS maps for the ecosystem studied have been plotted; these maps are used for determining excess critical loads. A space satellite (Google Company) image with a co-ordinate reference has been used as the GIS basis. All data on the ecosystem load as well the critical load are plotted on the digitized map using R program [R development, 2010]. The total territory area according to chosen co-ordinates is 0.54 ha. A part (%) of the area with excess critical loads is recognized by the software. A 95% ecosystem protection when the area with excess critical loads does not exceed 5% of the total value (0.54 ha) is taken as the permissible adverse storage impact on the environment [16]. The analyzed plotted GIS maps show that in a studied territory the area characterized by excess the critical AC is 61% (that exceeds the permissible value (5%)). So, the risk of waste storage effects on a surrounding terrestrial ecosystem should be characterized as inadmissible that highlights the unstable state in the territory studied within the next few years.

When assessing the ecological risk, some uncertainties have to be considered in such investigations. First, interspecies extrapolation, i.e. data for radionuclide effect on individual reference species have been extrapolated to the whole ecosystem. Second, in the critical load assessment the form of radionuclides found in soil has not been taken into account at the stage of data collection. Third, the process of radionuclide return into soil as a result of plant die-away and mollusc loss has not been considered.

4. Conclusions

Risk assessment for the population and the ecosystem allows us to conclude the following.

Radiation doses and risks for the population, or habitually resident in the territory adjacent to the waste repository show that the radioactive storage is "relatively" safe target for the population.

Hence, the environmental impact of this storage should be characterized as inadmissible and implies unstable conditions and the expected loss of current biocenosis in the studied territory in the near future.

Acknowledgments
This research was conducted with financial support from the Russian Foundation for Basic Research (Agreement No. MK-15-38-20142 \ 15, No. MK-15-38-20142 \ 16).

References

[1] ICRP Publication 26. The 1977 Recommendations of the International Commission on Radiological Protection. Ann.ICRP1977. V.1, No.3. p.1-53.
[2] ICRP Publication 60. The 1990 Recommendations of the international Commission on Radiological Protection. Ann.ICRP 1991.V.21. p.1-201.
[3] ICRP Publication 103. The 2007 Recommendations of the International Commission on Radiological Protection. Ann ICRP. 2007. V.37. No. 2-4, p. 1- 332.
[4] Alexakhin, R., Fesenko, S. 2004 Radiation biology. Radiology. V.44, No.1, p.93-103.
[5] Ecological Risk Assessment 2007 \ edited by Glenn W. Suter II. – 2nd edd.(ISBN-10: 1-56670-634-3).
[6] Biya E A, Kazakov S V, Linge I I 2003 Development of ecological approaches to regulating radiation effects on aquatic ecosystems. IBRAE preprint. M.: RAS Institute of safe development problems of nuclear power engineering, 24 pp.
[7] Lavrentyeva G V 2014 Journal of Environmental Radioactivity № 135, pp. 128-134
[8] IAEA-TECDOC-1380. 2003 Derivation of activity limits for the disposal of radioactive waste in near surface disposal facilities, 145 p.
[9] RSS: Radiation Safety Standards. RSS - 99/2009, 2009. Moscow. (in Russian).
[10] Instruction : Procedure of radiochemical strontium-90 assessment in soil and plant samples, 1994. Instruction of Rosleskhoz of Russia from 05.09.1994 № 192. Moscow (in Russian).
[11] www.r-project.org
[12] IAEA – International Atomic Energy Agency. The Principles of Radioactive Waste Management, Safety Series No.111-F, IAEA, Vienna, 1995.
[13] Gudkov D I, Derevets V V, Zub L N 2005 Rodioecology, Radiation biologyV.45, No.3, p 271-280.
[14] Frantsevich L I, Panjkov I V, Ermakov A A 1995 Ecology p. 57-62.
[15] Lavrentyeva G V, Mirzeabasov O A, Synzynys B I 2014 Int. J. Environ. Res., 8(4): p 961-970
[16] Vorobejchik E L 2004 Ecological rationing of toxic loads on terrestrial ecosystems: Doctor's thesis Ekaterinburg 362 pp.

Ultrasmall Quantum Dots: A Tool for in Vitro and in Vivo Fluorescence Imaging

Pavel Linkov[1,2,3], Kirill V Vokhmintcev[1], Pavel S Samokhvalov[1], Igor Nabiev[1,3]

[1]National Research Nuclear University MEPhI (Moscow Engineering Physics Institute), 115409 Moscow, Russian Federation
[2]Department of Chemistry, Moscow State University, 1 Leninskie gory, 119991 Moscow, Russia
[3]University of Reims Champagne-Ardenne, 51100 Reims, France

E-mail: valinkov@gmail.com

Abstract. Fluorescence bioimaging is an increasingly popular approach in biomedical research and diagnosis, where semiconductor nanocrystals or quantum dots (QDs) have proved to be excellent fluorescent labels. The use of ultrasmall QDs in nanoprobes extends the possibilities of bioimaging owing to an enhanced capacity for penetrating through cell membranes. However, the QDs synthesis is accompanied by the rapid growth of nanocrystals in colloidal medium what prevents obtaining sufficiently small QDs prepared by conventional approaches. Here, a one-pot injection technique of QD synthesis in an organic medium, with the reaction terminated at an early crystal growth stage and excess precursors eliminated by gel permeation chromatography, is proposed. This technique yields defect-free cadmium selenide QD cores about 1.5 nm in size emitting at the wavelengths less than 500 nm. Coating of these QDs with epitaxial shells of different compositions ensures a photoluminescence quantum yield approaching 100%. The resultant ultrasmall QDs are promising components of nanoprobes to be used for imaging intracellular and intranuclear events down to the molecular level.

1. Introduction

Quantum dots (QDs) are semiconductor nanocrystals (NCs) several nanometers in size capable of photoluminescence (PL) in a wide spectral range. Their energy structure and characteristics of light absorption and fluorescence strongly depend on the QD size because of the quantum confinement effect [1,2]. Advanced methods of one-pot injection synthesis with one component injected into a hot solution of the other component [3] allow obtaining QDs with controllable sizes and, hence, PL parameters by varying the reaction time, temperature, and/or types of precursors. Core/shell QDs consisting of a fluorescing core (usually, CdSe) and a thin epitaxial inorganic shell of a semiconductor with a higher band gap (usually, CdS or ZnS) emitting in the range from 500 to 650 nm, depending on the core size, are currently the most common type of QDs. The shell enhances charge carrier localization in the core, protects the core from the environment, and suppresses blinking [4,5].

QDs are increasingly widely used in various fields [6,7], including biomedical research and medicine [8,9]. A typical biomedical application of QDs is their use as fluorescent labels, which are advantageous over the routinely used organic dyes due to the unique optical properties and high stability. In this field, the QD shell provides and additional advantage in that it facilitates the attachment of hydrophilic ligands to the QD surface to make the QDs water-soluble. These QDs can serve as components of complex suspension microarrays or nanoprobes, whose PL may be excited in either the one-photon [10] or the two-photon [11,12] modes.

In recent years, the possibility of imaging intracellular events down to the molecular level by means of QDs has raised special interest [13]. In this case, of particular importance are small sizes of QDs and the absence of PL blinking [14]. In addition, ultrasmall CdSe cores fluorescing in the blue spectral region are promising components of next-generation light-emitting diodes. Therefore, the development of ultrasmall QDs with a shell effectively enhancing fluorescence is an urgent task.

The main difficulty with obtaining QD cores smaller than 2 nm [15] is that the NC growth at early stages is so rapid that larger crystals are formed almost instantaneously. To date, methods for obtaining the so-called magic-size clusters with a high thermodynamic stability and very narrow optical transitions [16] in the UV and blue spectral regions have been developed . However, this type of NCs is unsuitable for biomedical applications because attempts at coating them with inorganic shells effectively enhancing the PL quantum yield (QY) and protecting the core have thus far failed. Apparently, the reason is that these clusters are structurally different from common QDs [16], which makes the standard coating techniques ineffective.

In this study, we suggest a new procedures for synthesizing CdSe cores with diameters as small as 1.5 nm and coating them with one or several layers of an inorganic shell to obtain CdSe/ZnS and CdSe/ZnS/CdS/ZnS core/shell QDs with highly uniform sizes and a PL QY as high as 90% fluorescing in the region between 480 and 540 nm.

2. Results

Our procedure, described in detail elsewhere [17], is based on the high-temperature injection synthesis method. In this study, the growth of QD cores was monitored by recording their absorption and emission spectra. The PL QY was measured relative to Coumarin 102 and fluorescein dyes.

Our first objective was to obtain CdSe NCs of strictly controllable size, which was fulfilled by precise control of the reaction time and removal of all unreacted precursors.

First, we measured the kinetics of NC growth by monitoring of the position of the first excitonic transition in the light absorption spectrum, and, by using the relationship between the latter and the NC diameter [18] we plotted an approximated calibration curve. The measurement showed that the reaction had to be terminated within about 45 s to obtain QD cores smaller than 1.5 nm in diameter. However, the commonly used method of terminating the synthesis by cooling the reaction flask with an air flow takes more than 1 min to cool the mixture to a temperature below 100°C, during which time the crystal growth gradually slows down. This leads to size heterogeneity of the resultant NCs.

Therefore, we put the reaction mixture into liquid nitrogen after 45 s of synthesis to stop the reaction instantaneously. The frozen mixture was collected into excess isopropyl alcohol, and when it warmed to room temperature, the mixture was centrifuged to settle the cores. Then, the cores were purified by dispersing them in hexane and coagulating with acetone. The purification was repeated two times. The absorption spectrum of the resultant CdSe cores had a peak at 401 nm, which, according to Ref. 18, corresponds to a core diameter of 1.5 nm. Their PL spectrum had a peak at 442 nm and a full width at half maximum (FWHM) of 30 nm.

The next problem consisted in that the rapid termination of synthesis leaved considerable amounts of unreacted precursors in the mixture. Cadmium precursors, in the forms of cadmium alkyl phosphonates and carboxylates, cannot be totally removed by the dispersion/coagulation procedure, because they are adsorbed on the NC surface [19], the process being further facilitated by the small size and, hence, large relative surface area of the nanocrystals. Moreover, these remaining precursors are involved in reaction during the subsequent formation of the shell, either causing additional growth of the core or being included in the shell and disturbing its designed stoichiometry.

This problem was addressed by additional purification of the CdSe cores by gel permeation chromatography (GPC) [19] using the SX-1 cross-linked polystyrene beads (BIO-RAD) as the stationary carrier phase. In this method, QD cores rapidly migrate along the stationary phase, while low-molecular-weight precursors are removed as a result of retention in the pores of the carrier [19].

In order to estimate how this additional purification improves the quality of the resultant QDs, we compared the optical characteristics of the core-shell QDs grown with or without GPC-purification of the cores. The results are shown in Table 1 and Figure 1.

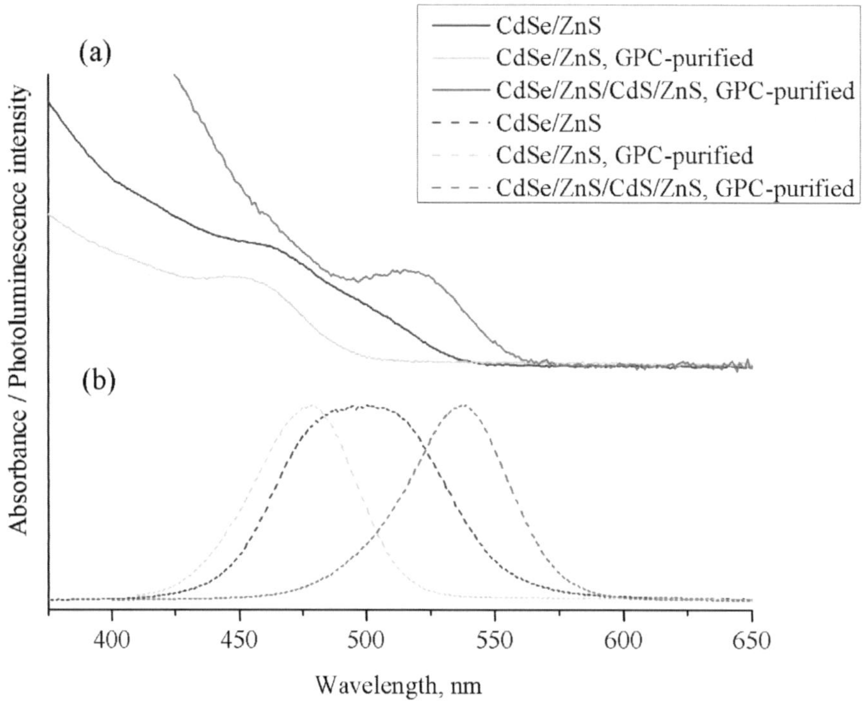

Figure 1. Absorption spectra (a) and fluorescence spectra (b) of quantum dots with different shell structures and procedures for nanoparticles isolation and purification.

The CdSe cores purified by the dispersion/coagulation method alone (Sample 1 in Table 1) or dispersion/coagulation followed by GPC (Sample 2) were coated with three monolayers of ZnS. We used the method of successive ion layer adsorption and reaction (SILAR) [20,21], because it allows layer-by-layer coating with a predetermined number of layers. Some GPC-purified cores were coated with a multicomponent (ZnS/CdS/ZnS) shell three monolayers in thickness (Sample 3). This type shell was earlier shown [17] to ensure an enhanced charge localization, thereby substantially increasing the PL QY. The final size of the CdSe/ZnS and CdSe/ZnS/CdS/ZnS QDs was less than 4 nm.

The incomplete removal of cadmium precursors from the solution of CdSe cores led to the formation of an inner CdS layer or a layer of a more complex and varying composition ($Cd_xZn_{1-x}S$) in the shells of QDs in Sample 1. As can be seen from Table 1 and Figure 1, this has "blurred" the excitonic transition due to the formation of several types of QDs, which is expressed in a wide FWHM of both absorbance and PL peaks. Sample 1 is also characterized by a 56-nm red shift of the PL peak. CdSe/ZnS QDs in Sample 2 are considerably more homogeneous, as evidenced by their more than 1.5-fold narrower PL band. In addition, the red shift of the PL peak is also smaller (37 nm). Thus, the additional GPC purification of CdSe cores has substantially improved the optical parameters of the resultant QDs.

As expected, CdSe/ZnS/CdS/ZnS QDs in Sample 3, which also are based on GPC-purified cores, were as homogeneous as Sample 2, as judged by the FWHM of their PL spectrum (Table 1). The multicomponent shell has ensured an improved charge carrier localization in the core and their better protection from the environment, which has resulted in a higher PL QY (90%). The large red shift of the PL band of these QDs (79 nm) is explained by the complex energy band structure of their multicomponent shell [3,17].

Table 1. Optical characteristic of quantum dots with different shell structures and procedures for nanoparticles isolation and purification.

No.	Sample	Absorbance peak, nm / PL peak, nm / FWHM, nm	PL QY, %
1	CdSe/ZnS	- / 498 / 73	63 (relative to Coumarin 102)
2	CdSe/ZnS, GPC-purified	449 /479 / 46	68 (relative to Coumarin 102)
3	CdSe/ZnS/CdS/ZnS, GPC-purified	517 / 537 / 46	90 (relative to fluorescein)

Thus, the experimental results confirmed the efficiency of our new procedure for the synthesis of ultrasmall CdSe cores with rapid termination of the reaction and additional GPC purification of the cores. We plan to use the CdSe/ZnS and CdSe/ZnS/CdS/ZnS QDs designed in this study as fluorescent labels in next-generation nanoprobes for *in vivo* and *in vitro* bioimaging [22].

Acknowledgements
This study was supported by the Ministry of Education and Science of the Russian Federation, State Contract no. 4.624.2014/K. We thank Mr. Vladimir Ushakov for the help in preparation of the manuscript.
This work was supported by Competitiveness Program of National Research Nuclear University MEPhI.

References
[1] Klimov V, Ivanov S, Nanda J, Achermann M, Bezel I, McGuire J and Piryatinski A 2007 Nat. 447 441-6
[2] Vokhmintcev KV, Samokhvalov PS and Nabiev I 2016 Nano Today 11 189-211
[3] Samokhvalov P, Artemyev M and Nabiev I 2013 Chemistry - Eur. J. 19 1534–46
[4] Chen Y, Vela J, Htoon H, Casson J L, Werder DJ, Bussian DA, Klimov VI and Hollingsworth JA 2008 JACS 130 5026–7
[5] Boldt K, Kirkwood N, Beane G A and Mulvaney P 2013 Chem. Mat. 25 4731–8
[6] Cao G 2004 Nanostructures and Nanomaterials: Synthesis, Properties, and Applications ed S Chong and L Narayanan (London: Imperial College Press) 2nd edn
[7] Anikeeva PO, Halpert JE, Bawendi MG and Bulović V 2009 Nano Lett. 9 2532–6
[8] Rakovich TY, Mahfoud OK, Mohamed BM, Prina-Mello A, Crosbie-Staunton K, Van Den Broeck T, De Kimpe L, Sukhanova A, Baty D, Rakovich A, Maier SA, Alves F, Nauwelaers F, Nabiev I, Chames P and Volkov Y 2014 ACS NANO 8 5682–95
[9] Bobrovsky A, Mochalov K, Oleinikov V, Sukhanova A, Prudnikau A, Artemyev M, Shibaev V and Nabiev I 2012 Adv. Mater. 24 6216–22
[10] Brazhnik K, Sokolova Z, Baryshnikova M, Bilan R, Efimov A, Nabiev I and Sukhanova, A 2015 Nanomedicine: NBM 11 1065–75
[11] Artemyev M, Ustinovich E and Nabiev I 2009 J. Am. Chem. Soc. 131 8061-5
[12] Hafian H, Sukhanova A, Turini M, Chames P, Baty D, Pluot M, Cohen JHM, Nabiev I and Millot J-M 2014 Nanomed.: NBM 10 1701–9

[13] Ishimoto T, Ozawa T and Mori H 2011 Bioconj. Chem. 22 1136–44
[14] Hollingsworth JA 2013 Chem.Mat.: publication of the ACS 25 1318–31
[15] Harrell S M, McBride J R and Rosenthal S J 2013 Chem. Mat. 25 1199–210
[16] Kudera S, Zanella M, Giannini C, Rizzo A, Li Y, Gigli G, Cingolani R, Ciccarella G, Spahl W, Parak WJ and Manna L 2007 Adv. Mat. 19 548–52
[17] Samokhvalov P, Linkov P, Michel J, Molinari M and Nabiev I 2014 Colloid. Nanopart. Biomed. Appl. IX. Proceed. SPIE 8955 89550S
[18] Jasieniak J, Smith L, Embden JV, Mulvaney P and Califano M 2009 J. Phys. Chem. C 113 19468–74
[19] Shen Y, Gee MY, Tan R, Pellechia PJ and Greytak AB 2013 Chem. Mat. 25 2838–48
[20] Blackman B, Battaglia DM, Mishima TD, Johnson MB and Peng X 2007 Chem. Mat. 19 3815–21
[21] Hao J, Zhou J and Zhang C 2013 Chem. Comm. 49 6346
[22] Linkov P, Laronze-Cochard M, Sapi J, Sidorov LN and Nabiev I 2015 Phys. Proc. 73 216–20

Nanomaterials from bacterial cellulose for antimicrobial wound dressing

E Liyaskina[1], V Revin[1], E Paramonova[1], M Nazarkina[1], N Pestov[1], N Revina[1], S Kolesnikova[2]

[1]National Research Mordovia State University, Ministry of Education and Science of the 430005, Saransk, Russian Federation
[2]National Research Nuclear University MEPhI (Moscow Engineering Physics Institute), 115409, Moscow, Russian Federation

E-mail: liyaskina@yandex.ru

Keywords: bacterial nanocellulose, *Gluconacetobacter sucrofermentans*, biocomposites, fusidic acid, antibacterial activity, wound dressing

Abstract. Bacterial nanocellulose (BNC) is widely used in biomedical applications. BNC has attracted increasing attention as a novel wound dressing material, but it has no antimicrobial activity. To get over this problem in the present study the BNC was saturated with antibiotic fusidic acid (FA). The subject of the experiment was BNC, produced by bacteria *Gluconacetobacter sucrofermentans* B-11267. The resulting biocomposites have high antibiotic activity against *Staphylococcus aureus* and can be used in medicine as a wound dressing. The structure of BNC was analyzed by atomic force microscopy (AFM) and Fourier transform infrared spectroscopy (FTIR).

1. Introduction

Cellulose is well known as one of the most abundant biodegradable materials in nature. It is traditionally extracted from plants but can also be produced by certain bacterial species by fermentation. Bacterial cellulose (BC) is produced mainly by microorganisms belonging to genera Gluconacetobacter (renamed as Komagataeibacter) [1]. BC is chemically identical with plant cellulose but it is not contaminated such products as lignin, pectin, and hemicelluloses [2]. BC has a unique reticulate network from thin fibers with diameter more then 100 times smaller than that of plant-derived fibers [2-4]. BC has better mechanical properties, higher water-holding capacity, higher crystallinity, higher porosity, larger surface area, ultrafine network structure and biocompatibility as compared with plant cellulose [3-7].

This unique nanostructured matrix is being widely explored for various medical and nonmedical applications. BNC is used extensively in many fields, including biomedical materials [1, 8-10], food industry [11], acoustic diaphragm, functional paper [12], drug delivery [13], tissue engineering [14], nanostructured biomaterials and biocomposites [15-17].

BC has many characteristics of an ideal wound dressing. It is known that BC reduce pain and accelerate granulation, ensuring proper wound healing [18, 19]. Further, BC helps in creation of a moist environment at the wound area, absorption of exudates. Thereby BC is considered as as physical barrier between the wound and the surrounding environment, preventing microbial infections. But BNC in its native state does not exhibit antimicrobial effects. There are several successful attempts to impart antimicrobial properties to BC. Bacterial cellulose-silver nanocomposites were successfully prepared and they exhibited excellent antibacterial activity [20-22]. The antibacterial composite based on BC was successfully prepared by in-situ synthesis of SiO_2 coated Cu nanoparticles ($Cu@SiO_2/BC$) [23]. The synthesis of BNC/chitosan composites with high mechanical reliability and antibacterial activity was reported [24]. The antiseptics povidone-iodine (PI) and polihexanide (PHMB) were loaded into BNC with the idea to supplement the superior material properties of the nanostructured biopolymer as a wound dressing with antimicrobial activity [25]. Antibiotics are used in combination with BC for the preparation of composite membranes with antibacterial activity. For example, an antibiotic drug tetracycline hydrochloride (TCH)-loaded bacterial cellulose composite membranes were prepared [26].

The aim of this study was to prepare antibiotic fusidic acid loaded BNC biocomposites and evaluate their antibacterial activities.

2. Materials and methods

2.1. BNC Preparation

BNC was prepared in a static culture medium by *Gluconacetobacter sucrofermentans* B-11267, which was isolated from Kombucha tea and identified by sequencing the amplified product of 16S rRNA [27]. For BNC production HS medium was used (g/L): glucose (20), peptone (5), yeast extract (5), citric acid (1.15), and disodium hydrogen phosphate (2.7), pH 6.0. Culture media was autoclaved at 121°C 20 min. The media was inoculated with 10 % (v/v) inoculum. To prepare the inoculum, *G. sucrofermentans* B-11267 from an agar plate was transferred aseptically into a 250 ml Erlenmeyer flask containing 100 ml of culture medium and incubated on a shaker incubator (Model ES-20/60, BIOSAN, Latvia) at 28 °C for 24 h at 250 g. BNC was produced in static conditions at 28 °C for 5 days. After incubation, BNC was collected, washed thoroughly with de-ionized water to remove medium components, and treated with 4 % (w/v) sodium hydroxide solution at 80 °C for 1 h to eliminate bacterial cells. Further, BNC was rinsed extensively with 6 % acetic acid and then with de-ionized water until pH became neutral. The purified BNC was dried to constant weight at 60 °C and then weighed.

2.2. Production of BNC/ FA composites

BNC films were cut into round shapes with 10 mm diameter and immersed in 10 mL aqueous sodium fusidate solution with concentrations of 0.1, 0.2, 0.3 and 0.4 g/L for 1-24 h. Then they were dried at 60 °C for 24 h. The final composite films were named as $BNC_{0.1}$, $BNC_{0.2}$, $BNC_{0.3}$ and $BNC_{0.4}$, respectively.

2.3. Characterization

The surface morphology of BNC was studied by contact atomic force microscopy (AFM) using an SPM 9600 (Shimadzu, Japan).

FTIR spectra of BNC were obtained using a Fourier transform infrared spectrometer IRPrestige-21 (Shimadzu, Japan).

2.4. Antibacterial activity

The antibacterial activities of BNC/FA composites were investigated by disk diffusion method against *Staphylococcus aureus* 209 P. Lawns of test bacteria (about 1×10^5 CFU/plate) were prepared on medium N1 GRM for bacteria cultivation. The sterilized samples were then carefully placed upon the lawns and sodium fusidate solution with concentrations of 0.1, 0.2, 0.3 and 0.4 g/L was used as control. The plates were placed in a 37°C incubator for 24 h. Then inhibitory action of tested samples on the growth of the bacteria was determined by measuring diameter of inhibition zone.

3. Results and discussion

3.1. Morphology of BNC

The macrostructure morphology of BNC varied depending on the different culture methods. In stationary culture, a gelatinous cellulose film is formed on the air/ liquid interface of the medium. In agitated culture, cellulose is synthesized in deep medium in the form of fibrous suspensions, pellets, irregular masses [28]. When cultivating bacteria *G. sucrofermentans* B-11267 in static conditions in HS growing medium for 5 days on the surface of the medium a BC gel film was formed. After treatment a gel film became colourless and transparent (figure 1).

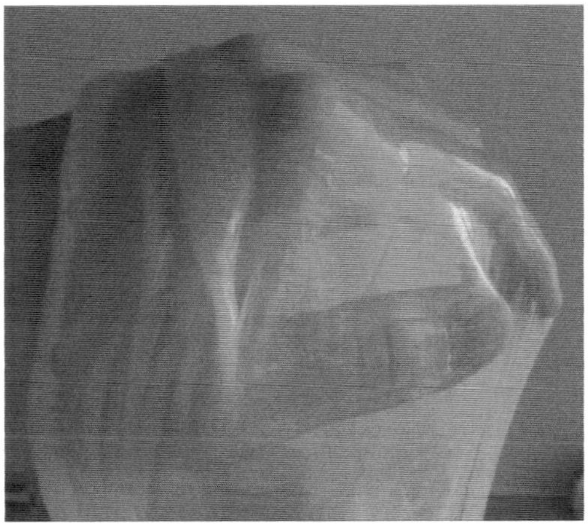

Figure 1. BC gel film.

To study the microscopic details of the biopolymer, atomic force microscopy (AFM) was used. Figure 2 shows the micro-morphology of BNC that exhibited a nanoporous three-dimensional network structure with a random arrangement of ribbon-shaped fibrils. The thickness of BNC fibrils formed on standard HS medium averaged 50-90 nm.

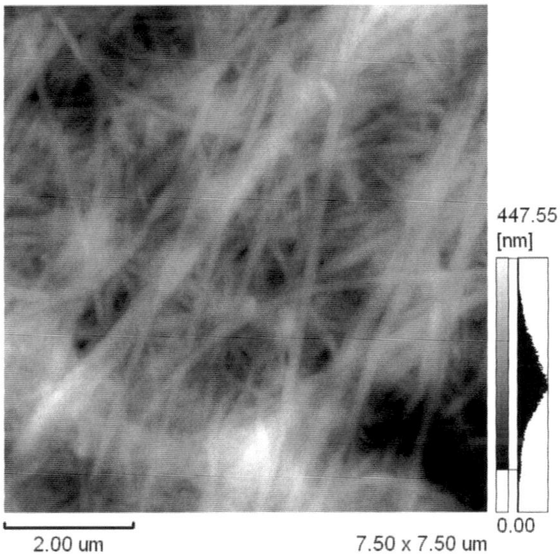

Figure 2. AFM image of the BNC.

3.2. FTIR spectroscopy

Figure 3 displays the FTIR spectra of BNC, FA and BNC $_{0.4}$ composite membranes. FTIR spectrum of BNC (figure 3a) was typical for this polymer. At the FTIR spectrum of the BNC the absorption at the area 3200-3500 cm^{-1} corresponds to the valence vibration of hydrogen bonded OH groups [26]. In the C-O stretching vibration region the peaks at 1164, and 1061 cm−1 correspond to the C-O-C asymmetric stretching and the C-C, C-OH, C-H ring and side group vibrations, respectively.

The infrared spectrum of pure FA exhibited three characteristic bands at 3420, 1725, 1260 cm^{-1} (figure 3c). Absorbtion at the 3420, 1725 cm^{-1} were caused by valence vibration of hydrogen bonded OH groups and carboxyl groups of the FA. In the case of BNC $_{0.4}$ composite the FTIR spectrum had typycal bands for BNC and FA (figure 3b). The band in the 3600-3100 cm^{-1} region became broader and more extensive. Absorption at this region can give considerable information concerning the hydrogen bonds. Probably broadening of this peak was caused by forming additional intermolecular hydrogen bonds between BNC and FA. At the FTIR spectrum of the BNC $_{0.4}$ composite the bands 1725 and 1260 cm^{-1}, that are typical for FA can be detected. Presence of these peaks at the FTIR spectrum of the BNC $_{0.4}$ indicates presence of the FA in the obtained composite.

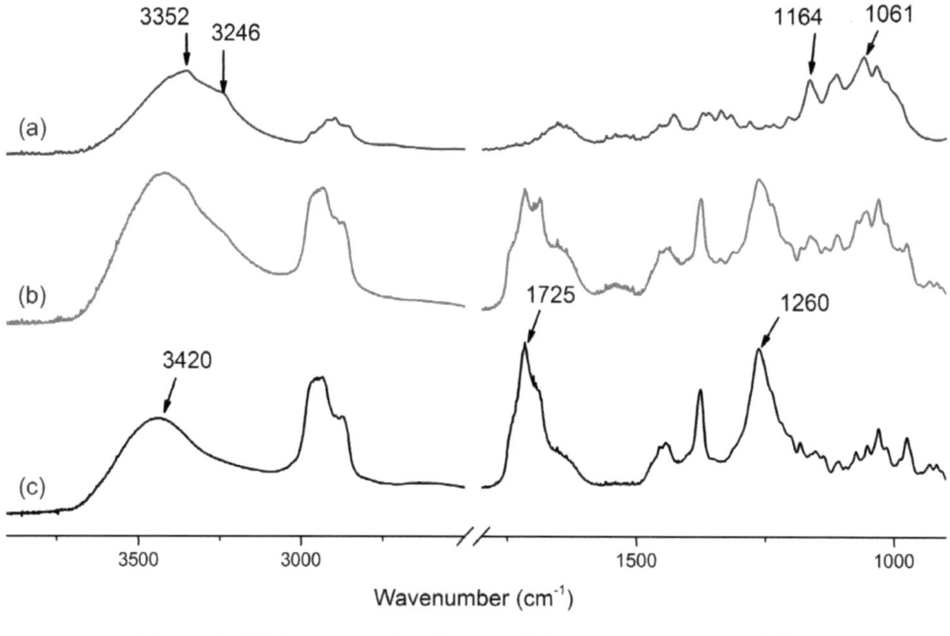

Figure 3. FTIR spectra of BNC (a); BNC $_{0.4}$ composite (b) and FA (c)

3.3. Antibacterial activity

Fusidic acid (FA), an antibiotic produced from the *Fusidium coccineum* fungus, belongs to the class of steroids, but has no corticosteroid effects [29]. FA was characterized as an *α, β*-unsaturated carboxylic acid; its molecular formula is $C_{31}H_{48}O_6$, containing one acetoxyl and two hydroxyl groups (figure 4).

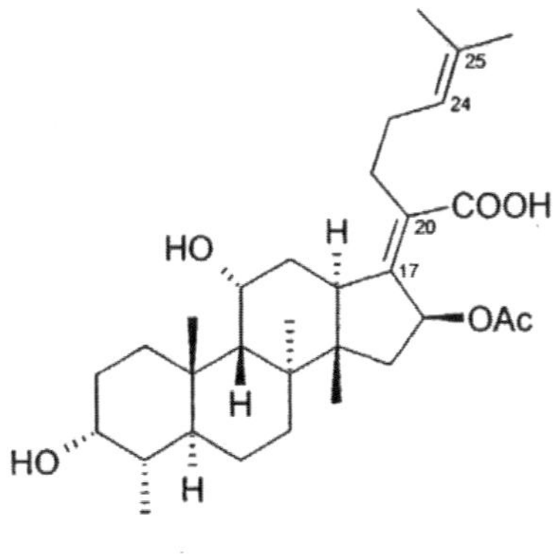

Figure 4. Chemical structure of fusidic acid.

FA has been widely used in the systemic and topical treatment of staphylococcal infections, including strains resistant to penicillin and other antimicrobials, making it an alternative for the treatment of diseases caused by MRSA strains [29, 30]. Furthermore, it is a hypoallergenic agent, has low toxicity, shows low resistance, and has no cross-resistance with other clinically used antibiotics [29].

In this work, the antibacterial activity of FA-loaded BC composites were investigated by disc diffusion method. The prepared composites were placed on a lawn of tested bacteria *S. aureus*. The efficiency of antibacterial activity of BNC/ FA composites was estimated by size of appeared clear zones of inhibition around the samples after 1-24 h of exposure (figure 4).

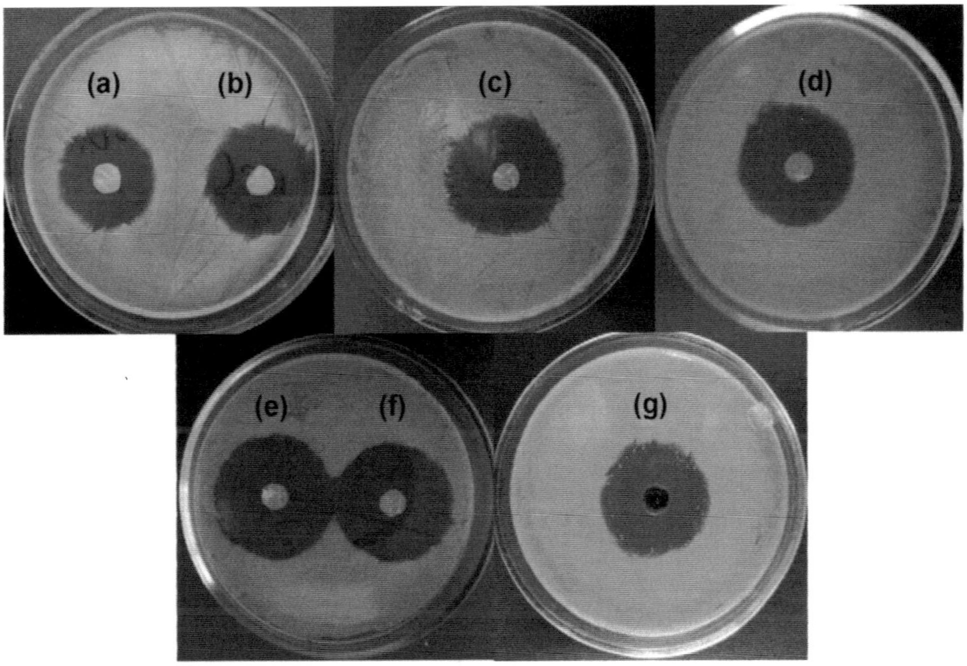

Figure 4. Antimicrobial activity of BNC/ FA composites against *S. aureus* after 24 h of exposure: BNC $_{0.1}$ (a); BNC $_{0.2}$ (b); BNC $_{0.3}$ (c); BNC $_{0.4}$(d); after 1 h of exposure BNC $_{0.4}$ (e); after 2 h of exposure BNC $_{0.4}$ (f); control FA $_{0.4}$ (g)

While within 24 h exposure, diameters of inhibition zones were 27 mm, 30 mm, 32 mm and 34 mm for BNC $_{0.1}$ (a); BNC $_{0.2}$ (b); BNC $_{0.3}$ (c); BNC $_{0.4}$, respectively. $BNC_{0.4}$ has the best antibacterial activities. This composite inhibited *S. aureus* grouth at the area with diametres 30 mm and 32 mm even after 1 (e) and 2 h (f) exposure of antibiotic with BC, respectively. The obtained results indicate that BNC/FA composite membranes have excellent antibacterial activities against *S. aureus*.

4. Conclusions

In summary, FA-loaded BNC biocomposites with antibacterial activity were prepared and investigated. FTIR spectrum of the BNC/FA composites indicates presence of the FA in the obtained composites. BNC/ FA biocomposites have high antibiotic activity against *Staphylococcus aureus* and can be used in medicine as a wound dressing.

Acknowledgments

The authors would like to thank Dr K.N. Nishchev for help in AFM studies. The work was supported by the Ministry of Education and Science of the Russian Federation (project 2913 No. 114112440236).

References

[1] Cacicedo M L, Castro M C, Servetas I, Bosnea L, Boura K, Tsafrakidou P, Dima A, Terpou A, Koutinas A and Castro G R 2016 Progress in bacterial cellulose matrices for biotechnological applications *Bioresour. Technol.* **213** 172–180

[2] Mohite B V and Patil S V 2014 A novel biomaterial: bacterial cellulose and its new era applications *Biotechnol. Appl. Biochem.* **61** 101–110

[3] Lee K Y, Buldum G, Mantalaris A and Bismarck A 2014 More than meets the eye in bacterial cellulose: biosynthesis, bioprocessing, and applications in advanced fiber composites *Macromol. Biosci.* **14** 10–32

[4] Dahman Y 2009 Nanostructured biomaterials and biocomposites from bacterial cellulose nanofibers *J. Nanosci. Nanotechnol.* **9** 5105–5122

[5] Numata Y, Sakata T, Furukawa H and Tajima K 2015 Bacterial cellulose gels with high mechanical strength *Mater. Sci. Eng. C Mater. Biol. Appl.* 147 57–62

[6] Andrade F K, Alexandre N, Amorim I, Gartner F, Mauricio A C, Luis A L and Gama M 2013 Studies on the biocompatibility of bacterial cellulose *J. Bioact. Compat. Polym.* **28** 97–112

[7] Avila M H, Schwarz S, Feldmann E-M, Mantas A, Von Bomhard A, Gatenholm P and Rotter N 2014 Biocompatibility evaluation of densified bacterial nanocellulose hydrogel as an implant material for auricular cartilage regeneration *Appl. Microbiol. Biotechnol.* **98** 7423–7435

[8] Lin S P, Loira Calvar I, Catchmark J M, Liu J R, Demirci A and Cheng K C 2013 Biosynthesis, production and applications of bacterial cellulose *Cellulose* **20** 2191–2219

[9] Rajwade J M, Paknikar K M and Kumbhar J V 2015 Applications of bacterial cellulose and its composites in biomedicine *Appl. Microbiol. Biotechnol.* **99** 2491–2511

[10] Kucińska-Lipka J, Gubanska I and & Janik H 2015 Bacterial cellulose in the field of wound healing and regenerative medicine of skin: recent trends and future prospectives *Polymer Bulletin* **72(9)** 2399–2419

[11] Shi Z, Zhang Y, Phillips G O and Yang G 2014 Utilization of bacterial cellulose in food *Food Hydrocolloids* **35** 539–545

[12] Basta A H and El-Saied H 2009 Performance of improved bacterial cellulose application in the production of functional paper *J. Appl. Microbiol.* **107** 2098–2107

[13] Abeer M M, Mohd Amin M C I and Martin C 2014 A review of bacterial cellulose-based drug delivery systems: their biochemistry, current approaches and future prospects *J. Pharm. Pharmacol.* **66** 1047–1061

[14] Andersson J, Stenhamre H, Backdahl H and Gatenholm P 2010 Behavior of human chondrocytes in engineered porous bacterial cellulose scaffolds *J. Biomed. Mater. Res.* **94** 1124–1132

[15] Hu W, Chen S, Yang J, Li Z and Wang H 2014 Functionalized bacterial cellulose derivatives and nanocomposites *Carbohydr. Polym.* **101** 1043–1060
[16] Ul-Islam M, Khan S, Ullah M W and Park J K 2015 Bacterial cellulose composites: Synthetic strategies and multiple applications in bio–medical and electro-conductive fields *Biotechnol. J.* **184** 47–61
[17] Kiziltas E A, Rhodes K, Emanetoglu N W, Blumentritt M and Gardner D J 2016 Electrically conductive nano graphite filled bacterial cellulose composites *Carbohydr. Polym* **138** 44–51
[18] De Olyveira G M, Manzine Costa L M, Basmaji P and Xavier Filho L 2011 Bacterial nanocellulose for medicine regenerative *J. Nanotechnol. Eng. Med.* **2** 34001–34009
[19] Sulaeva I, Henniges U, Rosenau T and Potthast A 2015 Bacterial cellulose as a material for wound treatment: Properties and modifications. A review *Biotechnol. Adv.* **33(8)** 1547–1571
[20] Yang G, Xie J, Hong F, Cao Z and Yang X 2012 Antimicrobial activity of silver nanoparticle impregnated bacterial cellulose membrane: Effect of fermentation carbon sources of bacterial cellulose *Carbohydr. Polym.* **87(1)** 839–845
[21] Berndt S, Wesarg F, Wiegand C, Kralisch D and Muller F 2013 Antimicrobial porous hybrids consisting of bacterial nanocellulose and silver nanoparticles *Cellulose* **20** 771–783
[22] Shao W, Liua H, Liub X, Wanga S, Wua J, Zhanga R, Minc H and Huang M 2015 Development of silver sulfadiazine loaded bacterial cellulose/sodium alginate composite films with enhanced antibacterial property *Carbohydr. Polym.* **132** 351–358
[23] Ma B, Huang Y, Zhu C, Chen C, Chen X, Fan M and Sun D 2016 Novel Cu,SiO2 / bacterial cellulose nanofibers: Preparation and excellent performance in antibacterial activity *Mater. Sci. Eng. C Mater. Biol. Appl.* **62** 656-661
[24] Zang P, Chen L, Zhang Q and Hong F F Using in sity dynamic cultures to rapidly biofabricate fabric-reinforsed composites of chitosan / bacterial nanocellulose for antibacterial wound dressings 2016 *Front Microbiol.* **7** 260
[25] Wiegand C, Moritz S, Hessler N, Kralisch D, Wesarg F, Muller F A, Fischer D and Hipler U C 2015 Antimicrobial functionalization of bacterial nanocellulose by loading with polihexanide and povidone-iodine *J. Mater. Sci. Mater. Med.* **26(10)** 245
[26] Shao W, Liu H, Wang S, Wu J, Huang M, Min H and Liu X 2016 Controlled release and antibacterial activity of tetracycline hydrochloride-loaded bacterial cellulose composite membranes *Carbohydr. Polym.* **145** 114–120
[27] Revin V V, Liyaskina E V Strain *Gluconacetobacter sucrofermentans* – producer of bacterial cellulose 2013 Patent RU 2523606
[28] Czaja W, Romanovicz D and Brown R M 2004 Structural investigations of microbial cellulose produced in stationary and agitated culture *Cellulose* **11** 403–411
[29] Curbete M M and Salgado H R 2016 A critical review of the properties of fusidic acid and analytical methods for its determination *Crit. Rev. Anal. Chem.* **46(4)** 352–360
[30] Wang J L, Tang H J, Hsieh P H, Chiu F Y, Chen Y H, Chang M C, Huang C T, Liu C P, Lau Y J and Hwang K P 2012 Fusidic acid for the treatment of bone and joint infections caused by meticillin-resistant *Staphylococcus aureus* *Int. J. Antimicrob. Agents* 2012 **40(2)** 103–107

IOP Conf. Series: Journal of Physics: Conf. Series **784** (2017) 012035 doi:10.1088/1742-6596/784/1/012035

Development of modern approach to absorbed dose assessment in radionuclide therapy, based on Monte Carlo method simulation of patient scintigraphy

Y V Lysak[1], V A Klimanov[1] and B Ya Narkevich[2]

[1]National Research Nuclear University MEPhI, 31 Kashirskoe highway, Moscow, 115409 Russia
[2]N. N. Blokhin Russian Oncological Research Center, 23 Kashirskoe highway, Moscow, 115478 Russia

E-mail: lysakyulia@yandex.ru

Abstract. One of the most difficult problems of modern radionuclide therapy (RNT) is control of the absorbed dose in pathological volume. This research presents new approach based on estimation of radiopharmaceutical (RP) accumulated activity value in tumor volume, based on planar scintigraphic images of the patient and calculated radiation transport using Monte Carlo method, including absorption and scattering in biological tissues of the patient, and elements of gamma camera itself. In our research, to obtain the data, we performed modeling scintigraphy of the vial with administered to the patient activity of RP in gamma camera, the vial was placed at the certain distance from the collimator, and the similar study was performed in identical geometry, with the same values of activity of radiopharmaceuticals in the pathological target in the body of the patient. For correct calculation results, adapted Fisher-Snyder human phantom was simulated in MCNP program. In the context of our technique, calculations were performed for different sizes of pathological targets and various tumors deeps inside patient's body, using radiopharmaceuticals based on a mixed β-γ-radiating (^{131}I, ^{177}Lu), and clear β-emitting (^{89}Sr, ^{90}Y) therapeutic radionuclides. Presented method can be used for adequate implementing in clinical practice estimation of absorbed doses in the regions of interest on the basis of planar scintigraphy of the patient with sufficient accuracy.

1. Introduction.

Application of internal radiation sources in radionuclide therapy attributes very special features to the methodologies of radiological control of absorbed doses. Direct measurement of absorbed dose value in tumor volume is extremely difficult because of the necessity of invasive intervention and the complexity of implementation of the procedure per se in routine clinical practice. At the same time, available reference data on the distribution of radiopharmaceutical (RP) in bodies of small laboratory animals, upon which the values of absorbed dose within the region of interest is directly dependent, do not reach satisfactory levels of correlation with respective indicators for human body [1]. Increased probability of radiation induced complications in normal tissues is the limiting factor both in RNT and in the external beam therapy imposing restrictions on the value of RP activity introduced in the body. Values of absorbed doses in the pathological targets and in normal tissues is extremely important for ensuring the required therapeutic doses in tumor under the condition of maintaining tolerant levels of irradiation of not affected areas.

Value of absorbed dose within the region of interest is directly dependent on the value of RP activity accumulated in tumor. Modern methods for accumulated activity measurement in the body include the following methodologies [2]:

– Measurement of count rates in region of interest using clinical radiometer;
– Producing planar images using gamma camera;
– Application of SPECT- and PET-tomography;
– Evaluation of activity in blood and urine samples.

For a long time and even nowdays obtaining planar images using gamma camera is the most convenient and easy to use method for quantitative assessment of RP activity accumulated in the region of interest in patient's body at different time moments. Relative simplicity of the measurement procedure, acceptable costs and low time consuming assured gamma cameras to be "workhorses" of clinical departments of radionuclide therapy.

Nevertheless, certain aspects associated with the specific design of gamma camera and method used for images obtaining must be taken into consideration in order to achieve required level of accuracy of activity estimation according to the data from obtained images. In particular, it is necessary to introduce corrections on the absorption of radiation, radiation scattering in the patient's body and in the elements of the gamma camera system and on the possibility of photon penetration through collimator septum walls. These and some other specific features impose restrictions on the energy and space resolution of the gamma camera. In particular, in radionuclide therapy radiation dose rate during early stages after RP introduction in the patient's body can be sufficiently high which may lead to the loss of counts in the camera and to potential possibility of errors in the identification of the position where the photon is generated. On the other hand, low dose rate is the cause of high levels of statistical uncertainties and noises in the images [3].

Spatial resolution of planar images depends, first of all, on the characteristics of the scintillator and the configuration of collimators of gamma camera. Design features of the collimator also play important role in the determination of geometrical sensitivity of the system. Parallel hole collimators are often used in clinical practices and proved themselves in the most positive way for the estimation of activity accumulated in the patient's body. Compared to other types of collimators (convergent, divergent and pinhole) parallel hole collimators ensure the smallest geometrical distortion of the obtained information and demonstrate relative independence of sensitivity of the image producing system on the source-to-collimator distance within the limits of certain area. Thus, determination of the calibration factor for the selected collimator configuration becomes a routine procedure.

At the same time the processes of radiation absorption along the source-to-collimator distance which is dependent on the RP distribution and on the specific constitution of the patient's body cannot be left un-noticed and, therefore, the value of the calibration factor in the given geometry becomes the value specific for each separate study and introduction of correction of absorption of radiation is strictly necessary.

The most widely used planar scintigraphy method in vivo for measuring the value of activity within the region of interest as well as for introduction of correction of the absorption of radiation in the patient's body became the method of obtaining two conjugated opposite scintillation camera images of the patient's body [4]. Such methodology ensures sufficient accuracy of evaluation of measurements for geometrically well separated sources of activity with insignificant activities of surrounding tissues. The main advantage of this method is the absence of the need to obtain information on the depth of positioning of the region of interest inside the patient's body which significantly simplifies the procedure of obtaining the required data.

The methodology in question assumes obtaining two opposing (usually the anterior and posterior view) planar images of the whole patient's body or of the selected region of interest. For point source located at depth d in the medium with absorption factor μ the gamma camera count rates obtained for the anterior C_A and posterior C_P images, respectively, are equal to

$$C_A = C_0 e^{-\mu d}, \tag{1}$$

and

$$C_P = C_0 e^{-\mu(L-d)}, \tag{2}$$

where L is the thickness of the patient for the particular section; C_0 is the detector count rate in the absence of absorption of radiation in tissues of the patient's body when detector head is positioned directly above the patient's body and after its rotation by 180°. The geometrical mean value calculated according to the following relation:

$$(C_A \cdot C_p)^{1/2} = (C_0{}^2 e^{-\mu L})^{1/2} \tag{3}$$

appears to be independent on the depth of positioning of the region of interest in the patient's body. Thus, the value of accumulated activity is equal taking into account the system's sensitivity C measured in the air to

$$A = C_0 / K = (C_A \cdot C_p)^{1/2} / (C \cdot e^{-\mu L/2}) \tag{4}$$

Absorption factor applied in the methodology is a weighted sum of absorption factors for all tissues of the body along the direction of projection of the image.

Subsequent calculations of activity accumulated within the region of interest using the described methodology require correction caused by radiation scattering along the source–to-detector path which implies the need to implement additional measurements and calculations.

However, application of this methodology in clinical practices results in the accumulation of uncertain-ties in the determination of the value of absorbed activities associated with the following factors:

– Significant simplification of mathematical description of the processes of radiation transfer along the source-to-detector path which may lead to the underestimation or overestimation of absorption and scattering factors in different tissues;

– Application of described methodology is limited by the cases of centers of RP accumulation located at significant geometrical distances from each other which appears to be rather the exclusion than the rule in the clinical practice;

– Absence of the contribution in the detector readings of results of penetration of high-energy photons through the collimator septum walls.

Moreover, application of the discussed methodology requires implementation of additional radiological measurements for correct estimation of absorption and scattering factors, as well as additional irradiation of the patient by external radiation source for estimating the effective value of attenuation factor in the region of interest.

All these factors complicate the procedure for accurate estimation of the value of accumulated activity of RP in tumor gives the stimulus for searching for the alternative methodology of tumor dose control in radionuclide therapy. Methodology of simulation of radiation transfer by Monte-Carlo method which deservedly becomes more and more popular in RNT allows discussing the possibility of significant enhancement of accuracy of estimation of activity in the patient's body [5 – 8].

Application of Monte-Carlo method in RNT for calculation of absorbed doses within the tumor and in organs at risk, as well as for estimation of radiation absorption and scattering in the body tissues found its reflection in a number of foreign and indigenous studies [9 – 15].

The purpose of the present study is to develop clinical method for estimation of RP activity accumulated in the region of interest on the basis of simulation by Monte-Carlo method of processes of radiation transfer in the system for obtaining planar scintigraphic images of the patient's body using gamma camera.

2. Materials and methods.

Therapeutic effect and tumor control assessment in the process of implementation of radionuclide therapy and the probability of development of complications in healthy tissues presupposes the need to accurately determine the doses absorbed in the regions of interest. However, significant dispersion in the values of RP accumulation in the affected organs and in the unaffected structures of the patient's body associated with individual features of RP dynamics and patient's metabolism rate serves as the indisputable proof of the need of implementation of individual control of tumor doses after administration of RNT treatment course [16 – 19]. It has to be noted that in connection with specifics features of RNT treatment, significant variability of biokinetic mechanisms of RP accumulation, distribution and washout from the body, as well as with specific features of implementation of radiological measurement procedures, determination of the value of absorbed dose in the region of interest to one order of magnitude is accepted as sufficiently good result, while estimation of uncertainty up to several tens percent is the maximum achievable result during the contemporary phase of RNT development [4].

The layout of determination of the value of accumulated RP activity in the patient's tumor site suggested in the present paper implies successive implementation of the following three steps.

1. Scintigraphic images are obtained of the vial containing already known activity of the radiopharmaceutical A_0 placed at the fixed source-to-collimator distance, following which estimation of the detector count rate k_0 within the specified region of interest of the vial image is undertaken.

2. Therapeutic activity A_0 is introduced in the patient's body, after a certain period of time specific for each RP type and characteristics the period of maximum RP accumulation in the tumor scintigraphic examination of the patient is performed. Estimation of the detector count rate k_{tum} in the region where the tumor is located and the value of tissue background k_{bg} in the close enough vicinity to the tumor is performed using the tools for contouring the region of interest on the obtained planar image provided using the software imbedded in the scintigraphic equipment.

3. Value of accumulated activity in the affected tumor is determined according to the following formula:

$$A = A_0 \cdot (k_{tum} - k_{bg}) / (k_0 \cdot p), \tag{5}$$

Where p is the correction factor calculated using Monte-Carlo method for specific clinical case for the geometry used in obtaining scintigraphic images which is identical to the conditions of measurement of activity in the vial and in the patient's body.

The obtained correction factor allows taking into account the combination of corrections on the radiation absorption and scattering in biological tissues of the patient, in the air gap between the patient's skin surface and the gamma camera collimator, as well as in the collimator of the imaging system per se taking into account penetration of high-energy photons through the collimator septum walls.

Fischer-Snyder phantom parameters of which correspond to the sizes of adult human body was simulated for calculating the above factor in the MCNP code [20]. Different target volumes, with different sizes approximated with spheres with radii varying from 1 cm to 5 cm were located inside the phantom's body at different depths.

Also model of the system for obtaining scintigraphic images including gamma camera (NaI crystal detector) and parallel hole collimator in which walls of apertures are made of tungsten and have hexagonal shape was developed for performing calculations using the MCNP code. The number of apertures, their lengths and cross-section geometrical characteristics were selected in according to the type and the energy distribution of radiation emitted by radionuclide in the respective RP. Parameters of gamma camera collimators used for simulation and calculations using the MCNP code, are presented in Table 1.

Table 1. Parameters of gamma camera collimators manufactured by Siemens.

	Medium energies	High energies	Super high energies
Isotope	^{67}Ga	^{131}I	^{18}F
Aperture shape	Hex	Hex	Hex
Number of apertures (×1000)	14	8	4
Aperture length, mm	40.64	59.7	50.5
Membrane thickness, mm	1.14	2	3.4
Aperture diameter, mm	2.94	4	2.5

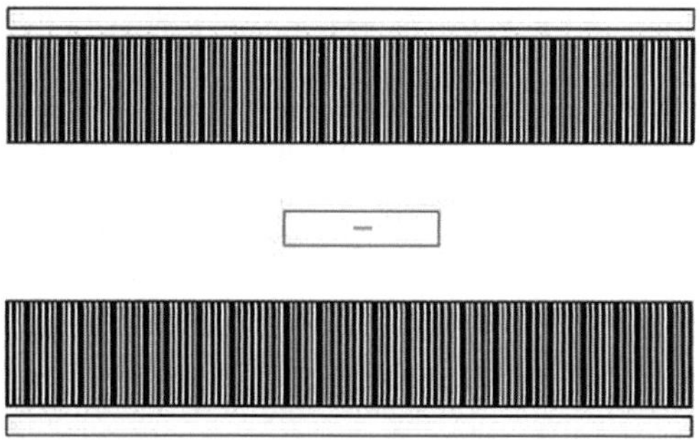

Figure 1. Layout for the determination of gamma camera count rate in the measurements of vial containing activity A_0.

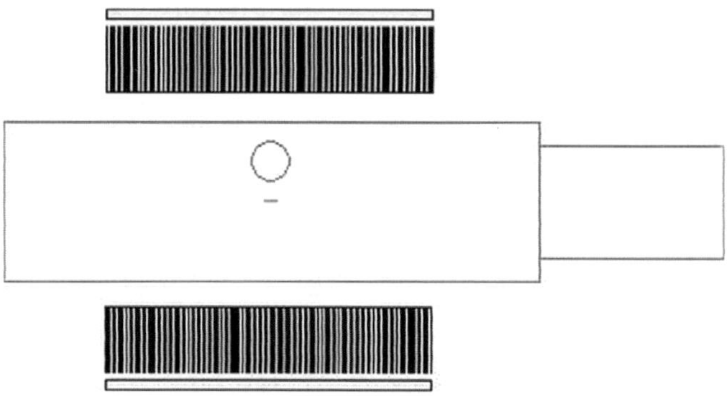

Figure 2. Layout for the determination of gamma camera count rate during scintigraphic study of the tumor located inside the patient's body and containing activity A_0.

Geometry of the experiment simulated in the calculations coincides with the conditions for obtaining planar scintigraphic images using gamma camera. Layout for the determination of the gamma camera count rate in the measurements of 5-ml vial containing activity A_0 administered to the patient is presented in Figure 1. Such model of the experiment corresponds to the realistic clinical situation. Planar images of the vial are obtained when the detector is in position above and below.

Figure 2 demonstrates the model of scintigraphic examination of the patients with RP introduced in the body. It is important to note that the value of activity of the radiopharmaceutical modeled in the target of pathology is equal to the value of A_0 originally contained in the vial. Thus, the required factor taking into account corrections on the radiation absorption and scattering along the whole source-to-detector path is determined in the experiment under modeling from the following relation:

$$p = k^{mc}{}_{amp} / k^{mc}{}_{tum} \quad (6)$$

In this relation $k^{mc}{}_{amp}$ is the gamma camera count rate for simulation by the Monte-Carlo method of the process of obtaining planar images of the vial containing activity A_0; $k^{mc}{}_{tum}$ is the count rate for the modeled scintigraphic study of the phantom of human body with respective sizes of the tumor and depth of its position. Position of the gamma camera during measurement corresponds to realistic clinical situation when in accordance with settings of the equipment the distance between the patient's body surface and the collimator of the system for obtaining the image is fixed to be equal to 3 cm.

3. Results and discussion.

Measurement of correction factor p according to the described methodology was implemented for RP on the basis of both mixed β-γ-emitters ^{131}I and ^{177}Lu and pure β-emitters ^{90}Y and ^{89}Sr for which brehmstrallung emission generated in the patient's body was registered.

In accordance with settings of the system for obtaining planar images using gamma camera applied in the clinical practices, energy range of the registered radiation was set in the code at the level corresponding to 20% deviation upwards and downwards from the most typical line in the spectrum of the investigated radionuclide.

Radius of tumors in the patient's body examined in the study was approximated by spherical geometry with sizes varying from 1 cm to 5 cm, while the examined depths of tumor localization in the patient's body changed from the central lateral line of the phantom to its surface with step equal to 1 cm. Indicators of gamma camera count rate used in the calculations of correction factor were normalized to unit activity in the region of interest.

Results of calculations of correction factors for radiation absorption and scattering obtained by the Monte-Carlo method in the geometry corresponding to the problem described in the present paper are shown in Tables 2 – 5.

Table 2. Calculated values of the correction factor p for pathological target volumes containing RP on the basis of ^{131}I

Distance from the central lateral axis of the phantom, cm (target radius 1 cm)									
0	1	2	3	4	5	6	7	8	9
0.423	0.532	0.593	0.665	0.732	0.762	0.811	0.832	0.901	0.944

Target radius, cm (distance from the axis 0 cm)				
1	2	3	4	5
0.423	0.458	0.484	0.514	0.538

Table 3. Calculated values of the correction factor *p* for pathological target volumes containing RP on the basis of ^{177}Lu

Distance from the central lateral axis of the phantom, cm (target radius 1 см)									
0	1	2	3	4	5	6	7	8	9
0.216	0.219	0.222	0.228	0.232	0.236	0.243	0.252	0.273	0.294

Target radius, cm (distance from the axis 0 cm)				
1	2	3	4	5
0.216	0.224	0.233	0.239	0.243

Table 4. Calculated values of the correction factor *p* for pathological target volumes containing RP on the basis of ^{90}Y

Distance from the central lateral axis of the phantom, cm (target radius 1 см)									
0	1	2	3	4	5	6	7	8	9
0.010	0.013	0.020	0.023	0.028	0.036	0.041	0.047	0.054	0.074

Target radius, cm (distance from the axis 0 cm)				
1	2	3	4	5
0,010	0,013	0,019	0,024	0,032

Table 5. Calculated values of the correction factor *p* for pathological target volumes containing RP on the basis of ^{89}Sr

Distance from the central lateral axis of the phantom, cm (target radius 1 см)									
0	1	2	3	4	5	6	7	8	9
0.013	0.015	0.017	0.022	0.024	0.027	0.029	0.034	0.037	0.046

Target radius, cm (distance from the axis 0 cm)				
1	2	3	4	5
0.013	0.019	0.022	0.023	0.026

Characteristic increase of the value of correction factor with increased size of tumor in the phantom body, as well as with its distance to the surface of the body is associated with decreasing thickness of the layer of biological tissue of the phantom absorbing radiation along the source-to-detector path. Reduction of numerical values for RP on the basis of ^{177}Lu as compared to RP on the basis ^{131}I is explained by significantly lower characteristic energy of the spectrum and reduction of photon contribution in percent in the total exposure dose created by radiation emitted by the radionuclide.

Determination of count rate of the gamma camera for the cases of ^{90}Y and ^{89}Sr required significant in-crease of computer processor time expenditures by the code for obtaining statistically reliable results which is associated with low probability of registration of brehmstrallung emission generated in the body tissues.

Calculated correction factor allows obtaining information about the accumulated value of RP activity in the region of interest. In this case determination of absorbed doses in tumor and in normal tissues does not cause any difficulties.

4. Conclusions.

The methodology of target doses control in the RNT on the basis of obtaining planar scintigraphic images of the patient and calculation of radiation transfer by Monte-Carlo method suggested in the present study allows determining the value of activity of the radiopharmaceutical accumulated in the tumor and in normal tissues taking into account radiation absorption and scattering in biological tissues of the patient and in the materials of the imaging system.

Values of factors determining corrections on radiation absorption and scattering calculated for RP on the basis of ^{131}I, ^{177}Lu, ^{90}Y, ^{89}Sr and presented in the paper for defined geometry and different depth of and different sizes of the affected volume, allowing implementing in clinical practice estimation of absorbed doses in the regions of interest on the basis of planar scintigraphy of the patient with sufficient accuracy.

Acknowledgments
This work was supported by Competitiveness Program of National Research Nuclear University MEPhI.

References

[1] Sgouros G 2005 *J. Nucl. Med.* **46** 18
[2] Fischer D R 2003 *Radiation Protection Dosimetry* **105** 581
[3] Plyku D, Loeb D M, Prideaux A R, Baechler S, Wahl R L, Sgouros G and Hobbs R F 2015 *Cancer Biotherapy and Radiopharmaceuticals* **30** 369
[4] Siegel J A, Thomas S R, Stubbs J B, Stabin M G, Hays M T, Koral K F, Robertson J S, Howell R W, Wessels B W, Fisher D R, Weber D A and Brill A B 1999 J. Nucl. Med. 40 37
[5] Klyopov A N, Kurachenko Yu A, Matusevich E S 2006 *Application of mathematical modeling methods in nuclear medicine* (Obninsk, SOCIN) p 204
[6] He B, Du Y, Song X, Segars W P, Frey E C 2005 Phys. Med. Biol. **50** 4169
[7] Taschereau R, Chatziioannou A F 2007 *Med. Phys.* **34** 1026
[8] Ljungberg M, Sjogreen-Gleisner K 2011 *Acta Oncol* **50** 981
[9] Saadzadeh E, Sarkar S, Tehrani-Fard A A, Ay M R, Khosravi H R, Loudos G 2012 *Radiation Protection Dosimetry* **150** 298
[10] Jonsson L, Ljungberg M, Strand S E 2005 *Journal of Nuclear Medicine* **46** 1679
[11] Vlasova O P, Klyopov A N, Garbuzov P I, Drozdovsky B J, Matusevich E S, Oleynik N A, Spychenkova O N 2007 *Journal of Med. Radiology and Rad. Safety* **52** 53
[12] Dolya O P, Matusevich E S, Klyopov A N, Kurachenko Yu A 2008 *Medicinskaya fizika* **2** 63
[13] Clairand I, Ricard M, Gouriou J, Di Paola M, Aubert B J. 1999 *Nucl. Med.* **40** 1517
[14] Wilderman S J, Dewaraja Y K 2007 *IEEE Trans. Nucl. Sci.* **54** 146
[15] Dewaraja Y K, Ljungberg M, Koral K 2001 *Eur. J. Nucl. Med.* **28** 900
[16] Kost S D, Dewaraja Y K, Abramson R G, Stabin M G 2015 *Cancer Biotherapy and Radiopharmaceuticals* **30** 1
[17] Song N, He B, Wahl R L, Frey E C 2011 *Phys. Med. Biol.* **56** 5503
[18] Sgouros G, Frey E, Wahl R, He B, Prideaux A, Hobbs R 2008 *Semin. Nucl. Med.* **38** 321
[19] Dewaraja Y, Wilderman S J, Ljungberg M, Koral K F, Zasadny K R, Kaminiski M 2005 J. Nucl. Med. 46 840
[20] Briesmeister J F 2000 MCNP – A General Monte Carlo N-Particle Transport Code, Version 4C (LA-13709-M) p 823

Assessment of organochlorine hydrocarbons transformation in contaminated agricultural products and foodstuffs under gamma-radiation

T V Mel'nikova[1], L P Polyakova[1], and A A Oudalova [1,2]

[1]National Research Nuclear University "MEPhI", Moscow, Russia
[2]Russian Institute of Radiology and Agroecology, Obninsk, Russia

E-mail: tritel2010@gmail.com

Abstract. The problem of an estimation of organochlorinated pollutants transformation (particularly organochlorinated pesticides (OCP) and polychlorinated biphenyls (PCB)) under gamma-irradiation has become important in connection with radiation technologies application in the food industry. According to earlier researches, small doses of OCP lead to serious damages of an organism, comparable with damages from high doses. Among radiolysis products of OCP in model solutions various substances on a structure have been found out. Though of trace concentration of each of them, in sum with the initial pesticides residue they make up significant of mass contamination (as shown earlier up to 90% from initial OCP). In this work fish samples (bream) containing OCPs (15.20 ng/g of hexachlorocyclohexane isomers and 87.10 ng/g of DDT and its metabolites), as well as PCB (18.51 ng/g) were studied. The minced fish was irradiated at dose of 10 kGy with dose rate of 1.35 Gy/sec. Then, by methods of gas-liquid chromatography (GLC) and gas chromatography-mass spectrometry (GC-MS), it was found that the OCPs degradation varied from 3 up to 61% and the PCB degradation – 24-52%. Significant complication of chemical composition was shown comparing to the primary biological sample contamination. As a result of fish irradiation, secondary pollution appeared that included residues of primary organochlorine hydrocarbons and their radiation-induced metabolites. Among the investigated OCPs the most stable proved to be alfa-hexachlorocyclohexane (alfa-HCH), the least stable – DDT which corresponds to the previous findings about the radiation stability of OCPs in model solutions. Mass spectra of the irradiated samples of minced bream showed the presence of radiation metabolites of OCPs, that had also been found at irradiation of model solutions of 2,2-di(4-chlorophenyl)-1-chlorethylene (DDMU), DDD and 1a, 2e, 3e, 4e, 5e-pentahlorcyclohexane. There was revealed a decomposition product formed during the deep destruction of chlorinated xenobiotics molecules – 1,1,2-trichloro-4-methyl-1-pentene ($C_6H_9Cl_3$); this substance was absent in the original fish samples. In the samples of irradiated fish there was not defined any OCP oxidation products or any intermediates that may be due to the conditions of sample preparation. To estimate toxicity of the secondary pollution, integral bioassays methods are proposed to apply.

1. Introduction

Currently, in order to optimize processes of agro-food production, new technologies that involve exposure of agricultural raw materials and prepared food products have intensively being implemented. This, in particular, allows improving crop yields, increasing shelf life and reducing the time during transportation. In our country these technologies have to meet requirements of special standard documentation [1–2].

The relevance of the study of composition and properties of irradiated foods is, first of all, determined by the need to substantiate their absolute safety. Based on the results of experiments carried out with biological objects [3–4], it is shown that the radiation treatment of food products, initially containing OCPs in residual concentrations, may initiate processes leading to their secondary contamination.

Studies carried out previously [5] showed that irradiation of the model solutions of OCPs (alfa-HCH, gamma-HCH, p,p-DDE and p,p-DDT) resulted in decreasing of organochlorine substances content in these solutions. This process of the OCPs degradation was dependent on chemical nature and initial concentration of pesticide, solvent properties (polarity and presence of dissolved oxygen) as well as irradiation dose (D) and dose rate ($\dot{D}$). Conditions for maximum degradation of OCP at low initial concentrations (C = 0.01 g/ml, D = 10 kGy, $\dot{D}$ = 0,43 Gy/sec) were determined [5].

In subsequent studies, the OCP radiolysis products in model solutions were identified. It has been shown that irradiation leads to the OCPs radiochemical degradation according to mechanisms of dechlorination, dehydrochlorination, destruction and oxidation [6]. It was also found that the transformation occurs due to isomerization of both initial and dechlorinated molecules [6]. It is possible to suggest that the negative biological effect may be caused by the total action of initial pesticide, radiolysis products and addition products.

Based on the previous findings, it appeared interesting to identify the features of radiation transformations process of OCP in a natural matrix of some biological object. With this purpose, foodstuff containing trace concentrations of organochlorine hydrocarbons were irradiated under optimal conditions found in [5].

2. Materials and methods

For the study, samples of minced fish – bream imported from the Astrakhan region for municipal market – were used. This foodstuff was chosen for irradiation according to the following requirements: a) high coefficient of OCPs bioconcentration [7]; b) significant contribution in population consumption [8] - preparation technology for the consumer may include radiation treatment stage [9-10].

According to our earlier studies [11] the total OCPs content in different samples of fish varied from 12.59 up to 87.10 ng/g. As shown in [12], in recent years, concentrations of some persistent organic pollutants (of limited application) decreased in comparison with the period of the 1990s. However, as a result of coastal enterprises and marine fleet activities, some PCBs continue to come into the water, and pesticides from previous incomes accumulate in sediments. This is why the OCPs and their metabolites are occasionally detected in water and in hydrobionts.

The OCPs, PCBs as well as their products of radiation transformations were determined in bream before and after irradiation. Procedures for samples analysis during the sample preparation stage were developed earlier and described in [11,13-14].

Sample preparation included: homogenization of test sample; extraction of organochlorine hydrocarbons from the sample (with hexane-acetone) and drying (with anhydrous sodium sulfate); pre-concentration of the extract by distillation; acid cleaning and separation of the extract on an adsorption column; final concentration (with nitrogen stripping solvent) of some chromatographic fractions, followed by their analysis with gas-liquid chromatography (GLC). Removal of the lipids was controlled by multiplicity of sulfuric acid treatment of extracts, which was applied 3-5 times depending on the fat content in the sample.

Qualitative and quantitative analysis of the OCP stable transformation products was carried out using gas chromatograph Varian 3400 (Varian, Inc. (USA)) and chromatography-mass spectrometer Varian Saturn 4D (Varian, Inc. (USA)).

The OCPs stability was estimated on the basis of their degradation under irradiation (P, %):

$$P = 100\% - \frac{C_{after}}{C_{before}} 100\%, \quad (1)$$

where P – degradation extent, %, determined by liquid gas chromatography as ratio of their concentration before and after irradiation; C_{after} – concentration of pesticide after irradiation, ppm; C_{before} – concentration of pesticide before irradiation, ppm.

State standard samples of OCPs were used for the calibration of instrumental techniques. The identification of the components was carried out with the full mass spectra library (NIST/EPA/NIH Mass Spectral Library 2011 5th Edition).

Fish (bream) was irradiated on "Issledovatel" (^{60}Co) installation at dose of 10 kGy with dose rate of 1.35 Gy/sec. Doses were estimated with the ferro-sulfate method using the Fricke dosimeter. An uncertainty of this dosimetry method does not exceed 10%. LET was about 0.2 keV/m [16].

3. Results and discussion

The results of the chromatographic analysis of fish showed (Table 1) that the content of the individual OCPs meet the requirements for maximum permissible concentration (MPC). Samples containing the highest amount of pesticides [17] were selected for further analysis. This provided the most informative results of comparative studies of OCPs (before and after irradiation of samples) since there are usually some losses during procedure of biomaterial sample preparation.

Table 1. Concentrations of OCP in bream (in terms of fat)

Sample	Concentration, 10^{-3} ppm				Σ concentrations, ppm	
	alfa-HCH	gamma-HCH	DDT	DDE	HCH	DDT+DDE
№1	-[a]	-[a]	72,90±0,70	14,20±13,40	-[a]	87,10
№2	1,80 ± 1,10	-[a]	1,10±0,70	12,30±4,30	1,80	13,40
MPC[b]					200	200

[a] Pesticide is not detected.
[b] [17].

The chromatograms obtained from the preliminary analysis on a packed column showed different content of pesticides in non-irradiated and irradiated samples №1. DDE and DDT contents in irradiated sample significantly decreased (Figure 1, A and B, respectively). This indicates the radiation degradation of these substances. At the same time, the peaks (1, 2, 3, 4 in Figure 1) missing in the chromatogram A were detected in the chromatogram B. It indicated the appearance of new substances in the irradiated sample. The peaks could be identified as following chemical compounds: 1 – ε-3a,4e,5e,6e-tetrachlorocyclohexene, 2 – 1e,2a,3a,4a,5e-pentachlorocyclohexane, 3 – 1e,2a,3e,4e,5e-pentachlorocyclohexane and 4 – p,p-DDD.

Stereochemical formulas of the OCPs detected before and after irradiation as well as possible ways of derivation of their transformation products are shown in Figure 2.

Gas chromatography-mass spectrometry analysis on capillary column of the irradiated fish samples (№1 and №2), initially differing in content of the OCPs, gave an additional information about radiation metabolites generated in these samples.

On the chromatograms of extracts obtained before and after irradiation of minced fish (Figure 3, *a* and *b*, respectively), compounds were marked that were possible to determine quantitatively. The OCPs concentrations and the radiation degradation degree (P) are presented in Table 2. There are identified 6 pesticides in sample №2 (unlike sample №1); their concentrations after irradiation decreased as it was in the first experiment. It followed from comparing the degradation degrees that the same degradation degree of the OCPs appeared in fish and in model solutions [5-6]. The most stable was alfa-HCH, the least stable – DDT. Organochlorine hydrocarbon p,p-DDD found in fish sample №2 had a stability comparable to the stability of p,p-DDT. The degradation degrees of the same OCPs were the same despite their different content in the samples. Apparently, the effect of the OCPs concentration in the fish on the degradation degree was impossible to reveal, using the specified method of sample preparation.

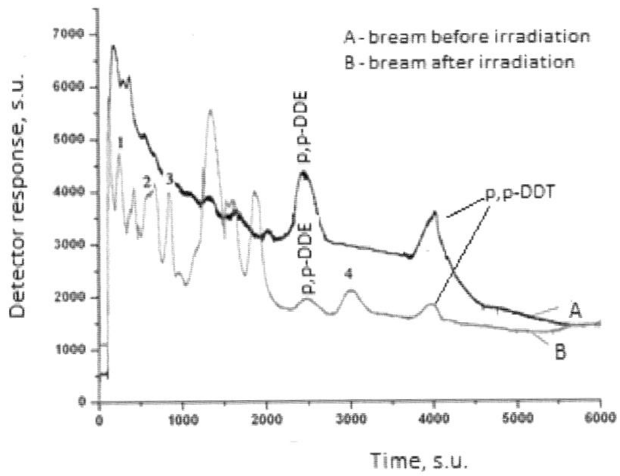

Figure 1. Chromatograms of bream extracts before and after irradiation (dose 10 kGy, dose rate 1.35 Gy/sec). Qualitative and quantitative analyses of the OCPs stable transformation products by gas chromatograph "Model 3700": packed column (2 m), stationary phase SE-54; electron capture detector (^{63}Ni); carrier gas – nitrogen. Isothermal mode. Sample volume - 4 µl.

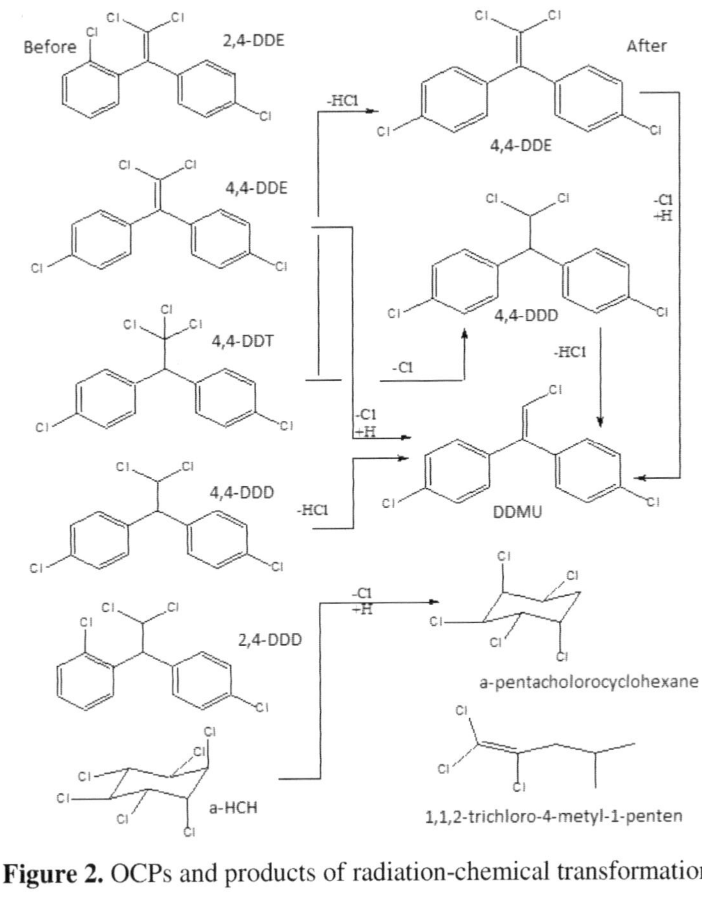

Figure 2. OCPs and products of radiation-chemical transformations.

a

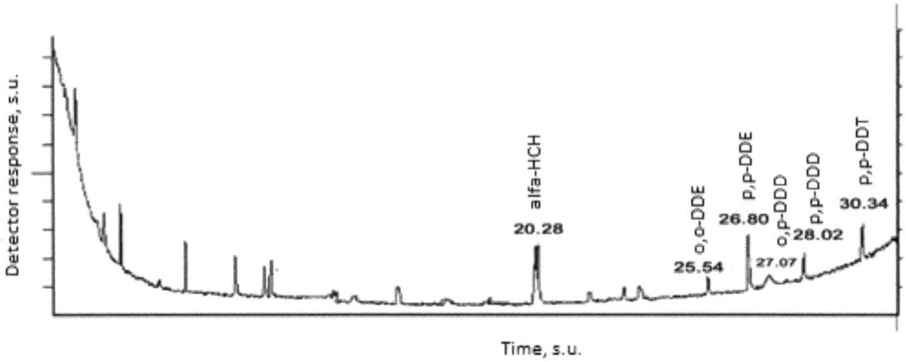

b

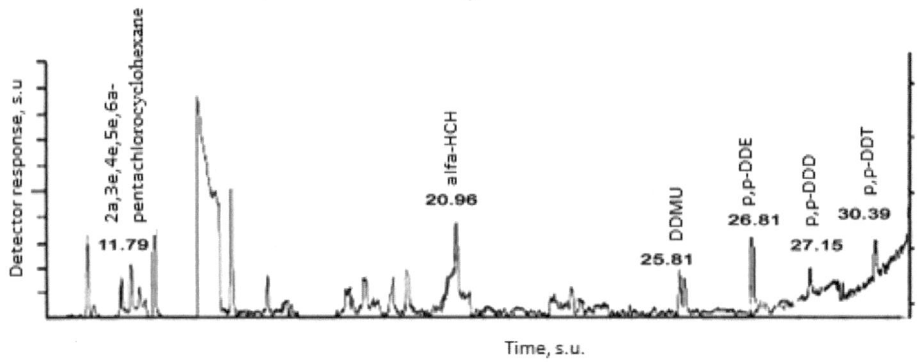

Figure 3. Chromatograms of bream extracts (sample № 2) before (*a*) and after (*b*) irradiation (dose 10 kGy, dose rate 1.35 Gy/sec). Analyses by gas chromatography-mass spectrometer Varian Saturn 4D.

Table 2. Concentrations and degradation degree of OCPs in bream (samples № 1, 2) before and after gamma-irradiation

Pesticide	Sample №1			Sample №2		
	Concentration, ng/g		*P*, %	Concentration, ng/g		*P*, %
	before	after		before	after	
α-HCH	-	-	-	1,80 ± 1,10	1,70 ± 1,21	2,70
2,4-DDE	-	-	-	0,035 ± 0,020	-	-
4,4-DDE	72,90 ± 0,70	44,15 ± 3,55	39,44	12,32 ± 4,11	7,72 ± 3,00	37,40
2,4-DDD	-	-	-	0,32 ± 0,18	-	-
4,4-DDD	-	-	-	13,64 ± 5,03	5,30 ± 1,71	61,40
4,4-DDT	14,20 ± 13,40	5,79 ± 1,52	59,23	1,64 ± 0,73	0,68 ± 0,29	58,53
ΣDDT metabolites	87,10	49,94	-	29,31	15,40	-

In sample №2 before and after irradiation by chromatography-mass spectrometric analysis there were found not only initial dechlorination products of pesticides, but also products of deep degradation of their molecules.

For example, among the newly formed after fish irradiation substances was found 1,1,2-trichloro-4-methyl-1-pentene ($C_6H_9Cl_3$). This compound can be formed as a result of the radiation transformation of any of the OCPs contained in bream. However, this substance can also be formed in the radiolysis of other chlorinated xenobiotics present in fish. It appeared that sample №2 was contaminated not only by the OCPs but also by polychlorinated biphenyls (PCB). In an extract from fish sample №2 before irradiation there were found 15 PCBs, containing from 3 up to 7 atoms of chlorine (Table 3). After irradiation all PCB undergone to degradation. The degradation extent varied from 24 up to 52%.

Table 3. Concentrations and degradation degree of PCBs in bream (sample № 2) before and after gamma-irradiation

PCB isomers	Concentration, ng/g		P, %
	before	after	
№28/№31 [CL3]	0,43	0,50	-
№52 [CL4]	0,36	0,28	24,7
№101 [CL5]	1,18	0,75	36,86
№99 [CL5]	1,23	0,82	33,33
№110 [CL5]	1,76	0,88	50,00
№118 [CL5]	2,05	1,18	42,54
№153 [CL6]	3,81	1,84	51,71
№105 [CL5]	0,73	0,36	49,79
№138 [CL6]	2,88	1,41	50,96
№187 [CL7]	0,76	0,37	51,06
№183 [CL7]	0,40	0,19	52,32
№128 [CL6]	0,81	0,41	49,57
№156 [CL6]	0,33	0,21	36,54
№180 [CL7]	1,09	0,54	50,92
№170 [CL7]	0,72	0,36	49,79
ΣPCB	18,51	10,08	45,55

The OCPs oxidation products and intermediates (compounds containing residues of OCPs molecules and molecules of the organic matrix material (e.g., fish fat)) was not detected in the irradiated samples. Obviously, this is due to losses during sample preparation. Products of chemical transformation of complex organic compounds are usually more polar than the original compounds themselves [3]. Therefore, in the adsorption treatment of the organic extract some transformation products of the OCPs are not eluting.

This study showed that organochlorine hydrocarbons in the biological object under irradiation were not completely destroyed and transformed into compounds with similar properties (including toxicological ability). In addition, metabolites with unstudied toxicological properties appear. That is a new contamination of biological object appears. The danger of such pollution will be determined by synergistic or antagonistic effects or the combined effects due to the initial and newly formed contaminants.

At present to estimate the toxicity of secondary pollution, an integral method of bioassays is often proposed to apply. These methods by indicators of substances biological activity are able of detecting deviation in the system from the equilibrium state. We also plan to use such approach in the future to assess toxicity of radiation metabolites of organochlorine hydrocarbons contaminating food.

As shown in [18], the OCPs keep stable even under irradiation at high dose of 100 kGy but with low dose rates (0.30 - 0.50 Gy/sec). At the same time, a wide range of doses and dose rates could be applied in up-to-date radiation technologies for foodstuffs irradiation [19]. Doses may vary between 1 and 60 kGy and dose rate – from 0.003 up to 0.25 Gy/sec [20-21]. The effect of these variations on the stability of the OCPs in biological objects has not been carefully studied so far. In our opinion, this issue also becomes very important in the case of food irradiation.

4. Conclusions

On the example of this study with irradiated fish it could be concluded that a detailed understanding of the composition and properties of radiation metabolites is necessary to optimize process of foodstuffs decontamination and avoid the potential deterioration of product quality due to the secondary contamination.

Acknowledgments
This work was supported by Competitiveness Program of National Research Nuclear University MEPhI.

References

[1] GOST ISO 14470-2014 «Food irradiation. Requirements for the development, validation and routine control of the process of irradiation using ionizing radiation for the treatment of food» (in Russia)
[2] Molin A A 2012 http://2012.atomexpo.ru/mediafiles/u/files/Present2012/Molin.pdf (in Russia)
[3] Lepine F L, Brochu F, Milot S, Mamer O A and Pepin Y 1994 *J. Agric. Food Chem.* vol 42 pp 2012-2016
[4] Lepine F L, Brochu F, Milot S, Mamer O A and Pepin Y 1995 *J. Agric. Food Chem.* vol 43 pp 491-494
[5] Mel'nikova T V, Polyakova L P, Koz'min G V 2001 *J.Radiatsionnaya biologiya i Radioekologiya* vol 41(6) pp 683-687 (in Russia)
[6] Mel'nikova T V, Polyakova L P, Koz'min G V 2003 *J.Radiatsionnaya biologiya i Radioekologiya* vol 43(6) pp 697-705 (in Russia)
[7] Fedorov L A, Yablokov A V 1999 (*Moscow*) 462 p
[8] Sazonova O V 2008 *Vestnik RUDN* vol 7 pp 476 – 479
[9] Safety and nutritional adequacy of irradiated food 1995 (*Moscow*) 209 p
[10] Technical Report Series 890 1997 (*Geneva*) FAO/ IAEA/ WHO 198 p
[11] Mel'nikova T V, Polyakova L P, Koz'min G V 1999 *J. Yadernaya energetika* vol 1 pp 66-74
[12] Il'in G V 2012 *Vestnik Yuzhnogo nauchnogo tsentra RAN* vol 8(1) pp 60–69 (in Russia)
[13] Voogt P, Klamer J C and Govers H 1986 *J. of Chromatography* vol 363 pp 407-411
[14] Barkatina E N, Zastenskaya I A, Shulyakovskaya O V and etc 2007 *J. of Analytical Chemistry* vol 62(9) pp 960-964
[15] Klisenko M A 1992 Kolos (*Moscow*) vol 2 304 p
[16] Pikaev A K 1985 Nauka (*Moscow*) 375 p
[17] SanPin 2.3.2.1078-01 2002 (*Moscow*) 165 p
[18] Mel'nikova T V, Polyakova L P, Koz'min G V and etc 2014 *J. Mezhdunarodnyy zhurnal prikladnykh i fundamental'nykh issledovaniy* vol 4(11) pp 210-215 (in Russia)
[19] Ershov B G 2013 *Vestnik RAN* vol 83(10) pp 885-895 (in Russia)
[20] Irradiation of Food Commodities: Techniques, Applications, Detection, Legislation, Safety and Consumer Opinion ed. Ioannis S. Arvanitoyannis, Amsterdam, Boston, Heidelberg, London, New York, San Francisco, Singapore, Sydney, Tokyo. Academic Press Elsevier 2010 710 p
[21] Radiation technologies in the agriculture and food industry ed. Koz'min G V, Geras'kin S A and Sanzharova N I 2015 RIARAE-RAAS 40 p (in Russia)

Biotesting of radioactively contaminated forest soils using barley-based bioassay

T V Mel'nikova[1], L P Polyakova[1] and A A Oudalova [1,2]

[1]National Research Nuclear University MEPhI, Moscow, Russia
[2]Russian Institute of Radiology and Agroecology, Obninsk, Russia

E-mail: tritel2010@gmail.com

Abstract. Findings from radioactivity and phytotoxicity study are presented for soils from nine study-sites of the Klintsovsky Forestry located in the Bryansk region that were radioactively contaminated after the Chernobyl accident. According to the bioassay based on barley as test-species, stimulating effect of the soils analyzed is revealed for biological indexes of the length of barley roots and sprouts. From data on ^{137}Cs specific activities in soils and plant biomass, the migration potential of radionuclide in the "soil-plant" system is assessed as a transfer factor. With correlation analysis, an impact of ^{137}Cs in soil on the biological characteristics of barley is estimated.

1. Introduction

As a result of the Chernobyl accident, the highest radioactive contamination of soil (in particular, forest soils) occurred in the Tula, Kaluga, Bryansk and Orel regions, representing more than 30% of the total area of forest in these regions. Contaminated soils have become one of the most important sources of radioactive impact on plants and other living organisms in forest ecosystems.

Nowadays there are a number of modern techniques to assess soil toxicity, but their application is usually limited due to their high labor intensity and cost. An effective and relatively low-cost approach to estimate toxicity of radioactively contaminated soils is based on an application of bio-testing methods using higher plants test-systems, such as barley.

Thus, this work was aimed at an estimation of biological effects in the test-species (barley) from the radioactively contaminated soils from the Klintsovsky Forestry located in the Bryansk region.

2. Materials and methods

Soil samples for the study were taken in 9 study-sites within the techniques forestry (the Bryansk region) that was radioactively contaminated after the Chernobyl accident [1]. As a control sample, forest soils were taken in the "Ugra" National park, near the Sukovka village (NL $54^0 51'$ HP $35^0 07'$).

The specific activities of ^{137}Cs in soils and in barley were measured by gamma spectrometry (table 1). Review of published data shows that in the south-west of the Bryansk region the specific activities of ^{137}Cs in soils decreased with a factor of 7-13 times comparing to 1986 [2].

Transfer factor (*TF*) was calculated as a ratio of the radionuclide specific activities in plant mass (A_{plant}, Bq/kg) to the specific activities of soil (A_{soil}, Bq/kg) [3]:

$$TF = \frac{A_{plant}}{A_{soil}} \quad (1)$$

Table 1. Specific activities of ^{137}Cs in forest soils and in barley and results of *TF* calculation.

№	Study-site	Soil type	Specific activity of $^{137}Cs \pm \Delta$, Bq/kg in soil	in barley	*TF*
1.	Baturovka 6	sand	8214 ± 220	1793 ± 661	0.22
2.	S/H 19	sand	274 ± 15	500 ± 41	1.82
3.	Petrova Buda	sand	1306 ± 116	729 ± 141	0.56
4.	Pravda 9	sandy loam	288 ± 14	500 ± 30	1.74
5.	Ushherp'e 21	sandy loam	2730 ± 180	548 ± 199	0.20
6.	Ushherp'e 81	sandy loam	1352 ± 118	500 ± 60	0.37
7.	Krasnaya gora 44	light loam	607 ± 50	400 ± 46	0.66
8.	Ushherp'e 10	light loam	1206 ± 114	650 ± 50	0.54
9.	Ushherp'e 64	light loam	1378 ± 113	1793 ± 661	0.29

For biotesting, soil samples (m = 500 g) are placed in special containers and moistened to 60% of the total field capacity. In the prepared soils, barley seeds (not less than 40-50) are planted. Containers are placed into a climate chamber (Panasonic MLS) with automatically controlled parameters: temperature (day time - 21 ^{0}C, night - 16 ^{0}C) and illumination (day - 16 hr, night - 8 hr). Exposure period is 14 days (Figure 1).

Figure 1. Sprouting barley seeds in climate chamber (Panasonic MLS).

At the end of exposure, plants are carefully pulled off the soil, washed with tap water, and length of barley roots and sprouts are measured. Toxicity of the soils tested is estimated from a difference between control and impacted samples; the Student t-test was used as a statistical criteria.

Soil phytotoxicity level (*PE*) was calculated as:

$$PE = \frac{TF_{experiment} - TF_{control}}{TF_{control}} \cdot 100\%, \quad (2)$$

where *PE* – index of phytotoxicity, %; $TF_{experiment,control}$ – test-function of biological indexes: length of barley roots and sprouts (mm) in the experimental and control soil samples, accordingly [4].

3. Results and discussion

The data obtained at barley seeds' germination on soils from the Klintsovsky Forestry are shown in Figure 2a. According to the bioassay, the biological indicator of barley rootlet length showed the stimulation effect compared to the control sample. Root length at the test-species germination on all the soil samples from the radioactively contaminated study-sites (except for Ushherp'e 81) was significantly higher ($\alpha<0.05$) than in control.

With second treat, the barley sprout length, the stimulation effect was also revealed for barley plants grown on radioactively contaminated soils as compared with the control soil sample (Figure 2b). Statistically significant increase in sprout growth was found in soil samples collected from Baturovka 6, Petrova Buda 3, Pravda 9 and Krasnaya gora 44 study-sites, in all other samples the differences appeared insignificant.

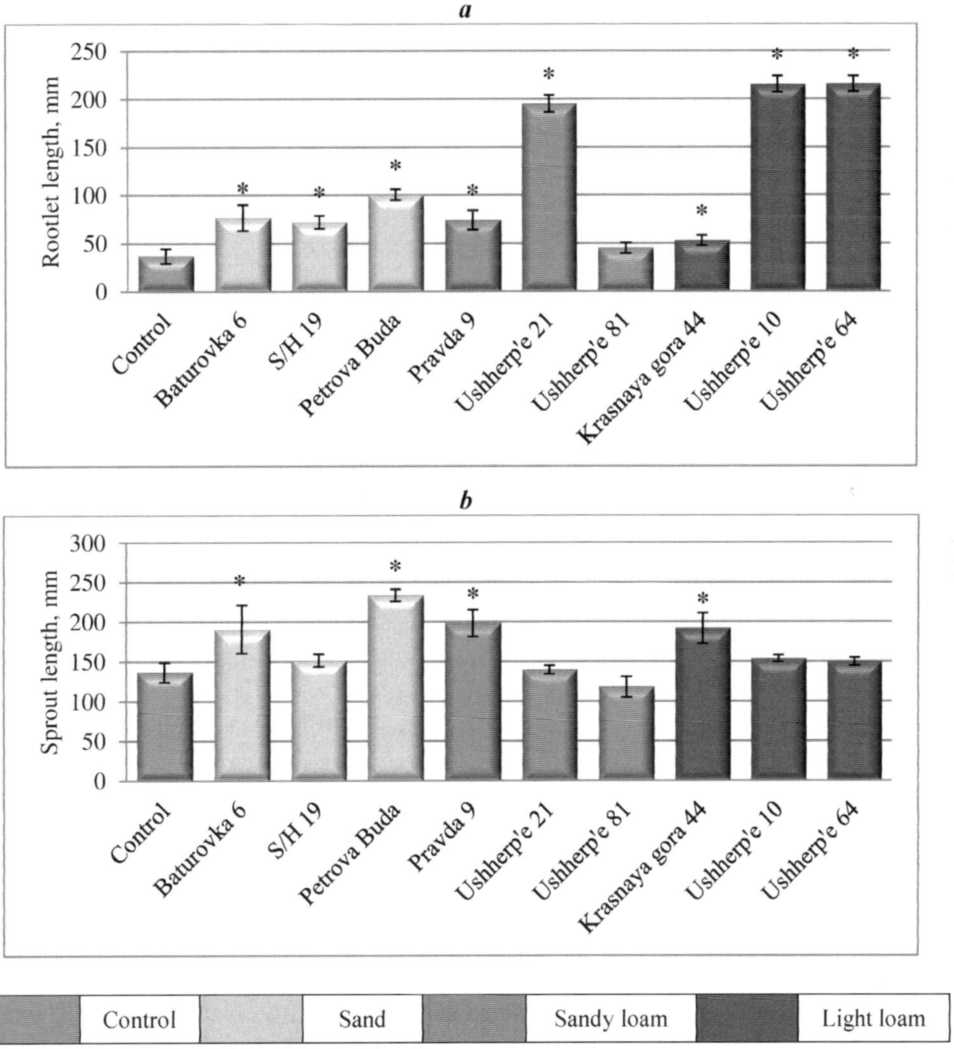

Figure 2. Length of roots (*a*) and sprouts (*b*) at barley seeds germination on radioactively contaminated soils from the Klintsovsky Forestry located in the Bryansk region.

* - difference from control is significant, $\alpha<0.05$

Data from phytotoxicity assessment are presented in table 2. Based on an integral assessment study of soils phytotoxicity with the "length of barley roots" test-function, a significant stimulation of root growth in terms of *PE* is concluded as this index increases by a factor of 1.1–4.8 (table 2). The soil samples tested can be ranged by a degree of phytotoxicity manifestation as follows: Pravda 9 < Baturovka 6 < Petrova Buda 3 < Ushherp'e 21 < Ushherp'e 10 < Ushherp'e 64. Phytotoxicity estimation with the "length of barley sprouts" test-function showed that the most pronounced stimulating effect is registered in the following samples: Baturovka 6, Petrova Buda 3, Pravda 9 and Krasnaya gora 44; in soils from all other study-sites phytotoxicity effect of stimulation is not significant.

Table 2. Integral assessment of soil phytotoxicity in samples tested.

Study-site	Root		Sprout	
	< length >± Δ, мм	*PE*, %	< length >± Δ, мм	*PE*, %
Control	36.9 ± 7.8		136.7 ± 12.4	
Baturovka 6	76.6 ± 13.5	107.5	190.6 ± 30.4	39.7
S/H 19	71.9 ± 6.8	94.9	151.5 ± 8.2	11.1
Petrova Buda	100.2 ± 5.6	171.5	233.3 ± 7.8	70.9
Pravda 9	44.9 ± 5.6	21.7	118.1 ± 13.0	-13.4
Ushherp'e 21	195.4 ± 8.9	429.1	139.7 ± 5.3	2.4
Ushherp'e 81	73.9 ± 10.1	100.2	198.3 ± 16.9	45.3
Krasnaya gora 44	52.8 ± 5.4	42.9	191.8 ± 19.3	40.5
Ushherp'e 10	215.8 ± 8.1	484.4	149. 8 ± 5.0	9.8
Ushherp'e 64	215.3 ± 8.5	483.3	153.4 ± 4.5	12.5

To reveal a relationship of biological indexes of barley (the length of roots and sprouts) from the ^{137}Cs specific activity in soil, a linear correlation analysis was performed. Findings obtained show that soil contamination with ^{137}Cs-137 has had a significant influence (α <0.05) on the length of barley sprout only in light loam soils. The correlation between the transfer factor of ^{137}Cs in the "soil-plant" system and the bioassay indexes is not revealed. An absence of the relationship between these indexes can be explained by limited data samples received for the radionuclide concentration in every type of soil tested.

4. Conclusions

In studies of soils sampled in the Klintsovsky forestry and laboratory barley-based bioassay, the specific activities of ^{137}Cs are measured in the "soil-plant" system and the transfer factor is calculated for the radionuclide uptake by the test-species, which amounts to 0.20-1.82. The biological effect of radioctively-contaminated soils for barley is analyzed with indexes of length of roots and sprouts as well as phytotoxity index. Findings obtained indicate that nearly for all soils tested (except for the Ushherp'e 81) the stimulating effect for growth processes in barley roots are detected, and also in many soil samples the stimulation of above-ground part of the plants is observed. A reliable correlation between the specific activity of ^{137}Cs in soil and biological indexes in barley is installed for only one type of the soils tested – light loam.

Acknowledgments

This work was supported by Competitiveness Program of National Research Nuclear University MEPhI.

References

[1] Reglament GU «Klincovskoe lesnichestvo» http://bryanskleshoz.ru/?p=40&parent=39 (in

Russian)
[2] ASPA Russia-Belarus ed. Izraehlya Yu A and Bogdevicha I M 2009 (Moscow-Minsk) 140 p
[3] Rosleshoz Prikaz № 81 March 16 2009 21 p (in Russian)
[4] Voronina L P, Maj Thu Lan Cheremnyh E G 2012 J. Problemy agrohimii i ehkologii vol 1 pp47-52 (in Russian)

Comparative study of cytotoxicity of ferromagnetic nanoparticles and magnetitecontaining polyelectrolyte microcapsules

O V Minaeva[1], E P Brodovskaya[1], M A Pyataev[1], M V Gerasimov[1], M N Zharkov[1], I AYurlov[1], O AKulikov[1], A A Kotlyarov[2], L A Balykova[1], A V Kokorev[2], A V Zaborovskiy[3], N A Pyataev[1] and G B Sukhorukov[4,5]

[1]National Research Mordovia State University
[2]National Research Nuclear University "MEPhI", Moscow, Russia
[3]Moscow State University of Medicine and Dentistry named after A.I. Evdokimov
[4]Queen Mary University of London
[5]Saratov State University

Main author email address: polinanew@mail.ru

Abstract. The cytotoxicity of magnetite nanoparticles (MNP) stabilized with citrate acidand polyelectrolyte multilayer microcapsules containing these particles in the shell is analyzed. Microcapsules were prepared by co-precipitation of iron (II) and (III) chlorides. Polyelectrolyte microcapsules synthesized by the layer-by-layer method from biodegradable polymers polyarginine and dextran sulfate. Cytotoxicity of the synthesized objects was studied on the L929 cells culture and human leucocytes. It was also investigated the phagocytic activity of leukocytes for the MNP and magnetite containing polyelectrolyte microcapsules (MCPM). A set of tests (MTT assay, neutral red uptake assay, lactate dehydrogenase release assay) was used to study the cytotoxicity in vitro. All the tests have shown that the magnetic nanoparticles have a greater cytotoxicity in comparison with microcapsules containing an equivalent amount of magnetite. In contrast to the mouse fibroblast culture, human leukocytes were more resistant to the toxic effects of magnetite. At the concentrations used in our studies no significant reduction in the viability of leukocytes has been registered. Both MNP and MCPM undergo phagocytosis, however, the phagocytic activity of leukocytes for these particles was lower than for the standard objects (latex microparticles).

Introduction

Polyelectrolyte microcapsules (PM) are a promising system for packaging and magneticaly driven delivery of verious substances. Microcapsules are commonly synthesized by layer-by-layer (LBL) technology when oppositely charged polyelectrolytes are applied in layers on a core of inorganic material which can contain different substances[1]. Then the core is dissolved by an acid and the substance remains in the PM. Inclusion of different functional groups into the core or shell of capsules allows to control the delivery and release of the drug.A method for the synthesis of PM from biocompatible and biodegradable polymers (polyarginine and dextran-sulphate) with nanoparticles of magnetite incorporated into the shell was proposed in paper [2]. It has been shown in in-vitro experimentsthat these capsules are more actively phagocytosed by cells [2].andcan act as intracellular anchors that allow manipulating cells by an external magnetic field [3].

However, there are several problems that limit the use of this technology. One of the main problems is the toxicity of magnetic nanoparticles. Iron is not a foreign substance to the body being a part of its major systems (hemoglobin, cytochrome p450, etc.). There are several systems involved in assimilation, transportation and excretion of iron in the human body. However, magnetically driven drug delivery requires parenteral administration of rather large amount of iron which can lead to some side effects [4]. In particular, the ionized iron can activate lipid peroxidation and be toxic for cell membranes [5]. However, the isolation of iron from contact with the blood cells and endothelium can decrease the cytotoxicity [6].

In the present paper, the cytotoxicity of magnetite nanoparticles (MNP), and magnetite containing polyelectrolyte microcapsules (MCPM) with these nanoparticles included in the shell was studied.

1. Materials and methods:

1.1. Reagents

Iron chloride (III) (FeCl3 • 6H2O) (99.99%, Sigma-Aldrich),
Iron chloride (II) (FeCl2 • 4H2O) (99.99%, Sigma-Aldrich),
Ammonium hydroxide 25% (99.8% Vekton, Russia)
Citric acid (99.99%, Sigma-Aldrich),
Sodium ethylenediaminetetraacetate (99.9% Vekton)
Sodium carbonate (99.9% Vekton)
Calcium chloride (99.9% Vekton)
Sodium dextran sulfate (DS) (M ~ 100 kDa, Sigma-Aldrich),
Polyarginine hydrochloride (PA) (M ~ 15-70 kDa, Sigma-Aldrich),
Sodium chloride (99.9% Vekton)
Hydrochloric acid 38% (99.8% Vekton).
Dulbecco's Modified Eagles Media (DMEM),Biolot (Russia).
Fetal Bovine Serum (FBS), Biolot (Russia).
Isopropanol (99.9% Vekton)
Neutral red, Sigma-Aldrich.
MTT (3- [4,5- dimethylthiazol-2-yl-2,5-difeniltetrazolium bromide), Sigma-Aldrich.
LDH Cytotoxicity kit, (The Thermo Scientific, USA).
The latex particles (PanEco, USA).
Trypan blue, Sigma Aldrich.
All reagents were used without further purification.

1.2. Synthesis of ferromagnetic nanoparticles

Magnetite nanoparticles were prepared by coprecipitation of iron (II) and (III) chlorides in a molar ratio of 1:2 [Volodkin] using citric acid as a stabilizer. Iron chlorides (2.35g of $FeCl_3{\bullet}6H_2O$ and 0.86 g of $FeCl_2{\bullet}4H_2O$) were dissolved in 40 mL of deionized water. The obtained solution was poured into three-necked flask and heated to 80°C under nitrogen. Then ammonium hydroxide (5 ml of 25% solution) were slowly added to a heated solution with vigorous stirring and kept for 15 minutes. Then citric acid (4 ml of25% solution) was added to the particles colloidand kept stirring for 1 hour at 90°C. The magnetite particles were purified by dialysis(cellulose membrane Q1210-55 F3 ("Orange scientific", Belgium) with 12-14 kDa pore, the eluent 0.15 MNaCl, the volume ratio suspension/eluent equal 1/1500, the treatment time 24 hours). Then the purified colloid was centrifuged at 16 000 min^{-1} for 5 minutes to remove large particles.

1.3. Synthesis of magnetite containing polyelectrolyte microcapsules

Polyelectrolyte microcapsules were prepared by the «layer-by-layer» method with the calcium carbonate core using the technique described by Volodkin et al [1]. Synthesis of core was performed by rapid mixing of calcium chloride and sodium carbonate solutions in equal volume proportions at room temperature. The cores were precipitated by centrifugation (2000 min^{-1} for 30 seconds) and washed three times with deionized water. Calcium carbonate core were added in a 1 mg/mlPA solution.The suspension was shaked for 15 minutes washed three times with deionized water and then transfered into DS solution. The same manipulation was repeated for subsequent adsorption layers. Magentic nanoparticles having a negative surface charge were deposited on the layer of PA. After deposition ofthe magnetite NP the layer was supplementedwithDSfor compensation of remaining cationic charges. The next layer of magnetite was deposited on the 3-th layer of PA in the same way. As a result, we have obtained the microcapsule shells with following composition:PA/DS/PA/Fe_3O_4NP+DS/PA/Fe_3O_4NP+DS/PA/DS.

In the last stage of synthesis, cores were dissolved with a solution of 0.2M EDTA (pH 6,5). After dissolving of the cores, the microcapsules were washed three times with deionized water.

Polyelectrolite microcapsules without magnetite (PA-DS-M) were synthesed by the similar procedurefor the control group in cytotoxity studies.

1.4. Characterization of the objects

1.4.1. Particle size. Determination of dimensional characteristics of Fe_3O_4 nanoparticles was performed by dynamic light scattering (Nano-Flex analyzer, Microtrac Inc., Germany).

1.4.2. Zeta potential. Measurement of zeta potential was carried out by electrophoretic mobility measurements (Stabino analyzer, Microtrac Inc., Germany).

1.4.3. Magnetite concentration. Quantitative determination of the magnetite in the samples (MNPand MCPM) was carried out by photocolorimetry. This method is based on the measurement of the optical density of colored iron (III) complexes with sulfosalicylic acid in an alkaline medium. Colloid of MNP (30 μl) or microcapsule suspension (30 μl containing capsules 260 mln / ml) was dissolved in 100 μl of 36% hydrochloric acid. A solution of hydrogen peroxide (200 μl of 6% solution) was added for the oxidation of iron to Fe^{3+}. The mixture was heated to 100°C and held at this temperature for 25 minutes. Then 1 ml of 17% sulfosalicylic acid and 1 ml of 10% ammonia solution were added. Then the mixture was resolved with deionized water to 50 ml and held for 10 minutes. The optical density of the solution was measured at λ = 430 nm (spectrophotometer Shimadzu UV2600). Iron concentration was determined with the calibration curve. The concentration of magnetite was obtained from the iron concentration by the formula:

$C (Fe_3O_4) = 1.38\ C(Fe)$.

Calculation of the content of magnetite in a capsule was carried by the formula:$S = C (Fe3O4)/n$, where C is the concentration of magnetite, and n is concentration of capsules in the suspension.

1.4.4. Visialization. Scanning Electron Microscopy (SEM). For SEM analysis, samples were prepared by applying a drop of the particle suspension to a glass slide and then drying overnight. Then samples were sputtered with gold and measurements were conducted using aGemini Leo 1550 instrument at an operation voltage of 3 keV (AFM, NT-MDT Ntegra system, Spectrum Instrument).

1.4.5. Magnetic properties. The magnetization of samples was measured by the induction method with a vibration magnetometer (EZ11, Microsense).

1.5. Cytotoxicity

Cytotoxicity of the synthesized nanoparticles and magnetite containing polyelectrolitemicrocapsules was investigated in accordance with the international standard operation procedure "Tests for in vitro cytotoxicity" (ISO 10993-5). [7].To investigate cells viability a series of tests were performed: MTT-test, neutral red uptake (NRU) test, lactate dehydrogenase (LDH) release test.The study was carried outaccording tothe Integrated Testing Strategies (ITS), because the median effective dose (ED_{50}) determined by asingle in-vitro toxicity test (such as MTT or NRU) have a low correlation (50-70%) with its real value (false negative results are frequently detected) [8].

1.5.1. Cell cultures and treatments. We used the culture of mouse fibroblast L929 obtained from tissue culture collections of D.I. IvanovskyInstitute of Virology. Cells were cultivated in flasks of DMEMsupplemented with 10% fetal bovine serum under the conditions: 5%atmosphere of CO_2, t = 37 ° C, humidity equal to 5% (incubator Sanyo, Japan). The cells in exponential growth phase were reseeded in a 96-well plate ($2 \cdot 10^4$ cells/well) containing 100 µl per well of DMEM supplemented with 10% FBS. Cells were incubated for 24 hours and then the mediumwas replaced with a fresh one with addition of MNP andMCPM in test series. The medium for the first control series (Control 1)contained 50 µl of phosphate buffer per well. The biodegradable capsules without iron oxide was added in the wells for second control series (Control 2).The concentration of objects were determined by the containing of magnetite. In every test series we used 6 levels of magnetite concentrations: 0.003; 0.032; 0.322; 0.644; 1.610;3.220 mg / ml. We used three values of exposure time: 24, 48 and 72 hours. This stage of the sample preparation was idetical for all tests.

1.5.2. MTT-test. MTT (3- [4,5- dimethylthiazol-2-yl-2,5-difeniltetrazolium bromide) is a water soluble salt of tetrazolium.It is transformed into an insoluble formazan which is accumulated in cytoplasm. The amount of formazan depends on activity of mitochondrial succinate dehydrogenase and is proportional to the number of living cells. The concentration of formazan can be determined by a colorimetric method. MTT test was performed as follows. After the sample preparation described above, we aspirated the medium and added 50 µl of 5 mg / ml MTT solution. Cells were incubated for 2 hours. After that the medium was removed. Then 100 µl of isopropanol was added in the well, and the plate was vortexed for 20 minutes until the formazan was dissolved completely. Optical density was measured by a microplate reader (EFOS 9305, Russia) at a wavelength of 492 nm with a reference wavelength of 620 nm. The cell viability was determined as the ratio of the optical density (OD) of the sample to the OD of the control solution, and expressed as a percentage.

1.5.3. Neutral Red Uptake test. The method is based on the original protocol proposed by Borenfreund and Puerner (1984) [9] and detects the accumulation of neutral red dye in the lysosomes of intact viable cells. It was realised in the following way. After incubation, medium was removed and 90 µl of fresh DMEM medium with 10 µl of 0.33% neutral red solution was added to each well. The cells were held in the incubator for 2 h. Then, the medium was carefully removed and the cells were fixed quickly with 0.5% formalin solution supplemented with $CaCl_2$. The fixing solution was removed by washing with a phosphate buffer. After washing,the cell monolayerwas dissolved by 1% acetic acid solution in ethanol. For solubilization of dye, plates were shaked for 10 min at room temperature, then supernatant was carefully aspirated and transferred to a clean plate for further measurements. Optical density was measured on a microplate ELISA reader (EFOS 9305, Russia) at a wavelength of 450 nm against a reference of 620 nm. The viability calculations were carried out similarly to the MTT test.

1.5.4 LDH release test. The method is based on determining of lactate dehydrogenase enzyme which is normally located in the cytoplasm and is released into the cultural medium whencells are irreversible damaged [10]. LDH content was determined using a colorimetric method with commercial kit (Pierce LDH Cytotoxicity Assay, The Thermo Scientific, USA). It was realised as follows. After the sample preparation, the medium was carefully aspirated and transferred to a clean plate.According to the procedureprotocol, reagents were added. Finally the stop-reagent was added and the absorbance was evaluated at 490 nm wavelengthagainst 680 nm reference.

1.6.Phagocytosis

Phagocytosis of particles was studied in the in vitro reaction with human whole blood [11]. The particles were added to the blood of healthy donors stabilized with sodium citrate. The latex particles (PanEco, USA) were added in the control series. Samples were incubated for 30 minutes at 37° C. After incubation, the samples were separated to two parts. In the first one, neutrophil viability was evaluated by the intravital staining with trypan blue. From the second part, smears were made and stained byRomanovsky-Giemsa.The activity of phagocytosis was evaluated by light microscopy. We calculateoverall phagocytic capacity (Percent phagocytosis (%P)) as the percentage of phagocytizing cells related to the total number of phagocytes. The phagocytic index (PI)was defined as the average amount oflatex particles internalized by one phagocyte.

1.7. Statistics

Statistical processing was made using t-test by MS Excel.

2. Results and discussion

2.1. Characterization ofMNP and MCPM

The size distribution for magnetite nanoparticles is shown in Figure 1.

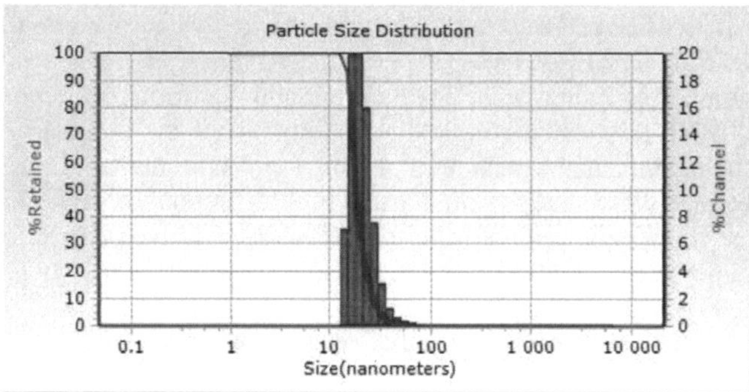

Figure 1. Size distribution of MNP.

The average size is 18±4 nm,zeta potential is equal to –39±2 mV at 7.0±0.1 pH. Suspension of MNP did not contain large aggregates and it was stable for 4 weeks. The SEM-images of MCPM are shown in Figure 2.

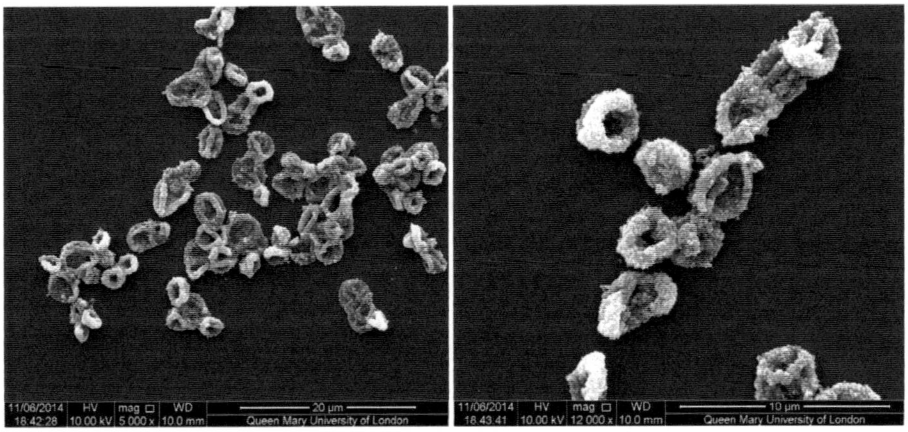

Figure 2. The SEM-images of MCPM.

One can see that the microcapsules have spherical or biconcave shape and average size of 2,9±0,7µm. The content of the magnetite in the colloid was3,24±0,05µg/ml or ≈0,012 ng per 1 capsule. These data is in agreement with the results obtained previously by other authors using the same synthesis method [12].

2.2. Magnetic properties

The hysteresis curves for MNP and MCPM samples are shown in Figure 3. The dependence of magnetization *M* on magnetic field *H* represented in the figure is typical for small magnetic particles in a liquid [13]. The saturation magnetization M_s (renormalised to the magnetite mass) was 55 emu/g and 70 emu/g for MNP and MCPM respectively. The saturation magnetization values for both samples are of the same order of magnitude as the value of M_s for bulk magnetite (about 90 emu/g). This result allows to conclude that magnetic properties of magnetite are basically preserved both in MNP and MCPM. Consequently, both the objects can be used in the development of new tools for magnetically driven target drug delivery.

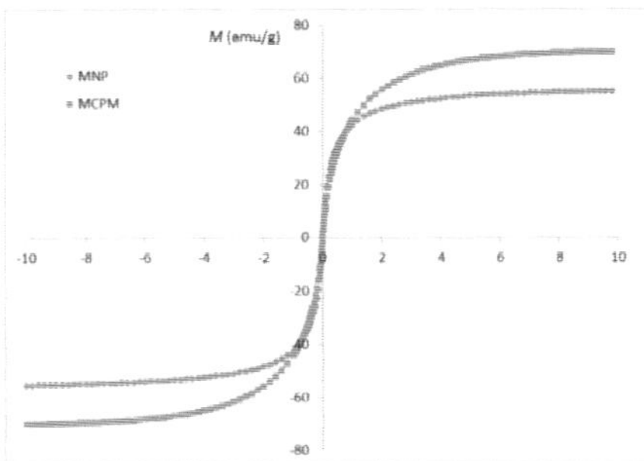

Figure 3. Magnetization hysteresis curves for MNP and MCPM samples. Magnetization is renormalized to the mass of magnetite in the sample.

For further studies of cytotoxicity, we have prepared solutions of colloid MNP and MCPM with identical concentrations of magnetite.

2.3 Cytotoxicity

Results of Cytotoxicity analysis of MNP and MCPM are shown in Figure4.

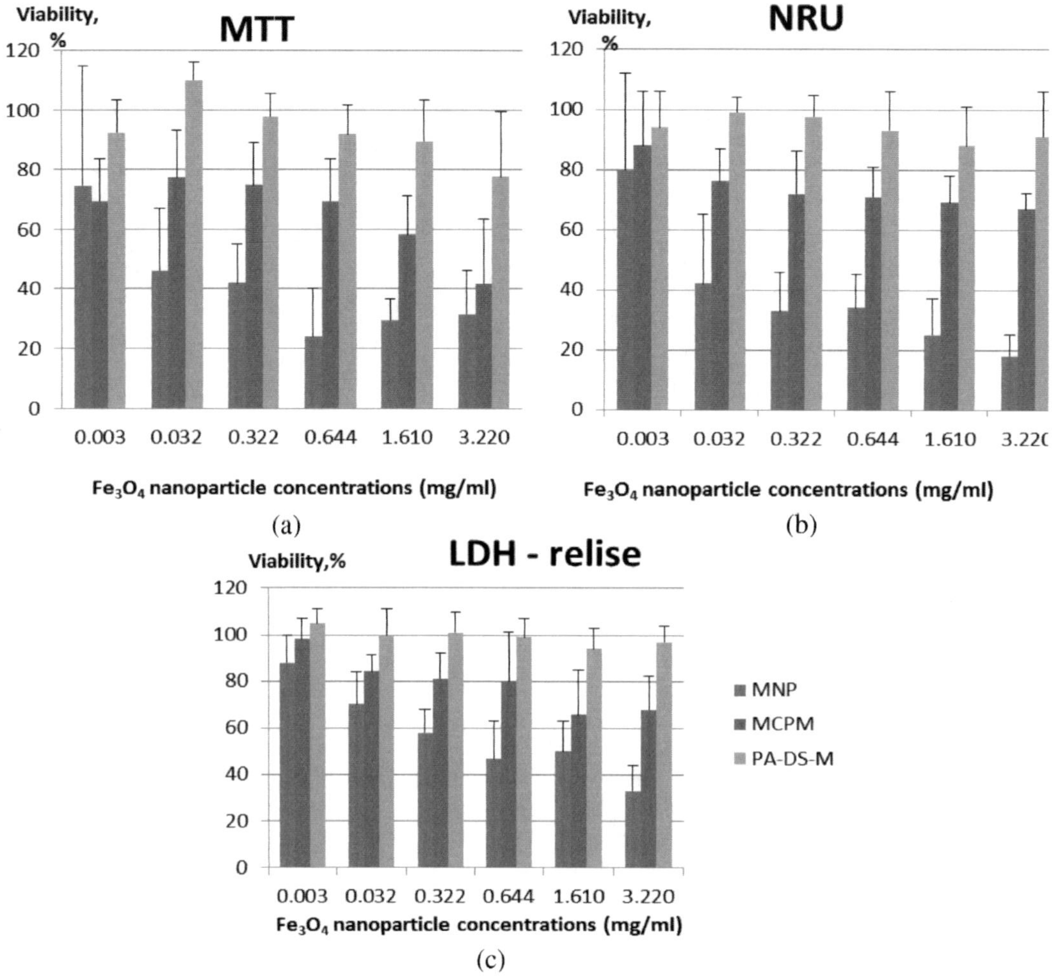

Figure 4. The viability of L929 cells after 24 h incubation with MNP and MCPM particles and PA-DS-M capsules according to the MTT test (a), NRU test (b) and LDH release test (c).

Analysis of MTT assay data (Figure 4, a) shows that a minimal concentration of MNP necessary for viability inhibition of L929 is 0.322 mg / ml. In the concentration range from 0.322 to 3.22 mg / ml, fraction of viable cells was 42-23%. Unlike MNP, polyelectrolyte capsules containing the equivalent amount of magnetite provided significantly less impact on the MTT assay results. In this case, the minimum concentration of magnetite causing a significant reduction in viability of cells was equal to 3.22 mg / ml. Polyelectrolitemicrocapsules without magnetite do not have a negative impact on the culture of L929.Similar results were obtained by NRU test (Figure 4, b). In the LDH release test,it wasalso found that thecytotoxicity of microcapsules is lower compared to the free magnetite nanoparticles. However, this method had significantly lesssensitivitythan the others. The minimal concentration of MNPcausing a cytotoxic effect determined by this testwas equal to 0.644 mg/ml. There was no significant cytotoxic effect observed by this method in the MCPM series as well as in the control series with the empty polyelectrolyte capsules.

Low sensitivity of LDH release method is explained by the fact that damage of cell membranes (deteced by this method)occureat the final stage of the cell death process. At the same time, MTT and NRU tests detect earlier and "finer" violations ofcell functions (thedamage of oxidative enzymes) [14].

Cytotoxicity studies have shown that the most sensitive test for estimation of the cells viability was MTT assay. According to this test, minimal doses providing significant effect on the cell viability was 0.322 mg/ml and 3.22 mg/ml of iron for MNP and MCPM respectively. To investigate the time-dependence of cytotoxicity the viability of L929 cells for these concentrations was estimated with MTT assay in 24, 48 and 72 hours after addition of MNP and MCPM to the incubation medium. The results are shown in Figure 5. It can be seen that at both concentrations cytotoxic effect reaches a maximum in 48hours and remains at approximately the same level for 72 hours.

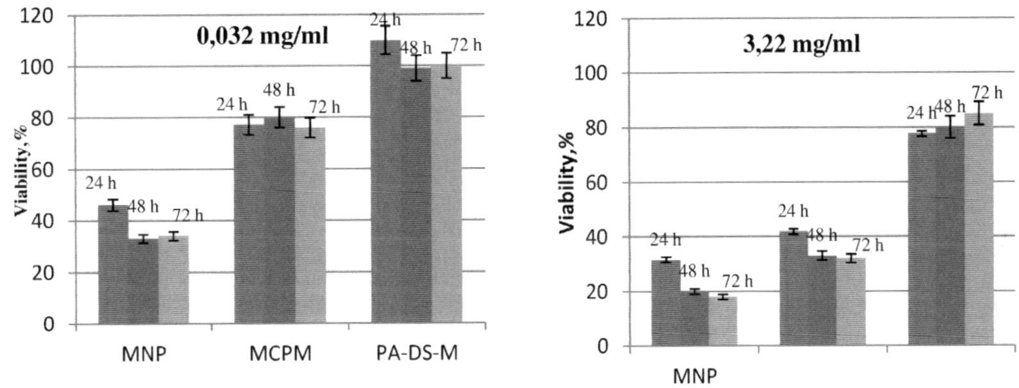

Figure 5. Cell viability studies at different time of exposureof MNP and MCPM particles and PA-DS-M capsules according to the MTT test.

The results of study of interaction between white blood cells and the magnetite containing objects are shown in Table 1. It is found that there is not significant reduction of phagocytes viability after incubation with MNP and MCPM.

Table 1. Impact of the magnetite containing objects on white blood cells viability and phagocytosis.

Parameter	Latex particles (control)	MNP	MCPM
Whitebloodcellsviability,%	93,1±3,3	95,1±1,7	92,9±0,7
Percent phagocytosis(%P), %	95±3,7	39±3,7 (p_{1-2}=0,002)[a]	46,7±3,1 (p_{1-3}=0,002)[b] (p_{2-3}>0,2)[c]
Phagocytic index (PI), n	8,7±0,9	-	3,5±1,5 (p_{1-3}=0,03)[b]

a - :p_{1-2} – statistically significant differences MNP vs Control (latex),
b - p_{1-3} –statistically significant differences MCPM vs Control (latex),
c - p_{2-3} - statistically significant differences MNP vs MCPM.

It was found by the light optical microscopy of leukocytes incubated with MNP and MCPM that both of these objects undergo phagocytosis. Neutrophils with inclusion of brown magnetite granules as well as cells with internalized microcapsule were observed in the blood smears. Phagocytosis index did not differ significantly in the groups with MNP and MCPM but it was two times lower compared to the control. The intensity of MNP phagocytosis was not determined because the particle size did not allow them to be identified using light microscopy. The magnetite containing microcapsules inside phagocytes were visualized clearly enough. The phagocytosis number for MCPM was 2 times lower in comparison with the one for latex.

The toxicity of iron oxide was discussed in detail in a number of studies. It was found that high concentrations of iron oxide nanoparticles induce cell necrosis [15,16]. Most authors proposed that the mechanism of necrosis inducing is associated with the generation of reactive oxygen species (ROS) [17]. Although antioxidant mechanisms present in the living systems, excess of ROS can cause lipid peroxidation and damage of cell membranes, proteins and DNA.

Autophagy can be another possible mechanism of toxicity. The majority of nanoparticles penetrating into the cell by endocytosis merges with lysosomes and form autophagosomes.[18, 19]. Internalization allows nanoparticles to reach subcellular structures and interact with various organelles (mitochondria, endoplasmic reticulum, nuclear membrane, chromosomes). The integration of these structures with autophagosomes blocks the cell functions. Critical increase in the autophagosome number leads to cell necrosis[20, 21].

The polymer shell of microcapsules blocks the toxic effects of ferromagnetic nanoparticles. The most likely mechanism for reducing of cytotoxicity is shielding of iron oxide nanocrystals from the intracellular environment. Thus incapsulation reduces their catalytic effect on the lipid peroxidation processes.

Tests with native blood did not show toxic effects of the magnetite containing objects in contrast to the tests with cell cultures. Both MNP and MCPM did not reduce the viability of neutrophils. This may be due to the presence of the protein in the environment. It is known that interaction of nanomaterials with many plasma proteins form so-called "protein crown" around the particles [22,23]. In this case, the particles interact substantially less intensive with cellular membranes and consequently produce less damage on them. This data illustrates the differences of in vitro and in vivo toxicity and suggests that the toxic effect for MNP and MCPM in the living organism is to be lower compared to cell cultures.

Concerning to the interaction of nanomaterials with immune cells, many questions still remain unresolved [24].It is known that nanoparticles having no antigenic determinants are not recognized by immune cells. In our study, we have shown possibility of phagocytosis of ferromagnetic nanoparticles and polyelectrolyte capsule in human blood but its activity was significantly lower than that of the latex which is the standard object for the study of phagocytosis. Most likely, this is due to low antigenicity and non-specific adsorption of plasma proteins. Even registered phagocytosis level can make a significant contribution to the biokinetics of nanoparticles and microcapsules. Furthermore, phagocytized magnetic microcapsules can be used for intracellular delivery of various compounds, as well as for remote control of phagocytes by external fields.

3. Conclusions

The interaction of magnetite nanoparticles and polyelectrolyte microcapsules containing these particles in the shell with L929 cells and human white blood cells was investigated. The dose-dependent effect of the substances on the viability of the L929 cells was observed. It is shown that iron oxide nanoparticles have higher cytotoxicity than microcapsules containing an equivalent amount of magnetite. The maximum cytotoxic effect was registered in 48 hours after administration of the substance in the incubation medium and persisted for 72 hours. It is found that nanoparticles and microcapsules do not affect significantly on the viability of neutrophils. It is shown that both the microcapsules and nanoparticles undergo phagocytosis, but the phagocytosis activity was lower for them than that for the latex. The results indicate the possibility for in vivo administration of magnetite containing microcapsules. Moreover, the capsules are the promising tool for the development of targeted delivery systems based on autologous or homologous blood cells.

Acknowledgements
The work was supported by Ministry of Education and Science of the Russian Federation as part of the State task for National Research Mordovia State University, project No. 2952 and the Government of the Russian Federation (grant №14.Z50.31.0004 to support scientific research projects implemented under the supervision of leading scientists).

References
[1] Volodkin D,Larionova N and Sukhorukov G 2004 Protein Encapsulation via Porous CaCO3MicroparticlesTemplatingBiomacromolecules5 1962-72
[2] Pavlov A, De Geest B, Louage B, Lybaert L, De Koker S, Koudelka Z, Sapelkin A and Sukhorukov G 2013 Magnetically Engineered Microcapsules as Intracellular Anchors for Remote Control Over Cellular Mobility Adv Mater25 6945-50
[3] Pavlov A, Gabriel S, Sukhorukov G and Gould D 2015 Improved and Targeted Delivery of Bioactive Molecules To Cells With Magnetic Layer-By-Layer Assembled MicrocapsulesThe Royal Society of Chemistry DOI: 10.1039/c5nr01261a
[4] McBain S, Yiu H and Dobson J2008 Magnetic Nanoparticles for Gene and Drug DeliveryInt J Nanomed, Jun 3(2) 169-80
[5] Schafer F, Qian S and BuettnerG 2000 Iron and Free Radical Oxidations In Cell MembranesCell MolBiol (Noisy-le-grand) May 46(3) 657-62
[6] Singh N, Jenkins G, Asadi R and Doaka S 2010 Potential toxicity of superparamagnetic iron oxide nanoparticles (SPION) Nano Rev1 10.3402/nano.v1i0.5358
[7] International Standard Operation (ISO) 10993–5 2009-06-01 "Tests for in vitro cytotoxicity" Third edition
[8] EUR 25946 - EURL ECVAM 2013 Recommendation on the 3T3 Neutral Red Uptake Cytotoxicity Assay for Acute Oral Toxicity Testing (Luxembourg: Publications Office of the European Union) p 47
[9] Borenfreund E andPuerner JA 1984 A simple quantitative procedure using monolayer culture for toxicity assays J Tissue Cult Meth9 7–9
[10] Chan FK-M, Moriwaki K, De Rosa MJ 2013 Detection of Necrosis by Release of Lactate Dehydrogenase (LDH) ActivityMethods MolBiol 979 65-70
[11] Reviews on Immunoassay Technology 1988 vol. 2. ed SB Pal (London: The macmillan Press LTD)
[12] Antipina MN and Sukhorukov GB 2011 Remote control over guidance and release properties of composite polyelectrolyte based capsules Adv Drug Deliv Rev63 716–29
[13] Shliomis M I 1974 Magnetic fluids SovPhysUsp17 153–69
[14] Fotakis G andTimbrell JA 2006 In vitro cytotoxicity assays: comparison of LDH, neutral red, MTT and protein assay in hepatoma cell lines following exposure to cadmium chloride Toxicology Letters160(2) 171-77
[15] Liu G, Gao J, Ai H and Chen X 2013 Applications and Potential Toxicity of Magnetic Iron Oxide Nanoparticles Small May 27 9(9-10) 1533-45
[16] Soenen S J, De Cuyper M, De Smedt S C and Braeckmans K 2012 Investigating the Toxic Effects of Iron Oxide NanoparticlesMethods in Enzymology509195-224
[17] Patil SU, Adireddy S, Jaiswal A, Mandava S, Lee BR andChrisey DB 2015 In Vitro/in Vivo Toxicity Evaluation and Quantification of Iron Oxide Nanoparticles Int J MolSci16 24417–50
[18] Vilella A, Ruozi B, Belletti D, Pederzoli F, Galliani M,Semeghini V et al 2015 Endocytosis of Nanomedicines: The Case of Glycopeptide Engineered PLGA Nanoparticles Pharmaceutics 7(2) 74-89

[19] Khan MI, Mohammad A, Patil G, Naqvi SA, Chauhan LK and Ahmad I 2012 Induction of ROS, Mitochondrial Damage and Autophagy in Lung Epithelial Cancer Cells by Iron Oxide NanoparticlesBiomaterials 33 1477-88
[20] Kayal S andRamanujan RV 2010 Anti-cancer Drug Loaded Iron-Gold Core-Shell Nanoparticles (Fe@Au) for Magnetic Drug TargetingJ Nanoscinanotechno10 5527-39
[21] Li JJ, Hartono D, Ong CN, Bay BH and Yung LY 2010 Autophagy and oxidative stress associated with gold nanoparticles Biomaterials31 5996-6003
[22] Rahman M, Laurent S, Tawil N, Yahia L andMahmoudi M 2013 Protein-Nanoparticle Interactions The Bio-Nano Interface p 86
[23] Lundqvist M, Stigler J, Cedervall T, Berggard T, Flanagan MB, Lynch I, Elia G and Dawson K 2011 The Evolution Of The Protein Corona Around Nanoparticles: A Test StudyACS Nano5 7503-09
[24] Hussain S, Vanoirbeek J A and Hoet P H 2012 Interactions of nanomaterials with the immune system Wiley Interdiscip Rev NanomedNanobiotechnol4(2) 169-83

Ways of development of compact coherent femtosecond X-ray sources for applications in nano- and biophotonics

L Mikheev[1,2]

[1]P.N. Lebedev Physical Institute, Russian Academy of Sciences, Leninsky prosp. 53, 119991 Moscow, Russia
[2]National Research Nuclear University "MEPhI" (Moscow Engineering Physics Institute), Kashirskoe sh. 31, 115409 Moscow, Russia

E-mail: mikheev@sci.lebedev.ru

Abstract. Ways of the development of compact coherent sources of soft X-ray femtosecond pulses are discussed, which meet the requirements for the implementation of the "diffraction-before-destruction" approach in the lensless X-ray Coherent Diffractive Imaging (CDI) technique enabling quantitative 3D mapping of material structure with the nanoscale spatial resolution. An innovative hybrid (solid/gas) approach to produce ultra-intense femtosecond laser pulses in the visible is described in the context of its applications for laser driven high harmonic generation (HHG) and soft X-ray generation in laser plasmas due to recombination mechanism of excitation.

1. Introduction

Successful development of new technologies in nano- and bio-photonics strongly depends on the availability of a modern instrumentation for structural analysis. The most powerful and informative tool in the study of matter structure is the lensless X-ray Coherent Diffractive Imaging (CDI) technique enabling quantitative 3D mapping of materials with the nanoscale spatial resolution determined by the wavelength of a radiation source (e.g., [1]). In this technique, a sample is illuminated with a spatially coherent beam of light to obtain far-field diffraction patterns allowing the reconstruction of sample images with a computational phase retrieval algorithm. X-ray CDI has been successfully applied for nanoscale imaging with coherent soft x-ray radiation from synchrotrons, X-ray free-electron lasers (XFELs), laser driven high harmonic generation (HHG) and with a partially coherent compact gas–discharge EUV light source at 17.3 nm in both single and multiple shot modes [1-3].

It is well known that biological structures are highly sensitive to radiation damage. To solve this problem, a "diffraction-before-destruction" approach is used, which enables diffraction signals to be collected from a sample in a single femtosecond-duration pulse of high intensity before the onset of significant radiation damage. Among the above mentioned X-ray and EUV light sources, nowadays only XFELs, providing intense femtosecond pulses in diffraction-limited soft–X-ray beams, meet the requirements for implementation of the "diffraction-before-destruction" method in CDI.

However, despite the clear potential for x-ray microscopy, XFELs have limited availability because of their high cost and the enormous scale of linear electron accelerators of several kilometers in length. This limits the wide spread of ultrafast X-ray CDI technique and makes it relevant to the development of alternative coherent X-ray sources for tabletop microscopy suitable for applications in academic and industrial laboratories.

One of the greatest challenges is the development of compact coherent X-ray sources radiating in the ''water window'' between the K absorption edges of carbon (4.37 nm) and oxygen (2.33 nm), where carbon is highly opaque, while water is largely transparent. This spectral range enables the observation of living biological structures in a natural aqueous environment with high contrast ratio and unprecedented spatial resolution.

2. High-order harmonic generation

Recently, intensive efforts have been made to develop such compact soft X-ray sources on the basis of laser driven HHG widely used to generate spatially coherent short wavelength radiation in gases and on the surface of solid targets (e.g., [4-6]). HHG can provide a single burst or a train of attosecond pulses which allow for applying X-ray CDI technique for ultrafast tabletop microscopy. HHG in gases is nowadays a routine operation. However, it has low up-conversion efficiency (typically less than 10^{-4}) and limited laser intensity scaling above 10^{16} W/cm^2 because of phase mismatch due to plasma generation [4, 6]. These features impede HHG in gaseous media to be used for the development of highly intense X-ray sources which would be suitable for the "diffraction-before-destruction" CDI microscopy of biological structures.

Other concept relies on the generation of high-order harmonics by reflecting ultra-intense lasers from a solid surface, which has no limitation on driving laser intensities and promises much higher conversion efficiencies. According to the simplest oscillating plasma mirror model [7] the laser field induces a relativistic oscillation of the overcritical plasma surface, which results in Doppler frequency upshift of the laser light reflected by the dense plasma. This gives rise to the incident frequency harmonics emerged as an attosecond pulse train. HHG driven by a few-cycle laser field produces an isolated attosecond pulse [8], that is required to implement the "diffraction-before-destruction" method in X-ray CDI. Particle-in-cell (PIC) simulations [9] demonstrate a significant enhancement of conversion efficiency up to 20% with shortening driving laser pulse down to single-cycle pulse-width.

Up-conversion efficiency of HHG from solids displays strong driving laser frequency scaling [5, 10] so that shorter wavelength lasers are more efficient in the harmonic generation.

3. Soft X-ray lasers

Only collisional and recombination mechanisms of active media excitation in laser plasmas are of practical interest for the generation of coherent soft X-ray radiation. The first one was realized in laser plasmas with high electron temperature, providing a population inversion on transitions between excited states of ions, which typically lie in the range 10-50 nm. There is only one experimental result showing the X-ray emission at shorter wavelengths of 4.3 and 4.5 nm [11], close to the edge of the "water window". However, pulse energies reached in collisional soft X-ray lasers are too low for CDI applications.

The most promising way to reach desirable energy within the ''water window'' with soft X-ray lasers lies in the realization of recombination scheme of laser action on transitions to the ground state of recombining fully stripped ions, which are of great interest from the view point of scaling to shorter wavelengths. Optical-field-ionization (OFI) in a strong field of a linearly polarized laser beam is the most effective technique to produce fully stripped plasmas consisting of nuclei and cold electrons in a time much shorter than the time of their recombination. Low temperature of electrons enables their rapid recombination with nuclei resulting in the formation of the population inversion of low-lying levels of ions.

Residual energy of the electrons produced via OFI is proportional to $I\lambda^2$, where I and λ are the intensity and wavelength of the pump beam, respectively. That is why the first gain observation in the recombination scheme was performed with the use of a KrF-laser [12] at the wavelength of 248 nm. In these experiments, quite a large gain of 20 cm^{-1} at the Lyman-alpha transition of the H-like LiIII ion (13.5 nm) was achieved. Subsequently, this result was confirmed in several studies performed at the

Berkeley National Laboratory [13] and Princeton University [14]. However, the gain saturation necessary for the efficient energy extraction from the gain medium was not demonstrated. This is due to the fact that the KrF-laser pulse-width lying in the subpicosecond time-domain is too long and leads to overheating electrons in the laser plasma. Pump pulse-widths shorter than 100 fs are unattainable for the KrF-lasers characterized by narrow gain bandwidth (~ 300 cm^{-1}).

Pulse-widths and intensities of the widespread Ti:Sapphire laser systems are much more suitable for the recombination scheme realization. However, the radiation of such systems lies in the near infrared (800 nm) and the electron residual energy turns out to be too high for the recombination scheme implementation. The most attractive for the excitation of lasing on the transitions to the ground state is the second harmonic of a Ti:Sapphire system that allows for increasing gain on these transitions by two orders of magnitude [15, 16]. In particular, the numerical simulation of the gain on the transition to the ground state of CVI H-like ions (3.4 nm falling within the "water window") shows that gain as high as 180 cm^{-1} can be attained at an intensity of $(0.5\text{-}1)\times10^{19}$ W/cm^2 in a 20-50 fs pulse [16].

Achieving these parameters in a Ti:Sapphire laser system with nonlinear frequency doubling encounters serious technical difficulties in the manufacture of thin nonlinear crystals of large diameter, which have not yet been resolved. Development of hybrid (solid/gas) femtosecond systems in the visible can solve this problem and promises the output parameters, which meet the requirements of recombination X-ray lasers excitation.

4. Hybrid (solid/gas) femtosecond systems in the visible

An innovative hybrid (solid/gas) approach to produce ultra-intense femtosecond laser pulses in the visible was proposed at P.N. Lebedev Physical Institute of RAS (LPI, Moscow, Russia) [17] and thereafter it was developed in the course of experimental and theoretical studies carried out in cooperation with the Lasers Plasmas and Photonic Processes (LP3) Laboratory (Marseille, France) and the Institute of High-Current Electronics (IHCE, Tomsk, Russia) (see [18, 19] and references cited therein). This approach is based on original methods, developed at the LPI, of photochemical excitation of active gas media on broadband electron transitions in molecules (XeF(C-A), Xe_2Cl and Kr_2F) and on amplification of the second harmonic of a Ti:sapphire femtosecond laser in these active gas media. As a result of these studies, two hybrid femtosecond systems in the visible (475 nm) have been built at the LPI and the IHCE: THL-30 with a design peak power of 5-10 TW and THL-100 designed for 50–100 TW peak power, respectively (figures 1,2). Both of them are based on application of the XeF(C-A) active medium in a boosting amplifier.

(a) (b)

Figure 1. THL-30: Photos of the Ti:Sapphier front-end (a) and XeF(C-A) amplifier (b).

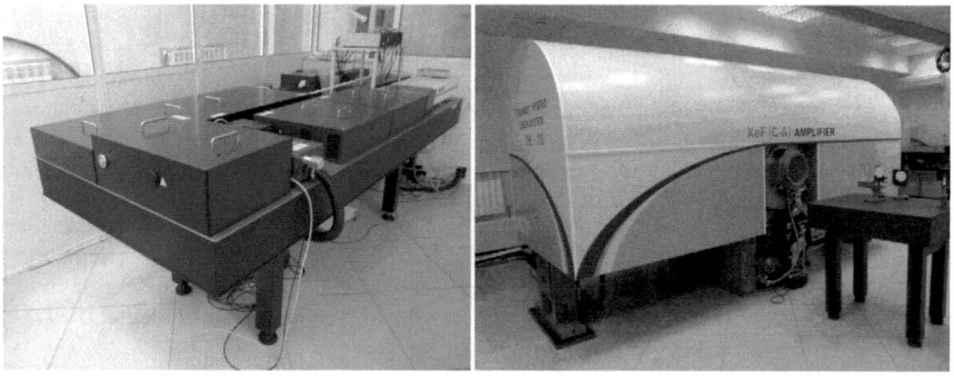

(a) **(b)**

Figure 2. THL-100: Photos of the Ti:Sapphier front-end (a) and XeF(C-A) amplifier (b).

Pump technique is based on the conversion of e-beam energy to the VUV radiation of xenon at 172 nm used to excite the XeF(C-A) active medium. Principle of the converter operation is explained in figure 3 for the case of THL-30 system in which four 120 cm long × 15 cm wide radially converging e-beams are injected through 40 μm Ti foils into the converter filled with xenon at pressure of 3 bars. Xenon radiation excited by electrons comes through CaF_2 windows to a laser cell housed into the e-beam converter along its axis and filled with XeF_2/N_2 mixture at 0.25 bar. The XeF(C-A) active medium is excited due to photodissociation of the XeF_2 vapor by the 172 nm radiation. Principle of the THL-100 operation is practically the same besides using six radially converging e-beams in the converter and twice higher pump energy. More details on the subject can be found elsewhere [18, 19].

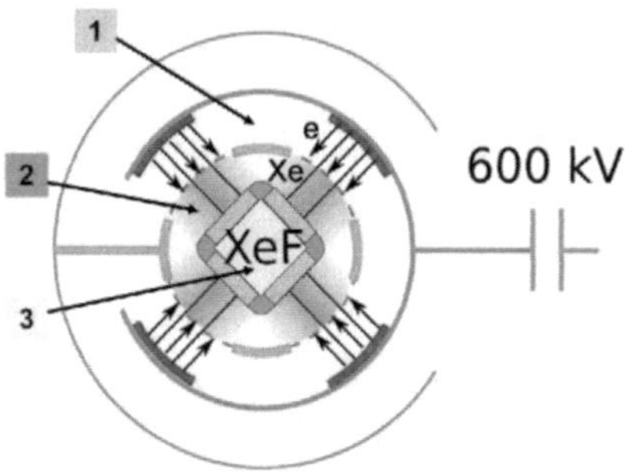

Figure 3. Cross sectional schematic diagram of the XeF(C-A) amplifier: (1) vacuum diode, (2) e-beam converter, (3) photolytic laser cell.

Laser XeF(C-A) transition is centered near 480 nm spectrally matching with the second harmonic of a Ti:Sapphire laser and has very broad bandwidth of $\Delta\lambda$=70 nm corresponding to the transform limited pulse duration of 5 fs. The wide emission continuum for the C–A transition is caused by strongly repulsive nature of the lower A state. The boosting XeF(C-A) amplifier is seeded by 50 fs pulses early down-chirped to 1-2 ps in a prism pair. Seed pulses of 0.1 TW peak power at 475 nm are produced with a solid-state Ti:Sapphire front-end (Avesta Project Ltd) consisting of a Ti:Sapphire oscillator operating at 950 nm, regenerative and multipass amplifiers, and KDP frequency doubler spectrally matching the front end to the XeF(C–A) transition. Down-chirped pulses amplified in the boosting amplifier are recompressed in bulk fused silica compressor.

In pilot proof-of-principle experiments made in the THL-100 system, 14 TW peak power was obtained in a 50 fs pulse at the output energy of 0.7 J [19]. In the hybrid system THL-30 the output energy of 0.25 J was achieved, which allows a peak power of 5 TW to be obtained after temporal pulse compression to 50 fs [19].

5. Conclusions

Results obtained in the course of the novel hybrid concept studies demonstrate the high potential of the approach relying on the optically driven broadband active media in the visible. The most remarkable achievement of these studies is the record breaking peak power of 14 TW in the visible, which seems to be promising for the development of compact coherent X-ray sources based on the laser driven HHG from solids and recombination mechanism of X-ray generation in laser plasmas.

Acknowledgments

This work was supported by the Competitiveness Program of NRNU MEPhI.

References

[1] Chapman H and Nugent K 2012 *Nature Photon.* **4** 833-9
[2] Seaberg M D *et al.* 2011 *Opt. Express* **19** 22470–9
[3] Dstrcil M O, Ussmann J B, Udolf D R, Resenitz R B, Jianwei Miao, Brocklesby W S and Juschkin L 2015 *Opt. Lett.* **40** 5574-7
[4] Eden J G 2004 *Progr. Quantum Electron.* **28** 197-246
[5] Teubner U and Gibbon P 2009 *Rev. Mod. Phys.* **81** 445-79
[6] Brabec T and Krausz F 2000 *Rev. Mod. Phys.* **72** 545-91
[7] Lichters R, Meyer-ter-Vehn J and Pukhov A 1996 *Phys. Plasmas* **3** 3425-37
[8] Heissler P 2012 *Phys. Rev. Lett.* **108**, 235003
[9] Ma G, Dallari W, Borot A, Krausz F, Yu W, Tsakiris G D and Veisz L 2015 *Phys. Plasmas* **22**, 033105
[10] Gibbon P 1996 *Phys. Rev. Lett.* **76** 50-3
[11] MacGowan B J et al. 1990 *Phys. Rev. Lett.* **65** 420-3
[12] Nagata Y 1993 *Phys. Rev. Lett.* **71** 3774-7
[13] Donnelly T D et al. 1994 *Proc. 4th Int. Conf. on X-Ray Lasers* (Williamsburg, VA) (New York: AIP) p. 106
[14] Korobkin D V, Nam C H, Suckewer S and Goltsov A 1996 *Phys. Rev. Lett.* **77** 5206-9
[15] Spence D J and Hooker S M 2005 *Opt. Commun.* **249** 501-13
[16] Avitzour Y and Suckewer S 2007 *J. Opt. Soc. Am. B*, **24** 819-28
[17] Mikheev L D 1992 *Laser Part. Beams* **10** 473-8
[18] Mikheev L D, Tcheremiskine V I, Uteza O P and Sentis M L 2012 *Progr. Quantum Electron.* **36** 98–142A reference
[19] Mikheev L D and Losev V F 2016 *Multiterawatt Hybrid (Solid/Gas) Femtosecond Systems in the Visible. In "High Energy and Short Pulse Lasers"*, ed. R Viskup (Croatia: InTech) chapter 6 pp 131-161

IOP Conf. Series: Journal of Physics: Conf. Series **784** (2017) 012040 doi:10.1088/1742-6596/784/1/012040

Health risk assessment of potable water containing small amount of tritium oxide

O A Momot[1], B I Synzynys[1], A A Oudalova[1,2]

[1] Ecology Department, Obninsk Institute for Nuclear Power Engineering of the National Research Nuclear University MEPhI, Obninsk, Russia

[2] Russian Institute of Radiology and Agroecology, Obninsk, Russia

E-mail: momotulya@gmail.com

Abstract: The problem of groundwater pollution with tritium in a vicinity of radiation-dangerous facilities in Obninsk is considered. The information on the specific activity of tritium in Obninsk water sources is provided. The formula for the calculation of the β-radiation absorbed dose from tritium ingestion is proposed, reflecting the biological behavior of tritium in a human body. To establish the extent of tritium effects on human, the health risk is assessed. It is shown that if the specific activity of tritium in drinking water amounts to 10 Bq/l, the risk of stochastic effects of radiation will not exceed the limit of the individual lifetime risk.

1. Introduction

Obninsk is situated in the Kaluga region 100 km south-west of Moscow on the left bank of the Protva river. Several enterprises utilizing nuclear energy are operating in the town, the main of which are State Scientific Centre of the Russian Federation – the "Institute of Physics and Power Engineering" (IPPE) and the Branch of JSC «Karpov Institute of Physical Chemistry». Development of nuclear technologies, the available centers of utilization, reprocessing and disposal of radioactive materials and wastes have caused the need for studying technogenic radionuclide behavior and migration in this region. The primary effects associated with Obninsk radiation-dangerous facilities are gas and aerosol releases into the atmosphere, liquid waste discharges into the Protva river and groundwater pollution. Special emphasis should be pointed at the problem of environmental pollution with tritium [1].

The aim of this work is to provide radiological characteristics of drinking water sources of Obninsk based on the tritium content and to evaluate the impact of tritium on the Obninsk citizens, using the risk assessment methodology.

2. Characterization of groundwater in Obninsk

Groundwater sources are used for Obninsk water supply. Municipal intakes involve 34 wells arranged 15 km apart in a sequence north-south along the foot of the left slope of the Protva river valley. Northern and southern water intakes are upstream and downstream of the town, respectively. They belong to river valley intakes with mixed feeding, i.e. precipitation infiltration is responsible for a smaller part of groundwater and a greater amount is being formed by overflowing from Protva. Water reserves are maintained by the Protva river runoff, the volume of which depends on the precipitation fallen out and watershed area.

Groundwater reserves in the intakes upstream of the town are much smaller as compared to those located downstream as feeding in the latter case is maintained also by a runoff of the Luzha river, a large tributary 1000 km^2 in area.

Actual water production is distributed between municipal intakes just "on the contrary", i.e. three fourth of water is drawn from the northern intakes and only a third from the southern one in Dobroe settlement. Such a "disrepancy" in groundwater production has caused more than 40 m lowering in a piezometric level of the industrial water bearing horizon for the very northern intake in Vashutino. Along the Obninsk intakes, a hydraulic slope is formed in an industrial water bearing horizon that is directed upstream opposite to the natural runoff of surface and groundwater. The inverse flow of river water ingressing into a groundwater bearing horizon, on the one hand, compensates the depletion of groundwater reserves in the northern intakes and, on the other hand, gives rise to the pollutant transfer from the Protva river into municipal water intakes [2].

Piezometric depression has essentially changed hydrodynamics and geochemistry of groundwater and is one of the reasons for dewatering of the upper water bearing horizons, the rocks of which turned out to be in the aeration zone. Rock oxidation in the aeration zone has stipulated the ingress of oxidation products into groundwater. Industrial and domestic wastes, the outlets for which are arranged in the south and downstream of the town, partially enter the northern intakes together with an inverse flow of river water.

The interaction of natural and anthropogenic factors led to a decline in the quality of produced water. Table 1 presents the results of laboratory testing of drinking water quality on organoleptic and chemical indicators of centralized drinking water supply systems of Obninsk, averaged for 2015 [3]. Among these indicators, only the content of ammonia nitrogen is above the established hygienic standards for drinking water (MPC), the rest are in the normal range.

Table 1. Results of laboratory analysis of drinking water in Obninsk

Sample	Indicator							
	Colour, degree	Turbidity, mg/l	Fe, mg/l	Dry residue, mg/l	General rigidity, mEq/l	Cl^-, mg/l	NH^{4+}, mg/l	Oil, mg/l
Average	5	0.5	0.21	448	5.77	2	7.25	0.009
MPC	20	1.5	0.3	1000	7	350	2	0.1

The previously described hydrodynamic and hydrochemical conditions are conducive for the radionuclides migration in Obninsk groundwater. Despite the nuclear specialization of large enterprises in Obninsk, groundwater pollution by radioisotopes is studied much worse and mainly in a course of initiative research surveys, because in frames of the Regional and National radiation monitoring networks there is no obligation to perform radiation monitoring of groundwater in monitor wells.

Monitoring of groundwater pollution on the IPPE industrial site is carried out by the external dosimetry service of the IPPE. In 2015, it was revealed that total β-activity in water of monitor wells in the sanitary-protective zone varied from 0.15 to 1.7 Bq/l, in the observation zone (5-km zone around IPPE) – from 0.15 to 14.4 Bq/l. Total α- and β-activity in water of Protva river was 0.1 and 0.15 Bq/l and did not exceed the reference levels for drinking water (0.2 Bq/l ($\Sigma\alpha$) and 1.0 Bq/l ($\Sigma\beta$)), according the RSS-99/2009 [4].

3. Contamination of Obninsk groundwater with tritium

Contamination of groundwater with tritium was first revealed in the sanitary-protective zone of IPPE by employees of Roshydromet performing the "Radiological monitoring program for territory of a nuclear energy and industry enterprise" [5].

A comprehensive survey of the IPPE's industrial site and adjacent territories has revealed that research nuclear reactors and accelerators where tritium targets are applied as well as radioactive waste storages are considered to be the sources of technogenic tritium. All the above sources are located within the sanitary controlled area of intakes. The expected release of tritium into the environment by this enterprise ranges from $2.5{\cdot}10^2$ to $2.5{\cdot}10^3$ Ci/year [6].

Table 2 presents data on tritium activity measurements in water of monitor wells located on-site of the solid radioactive waste (SRW) storage facility of IPPE and in water of the oxbow located in the observation zone of IPPE [7].

Table 2. Specific activity of tritium in monitoring points in 2010

Date of sampling	Specific activity of tritium in monitoring points, Bq/l			
	Well N-1	Well N-2	Well P-5	Oxbow
15.03	–	163	–	–
26.04	–	–	2472	–
17.05	–	–	14990	730
15.06	–	120	–	–
22.07	380	380	3020	1870
20.10	764	164	2436	1145
intervention level	7600	7600	7600	7600

As can be seen from table 2, in the monitoring wells of IPPE the increased levels of tritium are observed compared to background level (1.9 Bq/l), and even two-fold excess of the intervention level (IL=7600 Bq/l) is detected. The flow of tritium in groundwater have resulted from a leakage of the SRW storage at IPPE [6]. In this regard, SPA "Typhoon" of Roshydromet carries out periodic sampling of water from the intakes and drinking water communications in different sites of Obninsk and other nearby settlements (in Maloyaroslavets, Belousovo, Balabanovo, Mishkovo) to measure and control tritium levels.

A survey of drinking water in the city of Obninsk within the territory of SPA "Typhoon" in 2015 (table 3) showed that the content of tritium in drinking water from March to November ranged from 5.7 to 16.1 Bq/l with an average value of 10.6 Bq/l, which was about two orders of magnitude below the IL [3] and 5.6 times higher than the average content of tritium in fresh water rivers of the European part of Russia (1.9 Bq/l) [8]. In 2015, a decrease of the tritium concentration in drinking water was observed compared to 2014; this was due to the large amount of atmospheric precipitation in the study area. In 2014, precipitation was 288 mm less than in 2015 [9], which significantly affected the quality of groundwater. As a result, in 2015 the underground water was more watered, their dilution occurred, which resulted in a decrease in the tritium level.

Dynamics of an average annual content of tritium in drinking water in the territory of SPA "Typhoon" since 1997 [1] shows (figure 1) that the tritium content in water is gradually decreasing. During this period, it has reduced about 8 times. The maximum measured tritium concentration was recorded in 1999 and amounted to 150 Bq/l, minimum (4 Bq/l) – in 2005.

IOP Conf. Series: Journal of Physics: Conf. Series **784** (2017) 012040 doi:10.1088/1742-6596/784/1/012040

Table 3. Specific activity of tritium in Obninsk's municipal drinking water within the territory of SPA "Typhoon" [1]

2015		2014	
Date of sampling	Specific activity of tritium, Bq/l	Date of sampling	Specific activity of tritium, Bq/l
18.03	6.9	14.01	45.1
15.06	10.4	14.02	50.6
24.07	16.1	14.03	36.4
12.08	14.6	09.04	22.3
21.09	7.4	30.05	18.2
21.10	12.9	18.06	28.8
22.11	5.7	14.07	18.5
Average	10.6	Average	31.4

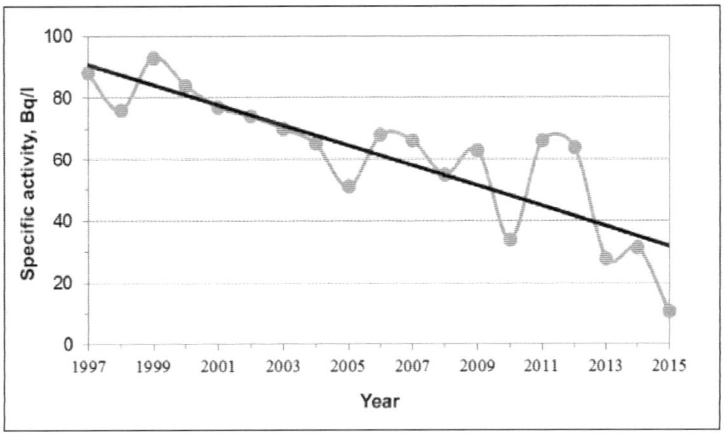

Fugure 1. Dynamics of tritium concentration in drinking water within the territory of SPA "Typhoon"

However, it should be noted that the intervention level for tritium in drinking water according to the recommendations of the US Environmental Protection Agency (US EPA) is 740 Bq/l, which is lower than the value of the same guidance level in Russia, and in the directives of the European Union it is only 100 Bq/l. This demonstrates a lack of attention to the tritium problem in our country, and perhaps requires a revision of the established standards for tritium. One of the criteria of hazard from a radionuclide such as tritium is a health risk assessment of its intake by a human body.

To assess human exposure from the above-mentioned concentrations of tritium, a corresponding risk assessment was carried out.

4. Radiological risk assessment at drinking water consumption

For the most complete assessment of harm that may be caused to health from tritium irradiation at low doses, damage (risk) is calculated. In accordance with the generally accepted linear no-threshold concept, the risk (r_i) is proportional to the effective dose of radiation (E, Sv) and is related to the dose in terms of linear coefficients of radiation risk (r_E, Sv^{-1}) [4]:

$$r_i = E \cdot r_E . \qquad (1)$$

At the risk assessment, the average tritium specific activity in municipal drinking water for 2015 was used which amounted to 10.6 Bq/L.

For the calculation of the absorbed dose of β-radiation D_β from tritium (rad) a formula has been proposed that reflects the biological behavior of tritium in a human body:

$$D_\beta(t) = 2{,}0 \cdot 10^{-3} \cdot \frac{\overline{E}_\beta \cdot A_v \cdot V \cdot f \cdot T_{eff}}{m} \cdot \left(t + \frac{T_{eff}}{0{,}693} \cdot e^{\frac{-0{,}693 \cdot t}{T_{eff}}} - \frac{T_{eff}}{0{,}693} \right) \quad (2)$$

where t – exposure time, day;
$\overline{E}_\beta$ – average energy of β-radiation decay, MeV;
Av – specific activity of tritium in drinking water, Bq/l;
V – intake rate of drinking water, l/day;
f – transfer coefficient in the critical organ (1);
$Teff$ – the effective half-life, s;
m – body weight, g.

Doses are calculated for a reference man weighing 70 kg and consuming 2 L of water per day. With daily use of water throughout the year, an Obninsk citizen receive an average absorbed dose of $1{,}2 \cdot 10^{-7}$ Gy, which is significantly below the annual radiation exposure limit.

In studies of biological effects of tritium, the relative biological effectiveness (RBE) was experimentally evaluated [10-12]. For carcinogenic potential, the RBE value for tritium β-radiation was set equal to 2.5. Based on these data the effective dose is calculated as $E = 3 \cdot 10^{-7}$ Sv.

Using the above approach, the risk of malignant tumors at consumption of drinking water with low tritium specific activity is calculated and it is equal to $1.6 \cdot 10^{-8}$. This level of risk assessed is 3000 times lower than an individual lifetime risk for the population.

5. Conclusion

During the study implementation, a holistic picture of natural water pollution with tritium in Obninsk was obtained. The data of the specific activities of tritium in tap water suggests that the intensity of inflow of tritium into groundwater from the industrial site of IPPE for the last years has decreased 8 times. Specific activities of tritium in water from monitor wells at the IPPE site in a vicinity of the RW repository are two times higher than the intervention level. Groundwater is essentially diluted along its transit to water intakes. The tritium content in drinking water supplied to local population is 10.6 Bq/l, which is 700 times lower than the intervention level. The radiological risk assessment at consumption of tritiated water did not reveal any excess of additional cases of radiation stochastic effects.

Acknowledgements

This work was supported by Competitiveness Program of National Research Nuclear University MEPhI.

References

[1] Radiation Situation in the Territory of Russia and Adjacent Countries in 2015, 2016. Annual book. Ed.by V.M. Shershakov, V.G. Bulgakov, I.I. Kryshev et al. (Obninsk: SPA "Typhoon") 348 p
[2] Silin I I 2003 Ecology and Economy of Natural Resources in the Protva River Basin (Moscow and Kaluga regions). (Kaluga: VIEMS) 315 p
[3] Information about the quality of drinking water and hot water systems centralized water supply // http://www.admobninsk.ru/obninsk/jkh/ecology/kachestvo-vodi-14/
[4] Russian Radiation Safety Standards RSS-99/2009: SRS 2.6.1.2.2523-09, 2009. (Moscow) 100 p

[5] Radiation Situation in the Territory of Russia and Adjacent Countries in 1999, 2001. Annual book. Ed.by K P Makhon'ko. (St.-Pet., Hydrometeoizdat).
[6] Starkov O V, Moiseeva O V 2002 *Nuclear and radiation safety in Russia* p 64-75
[7] Momot O A, Lavrentyeva G V, Synzynys B I, Silin I I 2012 *Regional Environmental Issues* vol 1 p 46-52
[8] Radiation Situation in the Territory of Russia and Adjacent Countries in 2014. 2015 Annual book. Ed.by V M Shershakov, V G Bulgakov, S M Vakulovskii et al. (Nizhny Novgorod) 322 p
[9] RIHMI – WDC. Hydrometeorological data // http://meteo.ru/data
[10] Review of Risk from Tritium 2007 *Document of the Health Protection Agency. Radiation, Chemical and Environmental Hazards* (November 2007)
[11] UNSCEAR 2016 *Reports to the General Assembly with Scientific Annexes. Biological effects of selected internal emitters* (New York) 320 p
[12] Akleyev A V, Azizova T V, Alexakhin R M et al. 2016 *Medical radiology and radiation safety* vol 61 No 1 p 60-72

Quantum Dot-Based Hybrid Nanostructures and Energy Transfer on the Nanoscale for Single- and Multi-Photon Imaging and Cancer Diagnostics

Igor Nabiev[1,2]

[1]Laboratoire de Recherche en Nanosciences, LRN - EA4682, Université de Reims Champagne-Ardenne, Reims, France
[2]Laboratory of Nano-Bioengineering, National Research Nuclear University MEPhI (Moscow Engineering Physics Institute), Moscow, Russian Federation

E-mail: igor.nabiev@univ-reims.fr

Abstract. An ideal single-photon (1P) or multiphoton fluorescent nanoprobe should combine a nanocrystal with the largest possible 1P or two-photon (2P) absorption cross section and the smallest possible highly specific recognition molecules conjugated with the nanoparticle in an oriented manner. However, the conditions used for conjugation of typical recognition molecules (conventional antibodies, Abs) with nanoparticles often provoke their unfolding and/or yield nanoprobes with irregular orientation of Abs on the nanoparticle surface. Conjugation of smaller Ab fragments, such as single-domain antibodies (sdAbs), with quantum dots (QDs) in an oriented manner can be considered as an attractive approach to engineering of ultrasmall diagnostic nanoprobes.

QDs conjugated to 13-kDa sdAbs derived from camelid IgG or streptavidin have been used as efficient 1P or 2P excitation probes for imaging of cancer markers. The 2P absorption cross sections (TPACSs) for some conjugates are higher than 49,000 GM (Goeppert–Mayer units), which is close to the theoretical value calculated for CdSe QDs and considerably exceeds that of organic dyes. A further step in advanced QD-based cancer diagnostics has been made through implementation of efficient FRET-based imaging with 2P excitation, which has been demonstrated for double immunostaining complexes formed on the surface of cancer cells from sdAb–QD conjugates (donor) and a combination of monoclonal Abs and secondary antibodies labeled with the AlexaFluor dye (acceptor). The proposed approach permits obtaining an exceptional contrast of 2P imaging of cancer biomarkers without any contribution of cell and tissue autofluorescence in the recorded images.

1. Introduction

Conventional immunoglobulins G (IgG) have a molecular weight of 150 kDa and average sizes of $14.5\times8.5\times4$ nm^3, which hampers their use as recognition molecules for targeted delivery of drugs, tumor imaging, and diagnosis [1]. In addition, the conditions of conjugation of monoclonal antibodies (mAbs) with labels often cause denaturation of IgG; another drawback is that mAb are attached to the nanoparticle carrier in a random orientation [2]. Smaller fragments of antibodies (Abs) conjugated with quantum dots (QDs) in a strictly predetermined orientation would be a good alternative to mAbs in designing nanoprobes of the smallest possible size [3, 4]. Single-domain Abs (sdAbs) or VhH fragments, with a molecular weight as small as 13 kDa, are the smallest known Ab fragments capable of recognizing the corresponding antigens with a binding constant compared to that of conventional Abs [5, 6]. SdAbs are also very little prone to aggregation; they exist in the monomeric form and better diffuse into tissues than full-size IgG do [7, 8]. These advantages make sdAbs optimal recognition molecules for the use in QD-based fluorescent nanoprobes for detecting biomarkers of diseases and immunodiagnostics.

We have recently published several papers on the development of stable, highly specific, ultrasmall next-generation nanoprobes based on highly oriented conjugates of sdAbs and semiconductor QDs and their employment in cancer diagnostic platforms using flow cytometry and immunohistochemistry [3, 9, 10], as well as in multi-photon imaging of tumor cells [4]. In the first proof-of-concept experiments with the novel nanoprobes, we used carcinoembryonic antigen (CEA) as a marker [3, 4, 9]. In addition, we carried out experiments with QD-conjugated sdAbs against human epidermal growth factor receptor 2 (HER2) as nanoprobes for detecting this marker of breast and lung cancers and compared the effectiveness of these probes with that of anti-HER2 mAbs conjugated with standard organic dyes, AlexaFluor488 and AlexaFluor568 [10]. The sdAb–QD conjugates exhibited better performance in experiments with a panel of lung cancer cells with different rates of HER2 expression. In general, the sdAb–QD conjugates effectively detected the CEA and HER2 proteins. Their comparison with the conjugates of the standard fluorophores with mAbs was performed with respect to both efficiency and specificity of labeling. Confocal microscopy was used to show that the sdAb–QD conjugates specific for HER2 were considerably more effective than the standard conjugates in detecting cancer cells with a low HER2 expression rate in a mixture with primary macrophages [10].

In multi-photon microscopy with excitation in the near-IR spectral region, the excitation band is remote from the spectral region of effective excitation of the autofluorescence of biological tissues, which makes it possible to achieve deep tissue imaging [11–13]. The main problem of multi-photon imaging is small two-photon absorption cross sections (TPACSs) of all known organic fluorophores; it is no larger than a hundred Goeppert–Mayer units (GM, 1 GM = 10^{-50} $cm^4 \cdot photon^{-1}$). Semiconductor QDs are characterized by a wide fluorescence excitation spectrum, a large absorption cross section, the possibility of exciting the fluorescence of QDs of different sizes at a single wavelength, and resistance to photobleaching [14–16]. The TPACSs of some QDs exceed 10^4 GM, which makes them optimal fluorescent labels to be used in nanoprobes for multi-photon imaging [12, 13]. Conjugates of QDs and sdAbs derived from llama IgG [17] have been used as effective probes with two-photon (2P) excitation for detecting CEA in the tissues of a normal human appendix vermiformis and colon carcinoma [3]. The TPACSs of some of the conjugates are higher than 49,000 GM, which substantially exceeds this value for organic dyes and is close to the theoretical value of 50,000 GM calculated for CdSe QDs.

Here, we summarize the results of our recent studies on single-photon (1P) and multi-photon optical characteristics of highly oriented sdAb–QD conjugates and consider their use under the conditions optimal for imaging cancer markers in terms of the absolute sensitivity of imaging and a high ratio of the sdAb–QD fluorescence signal to that of tissue fluorescence [13]. The conjugates developed in our studies are bright multi-photon nanoprobes with a high signal to autofluorescence ratio upon excitation at a wavelength of 800 nm, which allows the areas of tumor tissue overexpressing CEA to be clearly distinguished from normal tissue. The results of testing the sdAb–QD conjugates indicate their unprecedentedly high potential as new, more sensitive tools for early detection of cancer markers.

2. Materials and Methods

QDs were synthesized and stabilized in aqueous solution using the methods described earlier [14–16]. Then, they were solubilized in water according to the procedure described in [9]. In brief, the QDs were first transferred to the water phase by attaching DL-Cys (Sigma) to the surface. The resultant water-soluble QDs displayed bright-orange photoluminescence (PL) with an emission peak at 570 nm and a quantum yield close to 40% at room temperature. After that, DL-Cys on the QD surface was replaced with a mixture of thiol-containing polyethylene glycols (PEGs) with a carboxyl, hydroxyl, or amino end group. The mixtures of the polymers used for QD solubilization were selected in such a way as to minimize nonspecific binding of QDs by cells and tissues. The treatment of the QD surface with low-molecular-weight PEGs yielded water-soluble CdSe/ZnS QDs with the smallest possible diameters. Optimal results were obtained for organic shells consisting of hydroxy-PEG and mixtures of 90% of hydroxy-PEG and 10% of amino-PEG. The samples were incubated at 4°C overnight, preliminarily purified by centrifugation with blocks of Amicon Ultra-15 filters with a cutoff of 10 kDa (Millipore), and finally purified from excess ligand by means of exclusion gel chromatography in homemade columns containing Sephadex-25 (Sigma). All the QDs (and sdAb–QD conjugates) prepared were characterized by means of dynamic light scattering (DLS) and electrophoresis using a Zetasizer Nano ZS device (Malvern Instruments, United Kingdom). The samples suspended in ultrapure water were filtered through a filter with a mesh size of 0.1 μm, and the size distribution of the particles was measured at 25°C in a small-volume quartz cuvette (Hellma ZEN2112).

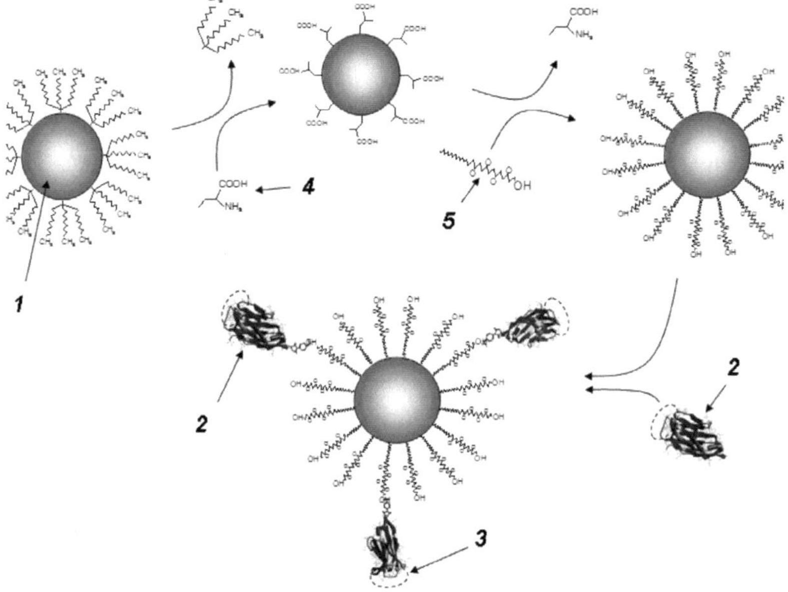

Figure 1. Schematic presentation of the main stages of obtaining water-soluble CdSe/ZnS QDs, their stabilization in aqueous solution, and highly oriented conjugation of sdAbs with these QDs, with the antigen-binding domain of each sdAb exposed to the outside and remaining functionally active.
(1) A CdSe/ZnS core/shell QD immediately after synthesis, with the surface covered with trioctylphosphine oxide (TOPO), a water-insoluble surfactant; (2) An sdAb serving as a recognition molecule; (3) The antigen-recognizing site of an sdAb (the part of the sdAb molecule that is capable of recognizing and specifically binding a disease marker); (4) A low-molecular-weight thiol-containing compound (here, cysteine) used at the first stage of nanoparticle solubilization, which is performed by replacing the hydrophobic surfactant (TOPO) on the nanoparticle surface with polar molecules; (5) Polyethylene glycol (PEG) derivatives.

The hydrodynamic sizes of the particles were calculated from the diffusion time using the Einstein–Stokes equation. The measurements were repeated 15 times for each sample (with a measurement intensity of 10 traces per measurement); the results were averaged and used for calculating the QD hydrodynamic diameter.

The sdAbs against CEA and HER2 were obtained as described earlier [5, 6]. The sdAb–QD conjugates were fabricated with the use of sulfosuccinimidyl-4-(N-maleimidomethyl)cyclohexane-1-carboxylate (Sulfo-SMCC) (Pierce) as a water-soluble heterofunctional crosslinker using the method described elsewhere [3]. The sdAb content of the sdAb–QD conjugates was determined spectrophotometrically using a reagent kit for protein assay (Bio-Rad Laboratories, United States); the QD content was calculated from the extinction coefficient of the first exciton band in the QD absorbance spectrum. The maximum sdAb to QD molar ratio in the conjugates was about 4:1.

Figure 1 shows the main stages of the preparation of water-soluble CdSe/ZnS QDs, their stabilization in aqueous solution, and highly oriented conjugation of sdAbs with these QDs, with the antigen-binding site of each sdAb exposed to the outside and remaining functionally active (see the caption to Figure 1 for details).

Flow cytometry of stained cells was performed by means of a FACStarPlus (Becton Dickinson, United States) or GuavaR EasyCyteTM Plus (Guava TechnologiesTM, United States) flow cytometer immediately after staining. An argon laser (488 nm) was used for exciting the fluorescence; the fluorescence intensity was measured in the spectral regions of 564–586 nm (FACStarPlus) and 570–596 nm (GuavaR EasyCyteTM Plus).

Tissue samples of the human appendix vermiformis and colon carcinoma were obtained from the Department of Pathology of the Robert Debré University Hospital Center (Reims, France). The study was not related to any diagnostic or therapeutic procedure; hence, the use of these samples did not require permission of the Committee for the Use of Clinical Material. In the control experiments corresponding to the "gold standard" of immunohistochemical labeling of tissue sections using DAB, the tissue specimens were incubated in the presence of mAbs and CEA from Ventana Medical Systems (France) (the TF 3H8-1 clone), washed with PBS, and then incubated in the presence of biotinylated anti-mouse sheep polyclonal immunoglobulin (Amersham, France) in PBS. The specimens were developed using a Dako REAL™ kit (Denmark) (peroxidase/DAB). After washing with PBS–Tween, the sections were examined under an optical microscope (Carl Zeiss, Germany).

A Chameleon Ultra II femtosecond mode-synchronization laser (Coherent, Germany) with a pulse duration of 140 fs, frequency of 80 MHz, and 2P excitation wavelength from 700 to 1000 nm was used for excitation of 2P fluorescence. The mean power of the laser was 3.3 W at an excitation wavelength of 800 nm. An Ar^+ laser (Coherent, Germany) was used for 1P excitation in the visible range (wavelength, 457.9 nm) at an excitation power of 10 mW. The fluorescence was recorded using an LSM-710 microscope (Carl Zeiss, Germany) with a 20× Plan Apo DIC II lens (0.45 N.A.) and an MBS760+ beam splitter. The image resolution was 1024 × 1024 pixels. The QD fluorescence from the stained specimens upon 2P laser excitation was absolutely stable throughout the irradiation period. The radiation intensity was permanently controlled during the experiments; photodegradation was not observed. In experiments with excitation in the UV spectral region (350–400 nm), the sections were examined under an Axiovert-200 fluorescence microscope (Carl Zeiss, Germany) with a low-frequency filter (450 nm).

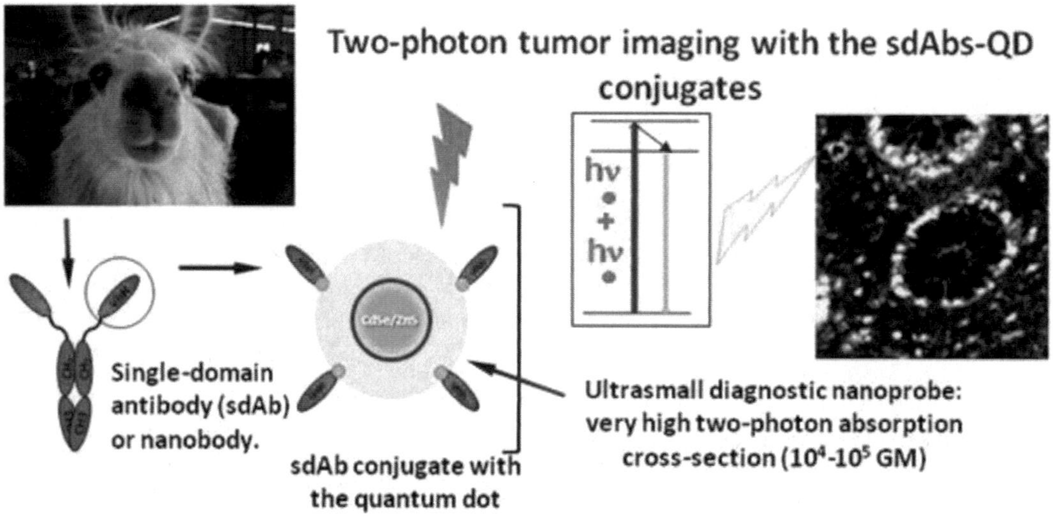

Figure 2. On the left, schematic description of the fabrication of a diagnostic nanoprobe by conjugating single-domain antibodies (sdAbs or VhH fragments) with semiconductor CdSe/ZnS quantum dots (QDs); on the right, the use of the nanoprobe for multi-photon imaging of a cancer marker.

3. Results and Discussion

SdAbs with high affinities for CEA and HER2 (KD $\approx$ 8.3 nM) were obtained by integrating an additional C-terminal cysteine residue [9] for subsequent site-specific oriented conjugation of the sdAbs with QDs. CdSe/ZnS QDs with a quantum yield of about 70% were transferred to water phase by replacing the TOPO/TOP ligands on their surface with DL-Cys. After that, we prepared a series of compact QDs with a controlled surface charge coated with three-functional low-molecular-weight PEG polymers. This procedure provides stable, compact PEG-functionalized QDs with a hydrodynamic diameter <9 nm and a quantum yield of about 50%.

The conjugation reaction using Sulfo-SMCC is the standard procedure for fabrication of batches of QD conjugates. This reaction links accessible amino groups on the QD surface with sulfhydryl groups of biomolecules. The PMPI conjugation reaction links hydroxyl groups on the QD surface with the exposed sulfhydryl group of the sdAb (Figure 1). Thus, both reactions ensures selective and oriented attachment of sdAbs to the QD surface specially functionalized for these reactions via the only accessible Cys residue introduced into their C-terminal domain. All the conjugated sdAb molecules are bound with the QDs via the same functional groups and have the same orientation relative to the QD surface.

DLS studies have shown that the conjugation of the QDs obtained in this study with sdAbs yields maximally compact nanoprobes with a very small hydrodynamic diameter (about 11.9 nm). They are only slightly larger than unconjugated QDs (8.84 nm) and are much smaller than IgG–QD conjugates (30.3 nm) [3, 9]. Thus, the ultrasmall diagnostic probes developed are nine times smaller than conjugates of full-size mAbs with nanoparticles of the same diameter.

The conjugates developed were originally used for flow cytometry analysis of mixtures of CEA-positive and CEA-negative MC38 cells in human blood serum, which demonstrated the possibility of detection of very small numbers of cancer cells in cell populations containing tumor and normal cells [3, 9]. The method used in our studies allows less than 1% of CEA-positive cells to be detected among a large excess (99%) of CEA-negative MC38 cells, which confirms a high specificity of flow cytometry using sdAb–QD conjugates. Figure 2 schematically shows the experimental procedure of multi-photon imaging with the use of sdAb–QD conjugates. We have found that excitation at 920 nm is a good compromise solution ensuring both the necessary intensity of QD fluorescence and strong suppression of tissue autofluorescence, with an ideal QD fluorescence to autofluorescence ratio higher than 50.

4. Conclusions

Recently, we have developed a next-generation ultrasmall diagnostic nanoprobe consisting of the smallest possible functionally active Ab fragments (13 kDa) highly affine for CEA (KD = 8.3 nM), which are termed single-domain antibodies (sdAbs) or "nanobodies", conjugated with a compact semiconductor CdSe/ZnS quantum dot (QD) in a strictly predetermined orientation [3–5, 9, 10]. The high degree of orientation is ensured by site-specific conjugation of the QD with sdAbs via an additional cysteine residue specifically integrated into the C-terminal domain of the sdAb amino acid sequence. The nanoprobes developed carry four identically oriented sdAb antibody molecules on the surface of each QD and have a hydrodynamic diameter smaller than 12 nm.

Experimental results have shown that the sdAb–QD nanoprobes specifically bind CEA, a cell receptor expressed by cells of many human tumors. We have also demonstrated that sdAb–QD conjugates are stable, retain their specificity for CEA-expressing tumor cells in human blood serum, and exhibit excellent sensitivity and specificity in quantitative flow-cytometric detection of cancer markers and rare biomarker-expressing cells. A high diffusion capacity of sdAbs used for fabrication of these ultrasmall nanoprobes makes them good candidate tools for immunohistochemical analysis of thick tissue sections impermeable for conventional IgG. Our data show that sdAb–QD conjugates label antigens more precisely than conventional Abs do, ensuring the quality of labeling in biopsies equal to or better than that of the "gold standard" immunohistochemical diagnosis using the DAB chromogen.

We have also demonstrated that the use of two-photon excitation in immunofluorescent diagnostics with the use of sdAb–QD conjugates provides the best results in detecting cancer markers in human histological specimens. This excitation mode allows the high brightness and stability of QD fluorescence to be combined with a high ratio of QD fluorescence to tissue autofluorescence. The sdAb–QD conjugates used in our studies [3, 7, 9] have proved to be a substantially better alternative to the conjugates of conventional monoclonal antibodies with organic fluorophores. These conjugates are bright and specific labels that make it possible to trace tumor cells in bulk tissues and detect target antigens at unprecedentedly low concentrations [2]. However, their effectiveness should be confirmed in model experiments on animals before the novel nanoprobes are used in clinical practice.

Acknowledgments

This study was supported by the Federal Targeted Program for Research and Development of the Ministry of Education and Science of the Russian Federation (grant no. 14.584.21.0012, contract no. RFMEFI58415X0012). We are grateful to Vladimir Ushakov for assistance in preparation of this manuscript.

References

[1] Amit, A., Mariuzza, R.A., Phillips, S.E., Poljak, R.J. // Science 1986. V. 233, P. 747.
[2] Mahmud, W., Rousserie, G., Reveil, B., Tabary, T., Millot, J.M., Artemyev, M., Oleinikov, V., Cohen, J., Nabiev, I., Sukhanova, A. // Anal. Biochem. 2011. V. 416, P. 180.
[3] Sukanova, A., Even-Desruleaux, K., Kisserli, A., Tabary, T., Reveil, B., Millot, J.M., Chames, P., Baty, D., Artemyev, M., Oleinikov, V., Pluot, M., Cohen, J., Nabiev, I. // Nanomedecine: NBM 2012. V. 8, P. 516.
[4] Hafian, H., Sukhanova, A., Turini, M., Chames, P., Baty, D., Pluot, M., Cohen, J., Nabiev, I., Millot, J.M. // Nanomedicine: NBM 2014. V. 10, P. 1701.
[5] Behar, G., Chames, P., Teulon, I., Cornillon, A., Alshoukr, F., Roquet, F., Pugnière, M., Teillaud, J.L., Gruaz-Guyon, A., Pèlegrin, A., Baty, D. // FEBS J. 2009. V. 276, P. 3881.
[6] Even-Desrumeaux, K., Fourquet, P., Secq, V., Baty, D., Chames, P. // Mol. Biosyst. 2012. V. 8, P. 2385.

[7] Saerens, D., Kinne, J., Bosmans, E., Wernery, U., Muyldermans, S., Conrath, K. // J. Biol. Chem. 2004. V. 279, P. 51965.
[8] Perruchini, C., Pecorari, F., Bourgeois, J.P., Duyckaerts, C., Rougeon, F., Lafaye, P. // Acta Neuropathol. 2009. V. 118, P. 685.
[9] Sukhanova, A., Even-Desrumeaux, K., Chames, P., Baty, D., Artemyev, M., Oleinikov, V., Nabiev, I. // Nature Protocols / Protocols Exchange 2012. DOI: 10.1038/protex.2012.042 http://www.nature.com/protocolexchange/protocols/2463
[10] Rakovich, T., Mahfoud, O., Mohamed, B., Prina-Mello, A., Crosbie-Staunton, K., Van Den Broeck, T., De Kimpe, L., Sukhanova, A., Baty, D., Rakovich, A., Maier, S., Alves, F., Nauwelaers, F., Nabiev, I., Chames, P., Volkov, Y. // ACS Nano 2014. V. 8, P. 5682.
[11] Denk, W., Strickler, J.H., Webb, W.W. // Science 1990. V. 248, P. 73.
[12] Larson, D.R., Warren, R.Z., William, R.M., Clark, S.W., Bruchez, M.P., Wise, F.W., Webb, W.W. // Science 2003. V. 300, P. 1434.
[13] Hafian, H., Sukhanova, A., Turini, M., Chames, P., Baty, D., Pluot, M., Cohen, J., Nabiev, I., Millot, J.M. // Nanomedicine: NBM 2014. V. 10, P. 1701.
[14] Samokhvalov, P., Artemyev, M., Nabiev, I. // Chemistry 2013. V. 19, P. 1534.
[15] Sukhanova, A., Devy, J., Venteo, L., Kaplan, H., Artemyev, M., Oleinikov, V., Klinov, D., Pluot, M., Cohen, J., Nabiev, I. // Anal. Biochem. 2004. V. 324, P. 60.
[16] Sukhanova, A., Venteo, L., Devy, J., Artemyev, M., Oleinikov, V., Pluot, M., Nabiev, I. // Lab. Invest. 2002. V. 82, P. 1259.

The influence of physical factors on recognizing blood cells in the computer microscopy systems of acute leukemia diagnosis

V G Nikitaev[1], A N Pronichev[1], E V Polyakov[1], V V Dmitrieva[1], N N Tupitsyn[2], M A Frenkel[2], A V Mozhenkova[2]

[1]National Research Nuclear University MEPhI (Moscow Engineering Physics Institute), Kashirskoe shosse 31, 115409, Moscow, Russia.
[2]N.N. Blokhin Russian Cancer Research Center, Kashirskoe shosse 23, 115478, Moscow, Russia

E-mail: VGNikitayev@mephi.ru

Abstract. The work investigated the effect of the choice of color space component on blood cell detection based on the calculation of texture attributes of blood cells nuclei in bone marrow. The study identified the most informative color space and texture characteristics of blood cells, designed for components of these spaces. Significance ratio was introduced to assess the quality of features. We offered features that have enabled to divide lymphocytes from lymphoblasts. The selection of the features was based on the results of the data analysis.

1. Introduction

Currently, study of hematopoietic cells (blood and marrow) are performed by a doctor–laboratory assistant through visual microscopy of stained smears. Differences in the quality of smears, variability of staining cells, professionalism level of the doctor in some cases may cause discrepancy in the interpretation of results. Data objectification of cells survey may be implemented by using computerized microscopy [1-2]. Automation research allows to enter new parameters of the cells characteristics, compare data obtained in different laboratories, and thus improve the level of diagnosis of haematopoiesis diseases [3-6].

An important feature for designing systems for digital processing of blood cells images is the spatial distribution of color in the region of the cell nucleus. The camera used in the computer systems of light microscopy generates three grayscale images of the incoming light flux by dividing it into light beams of three spectral bands: R(red spectrum area), G(green region of spectrum), B(blue spectrum area). When light photons are exposed to the photosensitive element CCD camera, photons knock electrons, creating a charge that is proportional to the light flux. The amount of charge is mapped to a digital code that represents the value of a color component of corresponding zone of the spectrum for a pixel in a digital image[7]. Pixel is the smallest possible element of a digital image. Pixel is assigned a color which describes the color model RGB. This model is the base for other color models. When solving the problem of blood cells recognition in conditions of close clusters of leukocytes in feature space, study of informativeness feature calculated for components of different color spaces is actual problem.

The aim of this work is to study the dependence of the information content of texture features from region of the spectrum of visible light and to determine the suitable components of color models in the task of hemopoiesis cells recognition.

2. Materials and methods

Investigation of the effect of the choice of the spectrum band (in the visible range of the electromagnetic radiation) and components color models in the calculation of textural features from the images of the nuclei of lymphocytes and lymphoblasts in the blood cells resulting image recognition performed on blood smears of bone marrow donors and patients with acute lymphoblastic leukemia (ALL). These preparations were fixed and stained by the method of May-Grunwald-Romanovsky. Morphological examination of blood smears and bone marrow aspirates were performed in the laboratory of immunology hematopoietic of N.N. Blokhin Russian Cancer Research Center by two experts. Acute lymphoblastic leukemia diagnosis was established on the basis of morphological, cytochemical, immunophenotypic studies. Blast cells prevailed on acute lymphoblastic leukemia preparation.

Image were obtained from stained smear preparations of peripheral blood and bone marrow preparations in the system of computer microscopy: microscope Olympus BX43 with motorized stage, camera Imperx IPX-IPX-4M1ST-GCFB (hereinafter Imperx). The resulting images are described by a format BMP with a depth coding by 24 bits per pixel 8 bits per component. 14 images of donor lymphocytes and 14 images of blasts cells of acute lymphoblastic leukemia patients were obtained.

Figure 1 shows examples of images of the nucleus blast cells in the components of different colors.

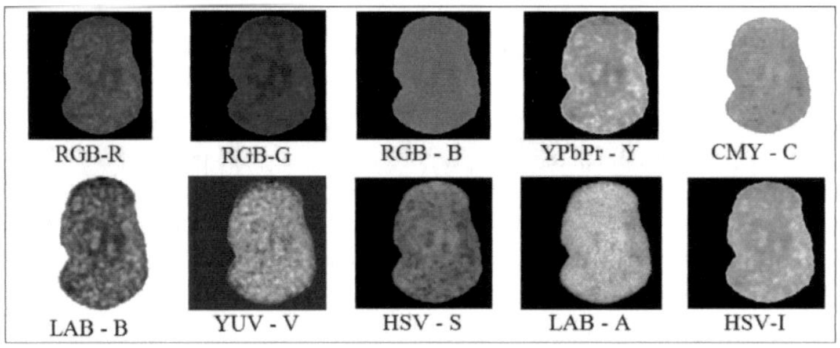

Figure 1. A sample of the transformations of the three (RGB) multispectral images of the nucleus of blast cells in the components of the various color models

Images presented in figure1 are characterized by different texture for the various components of color models. Therefore, the most informative features for the separation of blood cells and bone marrow should be looked for different color models.

3. Investigated physical characteristic

The paper discusses the characteristics of the three images for the three spectral bands. Recording is carried out with a digital camera Imperx, having three types of sensors, spectral sensitivity of which is shown in figure 2.

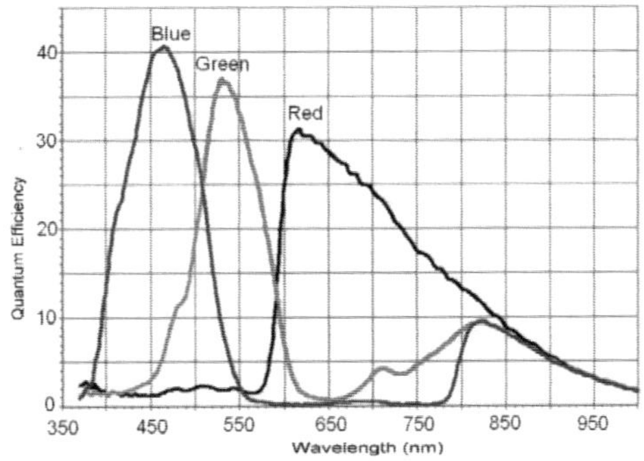

Figure 2. Spectral characteristics - monochrome quantum efficiency Imperx camera.

Images formed by these types of sensors correspond RGB color model.

4. Methodology of the experiment

To conduct the study was developed technique of the experiment, which allows to carry out a preliminary study of the impact of choice of spectral bands to the results of the classification of blood cells. Technique involves a series of steps:

1. Formation of digital sets of images of leukocytes in the computer system in the form of three microscopy multispectral images corresponding to zones R G B.
2. Image segmentation of the nucleus of cells of leukocytes.
3. Obtaining R, G, B, imaging nuclei of leukocytes, a transformation in color components models XYZ, HSL, Lab, Luv, LHC, HLS, HSV, YUV, YIQ, YCbCr, CMY. The resulting images from the components of color model are encoded 8-bit number and considered a halftone image [8].
4. Calculation textural features for halftone image. As textural features in this paper used features - energy, inertia, entropy, local homogeneity and maximum likelihood spatial adjacency matrix [9-10].
5. The index of significance of the sign was introduced to characterize the quality characteristic for solving the problem of separation of blood cells at the blasts and lymphocytes. It is calculated according to the formula
6. $I \dfrac{\bar{x}_b - \bar{x}_l}{2(S_b + S_l)}$, where $\bar{x}$ – is the arithmetic mean, S – standard deviation, index 'b' denotes the characteristics of the blast, 'l' – lymphocyte.
7. Search the most significant sign. The separation of leukocytes by type with using this sign.

5. Results of the study

The study calculated importance for textural features of spatial adjacency matrix, depending on the components of the color model. The figure shows the importance of the index for the different color components. The range [0, 0.4) correspond to the poor separability of different classes of cells [0,4; 0,7) - satisfactory [0.7; 1) - good, [1; + ∞) - excellent.

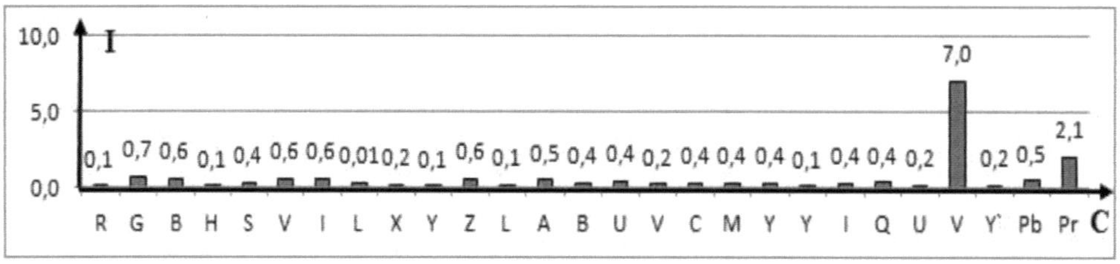

Figure 3. The dependence of the significance of the characteristic "Maximum likelihood" from components of a color model

As a result of analysis, according to the importance of the texture characteristic of the color model components, was detected chrominance component (V) color model (YUV), for which the characteristic value is maximum.

6. Analysis of the results

To confirm the adequacy of the results of the experiment carried out the separation of the types of white blood cells and lymphocyte blast on the found significance index. Separation of lymphocytes and blasts on the component (V) chrominance color model (YUV) for texture feature "maximum likelihood" matrix of spatial adjacency performed without error.

It was determined that the spectral zone G the most significant of the three zones of the spectrum of RGB, and composite component V (linear combination of spectral RGB component) is the most significant component of the considered (figure 3)

7. Conclusion

Comparative analysis of colour models showed that for the separation of blasts and lymphocytes by types, the most important is the component (V) color-difference signal of the color model (YUV).

A further stage of the research is to study the influence of other physical factors in the system of computer microscopy in the diagnosis of acute leukemia.

Acknowledgments

This work was supported by Competitiveness Program of National Research Nuclear University MEPhI.

References

[1] Reyes H et al. 2015 20th Symposium on Signal Processing, Images and Computer Vision (STSIVA) (IEEE Bogota Colombia) p1
[2] Amin M *et al.* 2015 Journal of Medical Signals & Sensors **5**(1) 49
[3] Putzu L *et al.* 2015 Artificial Intelligence In Medicine **62**(3) 179
[4] Chatap N and Sini S 2014 IOSR Journal of Computer Engineering **16**(5) 79
[5] Nikitaev V G *et al.* 2015 Proceedings of the 6th International Confererence on Bioscience and Bioinformatics (ICBB`15) Duba United Emirates February 22-24 p 120
[6] Computational Geometry and Artificial Vision (ISCGAV`15) Sliema Malta August 17-19 p 49
[7] Hsu D and Lee S H 2012 Laboratory Hematology Practice 707
[8] Nikitaev V G *et al.* 2015 WSEAS TRANSACTIONS on BIOLOGY and BIOMEDICINE 12 16
[9] Haralick R M, Shanmugam K and Dinstein I H 1973 IEEE Transactions on Systems, Man, and Cybernetics 6 610
[10] Nikitaev V G et al. 2011 (in Russia) Specialized machinery and communication 4-5 67

Study of iodine, gadolinium and bismuth quantification possibility with micro-CT IVIS spectrumct in vivo imaging system

V V Pervova[1], A A Lipengolts[1,2], A A Cherepanov[2], M A Abakumov[3]

[1] National Research Nuclear University «MEPhI»
[2] Blokhin Russian Oncological Scientific Center
[3] Pirogov Russian National Research Medical University

E-mail: pervova_ru@mail.ru

Abstract. The main task of radiotherapy is to create prescribed absorbed dose of irradiation in a tumor with minimal damage of healthy tissues. Contrast Enhanced Radiotherapy allows to achieve that by administration of a special drug into the tumor before irradiation, that increases the absorbed dose within the tumor volume. The concentration of the drug determines the value of the absorbed dose, therefore one of the major tasks in CERT is quantification of the drug concentration in the tumor during irradiation procedure. The present work deals with quantitative determination of iodine, gadolinium and bismuth water solutions by use of micro-CT IVIS Spectrum In Vivo Imaging System.

1. Introduction

Therapy of malignant tumors is one of the most actual problems in modern medicine. Each of the conventional methods of treatment – surgical removal, chemotherapy or radiation therapy – has its own advantages and disadvantages. In particular in radiation therapy it is not always possible to create the necessary ratio of the absorbed dose value in tumor and normal tissues. The main goal of radiation therapy is to deliver tumoricidal value of absorbed dose into the tumor without exceeding the tolerant dose of healthy tissues surrounding the tumor.

One of the new methods in radiation therapy is Contrast Enhanced Radiotherapy (CERT). The method consists of tumor saturation with dose enhancing agent (DEA) following x-ray irradiation. DEA is any chemical element with atomic number $Z>52$. The presence of DEA in the tumor leads to local increase of the absorbed dose caused by preferential external radiation photoabsorption by DEA atoms[1].

CERT allows to increase the value of the absorbed dose significantly within the irradiated target without exceeding tolerant dose in surrounding healthy tissue. Animal experiments have shown high efficacy of CERT in treatment of malignant tumors [2,3,4,5]. The value of the absorbed dose is determined by DEA concentration, so the determination of the DEA concentration in the tumor prior and during irradiation is one of the key issues arising in CERT implementation[6].

The aim of this work is to study the possibility of quantitative determination of iodine, gadolinium and bismuth by micro-CT IVIS Spectrum CT In Vivo Imaging System. Micro-CT IVIS is a widely used scientific instrument to study biodistribution of different substances labeled with fluorescent dyes. The CT-function is a not primary function of this device, therefore the quality of the CT image is inferior to specialized preclinical micro-CT facilities. However, the wide distribution of this instrument and radiation safety allowing it's installation in non-specialized rooms make the study of DEA quantification with micro-CT IVIS useful and practically significant for many research groups.

The following tasks were set to achieve the above mentioned goal:

1) Determine radiopacity of water in phantoms and air around them in arbitrary units of the micro-CT IVIS software using tomogram data obtained with IVIS.
2) Determine radiopacity of iodine, gadolinium and bismuth solutions with their different concentrations in microtubes inside a water phantom in arbitrary units of the micro-CT IVIS software using tomogram data obtained with IVIS.
3) Rescale obtained radiopacity data in arbitrary units to Hounsfield units considering water of 0 HU and air of -1000 HU.
4) Plot calibration curves for iodine, gadolinium and bismuth using rescaled data.

2. Materials and methods

The study was performed using micro-CT IVIS. Microtubes 250 ml containing different water solutions of iodine, gadolinium and bismuth were placed into 50 ml water filled cylinder with following dimensions 30x110 mm as visualized object. DEA concentration in microtubes was from 1 mg/ml to 50 mg/ml.

ImageJ software was used to measure the value of radiopacity in CT tomograms.

3. Results

Radiopacity of water along the longitudinal axis of the phantoms is shown in figure 1.

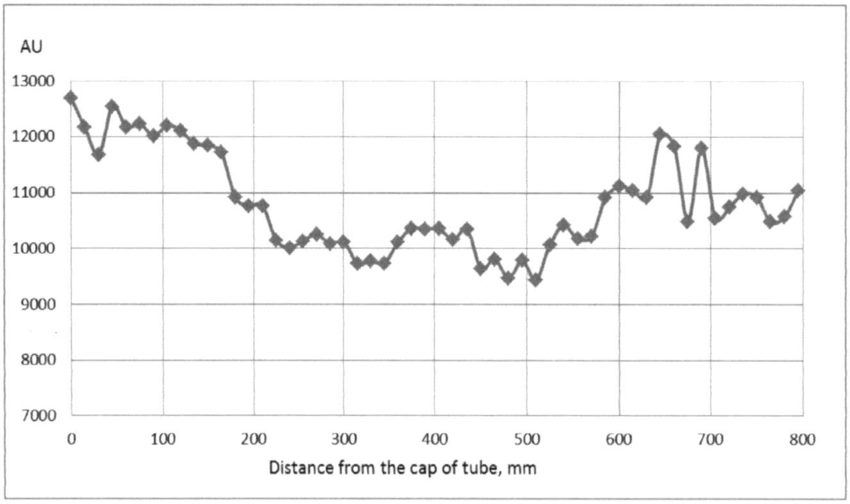

Figure1. The dependence of water radiopacity values in arbitrary units (AU) on the distance from the tube cap along the longitudinal axis of the phantom.

The maximum change of AU value is observed in the center of the tube and is about 25%.

Because of the longitudinal AU value significant change, the dependence of Hounsfield units (HU) on AU was built for each slice (Fig. 2).

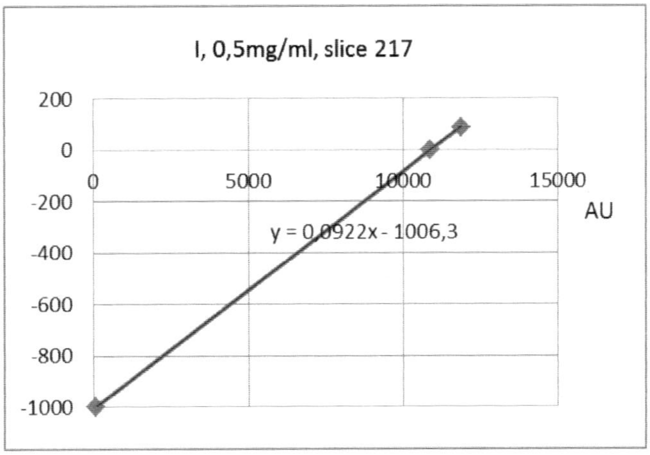

Figure 2. The example of a calibration curve.

Using obtained calibration curves for each slice of phantoms the dependence diagrams of the radiopacity values of iodine, gadolinium and bismuth water solutions in the Hounsfield's scale units on concentration of the elements were built (Fig. 3).

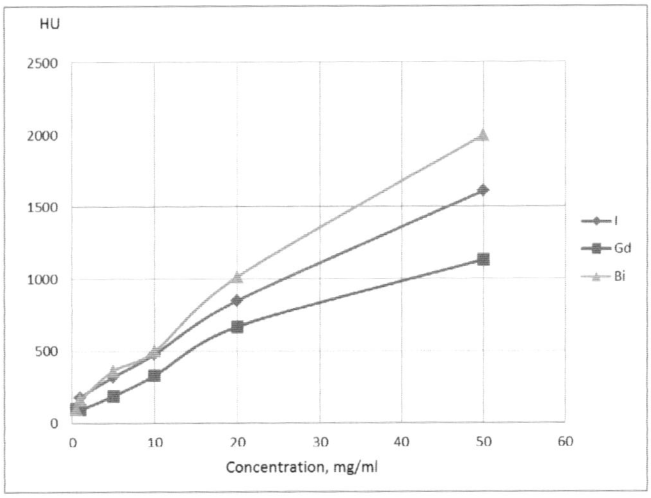

Figure 3. The dependence of radiopacity in HU units on the concentration of iodine (♦), gadolinium (■) and bismuth (▲).

Conclusion

The comparison of the radiopacity data of iodine, gadolinium and bismuth water solutions in different concentrations has revealed:

1) almost the linear dependence of the radiopacity conventional units which determined according to the IVIS system data (and recalculated in Hounsfield units), solutions of I, Gd and Bi in the water phantom on the concentration of elements;

2) significant change of radiopacity of the same DEA concentration along the longitudinal axis of the phantom.

Thus, for the experimental quantification of DEA in the body of laboratory mice using IVIS, a correction of the signal along the longitudinal axis of the object is necessary, namely, for each tomographic slice.

Acknowledgements
This work was supported by Competitiveness Program of National Research Nuclear University MEPhI.

References

[1] J. C. Roeske, L.Nunez, M. Hoggarth et al. 2007 Technol. Cancer Res. Treat, 6(5),395-401
[2] A. Norman, M. Ingram, R.G. Skillen et al. 1997, RadiatOncolInvestig, 5, 8–14
[3] J.F. Hainfeld, D.N. Slatkin, H.M. Smilowitz, 2004 Phys. Med. Biol., 49, N309–N315
[4] J.F. Hainfeld, H.M. Smilowitz, M.J. O'Connor et al. 2013, Nanomedicine, 8(10), 1601–1609
[5] Lipengolts A.A., Cherepanov A.A., Kulakov V.N., Grigorieva E.Y., Sheino I.N., Klimanov V.A. 2015. Antitumor efficacy of extracellular complexes with gadolinium in Binary Radiotherapy.Applied Radiation and Isotopes (2015)- Vol. 106.- Pp. 233-236
[6] V. N. Kulakov, A. A. Lipengol'ts, E. Yu. Grigor'eva, N. L. Shimanovskii. 2016. Pharmaceuticals for binary radiotherapy and their use for treatment of malignancies (a review). Pharmaceutical Chemistry Journal.-2016.- Vol. 50.- No. 6. Pp. 388-393

Pharmacokinetic properties of new antitumor radiopharmaceutical on the basis of diamond nanoporous composites labeled with rhenium-188

V M Petriev[1,2], V K Tishchenko[1], A A Kuril'chik[1] and V G Skvortsov[1]

[1]Tsyb Medical Radiological Research Centre – branch of the National Medical Research Radiological Centre of the Ministry of Health of the Russian Federation (A. Tsyb MRCC), Obninsk, Kaluga region, Russia, 249036, Korolev street, 4
[2] National Research Nuclear University MEPhI (Moscow Engineering Physics Institute), Moscow, Russia, 115409, Kashirskoe shosse, 31

E-mail: petriev@mrrc.obninsk.ru

Key words: composite materials, nanodiamonds, rhenium-188, cancer therapy, radionuclide therapy.

Abstract. Today the development of address therapeutic radionuclide delivery systems directly to tumor tissue is of current interest. It can be achieved by the design of drug containers of specific sizes and shapes from carbon-based composite materials. It will be allowed to enhance the efficacy of anticancer therapy and avoid serious side effects. In this work we studied the pharmacokinetic properties of nanodiamond nanoporous composite labeled with rhenium-188 in rats with hepatocholangioma PC-1 after intratumoral injection.
It was established that substantial part of injected radioactivity remained in tumor tissue. Within three hours after ^{188}Re-nanoporous composites injection activity in tumor constituted 79.1–91.3% of injected dose (ID). Then activity level declined to 45.9% ID at 120 hours. No more than 1.34% ID entered the bloodstream. In soft organs and tissues, except thyroid gland, the content of compound didn't exceed 0.3% ID/g. The highest activity in thyroid gland was 6.95% ID/g.
In conclusion, received results suggest ^{188}Re-nanoporous composites can be promising radionuclide delivery systems for cancer treatment.

1. Introduction

Modern technologies of cancer diseases treatment include so-called target therapy. One of its directions is the development of drug delivery systems (containers), which containing anticancer drug and/or radionuclide and acting directly on tumor cells. This approach will lead to an increase in selectivity of antitumor drugs or radiopharmaceuticals and reduce the frequency and severity of side effects. In addition, it will allow to overcome cancer drug resistance.

In this regard the various types of carbon-based materials (carbon nanotubes, nanodiamonds, fullerenes, etc.) are of particular interest due to its biocompatibility and low cytotoxicity [1 – 5]. It is possible to synthesize composite materials with defined mechanical properties and use them to create vector nanocontainers for address drug or radionuclide delivery.

Nanodiamonds possess unique properties so they can serve as promising compounds to obtain composite materials. The structure of diamond is characterized by superior hardness, Young's modulus, high thermal conductivity and electric resistivity, as well as the low friction [6]. Each particle consists of a chemically inert diamond core and a tunable surface, which can be modified with a wide variety of functional groups. Due to these features of the surface nanodiamonds have high adsorption capacity. It is possible to receive nanoporous diamond composites from disperse nanodiamond powder and carbon matrix [7, 8]. Then durable containers containing pharmaceutical and/or radionuclide can be created. This container may be surgically placed directly into a tumor. For example, these containers with synthetic antibacterial drug levofloxacinum suppressed the development of inflammatory processes in osteal tuberculosis in chinchilla rabbits. In control group of animals treated with intravenous injection of levofloxacinum inflammatory tuberculosis continued [9]. In addition, diamond nanoporous composites allow to compensate bone defect, arising from the inflammatory process in the bone [9]. There was also shown the possibility of the composites to target drug delivery to central nervous system [10].

Rhenium-188 (17h half-life, 84% beta with 2.12 MeV maximum energy, and 16% gamma 155 keV) is one of the most convenient radionuclide for cancer therapy. One potential advantage of ^{188}Re over other radionuclides is its routine availability from the generator ^{188}W/^{188}Re during 4-6 months [11, 12]. Beta radiation causes the therapeutic effect of ^{188}Re. The short half-life of ^{188}Re can effectively reduce the system irradiation secondary to escape from the treated target. It also has a gamma-line at 155 keV allowing imaging the distribution of pharmaceuticals in the body by gamma-camera. It is expected that using carbon-based carriers labeled with ^{188}Re will lead to high levels of radioactivity directly in tumor. The present work is devoted to study of pharmacokinetic properties of diamond nanoporous composites labeled with ^{188}Re after single intratumoral injection in a PC-1 hepatocholangioma rats model.

2. Methods and materials

Nanoporous composites were developed by Gordeev S.K. in Central research institute of structural materials (Saint-Petersburg, Russia) [7] and delivered to us for further investigation. Nanoporous composites represented cylinders with the length of 2 mm and the diameter of 1 mm. The embedding of ^{188}Re was performed at room temperature. Over 99% of ^{188}Re embedded in nanoporous structure of composites after 10 min from the beginning of adsorption. The retention of ^{188}Re by nanoporous structure was higher than 99% during 48 h in a temperature range from 20 ^{0}C to 75 ^{0}C.

Biodistribution study of diamond nanoporous composites labeled with ^{188}Re was carried out in rats with hepatocholangioma PC-1. The weighs of rats were 150 ± 20 g. To get a solid version of hepatocholangioma PC-1 the donor rat with tumor was killed by cervical disruption, and the tumor tissue was isolated. Then the tumor tissue was ground up, diluted in physiological saline and implanted subcutaneously in rats. Pharmacokinetic studies were performed after 8-10 days, when the tumor volume reached 1.0-1.2 cm^{3}.

Each animal was implanted only one composite with activity of 0.370 ± 0.074 MBq (10 ± 2 mCi) directly in tumor. For this purpose composite was placed in needle, which was injected in the core of the tumor. Then the composite was pushed out with mandrin.

The rats were sacrificed at 5 min, 1 h, 3 h, 24 h, 48 h and 120 h (four animals at each time), their organs and tissues were isolated, placed in plastic test-tubes and weighed on electronic scales "Sartorius" (Germany). Then samples of different organs and tissues were counted in automatic gamma counter "Wizard" version 2480 (PerkinElmer/Wallac, Finland). The activity of ^{188}Re in organs and tissues was calculated as percentage of injected dose per gram (%ID/g) and per organ (%ID). The results of radiometry were presented as mean values and mean-squared errors (M ± m).

3. Results and discussion

The results of studying of pharmacokinetic properties of nanoporous composites labeled with rhenium-188 are presented in tables 1 and 2. The biodistribution data of ^{188}Re-nanoporous composites indicated high tumor retention. In fifteen minutes after injection the level of specific activity in tumor tissue was 74.3% of injected dose per gram (ID/g). Subsequently, the level of radioactivity increased and varied from 90.4% ID/g to 100.1% ID/g (table 1).

Table 1. Pharmacokinetics of ^{188}Re-nanoporous composites in rats with transplanted hepatocholangioma PC-1 after intratumoral administration (in % of injected dose per gram).

Name of organ or tissue	Time after administration					
	15 min	1 h	3 h	24 h	48 h	120 h
Blood	0,06±0,01	0,89±0,12	0,07±0,01	0,06±0,01	0,08±0,01	0,06±0,01
Thyroid gland	0,53±0,11	6,95±1,14	1,12±0,33	1,60±0,37	0,91±0,06	2,69±0,93
Lungs	0,006±0,001	0,14±0,01	0,042±0,010	0,032±0,001	0,020±0,001	0,010±0,001
Liver	0,003±0,001	0,12±0,02	0,013±0,001	0,027±0,001	0,014±0,001	0,012±0,001
Kidney	0,003±0,001	0,27±0,02	0,045±0,010	0,042±0,010	0,024±0,001	0,022±0,001
Heart	0,005±0,001	0,21±0,02	0,015±0,001	0,010±0,001	0,021±0,001	0,023±0,001
Spleen	0,002±0,001	0,12±0,02	0,011±0,001	0,020±0,001	0,014±0,001	0,009±0,001
Stomach	0,008±0,001	0,21±0,04	0,18±0,03	0,21±0,05	0,029±0,001	0,016±0,001
Muscle	0,008±0,001	0,19±0,05	0,010±0,001	0,012±0,001	0,021±0,001	0,029±0,010
Femur	0,001±0,001	0,019±0,001	0,013±0,001	0,015±0,001	0,018±0,001	0,023±0,010
Tumor	74,3±7,78	90,4±3,39	100,1±18,9	91,8±7,98	95,8±1,11	90,5±28,0
Brain	0,002±0,0001	0,083±0,010	0,005±0,001	0,006±0,001	0,010±0,001	0,006±0,001

Assessment of total radioactivity showed that during the first 3 h after injection the level of activity in the tumor was very high: from 79.1% to 91.3% of injected dose (ID). Later the content of ^{188}Re-nanoporous composits went down to 45.9% ID in 120 h (table 2).

In the blood the highest activity was just 0.89% ID/g at 1 h after injection. At the other dates the content of ^{188}Re-nanoporous composites didn't not exceed 0.06-0.08% ID/g (table 1). The peak concentration of radioactivity in blood was 1.34% ID and was monitored at 1 h after intratumoral injection (table 2).

Relatively high radioactivity concentration was found in thyroid gland (6.95% ID/g at 1 h), but remained significantly lower than that in tumor (table 1). It is known that free rhenium-188 rapidly accumulates in thyroid gland mediated by the sodiumiodide transporter [13]. So we could conclude that ^{188}Re-nanoporous composites had high stability in vivo.

Low amounts of activity were observed in other soft organs and tissues (table 1 and 2). It is an advantage of ^{188}Re-nanoporous composites as the escape of the radiopharmaceutical from the treated target can result in severe adverse events. The peak concentrations were revealed at 1 h after intratumoral injection; this corresponded with change of ^{188}Re-nanoporous composites content in blood. Some radioactivity was also detected in stomach (0.18% ID/g at 3 h; 0.21 ID%/g at 1 h and 24 h). Extremely low activity was in the brain (no more than 0.083% ID/g at 1 h). It bore evidence of ^{188}Re-nanoporous composites didn't come through blood brain barrier and didn't irradiate brain.

Table 2. Pharmacokinetics of ^{188}Re-nanoporous composites in rats with transplanted hepatocholangioma PC-1 after intratumoral administration (in % of injected dose).

Name of organ or tissue	Time after administration					
	15 min	1 h	3 h	24 h	48 h	120 h
Blood	0,74±0,12	1,34±0,18	0,98±0,15	0,81±0,10	1,00±0,07	0,85±0,17
Thyroid gland	0,012±0,002	0,19±0,03	0,029±0,007	0,044±0,011	0,022±0,001	0,041±0,011
Lungs	0,006±0,002	0,16±0,01	0,050±0,011	0,039±0,005	0,022±0,005	0,011±0,002
Liver	0,013±0,004	0,65±0,09	0,063±0,014	0,14±0,03	0,066±0,004	0,062±0,014
Kidney	0,004±0,001	0,19±0,02	0,029±0,006	0,031±0,006	0,016±0,002	0,014±0,003
Heart	0,004±0,001	0,17±0,01	0,013±0,001	0,007±0,001	0,016±0,001	0,020±0,003
Spleen	0,002±0,001	0,16±0,02	0,014±0,002	0,033±0,006	0,021±0,001	0,014±0,003
Stomach	0,010±0,001	0,26±0,05	0,24±0,02	0,27±0,08	0,034±0,004	0,020±0,005
Muscle	0,69±0,20	2,03±0,57	0,98±0,11	1,23±0,14	1,88±0,43	3,08±0,67
Femur	0,001±0,0001	0,016±0,004	0,010±0,003	0,014±0,001	0,017±0,002	0,023±0,006
Tumor	88,7±13,7	79,1±9,75	91,3±9,50	72,3±10,3	63,7±10,6	45,9±8,12
Brain	0,001±0,0001	0,14±0,03	0,010±0,003	0,041±0,005	0,059±0,006	0,044±0,009

4. Conclusion

Today the development of address therapeutic radionuclide delivery systems directly to tumor tissue is of current interest. It can be achieved by the design of drug containers of specific sizes and shapes from carbon-based composite materials. It will be allowed to enhance the efficacy of anticancer therapy and avoid serious side effects. In this work we studied the pharmacokinetic properties of nanodiamond nanoporous composite labeled with rhenium-188 in rats with hepatocholangioma PC-1 after intratumoral injection.

It was established that substantial part of injected radioactivity remained in tumor tissue. Within three hours after ^{188}Re-nanoporous composites injection activity in tumor constituted 79.1–91.3% of injected dose (ID). Then activity level declined to 45.9% ID at 120 hours. No more than 1.34% ID entered the bloodstream. In soft organs and tissues, except thyroid gland, the content of compound didn't exceed 0.3% ID/g. The highest activity in thyroid gland was 6.95% ID/g.

In conclusion, received results suggest ^{188}Re-nanoporous composites can be promising radionuclide delivery systems for cancer treatment.

Acknowledgements

This work was supported by Competitiveness Program of National Research Nuclear University MEPhI.

References

[1] Lim D.J., Sim M., Oh L., Lim K., Park H. Carbon-based drug delivery carriers for cancer therapy. Arch. Pharm. Res., 2014, V. 37, P. 43–52.

[2] Schrand A.M., Dai L., Schlager J.J., Hussain S.M., Osawa E. Differential biocompatibility of carbon nanotubes and nanodiamonds. Diam. Relat. Mater., 2007, V. 16, No. 12, P. 2118–

2123.
[3] Schrand A.M., Huang H., Carlson C., Schlager J.J., Osawa E., Hussain S.M., Dai L. Are diamond nanoparticles cytotoxic? J. Phys. Chem. B., 2007, V. 111, P. 2–7.
[4] Liu K.K., Cheng C.L., Chang C.C., Chao J.I. Biocompatible and detectable carboxylated nanodiamond on human cell. Nanotechnology, 2007, V. 18, P. 325102.
[5] Zakrzewska K.E., Samluk A., Wierzbicki M., Jaworski S., Kutwin M., Sawosz E., Chwalibog A., Pijanowska D.G., Dariusz Pluta K.D. Analysis of the cytotoxicity of carbon-based nanoparticles, diamond and graphite, in human glioblastoma and hepatoma cell lines. PLoS ONE, 2015, V. 10, No. 3, P. e0122579.
[6] Mochalin V.N., Gogotsi Y. Nanodiamond-polymer composites. Diamond & Related Materials. 2015, V. 58, P. 161–171.
[7] Gordeev S.K. Trehmernye uglerodnye materially [Three-dimensional carbon-based materials]. Voprosy materialovedeniya – The questions of materials science, 2008, V. 2, P. 163–174. [Article in Russian].
[8] Gordeev S. Nanoporous and nanofragmental carbon composite materials. Nanostructured Carbon for Advanced Applications, Kluwer Academic Publishers, 2001, P. 71–88.
[9] Gordeev S.K. Carbon nanostructured drug containers based on a nanodiamond composite material. Metal Science and Heat Treatment. 2015, V. 56, No. 9–10, P. 559–563.
[10] Volnova A., Gordeev S., Lenkov D.. Targeted delivery of 4-aminopyridine into the rat brain by minicontainers from carbon-nanodiamonds composite. J. Neurosci. Neuroeng. 2013, V. 2, P. 569–573.

A comparative analysis of pharmacokinetics properties of diagnostic bone-seeking radiopharmaceuticals on the basis of phosphonic acids and technetium-99m

V K Tishchenko[1], **V M Petriev**[1, 2], **O A Smoryzanova**[1], **I N Zavestovskaya**[2]

[1] Tsyb Medical Radiological Research Centre - branch of the National Medical Research Radiological Centre of the Ministry of Health of the Russian Federation (A. Tsyb MRRC), 4, 249036, Korolev street, 4, Obninsk, Kaluga region, Russia
[2] National Research Nuclear University MEPhI (Moscow Engineering Physics Institute), 115409, Kashirskoe shosse, 31, Moscow, Russia

E-mail: petriev@mrrc.obninsk.ru

Key words: technetium-99m, 1-hydroxyethylidene-1,1-diphosphonic acid, oxa-bis-(ethylenenitrilo)tetramethylenephosphonic acid, ethylenediaminetetramethylenephosphonic acid (99mTc-EDTMP) or 99mTc-diethylenetriaminopentakis(methylphosphonic acid.

Abstract. This work is devoted to comparative research of pharmacokinetics properties of four bone-seeking radiopharmaceuticals (RPP) on the basis of bi-. tetra- and penta-phosphonic acids. Biodistribution studies were performed in intact rats after intravenous injections of ^{99m}Tc-hydroxyethylidenediphosphonic acid (^{99m}Tc-HEDP), ^{99m}Tc-oxabiphor (^{99m}Tc-OXB), ^{99m}Tc-ethylenediaminetetramethylenephosphonic acid (^{99m}Tc-EDTMP) or ^{99m}Tc-diethylenetriaminopentakis(methylphosphonic acid) (^{99m}Tc-PPA). In the structure of the HEDP contains two phosphonic groups, OENTMP and EDTMP – four phosphonic groups, PPA – five phosphonic groups. Radiochemical yield of labeled ^{99m}Tc HEDP, OENTMP, EDTMP, PPA is not less than 95%, the radiochemical impurities does not exceed 5%.
The investigated compounds have high stability in vivo and selective accumulation in osseous tissue. The highest concentrations of labeled compounds is reached in 3–24 hours after their intravenous injections. The investigated compounds are rapidly excreted from blood and soft organs and tissues mainly through the urinary routes.
So present study has showed that these RPP have properties, which making them promising candidates as a diagnostic pharmaceuticals of bone metastases.

1. Introduction

The success of cancer treatment depends on exact and well-timed tumor detection, and also on the evaluation of its dissimination in the body. A serious complication of many solid and hematological cancers is the development of bone metastases. The primary method of screening metastatic bone lesions is scintigraphy of skeleton using phosphonates labeled with technetium-99m (^{99m}Tc). To date, ^{99m}Tc-MDP represents the working horse for diagnostic bone scanning in conventional nuclear medicine due to the appropriate nuclear and physical characteristics of ^{99m}Tc (half-life 6 h, gamma-energy 140 keV). Various derivatives of phosphonates, containing from one to six phosphonic groups, are ideal ligands to react with ^{99m}Tc. Phosphonates have high affinity for hydroxyapatite crystals of bone and accumulate predominantly in inflammatory, destructive and metastatic sites of skeleton so these regions can be imaged as "warm foci". In addition to bone metastases imaging scintigraphy can be used to assess the efficacy of chemotherapy.

The mechanism of action of phosphonates depends on their structure. So, nonnitrogen-containing bisphosphonates can be metabolically incorporated into nonhydrolyzable analogues of adenosine triphosphate (ATP) [1 – 3]. Nitrogen-containing phosphonates inhibit enzymes of the mevalonate pathway, e.g. the farnesyl diphosphate synthase and geranylgeranyl diphosphate synthase [1 – 4]. They are necessary for posttranslational modification (prenylation) of small GTPases. As a result, this leads to the loss of osteoclast activity as well as the induction of tumor cells apoptosis [4, 5].

There are a lot of data of experimental researches of different phosphonates labeled with ^{99m}Tc in scientific literature [6 – 8]. For example, ^{99m}Tc-EDTMP had high affinity for bone tissue, but also accumulated in the kidneys: up to 2% of the injected dose at 90 min after intravenous injection [9]. In another study the possibility of bone metastases imaging with ^{99m}Tc-EDTMP and dicarboxypropane-diphosphonate labeled with ^{99m}Tc in patients clinical was assessed [10]. The results showed that the ratio of activity in metastatic sites of bone and soft tissue was higher for dicarboxypropane-diphosphonate labeled with ^{99m}Tc ($p < 0.05$), whereas the ratio of activity in metastatic and healthy areas of the skeleton for both drugs had no significant differences [10]. Láznícek M et al. [11] synthesized calcium salt of diethylenetriaminopentamethylphosphonic acid (PPA) labeled with ^{99m}Tc and compared its pharmacokinetic properties with ^{99m}Tc-MDP. Stability of ^{99m}Tc-PPA was higher than ^{99m}Tc-MDP. The levels of accumulation ^{99m}Tc-MDP and ^{99m}Tc-PPA in osseous tissue were of nearly equal.

The purpose of this work is to perform the comparative studies of pharmacokinetic properties of four radipharmaceuticals based on phosphonic acids with two, four and five phosphonic groups in intact rats after intravenous injection. ^{99m}Tc-1-hydroxyethylidene-1,1-diphosphonic acid (^{99m}Tc-HEDP) and ^{99m}Tc-oxa-bis(ethylenenitrilo)tetramethylenephosphonic acid (^{99m}Tc-OENTMF) have two phosphonic groups, ^{99m}Tc-N,N,N',N'-ethylenediaminetetrakis(methylenephosphonic acid) (^{99m}Tc-EDTMP) has four phosphonic groups and ^{99m}Tc- diethylenetriaminopentakis(methylphosphonic acid) (^{99m}Tc-PPA) has five phosphonic groups.

2. Materials and methods

For obtaining labeled radiopharmaceuticals monopotassium salt of 1-hydroxyethylidene-1,1-diphosphonic acid (officinal name – Xydifon, Semashko FSUC «Moschimpharmpreparaty», Moscow, Russia) was used. OENTMF was obtained from JSC «Karpov institute of Physical Chemistry» (Obninsk, Russia), EDTMP and PPA were manufactured by Sigma-Aldrich (England) and Fluka (Germany), respectively. ^{99m}Tc was received as $Na^{99m}TcO_4$ solution by elution from ^{99}Mo/^{99m}Tc-generator in isotonic saline. ^{99}Mo/^{99m}Tc-generator was purchased from JSC «Leypunskiy institute of Physics and Power Engineering» (Obninsk, Russia).

Labeling of phosphonic acids was carried out according to the methods, which were described in the articles [12, 13]. Labeling was performed at room temperature by the addition of 2.0 ml of $Na^{99m}TcO_4$ solution with volume activity 3.7 MBq/ml (0.1 mCi/ml). Labeling yield and radiochemical impurities were determined using paper chromatography on Whatman paper-1 manufactured by Sigma (England). Acetone and a mixture of C_2H_5OH, 25%NH_4OH and H_2O at a ratio of 2:5:5, respectively, were used acetone as mobile phase. Quantitative analysis of each radiolabelled compound and radiochemical impurities ($^{99m}TcO_4^-$ and $^{99m}TcO_2$) was performed by activity measurement of strips of chromatographic paper.

Pharmacokinetic studies were carried out in 64 white outbred male rats, weighing 180 ± 40 g. All animals were divided into four groups (16 rats in each group). Pharmacokinetics of ^{99m}Tc-HEDP, ^{99m}Tc-OENTMP, ^{99m}Tc-EDTMP and ^{99m}Tc-PPA was studied in the animals of the first, the second, the third and the fourth groups, respectively.

A dose of 0.37 MBq (0.01 mCi) in a volume of 0.1 ml of each radiolabelled compound was injected intravenously via a tail vein. The amount of injected compound was 2.27-3.57 mg/kg of animal weight. Then at various times after injection (5 min, 1, 3 and 24 h) the animals were sacrificed, and samples of various tissues were collected, placed in plastic tubes, weighed on electronic scales and counted in an automated gamma-counter with a standard of the injected dose. Then the amounts of radioactivity in each injected dose were calculated as percentage injected dose per gram of tissue (%ID/g) and as percentage injected dose per tissue (%ID/tissue). The results of radiometry were processed by estimation of standart error of mean value (M ± m). In addition, bone/blood and bone/muscle ratios were calculated.

3. Results

The chemical structure of various phosphonates derivatives almost didn't affect the nature of complex compounds with ^{99m}Tc. The efficacy of labeling was greater than 95%. Radiochemical impurities ($^{99m}TcO_4^-$ and $^{99m}TcO_2$) didn't exceed 5%.

The results of pharmacokinetics properties of ^{99m}Tc-HEDP, ^{99m}Tc-OENTMP, ^{99m}Tc-EDTMP and ^{99m}Tc-PPA in organs and tissues of intact rats are presented in the table 1. They indicate that after intravenous injection all labeled compounds accumulated mainly in the bone tissue.

Table 1. Pharmacokinetics of ^{99m}Tc-HEDP, ^{99m}Tc-OENTMP, ^{99m}Tc-EDTMP and ^{99m}Tc-PPA in intact rats after intravenous administration (in % of injected dose per gram).

№	Organ/tissue	Compound	Time after administration			
			5 min	1 h	3 h	24 h
1	Blood	^{99m}Tc-HEDP	0,78±0,17	0,19±0,03	0,048±0,004	0,043±0,009
		^{99m}Tc-OENTMP	0,56±0,09	0,24±0,04	0,11±0,01	0,050±0,004
		^{99m}Tc-EDTMP	0,54±0,03	0,10±0,02	0,019±0,002	0,033±0,009
		^{99m}Tc-PPA	0,71±0,13	0,10±0,01	0,020±0,001	0,010±0,002
2	Thyroid gland	^{99m}Tc-HEDP	0,32±0,07	0,10±0,03	0,12±0,02	0,23±0,05
		^{99m}Tc-OENTMP	0,31±0,05	0,22±0,05	0,30±0,10	1,48±0,77
		^{99m}Tc-EDTMP	0,81±0,14	0,09±0,02	0,10±0,02	0,48±0,09
		^{99m}Tc-PPA	0,70±0,11	0,17±0,04	0,11±0,03	0,23±0,05
3	Lungs	^{99m}Tc-HEDP	0,56±0,15	0,17±0,02	0,053±0,004	0,049±0,018
		^{99m}Tc-OENTMP	0,40±0,05	0,21±0,02	0,073±0,011	0,033±0,004
		^{99m}Tc-EDTMP	0,32±0,02	0,06±0,01	0,023±0,001	0,018±0,004
		^{99m}Tc-PPA	0,52±0,03	0,08±0,01	0,072±0,051	0,006±0,001
4	Liver	^{99m}Tc-HEDP	0,013±0,003	0,003±0,001	0,002±0,001	0,003±0,001
		^{99m}Tc-OENTMP	0,013±0,002	0,008±0,002	0,003±0,001	0,003±0,001
		^{99m}Tc-EDTMP	0,13±0,02	0,044±0,001	0,025±0,003	0,014±0,004
		^{99m}Tc-PPA	0,17±0,02	0,060±0,001	0,020±0,001	0,011±0,003
5	Kidney	^{99m}Tc-HEDP	1,59±0,34	0,74±0,10	0,60±0,04	0,59±0,20
		^{99m}Tc-OENTMP	1,59±0,14	1,46±0,17	1,49±0,12	0,93±0,07
		^{99m}Tc-EDTMP	1,78±0,23	1,25±0,05	1,35±0,09	0,96±0,18
		^{99m}Tc-PPA	2,83±0,19	0,63±0,03	0,50±0,03	0,27±0,01
6	Spleen	^{99m}Tc-HEDP	0,15±0,04	0,055±0,001	0,034±0,002	0,033±0,004
		^{99m}Tc-OENTMP	0,11±0,01	0,074±0,005	0,045±0,006	0,035±0,004
		^{99m}Tc-EDTMP	0,11±0,01	0,025±0,002	0,015±0,001	0,010±0,003
		^{99m}Tc-PPA	0,10±0,02	0,020±0,001	0,020±0,003	0,007±0,002
7	Stomach	^{99m}Tc-HEDP	0,30±0,06	0,11±0,01	0,15±0,02	0,21±0,08
		^{99m}Tc-OENTMP	0,23±0,03	0,15±0,02	0,09±0,02	0,040±0,007
		^{99m}Tc-EDTMP	0,25±0,02	0,042±0,006	0,022±0,003	0,019±0,003
		^{99m}Tc-PPA	0,26±0,06	0,060±0,001	0,070±0,010	0,010±0,001

8	Muscle	^{99m}Tc-HEDP	0,17±0,03	0,050±0,008	0,016±0,001	0,014±0,001
		^{99m}Tc-OENTMP	0,15±0,05	0,052±0,003	0,015±0,002	0,006±0,001
		^{99m}Tc-EDTMP	0,13±0,02	0,017±0,003	0,006±0,001	0,018±0,005
		^{99m}Tc-PPA	0,21±0,02	0,014±0,003	0,006±0,002	0,005±0,001
10	Femur	^{99m}Tc-HEDP	0,62±0,14	1,89±0,26	2,23±0,22	2,30±0,15
		^{99m}Tc-OENTMP	0,30±0,07	1,68±0,23	2,06±0,07	2,09±0,21
		^{99m}Tc-EDTMP	1,02±0,17	1,77±0,20	1,31±0,18	0,82±0,04
		^{99m}Tc-PPA	0,87±0,10	2,12±0,24	1,84±0,31	1,41±0,07
11	Skull	^{99m}Tc-HEDP	0,54±0,07	1,19±0,20	1,59±0,17	1,44±0,13
		^{99m}Tc-OENTMP	0,35±0,05	1,35±0,12	1,60±0,14	1,43±0,15
		^{99m}Tc-EDTMP	0,88±0,02	1,63±0,10	1,19±0,02	0,82±0,05
		^{99m}Tc-PPA	0,64±0,14	1,13±0,23	1,23±0,15	0,91±0,04
12	Rib	^{99m}Tc-HEDP	0,40±0,08	0,29±0,13	1,04±0,15	0,64±0,06
		^{99m}Tc-OENTMP	0,19±0,04	0,52±0,08	0,55±0,05	0,58±0,11
		^{99m}Tc-EDTMP	1,10±0,15	1,40±0,20	1,22±0,04	0,72±0,05
		^{99m}Tc-PPA	0,63±0,08	1,27±0,21	1,53±0,30	0,82±0,07
13	Spine	^{99m}Tc-HEDP	0,65±0,07	1,38±0,17	1,85±0,16	1,68±0,16
		^{99m}Tc-OENTMP	0,39±0,06	1,51±0,08	1,73±0,12	1,60±0,11
		^{99m}Tc-EDTMP	1,11±0,12	1,52±0,08	1,21±0,06	0,76±0,10
		^{99m}Tc-PPA	0,83±0,09	1,42±0,11	1,04±0,08	0,99±0,09

The highest level of radioactivity in skeleton was observed in femur: up to 2.30 %/g, 2.12 %/g, 2.09 %/g and 1.77 %/g for ^{99m}Tc-HEDP, ^{99m}Tc-PPA, ^{99m}Tc-OENTMP and ^{99m}Tc-EDTMP, respectively. In skull, ribs and spine the accumulation of labeled compounds was slightly lower than in femur.

In addition to the analysis of activity levels in bone tissue it is necessary to consider the values of ratios bone/blood and bone/muscle, which are particularly important for bone scintigraphy imaging. These values are presented in table 2. At 5 min after intravenous administration some of these ratios

Table 2. Ratios of ^{99m}Tc-HEDP, ^{99m}Tc-OENTMP, ^{99m}Tc-EDTMP and ^{99m}Tc-PPA in bones compared to those in blood and muscle in intact rats after intravenous administration

№	Ratio		Time after administration			
			5 мин	1 ч	3 ч	24 ч
1	Femur/blood	^{99m}Tc-HEDP	0,87±0,19	10,0±0,79	47,2±6,62	57,5±6,82
		^{99m}Tc-OENTMP	0,54±0,08	7,91±1,81	20,8±3,36	43,1±6,50
		^{99m}Tc-EDTMP	1,84±0,20	22,2±5,83	70,0±12,3	29,6±5,83
		^{99m}Tc-PPA	1,30±0,15	22,2±5,20	86,9±22,0	174,2±56,3
2	Femur/muscle	^{99m}Tc-HEDP	3,76±0,94	38,9±3,26	139,9±20,7	175,0±27,3
		^{99m}Tc-OENTMP	2,27±0,40	32,0±2,85	151,8±24,2	373,8±46,8
		^{99m}Tc-EDTMP	7,70±0,79	119,0±33,1	247,5±48,2	53,4±10,4
		^{99m}Tc-PPA	4,29±0,69	156,3±21,1	353,0±68,5	297,9±48,9
3	Skull/blood	^{99m}Tc-HEDP	0,78±0,15	6,35±0,84	32,9±3,33	35,6±4,06
		^{99m}Tc-OENTMP	0,65±0,05	6,44±1,55	16,2±2,88	29,4±4,51
		^{99m}Tc-EDTMP	1,64±0,08	19,6±4,11	63,5±7,20	28,6±5,03
		^{99m}Tc-PPA	0,95±0,23	12,0±3,78	54,9±3,79	111,3±35,8
4	Skull/muscle	^{99m}Tc-HEDP	3,36±0,66	23,9±1,43	97,9±10,9	110,5±20,5
		^{99m}Tc-OENTMP	2,89±0,64	26,2±2,51	114,3±14,9	254,9±31,9
		^{99m}Tc-EDTMP	7,00±0,82	103,5±17,5	222,5±25,5	54,7±13,5
		^{99m}Tc-PPA	3,36±0,90	83,1±15,7	253,9±72,4	188,4±25,7

5	Rib/blood	^{99m}Tc-HEDP	0,57±0,04	1,87±0,94	21,9±3,77	16,2±2,54
		^{99m}Tc-OENTMP	0,34±0,04	2,38±0,49	5,71±1,24	12,0±2,68
		^{99m}Tc-EDTMP	2,01±0,18	18,07±6,21	65,1±7,40	25,8±5,13
		^{99m}Tc-PPA	0,95±0,17	13,3±3,60	75,9±19,5	100,6±29,3
6	Rib/muscle	^{99m}Tc-HEDP	2,30±0,05	6,17±2,73	65,1±11,3	49,5±8,98
		^{99m}Tc-OENTMP	1,55±0,47	9,92±1,09	40,4±7,68	100,4±17,2
		^{99m}Tc-EDTMP	8,50±0,90	89,1±17,8	228,6±28,2	46,7±10,0
		^{99m}Tc-PPA	3,20±0,65	97,5±20,5	368,2±139,1	180,4±40,3
7	Spine/blood	^{99m}Tc-HEDP	0,92±0,14	7,36±0,57	38,5±3,23	41,5±5,13
		^{99m}Tc-OENTMP	0,71±0,03	7,11±1,43	17,4±2,68	32,8±4,13
		^{99m}Tc-EDTMP	2,02±0,11	18,5±4,41	64,8±8,42	26,0±5,23
		^{99m}Tc-PPA	1,22±0,10	14,6±2,56	49,7±10,6	113,1±23,5
8	Spine/muscle	^{99m}Tc-HEDP	3,91±0,47	28,4±2,41	114,0±10,4	129,2±24,5
		^{99m}Tc-OENTMP	3,29±0,90	29,2±0,82	124,1±14,7	290,1±38,1
		^{99m}Tc-EDTMP	8,70±1,09	95,2±13,5	226,0±27,6	46,9±8,45
		^{99m}Tc-PPA	4,14±0,72	106,4±16,7	216,5±56,8	214,8±51,6

were less than 1. Later, however, these values became significantly higher than 1 and reached at 3-24 hours. The highest values were noticed for ^{99m}Tc-PPA and ^{99m}Tc-OENTMP. For example, the ratio femur/blood for ^{99m}Tc-PPA reached 174.2 ± 56.3 and femur/muscle for ^{99m}Tc-OENTMP was 373.8 ± 46.8 (table 2). All these findings suggest the possibility of these radiopharmaceuticals for nuclear imaging of bone tissue.

It should be noted that all compounds were rapidly excreted from blood. During the early of the investigation there weren't significant differences between the radioactivity of labeled compounds. Further the content of compounds declined; meanwhile, in blood the activity of ^{99m}Tc-EDTMP and ^{99m}Tc-PPA was lower than ^{99m}Tc-HEDP and ^{99m}Tc-OENTMP.

4. Discussion

Some differences in the pharmacokinetic properties of ^{99m}Tc-HEDP, ^{99m}Tc-OENTMP, ^{99m}Tc-EDTMP and ^{99m}Tc-PPA can be explained by various binding of these complexes to plasma proteins [11].

The disappearance of many labeled phosphonates from blood is occurred through kidneys [13 – 17], so studied compounds accumulated and kept here in significant quantities during 24 h (table 1). Particularly high activity in the kidneys was observed for ^{99m}Tc-PPA: up to 2.83 ± 0.19 %/g at 5 min after intravenous injection. This caused a high radiation exposure to kidneys, but led to drop of radioactivity level in non-osseous tissues.

The stability of radiopharmaceuticals labeled with ^{99m}Tc can be assessed at their distribution in thyroid gland, since free ^{99m}Tc has a high affinity to in thyroid gland. The concentration of complexes in the thyroid gland was low. The peak levels of ^{99m}Tc-HEDP, ^{99m}Tc-EDTMP and ^{99m}Tc-PPA were noted immediately after intravenous injection. On the contrary, the amount of ^{99m}Tc-OENTMP reached 1.48 ± 0.77 %/g at 24 h. After 1 h there were no significant differences between the activity levels of compounds (table 1). Therefore, all radiopharmaceuticals were characterized by high stability in vivo.

The content of all complexes in non-osseous organs and tissues was low. ^{99m}Tc-EDTMP and ^{99m}Tc-PPA eliminated faster from all organs except the liver. Concentration of ^{99m}Tc-EDTMP and ^{99m}Tc-PPA in the liver was more than 10 times higher in comparison with ^{99m}Tc-HEDP and ^{99m}Tc-OENTMP (table 1).

5. Conclusion

The results of pharmacokinetics of ^{99m}Tc-HEDP, ^{99m}Tc-OENTMP, ^{99m}Tc-EDTMP and ^{99m}Tc-PPA in intact rats showed that all compounds exhibited high bone uptake and retention of radioactivity in the skeleton after their intravenous administration. The highest amounts of radioactivity were found in femur. In other bones (skull, ribs and spine) concentration of complexes was slightly lower than in femur. The retention of activity in bone tissue was high, whereas all complexes rapidly eliminated from soft organs and tissues. This was proved by the ratios of doses accumulated in bones compared to those found in other organs and tissues. The greatest values were observed after 3-24 h of injection. In addition, the studied complexes had high stability *in vivo*. Intensive removal of labeled compounds through the urinary system didn't lead to the significant accumulation of radioactivity in other organs and tissues.

The main result of the carried out research was the fact that the number of phosphonic groups in phosphonic acid derivate structures did not significantly affect the pharmacokinetic properties of radiopharmaceuticals, their uptake and retention in bone tissue, as well as the excretion from soft organs and tissues. These results suggested that ^{99m}Tc-HEDP, ^{99m}Tc-OENTMP, ^{99m}Tc-EDTMP and ^{99m}Tc-PPA had the potential to become suitable bone-imaging agents.

Acknowledgements
This work was supported by Competitiveness Program of National Research Nuclear University MEPhI.

References
[1] Ebetino F.H., Hogan A.M., Sun S. et al. 2011 *The relationship between the chemistry and biological activity of the bisphosphonates.* Bone., vol **49**, No. 1, p. 20–33
[2] Rogers M.J. 2003 New insights into the molecular mechanisms of action of bisphosphonates. *Curr. Pharm. Des.*, v. **9**, No. 32, p. 2643–2658
[3] Russell R.G. Bisphosphonates: the first 40 years. Bone., 2011, V. 49, No. 1, P. 2–19.
[4] Coxon F.P., Thompson K., Roelofs A.J. et al. Visualizing mineral binding and uptake of bisphosphonate by osteoclasts and non-resorbing cells. Bone., 2008, V. 42, No. 5, P. 848–860.
[5] Dunstan C.R., Felsenberg D., Seibel M.J. Therapy insight: the risks and benefits of bisphosphonates for the treatment of tumor-induced bone disease. Nat. Clin. Pract. Oncol., 2007, V. 4, No. 1, P. 42–55.
[6] Asikoglu M., Durak F.G.. The rabbit biodistribution of a therapeutic dose of zoledronic acid labeled with Tc-99m. Appl. Radiat. Isot., 2009, V. 67, No. 9, P. 1616–1621.
[7] Datta A., Panwar P., Chuttani K. et al. Synthesis of 99mTc-DOTMP and its preclinical evaluation as a multidentate imaging agent for skeletal metastases. Cancer Biother. Radiopharm., 2009, V. 24, No. 1, P. 123–128.
[8] Kumar V., Kumar D., Howman-Giles R.B. et al. Is (99m)Tc-labelled pamidronate a better agent than (99m)Tc-medronate for bone imaging? Nucl. Med. Commun., 2007, V. 28, No. 2, P. 101–107.
[9] Garnuszek P., Pawlak D., Licińska I. et al. Evaluation of a freeze-dried kit for EDTMP-based bone-seeking radiopharmaceuticals. Appl. Radiat. Isot., 2003, V. 58, No. 4, P. 481–488.
[10] Fueger B.J., Mitterhauser M., Wadsak W. et al. Bone lesion detection with carrier-added 99mTc-EDTMP in comparison with 99mTc-DPD. Nucl. Med. Commun., 2004, V. 25, No. 4, P. 361–365.
[11] Láznícek M., Láznícková A., Budský F. 99Tcm-DTPMP as a skeletal scintigraphy agent: distribution in rats in comparison with 99Tcm-MDP. Nucl. Med. Commun., 1996, V. 17, No. 12, P. 1016–1020.

[12] Petriev V.M., Skvortsov V.G., Ryzhikova T.P. Influence of Sn^{2+} concentration on complexation of ^{99m}Tc with potassium sodium hydroxyethylidenediphosphonate. Radiochemistry., 2007, V. 49, No. 3, P. 293-296.

[13] Petriev V.M., Tishchenko V.K., Siruk O.V. et al. Pharmacokinetic study of the new diagnostic radiopharmaceutical ^{99m}tc-pentaphosphonic acid in rats with an experimental bone-fracture model. Pharmaceutical Chemistry Journal., 2014, V. 48, No. 6, P. 357–362.

Radiopharmaceutical development based on human blood albumin microspheres and 90Y

V M Petriev[1,2], O P Vlasova[2], A A Postnov[2], N B Epstein[2]

[1]Tsyb Medical Radiological Research Centre – branch of the National Medical Research Radiological Centre of the Ministry of Health of the Russian Federation (A. Tsyb MRCC), Obninsk, Kaluga region, Russia
[2]National Research Nuclear University MEPhI, Moscow, Russia

E-mail: petriev@mrrc.obninsk.ru

Abstract. New radiopharmaceutial (RP) based on human serum albumin microspheres (MSA) and ^{90}Y was developed for treatment of liver cancer. The optimized synthesis using chelation resulted in approximately 80% yield with high specific activity. The RP developed was tested in mice with inoculated sarcoma-37. In two weeks the tumor size reduced by 43% after the treatment with the dose of 500 μCi injected into the tumor site.

1. Introduction

The current investigation is focused on the design of the new radiopharmaceuticals (RP) for the treatment of oncology. Cancer research is socially important especially keeping in mind the increase of malignant tumor diagnosis in the last years [1].

Oncological diseases remain one of the main cause of death in the developed countries. In Russia 280.000 men die from cancer each year. Cancer is responsible for the 15.3% of all mortality in the country next to cardiovascular diseases (50.1% in 2014) being ahead of trauma and poisoning (8.0% in 2014). 80641 men died in the working age 15-59 years as a result of cancer (15.4% for this group).

Cancer database of the P. Hertsen Moscow Oncology Research Institute reports that the number of primary liver cancer diagnosis changed from 6570 in 2004 to 7252 in 2014 with 10% increase.

Liver cancer can be primary namely originating from the liver cells themselves and secondary (metastatic) which is characterized by the growth of the tumor cells originated in the other organ and transported later to the liver [3].

Metastatic liver cancer is observed much more frequently than the primary one because of the liver function in the body and because of the high vascularization and perfusion.

Patients with the primary cancer of intestine, pancreas, stomach, lungs and reproductive system are prone to develop liver metastasis. Generally liver metastases are developed in 1/3 of all cancer patients. Metastatic cancer is characterized by the fast progression with no specific clinical observations [3]. The prognosis is negative predicting survival rate of 50% after one year [1].

Therefore the treatment of the liver malignant tumor remains of a key importance in clinical oncology. Cancer treatment depends on the stage of the disease and it can include surgery, chemotherapy and radiotherapy. Still the best results that allow the long survival time are achieved with the surgery. But at the time of diagnosis the radical resection of the tumor is possible with 5-15% of the patients only. Recurrent tumor is observed with 70-90% of the patients in 3-5 years. The second resection is possible with no more than 10% of the subjects. Hence the non-surgical treatment is of a vital importance for more than 80% of the liver cancer patients [4].

Currently in case of no surgery possible the systemic chemotherapy is used. The alternative is the locoregional chemo infusion and chemoembolization of the liver artery.

Last years the new approach of arterial radioembolization (RE) –a combination of the embolization and the radiotherapy was suggested. The method is based on the microspheres filled with isotope ^{90}Y (Theraspheres, MDS Nordion, Canada) or polymer microspheres again with ^{90}Y (SIR-spheres, Sirtex Medical, Australia) that are injected in the liver artery. Therapeutic dose is achieved in two weeks post injection [5,6].

RE implementation demonstrates good prospective of the method. However up to now there were only 6 patients in Russia treated with RE. Glass microspheres doped with ^{90}Y were used in 2009 in Russian Research Center for Radiology and Surgical Technologies (St.-Petersburg) with 4 patients. In 2012 another 2 patients received RE at Russian Cancer Research Center

(Moscow). The limiting factor remains the high price of the procedure and the short lifetime of ^{90}Y [6] especially taking into account the transportation time to Russia.

Development of a new highly efficient radiopharmaceutical is a multidisciplinary task with two main backbones: a selection of a proper radionuclide and a carrier among which human serum albumin plays an important role [7].

Microspheres of human serum albumin (MSA) are the unique carriers for the selective delivery of a radionuclide to the malignant tumors. Radiopharmaceuticals based on human serum albumin are highly physiological and easy to produce. The micro particles can be created with the controlled size and a given proteolysis speed in the body, ability to incorporate almost any isotope, keep them bound and release simultaneously with the proteolysis of the denatured protein [8–10].

The goal of the current study is to develop a radiopharmaceutical based on HBA and ^{90}Y to ally for the primary and metastatic liver cancer.

2. Materials and methods

The study on optimal fixation of ^{90}Y with human blood albumin microspheres 25-40 μm in diameter both non-modified and modified by the chelator were performed in the laboratory of Experimental Nuclear Medicine at A.Tsyb Medical Radiological Research Center – a branch of The Federal State Budgetary Institution the National Medical Research Radiological Center of the Ministry of Health of the Russian Federation. (FSBI NMRRC)

The following reaction conditions were studied: the acidity of the reaction mixture, heating time and the reaction temperature, influence of the other metals impurity on the fixation rate with ^{90}Y.

The results of ^{90}Y and MSA binding (both chelator modified and non-modified) are presented in table 1. It is clear modified MSA binds ^{90}Y two times better than non-modified one. This result is confirmed by serial experiments. Therefor only modified MSA was taken for further experiments.

After the extra purification of the metals impurity from the ^{90}Y solution the fixation rate of ^{90}Y with MSA increased up to 83% (the second and the third preparations). The fixation rate of ^{90}Y with MSA depends on the amount of inactive impurities of Fe, Al and Zn and we suppose that ^{90}Y chloride solution of a high quality will bind to MSA with 95% rate which is a good prerequisite for the development of a "cold kits" of radiopharmaceuticals for the liver cancer therapy.

Table 1. Fraction of ^{90}Y – MSA binding in %.

Substance	Time after synthesis, hours			
	0,1	1	2	24
First preparation of ^{90}Y chloride				
^{90}Y-MSA	23	26	30	30
^{90}Y-chelator-MSA	43	69	65	84
Second preparation of ^{90}Y chloride				
^{90}Y- MSA	39	42	43	38
^{90}Y- chelator-MSA	76	81	82	80
Third preparation of ^{90}Y chloride				
^{90}Y- chelator-MSA	68	83	70	75

The therapeutic efficacy on mice with inoculated tumor was also studied at FSBI NMRRC (figure1). The study was performed on female laboratory mice (N=30) divided into three groups 10 animals each. All mice were inoculated with sarcoma-37. In order to obtain a solid tumor the ascites fluid taken from donor mice was transplanted at day 8-10 of the tumor growth. The suspension of the tumor cells was injected under the hip skin. The study of the therapeutic efficacy started at day when the tumor size reached 0,8 – 1,0 cm^3.

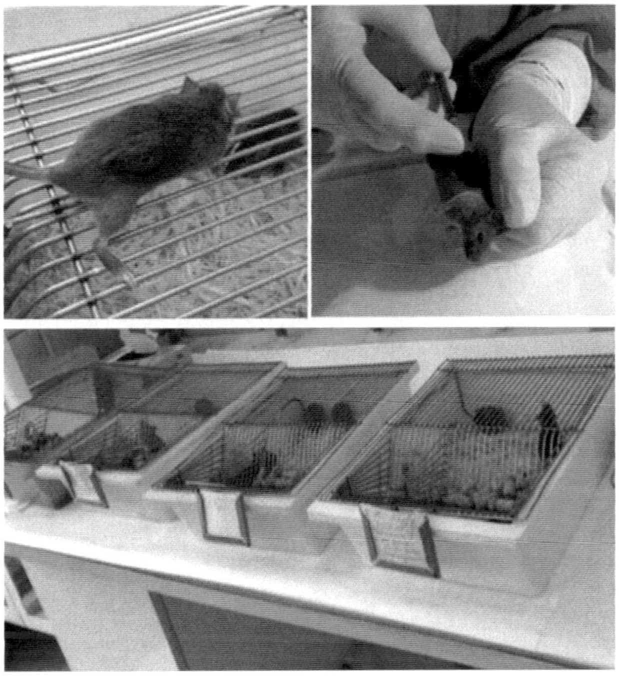

Figure 1. The study of the therapeutic efficiency of the designed radiopharmaceutical.

The first and the second groups received ^{90}Y-chelator-MSA 9,25 MBq (250 μCi) and 18,5 MBq (500 μCi) respectively injected in the tumor. The third group received the physiological solution injection of the same volume and served as control. The tumor volume was measured in all groups during 14 days next to transplante. All data in between was interpolated.

3. Results

In this experiment an active radiopharmaceutical was obtained with the yield of 82%. The specific volume activity was 126 mCi/ml which is high enough for the clinical applications.

The synthesis of MSA both chelated and pure was efficiently optimized to achieve the required characteristics.

The study of the therapeutic effect of the developed drug revealed the maximum retardation of the cancer growth achieved on day 14. compound compared to the control group (figure 2).

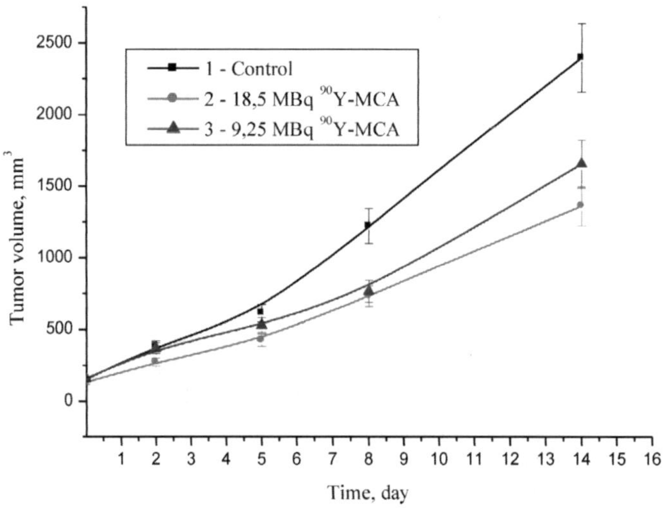

Figure 2. The retardation of a cancer growth (sarcoma-37) in mice after intratumor injection of ^{90}Y-albumin microspheres with the diameter of 25–40 μm.

4. Conclusions

The conducted radiochemical and biological studies concludes that ^{90}Y-albumin microspheres with the diameter of 25–40 μm is a prospective radiopharmaceutical for malignant tumors.

Acknowledgements

This work was supported by Competitiveness Program of National Research Nuclear University MEPhI.

References

[1] Statistics on malignant tumors in Russia and CIS in 2012 (editors Davydov M.I. and Axel E.M.) 2014 (in Russian) Moscow, 226 p.
[2] Malignant tumors in Russia in 2014 (morbidity and mortality), (editors Kaprin A.D., Starinsky

V.V., Petrova G.V.) 2016, Moscow, 250 p.

[3] Abdelmaksoud M.H.K., Louie J.D., Kothary N., et al. 2011 Consolidation of hepatic arterial inflow by embolization of variant hepatic arteries in preparation for yttrium-90 radioembolization, J. Vasc. Interv. Radiol, 22, 1364–1371.

[4] Tarazov P.G. 2013 Arterial radio embolization of the liver tumor by Y-90 microspheres, Voprosy oncologii, 59(4), 428-434.

[5] Ezziddin S, Meyer C, Kahancova S, Haslerud T, Willinek W, Wilhelm K, Biersack H J and Ahmadzadehfar H 2012 90Y Radioembolization After Radiation Exposure from Peptide Receptor Radionuclide Therapy, J Nucl Med 53, 1663–1669

[6] Salem R, Lewandowski R J, Gates V L, Nutting C W, Murthy R, Rose S C, Soulen M C at all., 2011 Research Reporting Standards for radioembolization of Hepatic Malignancies, J. Vasc. Interv. Radiol., 22 265–278.

[7] Tsyb A.F., Khachirov D.G., Strigunov V.I., Ispenkova L.I., Petriev V.M. 1983 Medical Radiology 28(2) 18–24.

[8] Petrova G.A., Petriev V.M., Skvortsov V.G. 2003 Problems of Biological, Medical and Pharmaceutical Chemistry. 3 30–34.

[9] Petriev V.M., Podgornichenko V.K., Ryzhikova T.P., Demidova N.A., Tarasova T.A., Smoryzanova O.A., 2007 Proceedings of the regional contest of the scientific projects in natural science, KNC AKO, RFFI, Kaluga 12 195–204.

[10] Petriev V.M., Shiryaev V.K., Smahtin L.A., Skvortsov V.G. 2010, Radiochemistry 52(2) 177–180.

The influence of protein fractions and electrolyte imbalance on refractive index of serum in patients with multiple myeloma

L Plotnikova[1, 4], A Polyanichko[2], M Kobeleva[1], M Uspenskaya[1, 5], A Garifullin[3] and S Voloshin[3]

[1] Saint Petersburg National Research University of Information Technologies, Mechanics and Optics, Kronveksky pr. 49, Saint Petersburg, 197101, Russia
[2] Saint Petersburg State University, Universitetskaya nab. 7/9, Saint Petersburg, 199034, Russia
[3] Russian Scientific Research Institute of Hematology and Transfusiology, 2nd Sovetskaya str. 16, Saint Petersburg, 191024, Russia

E-mail: [4] ljusja@mail.ru, [5]mv_uspenskaya@mail.ru

Abstract. Refractometric analysis is very rapid, accurate and simple method of analysis measuring refractive index of biological liquids such as serum, plasma, spinal fluid, urine. This method can be used for definition total protein and solids concentrations in serum. The value of refractive index depends on all substances in serum including proteins, lipids as well as low molecular weight compounds, for example ions of different metals.
Refractometric analysis shows strong correlations between protein concentrations in serum of patients with multiple myeloma and its(serum) refractive index which depends on protein concentration and doesn't depends on electrolyte disbalance.

1. Introduction

Multiple myeloma (MM) is a lymphoproliferative neoplasm with clonal plasma cells substrate. Multiple myeloma refers to a group of paraproteinemic hemoblastosis with typical produce and secrete a monoclonal immunoglobulin and/or its free light chains in serum and/or urine, seldom in cytoplasm of clonal plasma cells only [1, 2, 3]. Multiple myeloma accounts for about 1% of all types of malignancy and slightly more than 10% of hematologic malignancies [4]. The disease is considered incurable despite of advances in treatment and diagnostics [4, 6, 7].

There are different immunochemical types of MM which defined by types of involved secreting immunoglobulin (Ig): IgA-MM, IgG-MM, IgD-MM, IgE-MM, IgM-MM, Bence-Jones myeloma (BJ-MM), diclonal myeloma, nonsecretory myeloma and light chain myeloma [1, 2, 8, 11].

The increasing of calcium (total and ionized) and phosphorus concentrations are markers of active bone demineralization and bone destruction in MM patients with high monoclonal protein, especially for III stage of disease [8, 11].

Our research demonstrates differences in refractive index between control (donors) and experimental (patients with MM) groups with typical protein and electrolyte disturbance.

2. Materials and methods

2.1 Blood samples.

S-Monovette tubes were used (Sarstedt, Germany) with clotting activator for receiving of serum samples. The blood samples collected to tube and put in a support in room with +18-24°C within 20-30 min. After that serum was received by centrifugation within 15 min at 3,000 rpm (Thermo Scientific Heraeus Labofuge 200, USA). The samples were frozen (–30°C) and kept for refractometric analysis.

2.2 Refractometric analysis.

Serum were researched on refractometer (Anton Paar Abbemat 200, Austria) for dimension of total solids and protein concentrations in control group and patients with MM. Refractive index was measured on wavelength λ=589 nm, temperature T=17,5°C by experimental method. The accuracy of refractive index measurement is ±0,0001.

The sample was applying on the surface of prism and lit by LED illumination through an interference light filter with use different light angles and directions. The reflected light beam was recorded by matrix sensors. Information was processed by refractometer and displayed in automatic mode.

3. Results and discussions

Serum samples was researched from 14 patients with MM (stage II and III, age 42-79 years old), who was observed in Hematology Clinic of Russian Scientific Research Institute of Hematology and Transfusiology. Fifty-five serum samples from healthy donors entered into control group.

Refractometric analysis of total protein concentration in serum based on extent of change of light beam. The value of refractive index of serum depends on quantity and physic-chemical characteristics of being proteins in serum. It is bound to low depend of electrolytes and non-protein organic substances on refractive power because it has constant concentrations in healthy donors serum. Imbalance in calcium-phosphorus system is typical for MM patients and it leads to increase of concentrations these elements in serum and urine [8, 11]. Patients with MM and control group can have strong difference in concentration of electrolytes (calcium and phosphorus) owing to bone destructions in MM patients. Calcium ions can bound to monoclonal protein, that decrease differences between protein and solids concentrations in serum for some patients. [15]. This feature is shown on Figure 1-b (refractive index 1,3544 D and 1,3556 D). Never the less there is still the reliable differences remain between two groups (donors and MM patients) in concentrations of electrolytes and total protein.

M-protein is not secreted in some patients, who have specific forms of MM such as "nonsecretory" myeloma, Bence-Jones myeloma and light chain myeloma [1, 2, 11]. Aforementioned types of MM have not M-protein in serum. Our experimental data confirm this feature (see the points with refractive index 1,3462 D, 1,3483 D and 1,3521 D in Figure 1 (a, b)). M-protein is absent in serum of MM patients with nonspecific types (Figure 2).

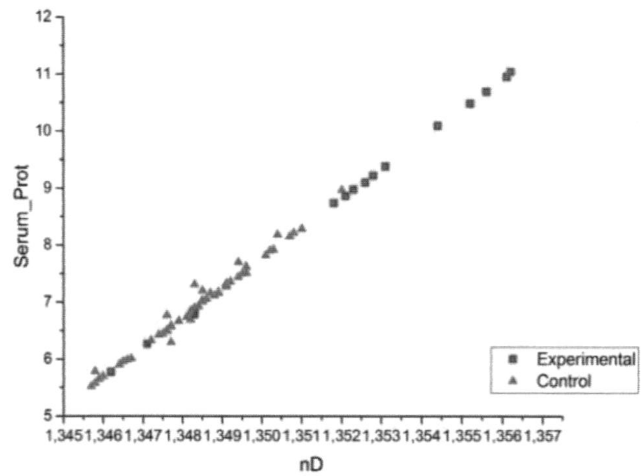

a

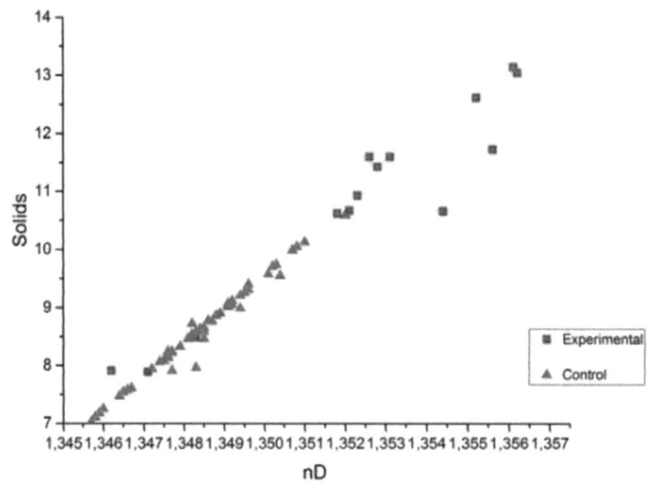

b

Figure 1. Dependence serum protein (a) and total solids (b) concentration on the refractive index. *Red squares* show MM patient's data and *green triangles* demonstrate donor's data

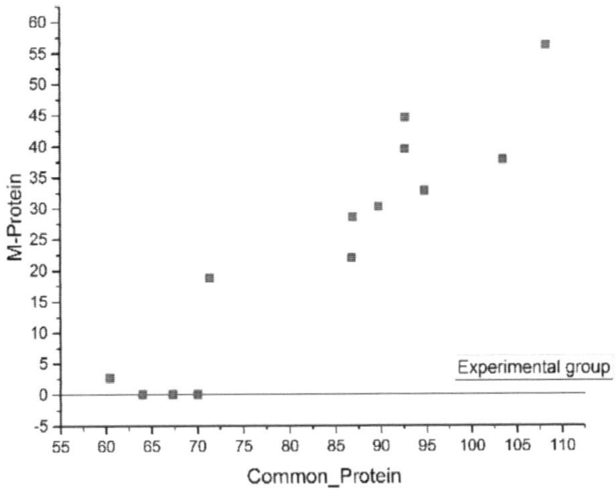

Figure 2. Total protein and M-protein concentrations in serum of patients with multiple myeloma

4. Conclusion.

The data of refractometric analysis show differences between patients with MM and control group (donors) in protein and electrolytes (calcium and phosphorus) concentrations in serum and their refractive index. Patients with MM have higher scores on all aforementioned features. The value of refractive index primarily depends on total protein and M-protein concentrations for patients with multiple myeloma as well as for healthy donors. Refractometric analysis can be use as one of suitable method of detection changes in protein concentration in MM patient's serum.

References

[1] M Volkova 2001 *Clinical oncohematology (Manual for Physicians)* ed M Volkova (Moscow: Medicine) p 576
[2] N Saraeva 2009 *Leukemias* (Irkutsk) p 122
[3] V Serov, M Paltsev 1998 *Pathological Anatomy (lectures)* ed V Serov, M Paltsev (Moscow: Medicine) p 640
[4] M Gertz, S Rajkumar 2014 *Multiple Myeloma. Diagnosis and Treatment* ed M Gertz, S Rajkumar (Springer) p 311
[5] N Munshi, K Anderson 2013 *Advances in Biology and Therapy of Multiple Myeloma* (*Basic Science* vol 1) ed N Munshi, K Anderson (Springer) p 319
[6] B Devenney, ChErickson 2004 Multiple myeloma: an overview *Clin. J. of Onc. Nurs.* **8(4)** 401
[7] M Mohty, J-L Harousseau 2015 *Handbook of Multiple Myeloma* ed M Mohty, J-L Harousseau (Adis) p 90
[8] O Rukavitcin, V Pop 2016 *Multiple myeloma and its related diseases (3-d edition)* ed O Rukavitcin, V Pop at al. (Moscow: HEOTAR-Media) p 224
[9] S Bessmeltsev 2013 Multiple myeloma (Part I) *Clin. Oncohem.* **6(3)** 237
[10] V Ryzhko, A Klodzinsky, E Varlamova, M Sataeva, I Kaplanskaya, I Nakastoev and O Paraskevova 2010 "Nonsecretory" multiple myeloma *Clin. Oncohem.* **3(3)** 270
[11] A Svistunov, A Ruta and O Shevchenko 2010 Characteristic of bone metabolism of patients with multiple myeloma *Sar. Sci-Med. J.* **6(1)** 48

[12] A Vorobiev 1985 *Guidelines for Hematology (vol 1)* ed A Vorobiev (Moscow: Medicine) p 429

[13] S Bessmeltsev 1998 Diagnosis of complications in multiple myeloma and methods of their correction (manual for physicians) ed S Bessmeltsev, K Abulkadyrov and U Katsadze (Saint-P.) p 17

[14] D Kravchenko, S Hoduleva and D Novik 2016 *Multiple myeloma (Practical Guide for Physicians)* ed D Kravchenko, S Hoduleva and D Novik (Gomel: government institution "Republic scientific and practical center of radiation medicine and human ecology") p 83

[15] T Annesley M Burritt and R Kyle 1982 Artifactual hypercalcemia in multiple myeloma *Mayo Clin Proc.* **57(9)** p 572

The bioimpedance analysis of a parenchyma of a liver in the conditions of its extensive resection in experiment

D Y Agibalov[1], D N Panchenkov[2], V B Chertyuk[3], S D Leonov[2] and D A Astakhov[4]

[1]National Research Nuclear University MEPhI, Moscow, Russia
[2]Department of endoscopic surgery of SEI HPE MSMSU them. A. I. Evdokimov, Moscow, Russia
[3]Institute of Emergency children's surgery and traumatology, Moscow, Russia
[4]Federal scientific clinical center of FMBA of Russia, Moscow, Russia

E-mail: agibalovd@bk.ru

Abstract. The liver failure which is result of disharmony of functionality of a liver to requirements of an organism is the main reason for unsatisfactory results of an extensive resection of a liver. However, uniform effective criterion of definition of degree of a liver failure it isn't developed now. One of data acquisition methods about a morfo-functional condition of internals is the bioimpedance analysis (BIA) based on impedance assessment (full electric resistance) of a biological tissue. Measurements of an impedance are used in medicine and biology for the characteristic of physical properties of living tissue, studying of the changes bound to a functional state and its structural features. In experimental conditions we carried out an extensive resection of a liver on 27 white laboratory rats of the Vistar line. The comparative characteristic of data of a bioimpedansometriya in intraoperative and after the operational period with the main existing methods of assessment of a functional condition of a liver was carried out. By results of the work performed by us it is possible to claim that the bioimpedance analysis of a liver on the basis of an invasive bioimpedansometriya allows to estimate morphological features and functional activity of a liver before performance of an extensive resection of a liver. The data obtained during scientific work are experimental justification for use of an impedansometriya during complex assessment of functional reserves of a liver. Preliminary data of clinical approbation at a stage of introduction of a technique speak about rather high informational content of a bioimpedansometriya. The subsequent analysis of efficiency of the invasive bioimpedance analysis of a liver requires further accumulation of clinical data. However even at this stage the method showed the prospect for further use in clinical surgical hepathology.

1. Introduction

At the present stage of development of surgical hepatology and anesthesiology of the indication to accomplishment of extensive resections of a liver considerably extended. However, the frequency of postoperative complications and a lethality remains high and makes, according to different data, 40% and 16% according to [1,2,3].

The main reason for unsatisfactory results in extensive liver resection is liver failure, which is the result of the functional capacity of the liver to the body's needs [4].

To date, there are a large number of methods of assessing the degree of hepatic dysfunction. The most informative indicators of liver function available in clinical practice are: total bilirubin, albumin, albumin-globulin index, leukocyte index of intoxication, the level aspartataminotransferaza, alanine aminotransferase and duration of cholestasis before operation, indices of protein, pseudocholinesterase, fibrinogen, fibrins, prothrombin index, clearance test, indocyanine green [4,5,6].

However, a single effective criterion for determining the degree of hepatic insufficiency is not currently developed.

One of the methods of obtaining data on morpho-functional state of internal organs is bioimpedance analysis (BIA), based on the estimated impedance (full electric resistance) of biological tissues [7,8]. Biological tissue consists of a cellular component and communication routes, including the vascular and interstitial slits, each of which is characterized by different resistance and a different ability to conduct alternating electric current.

Impedance measurements are used in medicine and biology to characterize the physical properties of living tissues, study of the changes associated with the functional status and structural features [9].

Although the research underlying the impedance were carried out at the beginning of the last century, the method is constantly evolving and finding more and more new diagnostic applications.

Work purpose: an assessment of biophysical characteristics of a parenchyma of a liver after its extensive resection.

2. Materials and methods.

The study was conducted on 27 white rats of Vistar line of both sexes weighing 180-230 g, which under ether anesthesia performed laparotomy, and measured the electrical impedance of the liver. Then we conducted extensive resection of the organ being examined (70%) and re-measured the electrical impedance of the parenchyma of the remaining lobe of the liver, then made a revision and suturing of the abdominal cavity.

After 72 hours was performed a relaparotomy, and measured the electrical impedance of the liver parenchyma. Experimental animals also conducted biochemical analysis of blood. Blood sampling was performed up to simulation of a pathological process and in 3 days after extensive liver resection. Animals were taken from experiment and a fragment of the liver were sent for histological examination.

At histologic research degree of dystrophy of hepatocytes was estimated in 25 viewing fields on a ball scale where 0 – normal hepatocytes; 1 – slight intracellular dystrophy of hepatocytes (mainly proteinaceous); 2 – moderate intracellular dystrophy of hepatocytes (small-drop fatty dystrophy); 3 – the expressed intracellular dystrophy of hepatocytes (largedropped fatty dystrophy); 4 – necrosis of hepatocytes. Expected a median of the sum of balls to ORP and the 3rd days after a liver resection.

The invasive bioimpedance was carried out by means of bipolar needle electrodes and the original device for measurement of total electrical resistance of biological tissues – BIM-II (The Russian Federation patent No. 2366360, OOO "Center for transfer of innovative technologies", Smolensk). Measurement was performed at frequencies of 2 kHz, 10 kHz, 20 kHz and voltage less than 1 volt with the subsequent calculation of coefficients of dispersion (D).

$$D = Z_{nch} / Z_{vch}$$

where D – coefficient of dispersion of an electric impedance, Z_{nch} – the electric impedance measured at a low frequency, Z_{vch} – the electric impedance measured at a high frequency.

Biochemical indicators of serum of blood investigated on the semi-automatic MiniScreen analyzer (HospitexDiagnostix, Italy).

3. Results and discussion

In the postoperative period the lethality made 52% (14 animals).

The electric impedance of an intact liver was equal 3,2±0,2 kOhm at a frequency of 2 kHz, 2,3±0,1 kOhm at a frequency of 10 kHz and 1,7±0,11 at a frequency of 20 kHz (n=27).

Table 1. An electric impedance of a parenchyma of a liver at experimental animals.

Measurement period	2 kHz		10 kHz		20 kHz	
	Z, kOhm	n	Z, kOhm	n	Z, kOhm	n
To ORP	3,2±0,2	27	2,3±0,1	27	1,7±0,11	27
After ORP	3,1±0,16	27	2,2±0,11	27	1,8±0,08	27
For the 3rd days after ORP	4,0±0,13 *	13	2,5±0,07	13	1,7±0,08	13

* - p <0,05 in comparison with data to and right after ORP

After an extensive resection the studied indicator at 20 rats decreased at a frequency of 2 kHz and 10 kHz a little and averaged at the corresponding frequencies: 3,1±0,16 kOhm, 2,2±0,11 kOhm, 1,8±0,08 kOhm (n=27, distinctions are not reliable, p> 0,05) that, is possible, knitted with increase in a blood-groove in the remained liver share.

In 72 hours after surgery the electric impedance of a parenchyma of a liver authentically increased to 4,0±0,13 kOhm (n=13, p <0,05) at a frequency of measurement of 2 kHz.

Coefficients of dispersion of an electric impedance were calculated only for the survived rats (n=13) and reduced in table 2.

Table 2. Coefficient of dispersion of an electric impedance of a liver of experimental rats.

Measurement period	Dispersion coefficient		
	D2kgts/20kgts	D2kgts/10kgts	D10kgts/20kgts
To ORP	1,95±0,12	1,43±0,10	1,40±0,10
After ORP	1,79±0,11	1,4±0,06	1,23±0,07
For the 3rd days after ORP	2,45±0,10 *	1,6±0,07	1,55±0,10

* - p <0,05 in comparison with data to and right after ORP

Coefficients of $D_{2kHz/10kHz}$ and $D_{10kHz/20kHz}$ authentically did not change throughout experiment. The liver parenchyma $D_{2kHz/20kHz}$ coefficient to a resection was equal 1,95±0,12, and then decreased right after a resection to 1,79±0,11 a little (n=13, distinctions are not reliable, p> 0,05). For the 3rd days after ORP $D_{2kHz/20kHz}$ coefficient authentically increased to 2,45±0,10 in comparison with initial indicators (n=13, distinctions are not reliable, p <0,05).

The value of the coefficient D depends on the structure, functions and status of biological tissue. Moreover, the higher it is, the more intense metabolism in the test body.

In the course of biochemical studies in experimental animals for 3 days there was a statistically significant increase in AST levels with 243,69±20,47 to 377,18±25,43 u/l (p<0.05) (table 3).

The concentration of total protein and albumin in the serum after modeling pathological process has also decreased slightly compared to the norm: total protein content from 84.1±7,92 g/l to 71,58±7.15 g/l (p>0.05), albumin – from 23.86±2.3 g/l to 21.85±2.8 g/l (p>0.05). Statistically significant differences between indicators were not observed.

Increase of concentration of nuclear heating plant in blood serum at the 3rd day after accomplishment of an extensive resection of a liver demonstrates existence of moderately expressed syndrome of a cytolysis.

At the time of Histological research of all surviving animals showed characteristic morphological picture. The liver parenchyma was represented by hepatocytes with intact girder structure, uneven hyperemia in the sinusoids and blood-filled veins big. Observed small focal perivascular lymphoid infiltrates. Hepatocytes were manifestations of fatty degeneration more pronounced in the periphery of the hepatic lobules, and in the center of lobules was determined by protein deposits. The nuclei of hepatocytes was with lumpy chromatin, clear nucleolus, sometimes two.

Table 3. Biochemical indicators of serum of blood of experimental rats before and after ORP.

Terms supervision	Biokhimicheskiyepokazateli			
	ALT, El	Nuclear heating plant, El	General protein, g/l	Albumine, g/l
	M±m	M±m	M±m	M±m
Before carrying out ORP	264,73±18,53	243,69±20,4784,1±7,92[1]		23,86±2,3
For the 3rd days after carrying out ORP	225,23±18,66	377,18±25,4371,58±7,15[1]		21,85±2,8

Note: distinctions are statistically significant at $p < 0{,}05$
[1] – between two groups

In the area of resection was observed formed a zone of demarcation with a zone of large-focal necrosis fibrinoidous and drain lymph leukocyte infiltration, sometimes extending to the liver parenchyma. Also was determined the drain hemorrhages with hemosiderin deposits and hemosiderofagus.

During the morphometry revealed that degeneration of hepatocytes, resected liver tissue was at the level of 9 points (min=7; max=15). However, 3 days later the PPR was observed severe degeneration of hepatocytes at the level of 60 points (min=50; max=68).

During surgical intervention on the liver in the hepatic parenchyma along with the processes of destruction of liver cells at the same time, the processes of regeneration. According to the literature, after performing a right hemihepatectomy for living-related liver transplantation within 10 days of the filling volume of the hepatic parenchyma is increased by 75% of the initial volume of the parenchyma and the regeneration of liver cells is triggered by increasing metabolic process [10].

Compensation of the disturbed functions of the liver is provided in all the variety of forms of regenerative reaction cell (mitotic division) and intracellular (compensatory hyperplasia ultrastructure and the actual intracellular regeneration).

During the post-resection regeneration actively divide by mitosis, not only hepatocytes, but also copperhouse cells and endothelial cells of blood capillaries. The maximum level of proliferation of hepatocytes largely depends on the scope of resection: the major part of the liver removed, the higher is the mitotic index in the period of the highest rise of mitotic activity [11].

Apparently, after the PPR in rats by 3 days of the experiment along with the phenomena of degeneration of hepatocytes, as evidenced by the increasing concentartion transamine blood, began an active process of cell regeneration of the liver, which was accompanied by intensive metabolism and increased functional activity of liver parenchyma. In this regard, there was a significant increase in the dispersion coefficient $D_{2kHz/20kHz}$ and the absolute value of electrical impedance at the frequency of 2kHz.

Thus, the electrical impedance parameters of liver reflect changes in the functional activity of the parenchyma and can be used to develop diagnostic technologies in surgical hepatology.

Acknowledgments
This work was supported by Competitiveness Program of National Research Nuclear University MEPhI.

References

[1] Sinenchenko G. I. 2008 Structure of postoperative complications of surgical treatment of locally-spread and metastatic cancer of stomach / G. I. Sinenchenko, L.D. Roman, A. M. Karachun, etc.//*Messenger of surgical gastroenterology,* No. 4., p. 65.
[2] Darwin V. V. 2012 Early postoperative complications in surgery of metastasises of a colorectal cancer in a liver / S. V. Onishchenko, M. M. Lysak//*the Collection of materials XIX of the International conference of surgeons-hepatologists of Russia and the CIS countries.* – Irkutsk, P. 45-46.
[3] Kemeny N. E. 2005 Hepatic arterial infusion after liver resection/N. E. Kemeny, M. Gonen//*N. Engl. J. med.*, Vol. 352, No. 7, p.734-735.
[4] Galperin E. I. 1978 Insufficiency of a liver / E. I. Galperin, E. A. Neklyudova, M. I. Semendyaeva, (M.: Medicine). – 327 pages.
[5] Metwally M. A. 2007 Predictors and noninvasive identification of severe liver fibrosis in patients with chronic hepatitis C/M. A. Metwally, C. O. Zein, N. N. Zein//*Dig. Dis. Sci.*, Vol. 52, No. 2, p. 582-588.
[6] Kotiv B.N., Dzidzava I. I., Cat's D.P., Kochatkov A. A., Bugaev S.A., Smorodsky A. V., Slobodyanik A. V. 2009 A quantitative assessment of function of a liver by method a clearance test with Indo-cyanine green//*Transplantology*, No. 1, p. 30-36.
[7] Hastsayev B. D. 1996 An impedance method in medicobiological researches and its instrument equipment *Medical technical*; , p. 34-40.
[8] Byelik D. V., Byelik K. D. 2007 Increase of informational content at determination of small arrays of oncological tumors a multifrequency impedansometriya. *Medical equipment*; **4**: 13-17.
[9] Tornuyev Yu. V., Hachatryan R. G., Hachtryan A.P. *et al* 1990 Electric impedance of biological fabrics (M.: VZPI).
[10] Istomin N. P. 2011 Features of a pancreatic and postoperative liver failure in surgery / N. P. Istomin, Yu. V. Ivanov, D. N. Panchenko. (Moscow) - 206 pages.
[11] Chikoteev S. P. 2001 Modern views on regeneration of a liver / S. P. Chikoteev, A. N. Plekhanov, N. G. Kornilov//*Surgery*, No. 6, p. 59-62.

Efficiency evaluation of mud applications of laser doppler of skin

S K Kasimova[1], E I Kondratenko[1], O M Alykova[1], N A Lomtieva[1] and A F Alykova[2]

[1]Astrakhan state university, Astrakhan, Russia
[2]National Research Nuclear University MEPhI Moscow Engineering Physics Institute), Moscow, Russia

E-mail: saule_kasimova@mail.ru

Abstract. The mechanism of the microcirculation's change of the face skin of the women under the influence of the sulfur silt mud application of the lake Karantinnoe of Astrakhan region was studied. The age particularities of vasorelaxation's peloid action on the microcirculation of the face skin was installed. Peloid promotes the influx of arterial blood, the improvement of the tissue's feeding and the reduction of the stagnant events. The prolonged action of the sulfur silt mud application reveals at more mature age.

1. Introdution

The Laser Doppler flowmetry (LDF) noninvasive method is highly effective in diagnostics of a condition of the microcirculator course of skin including studying changes of microcirculator processes using mud applications [3]. The studied sulphidic and silt therapeutic mud is extracted from the lake Karantinnoe located in the western part of the delta of the Volga River, in Limansky district of the Astrakhan region and according to the Balneological conclusion of FSU «RNCVM&KRossdrav» № 14/186 from 4/20/2007 is recognized as medical.

2. Methods

Examinees were divided into 4 age groups [1]. In total, 40 people participated in an experiment. Studying of microcirculation of face skin was carried out by means of the laser analyzer of rate of a superficial capillary blood flow "LAKK-01" (production of NPP Lazma, Russia, the registration certificate № 29/030200703/5555-03 of 11.09.2003). Measurements were performed in the center of the right cheek and in the center of a forehead (at the height of 1 cm from brows). Each measurement was performed within 4 min. that is sufficient for registration of necessary quantity of cycles of change of PI (perfusion index). Measurements were taken before drawing mud application, through 5 and in 15 min. after mud application. Application time – 10 min., temperature of the mud used for application, 30°C. In the first age group studying of change of microcirculation of a skin in 5 min. after processing of a skin by normal saline solution for assessment of mechanical impact on it was carried out. In this work the analysis of functioning of the active mechanisms caused by myogenetic activity of vazomotor (AmaxLF/M) and neurogenic influences (σ/AmaxLF) and also the passive mechanisms caused by cordial (AmaxCF/σ) and respiratory rhythms of a regulation of a blood flow

(AmaxHF/σ) was carried out. Calculated the index of efficiency of microcirculation (IEM) – the integrated characteristic of a blood flow between active and passive mechanisms of a regulation of a blood flow in system of microcirculation: IEM=AmaxLF/(AmaxHF+AmaxCF). Also the rheologic factor (intravascular resistance, AmaxCF/M) was considered.

3. Results

Wiping of face skin physiological solution had no significant effect on perfusion indicators of fabric, but increase in a tone of vessels has been testified. Thus, it is possible to exclude influence of manipulations with skin on microcirculation change.

At registration of an indicator of microcirculation of face skin after mud application at first the tendency to increase in perfusion (in 5 min.), then to reduction was observed (in 15 min).

Table 1. Change of microcirculation of face skin in a cheek at women of different age groups

Indicators	Before mud application	In 5 min. after mud application	In 15 min. after mud application
first age group, 16-25 years, n=10			
M, perfusion units	18,08±1,662	19,20±1,789	18,96±0,981
σ, perfusion units	1,02±0,124	1,40±0,165	1,33±0,196
Kv, %	5,36±0,558	7,50±0,627*	6,01±0,598
IEM, %	1,42±0,127	1,40±0,093	1,17±0,120
Intravessel. resistance,%	1,11±0,169	1,60±0,256	1,53±0,252
Second age group, 26-35 years			
M, perfusion units	17,68±1,017	19,82±1,931	19,41±1,233
σ, perfusion units	1,36±0,162	1,04±0,112	1,03±0,056
Kv, %	6,01±0,531	4,76±0,139	5,37±0,375
IEM, %	1,31±0,128	1,17±0,085	1,37±0,133
Intravessel. resistance, %	1,12±0,122	0,94±0,075	1,28±0,086
Third age group, 36-45 years			
M, perfusion units	18,6±1,69	18,2±1,54	23,0±2,31
σ, perfusion units	0,8±0,11	1,2±0,18	1,2±0,16 *
Kv, %	4,6±0,53	6,8±0,77 *	5,6±0,68
IEM	1,4±0,13	1,2±0,14	1,4±0,12
Intravessel. resistance, %	0,92±0,138	0,92±0,061	0,86±0,058
Senior age group, 46-55 years			
M, perfusion units	18,6±1,69	18,2±1,54	23,0±2,31
σ, perfusion units	0,8±0,11	1,2±0,18	1,2±0,16 *
Kv, %	4,6±0,53	6,8±0,77 *	5,6±0,68

IEM	1,4±0,13	1,2±0,14	1,4±0,12
Intravessel. resistance, %	0,92±0,138	0,92±0,061	0,86±0,058

*Notes: * - differences in comparison with measurements before mud application; * - p≤0,05, ** - p≤0,01,*** - p≤0,001; # - differences in comparison with measurements in 5 min. after before mud application; #- p<0,05, ## - p≤0,01, ### - p≤0,001*

Practically the efficiency index didn't change, it was only slightly decreasing in 15 min. The amplitude-frequency analysis showed that in case of the normal state of face skin the active mechanism of regulation prevails. Increase of a contribution of vazomotion in 5 min. ($p<0,05$) with its further decrease is fixed ($p\leq0,01$ – in 15 min.). The same tendency was observed in case of change of a vascular tone. In that time, there was a decrease in the regulating influence of the passive mechanism.

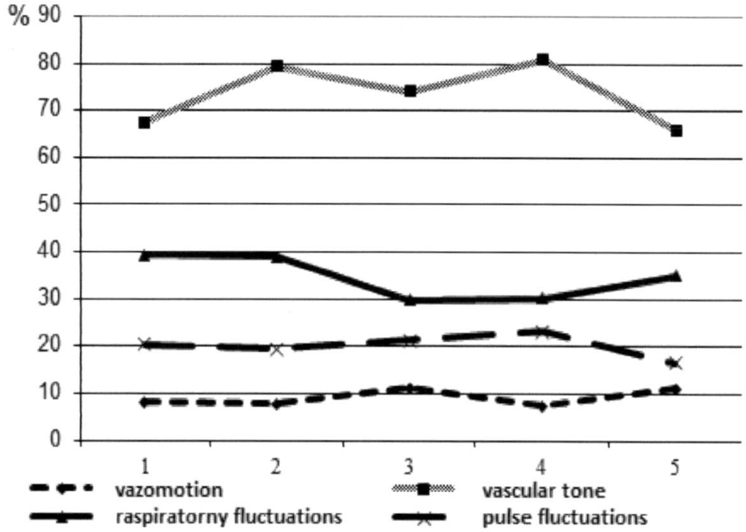

Figure 1. Change of microcirculation of face skin in a cheek at girls volunteers of 16-25 years (1 – before drawing before mud application, 2 – after physiological solution, 3 – in 5 min. after before mud application, 4 – in 15 min. after before mud application, 5 – after course application of before mud application, here and to figure 2)

In this case, if in 5 min. the active mechanism exerted impact, then in 15 min. it was replaced by action of passive. Intra vascular resistance after mud application had a downward tendency.

In the second age group after drawing mud application in a cheek the tendency to growth of an indicator of microcirculation was shown that was followed by decrease in mean square value and coefficient of a variation (tab. 1). At the same time in case of the first measurement after application decrease in an index of efficiency of microcirculation which by 15th minute was recovered to reference values is noted.

In the analysis of a contribution of various mechanisms to change of microcirculation the tendency similar to the previous age group in case of which in the fifth minute decrease in influence of vazomotion, a vascular tone and intra vascular resistance while compound the passive mechanism strengthen the influence is fixed is traced.

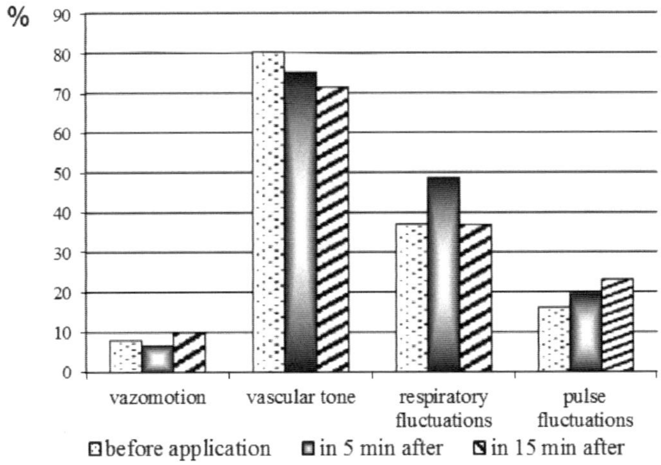

Figure 2. Change of microcirculation of face skin in a cheek after mud application at female volunteers of 26-35 years

In the 15th minute a little ambiguous picture as there is a further decrease in a vascular tone and respiratory fluctuations comes to light, and influence of pulse fluctuations and vazomotion (fig. 2) in parallel increases. When studying microcirculator processes of face skin in cheeks at women of 36-45 years after 5 min. after application changes of the main indicator of microcirculation of M (tab. 1) weren't revealed. In 15 min. after mud application the tendency to increase in this indicator was observed that was followed by increase of a mean square deviation ($p<0,05$).

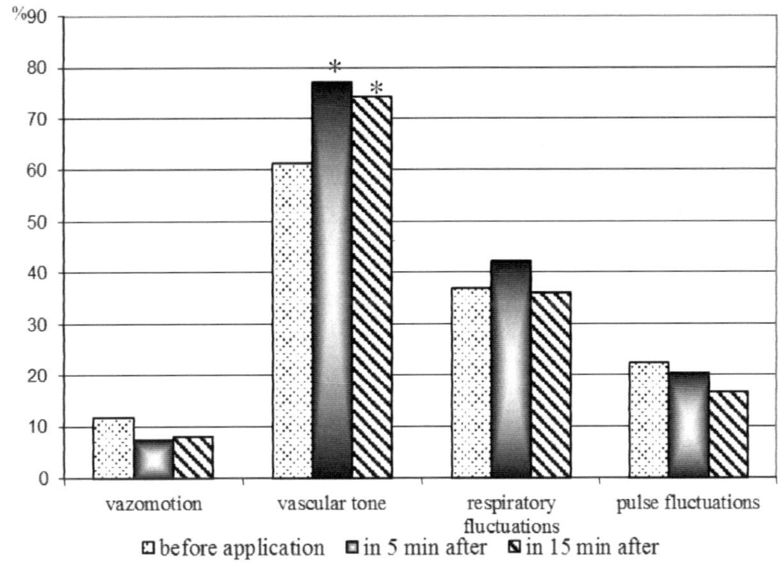

Figure 3. Change of microcirculation of face skin in a cheek after mud application at female volunteers of 36-45 years.

The variation coefficient in the fifth minute increased for 48% ($p<0,05$), further there was a decrease in this indicator. IEM practically didn't change throughout measurements. By means of the amplitude-frequency analysis the tendency to decrease in influence of the passive mechanism after mud application has been revealed. At the same time decrease of the activity of the smooth muscle cells and intra vascular resistance was observed (fig. 3). Obviously, the decisive factor explaining the increasing fabric perfusion is strengthening of a vascular tone which in 5 min. increases for 26% ($p<0,05$) and in 15 min. remains at the level, exceeding control values for 21%.

Changes of microcirculation of face skin in a cheek of women of the senior age group (46-55 years) as a result of mud applications had the following character: already on the 5th min. the tendency to growth of an indicator of M which continued to increase has been recorded, and on the 15th min. exceeded control values for 34% ($p<0,05$).

These changes were followed by increase of values of such indicators as σ, IEM (tab. 1). At the same time, the intra vascular resistance and coefficient of a variation tended to decrease in rather previous measurements.

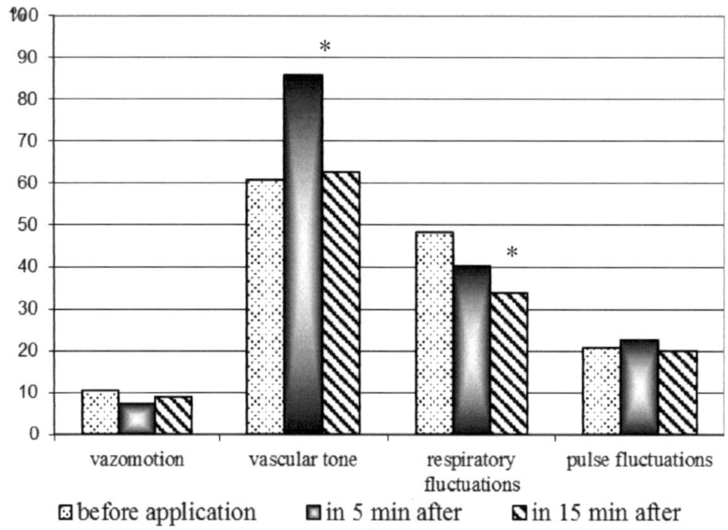

Figure 4. Change of microcirculation of face skin in a cheek after mud application at female volunteers of 46-55 years

As we can see from the fig. 4, change of microcirculation of face skin in a cheek at women of this age group in inverse proportion to change of respiratory fluctuations, i.e. their level decreased after mud applications. At the same time the nature of change of a vascular tone and pulse fluctuations had an identical tendency: increase on the 5th mines with further reduction. Exactly the opposite there was an activity change of the smooth muscle of cells.

Having carried out the comparative analysis of nature of change of microcirculation of face skin of female volunteers of different age after drawing mud applications (fig. 5), it is possible to note the following:

1. in the first age group (16-25 years) there is an insignificant increase in an indicator of M after mud applications;

2. this indicator increases in the second age group (26-35 years) for 12, 4% in 5 min. after mud application and 9,79% in 15 min. rather initial measurements;

3. in the third age group (36-45 years) of change are noted in 15 min., at the same time force of influence of mud application directed to increase of an indicator of M has made 11% ($p<0,05$);

4. mud application had the most expressed effect in the senior age group (46-55 years) where the M indicator in 5 min. increased for 21,66% (force of influence of mud application has made 13%, p<0,05) and for 33,77% – in 15 min. after application (force of influence has made 31%, p<0,001).

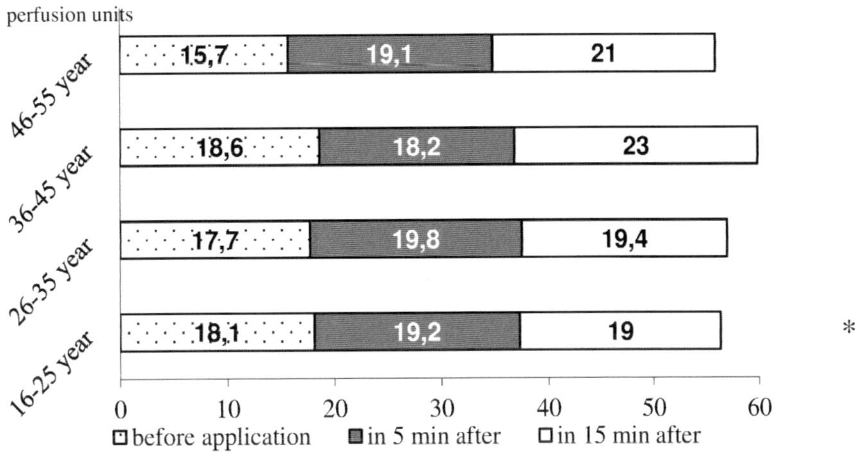

Figure 5. Change of microcirculation of face skin of female volunteers of different age groups as a result of mud applications

4. Conclusions

Thus, effect of sulphidic mud of the lake Karantinnoe is shown as improving microcirculation and in redistribution of influence of factors of active and passive mechanisms. Applications of a peloid promote inflow of arterial blood to the nutritive course of face skin and outflow venous, i.e. to decrease in probability of developments of stagnation. These changes are more expressed at measurement in 5 min. after mud application at volunteers of younger age group. With age, as well as in the analysis of indicators of thermometry, the prolonged action of mud application is shown, i.e. change of microcirculation is fixed later (in 15 min. after mud application) [2]. On the basis of the conducted researches it is possible to draw the following conclusions:

1. strengthening of microcirculation noted by reliability of distinctions before mud applications happens in the senior age groups (36-45 and 46-55 years);
2. these changes are connected, first of all, by increasing of muscular activity and respiratory fluctuations.

References

[1] Ivanova E. V. 2007 Evaluation of a morphological functional condition of the growing old face skin by noninvasive methods (parts I) // *Experiment. and wedge. Dermatocosmetology* N 6 p. 37-42.
[2] Kasimova S. K., Kondratenko E. I., Lomteva N. A. 2008 Age dynamics of change of temperature of face skin of women after mud applications // *Natural sciences* N 2, p. 61-63.
[3] Kozlov, V. I. 2006 Sistem of microblood circulation: clinical and morphological aspects of studying / A regional blood circulation and microcirculation **1,** N 1 p. 84-101.

The methods of optical physics as a mean of the objects' molecular structure identification (on the base of the research of dophamine and adrenaline molecules)

M D Elkin, O M Alykova, V V Smirnov, G P Stefanova

Astrakhan state university, Astrakhan, Russia

E-mail: olga-alykova@mail.ru

Abstract. Structural and dynamic models of dopamine and adrenaline are proposed on the basis of ab initio quantum calculations of the geometric and electronic structure. The parameters of the adiabatic potential are determined, a vibrational states interpretation of the test compound is proposed in this work. The analysis of the molecules conformational structure of the substance is made. A method for calculating the shifts of vibrational excitation frequencies in 1,2,4-threesubstituted of benzole is presented. It is based on second order perturbation theory. A choice of method and basis for calculation of a fundamental vibrations frequencies and intensities of the bands in the IR and Raman spectra is justified. The technique for evaluation of anharmonicity with cubic and quartic force constants is described. The paper presents the results of numerical experiments, geometric parameters of molecules, such as the valence bond lengths and angles between them. We obtain the frequency of the vibrational states and values of their integrated intensities. The interpretation of vibration of conformers is given. The results are in good agreement with experimental values. Proposed frequency can be used to identify the compounds of the vibrational spectra of molecules. The calculation was performed quantum density functional method DFT/B3LYP. It is shown that this method can be used to modeling the geometrical parameters molecular and electronic structure of various substituted of benzole. It allows us to construct the structural-dynamic models of this class of compounds by numerical calculations.

1. Introdution

Detecting the features of compounds spectral identification, their interpretation or their relation to a certain molecular fragment are both fundamental target of mathematical modeling in the theory of molecular system and applied one according to the quantitative and qualitative composition of bioobjects i.e. relations "State-Structure- Spectrum". The achievement of the given problem poccesses its own special features for each composition. The present work demonstrates the possibility of the use of the up-to-date programming software to create structural and dynamical models of the substance and accomplishment on the base of the spectrum identification of the substance. Optical physics methods are the most informative in the studied structures of the discussed types of the molecular objects.

Solving problems of the discussed types is possible due to the high techs of the molecular modeling i.e. the use of quantum method of analyses of parameters of adiabetic potential. These methods in the beginning of the 21 century completely replaced the classical methods of molecular dynamics which are based on the soluteon of inverse problems in theoretical physics.The applied methods are implemented as applied programming lay-outs. Dominating one is "Gaussian" according to the frequence use. Its particular feature is the implementation of the method of functional density DFT/B3LYP allowing to model effectively the problems discussed in article.

The interest to the study of the structure and spectrum of molecule of dofamine and adrenaline is understandable (Figure 1). As an example we refer to the works [1,2,3] where just one of the possible conformers' combinations. Интерес к изучению структуры и спектра молекул дофамина и адреналина (рис.1) понятен. Сошлемся, к примеру, на работы [1,2,3], где рассмотрен лишь один (A) из возможных конформеров соединений. In the work [3] such interpretation for high frequency range is differ from data given in study [4]. In the work [1] there used a classical approach to evaluate the parameters of harmonic force field . In the work [2] the authors researched the range of 570-1880 sm-1. The spectrum modeling calculation are limited by harmonic approximation of the molecular vibrations theory. To measure anaharmonic displacement there used the procedure of scaling [5].

Structurally dopamine and adrenaline are related to 1,2,4 – benzol substituents(C_6H_3XYZ).According to the study[6] the number of fundamental vibrations of benzol trisubstitution could be divided into following groups: characterized by frequency,mode and intensity,characterized by frequency and mode , characterized by mode. Theoretical interpretation of fundamental vibrations of spectrum suggested in the work [6] completely agrees with experimental fundamental vibrations spectrum allocation 1,2,4 – benzol substituted (X=F, CH_3), suggested in the work [7]. This gives the possibility to assume that the given interpretation of spectrum skeleton will possess a space for both dopamine molecules and (X,Y=OH, Z=CH_2–NH_2)and adrenaline (X,Y=OH, Z=CHOH–$NHCH_3$). The accuracy of such idea could be confirmed by quantum calculations.

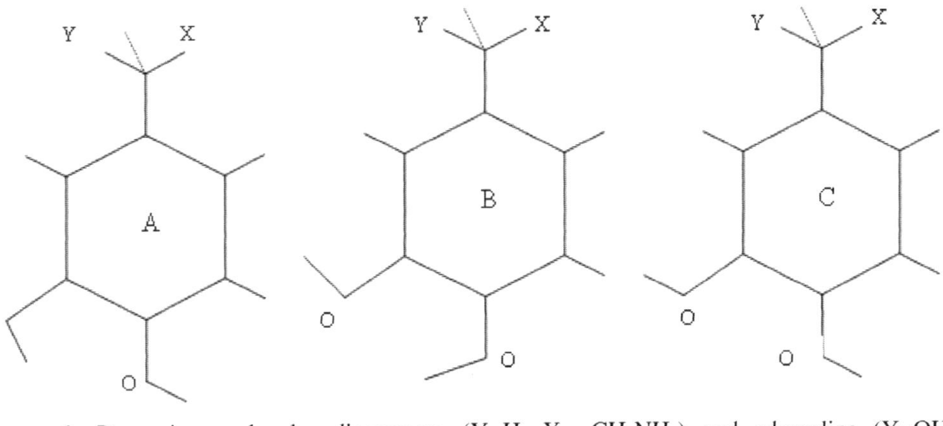

Figure 1. Dopamine molecular diagramms (Y=H, X= CH_2NH_2) and adrenaline (Y=OH, X= CH_2NHCH_3)

The given work suggests structural and dynamical models of possible dopamine and adrenaline conformers on the base of model quantum calculations.All calculations are carried out according to the density functional method DFT/B3LYP [8] with the use of the basis 6-311G*, 6-311G**, 6-311+G**.

2. Methods

The evaluation of the bands' state was carried out by means of the following ratio [9]

$$E_v = v_s(n_s + 1/2) + \chi_{sr}(n_s + 1/2)(n_r + 1/2). \quad (1)$$

There used the equation from the work [10] for anaharmonic constants χ_{sr}.

$$\chi_{ss} = \frac{1}{16}F_{ssss} - \frac{5}{48}\frac{(F_{sss})^2}{\nu_s} + \frac{1}{32}(F_{ssr})^2(\Omega(s;s;-r) - \Omega(s;s;r) - 12\Omega(r;r;r))(1-\delta_{sr}) \quad (2)$$

$$\chi_{sr} = \frac{1}{16}F_{ssrr} - \frac{1}{8}(F_{ssr})^2(\Omega(s;s;-r) + \Omega(s;s;r)(1-\delta_{sr})) + \frac{3}{8}(F_{srt})^2(\Omega(s;r;t) - \Omega(s;r;-t) + \Omega(s;-r;t) - \Omega(s;-r;-t)) \times$$
$$\times (1-\delta_{sr})(1-\delta_{st})(1-\delta_{rt}) + L(a;sr)^2\frac{\Omega(s;r;0) + \Omega(s;-r;0)}{2} \quad (3)$$

Here $P_\alpha = L(\alpha;sr)Q^s \cdot P_r$; $L(\alpha;sr)$ – constants Coriolis , ν_s –harmonic vibrations frequency(in sm^{-1}); Q^s – dimentionless normal vibration coordinates ; F_{srt} и F_{srtu} – cubic and quatric force constants ,$\Omega(s;\pm r;\pm t) = (\nu_s \pm \nu_r \pm \nu_t)^{-1}$ – resonance functions , n_s —*the quantum number of vibration state* .

When testing the method DFT/B3LYP resources , during the primarily calculations of geometrical and electronic structures 1,2,4- trisubstituted benzol , there used relevant difluoro and dihydroxytoluene .The given approach was used in the works [11-12] when constructing structural and dynamical models of a number of trisubstituted benzol.

3. Results

According to the calculations the geometrical structure of benzol ring when being replaced is not changed greatly. The length of the valence bonds are distinguished by the amount of ~ 0.008 Å, values of the bond angles ~ 1.6°.

The results of model calculations of the vibrational state of the benzol skeleton of the possible conformers of the given test compositions completely agree with the experimental fundamental frequency assignement suggested in the works [1,6,7].

In studied conformers of 1,2-digydro-4-methyl benzol, the great interest is in the band position which are interpreted as valent (q_{OH}) and non- plane deformational vibrations (χ_{OH}) of hydroxyl sections. These are the doublets ~ 3710 and 3650 sm^{-1} in highly frequency range of vibrational spectrum for conformers A and B . The bands are identical for the third conformer (~ 3700 sm^{-1}). We can notice that in the works [1,3] the value of frequency being interpreted as valent frequency bonds OH (q_{OH}) is transformed into long waved area of the frequency spectrum for ~ 300 sm^{-1} in comparison with data given above and suggested assignment for the mentioned frequency from the work [8]. Due to the fact that the relevant experiment introduced in the works [1,3] corresponds to the condensing state of dopamine and adrenaline.

The band state interpreted as torsion vibrations χ_{OH}, is characterized by the following doublets : 410 и 210 sm^{-1}, 420 и 170 sm^{-1}, 360 и 310 sm^{-1}. According to the model calculations the bands are the most intensive in a low frequency range of vibrational spectrum .

The most important fact is the vibrational characteristics of the benzol skeleton according to the frequency. There can be observed characteristics according to the intensity for the given type of benzol trisodium while interfusing according to the form for deformational vibrations of hydroxide fragments. При отсутствии смешивания по форме с деформационными колебаниями гидроксильных фрагментов, наблюдается характеристичность по интенсивности для данного типа тризамещенных бензола.The testing results confirm the suggested theory.The intensity band changing values is connected either with the varied electronic substitutes' properties or with the interfusion of ordinary vibrations forms interpreted as deformational vibrations of the relations CH (H=H, F, O) in the model calculations. It can be seen in 1,2 dihydro- 4 methyl benzol in the range of e 1550-1100 sm^{-1}.

The interpretation of vibrational spectrum of dopamine is introduced in the table 1. It was supposed that a molecule belongs to the symmetry group C_s (the carboatomic plane is the molecule symmetry plane). Torsional vibrations were produced when the fragments CH_2C и CH_2NH_2 of substitutes are in the transconfiguration relatively each other .

Table 1. Fundamental dopamine vibration interpretation

Form of vibration	$\nu_{эксп}$ [5]	$\nu_{г}$	Model_A			Model_B			Model_C		
			$\nu_{анг}$	IR	CR	$\nu_{анг}$	IR	CR	$\nu_{анг}$	IR	CR
Flat vibrations of 1,2,4 trisubstituted benzol fragment фрагмента (symmetryA′)											
Q,γ,β	1621	1663	1621	9.6	37	1622	20	8.7	1618	4.1	32
Q,γ,β	1596	1645	1605	178	6.1	1601	39	8.5	1601	35	12
β,Q	1522	1559	1521	146	2.0	1511	188	1.1	1526	138	1.5
β_{OH}.β	1382	1404	1376	38	6.7	1369	6.2	3.7	1364	14	5.1
β_{OH},β	1326	1354	1319	13	1.0	1321	140	1.9	1324	7.6	1.2
Q_{CC},β	1277	1314	1279	114	13	1281	126	13	1286	104	11
Q_{CO},Q	1234	1292	1259	132	1.1	1256	48	0.6	1266	30	2.6
Q_{CO},β_{OH}	-	1284	1247	73	3.7	1255	59	0.3	1257	36	5.3
β_{OH},β	1188	1214	1193	54	8.3	1188	31	5.3	1174	309	1.0
β_{OH},β	1176	1185	1149	75	3.4	1153	102	4.1	1169	67	3.2
β,Q	1144	1166	1135	7.2	2.4	1135	7.2	2.6	1144	5.2	3.4
γ,β_{OH}	1099	1128	1103	107	1.0	1101	92	4.3	1100	82	1.3
γ	922	958	931	2.7	5.8	937	13	5.0	939	12	4.7
γ	771	794	779	39	18	775	48	18	774	22	18
γ	728	758	741	5.8	8.4	741	10	7.3	742	7.2	8.7
β_{CO}. γ	571	588	579	19	2.8	578	9.3	5.4	567	10	3.3
β_{CC},$\beta_{CC'}$,	445	467	456	3.4	1.0	456	5.7	0.9	460	4.2	1.5
β_{CC},$\beta_{CC'}$	340	337	330	9.5	0.3	330	1.0	0.3	330	0.6	0.6
β_{CO}.γ	312	316	310	3.2	3.6	309	6.3	3.3	310	16	2.2
β_{CO}.γ	285	297	291	2.2	4.9	292	3.7	4.3	290	1.8	6.1
Non-flat vibrations of 1,2,4 trisubstitute fragment (symmetry A)											
ρ	932	936	914	1,6	0,2	890	3.5	0.1	887	1.0	0.1
ρ	854	849	818	29	0.6	851	11	0.7	820	24	0.8
ρ	804	807	806	13	0.7	780	29	0.4	783	22	0.7
χ,ρCO	688	706	690	0.1	0.1	689	0.2	0.1	691	0.2	0.1
ρ_{CC},χ	603	600	584	9.9	0.3	588	4.4	0.1	588	7.6	0.2
χ,χ_{OH}	434	456	445	2.9	0.9	448	11	1.3	446	1.6	0.2
ρ_{CO},χ	330	343	350	7.7	2.2	352	4.3	2.5	306	6.7	4.8
χ	237	223	220	2.5	1.2	224	19	0.3	213	4.9	1.4

χ	152	157	155	4.7	1.0	156	0.1	0.5	152	0.1	0.5
Vibrations of fragment CH_2-CH_2NH_2											
q_{OH}	3615	3844	3712	69	103	3714	79	125	3698	42	152
q_{OH}		3803	3658	101	77	3654	102	78	3696	72	75
q_{NH}	3427	3553	3432	0.1	60	3431	0.1	61	3431	0.1	60
q_{NH}	3360	3478	3361	1.1	110	3360	1.0	111	3360	1.1	110
γ_{HNH}	1623	1666	1622	27	5.2	1624	5.0	24	1622	24	5.6
α_{CNH}	1430	1388	1364	10	6.2	1349	8.0	11	1350	52	16
Q_{CN}	1059	1092	1067	8.2	6.6	1067	3.7	5.4	1066	3.9	6.0
Q_{C-C}	1040	1049	1025	34	1.8	1024	74	2.4	1025	44	2.3
β_{CNH}	780	861	840	223	3.8	844	198	3.1	843	217	3.3
χ_{OH}	-	-	415	70	1.2	424	66	1.0	362	63	2.4
χ_{OH}	303	-	199	141	3.0	190	131	4.3	311	137	0.1
χ_{CN}	-	293	287	41	1.9	287	42	1.9	287	49	1.8

Note: Vibrational frequency in sm^{-1},Intensity in IR spectrum in km\mole , in spectrum CR Å^4/ a.e.м.

For the spectrum identification of conformers there can be used IR bands intensity in the range of 1200-1400 cм$^{-1}$ and the bands interpreted as deformational non-flat vibrations of the bonds OH (χ_{OH}).The use of the basis 6-311G+** [7] leads to the displacement of calculated values of vibrations (Δ~ 50 cм$^{-1}$). But the significant intensity of the given bands in IR spectrum in the range of lower than 400 sm^{-1} does not create difficulties in the interpretation of low frequency range

The determined conformity related to the spectrum bands of the benzol fragment of 1,2,4- trisubstituted benzol allows to interpret the fundamental vibrations of the fragment of CH_2-CH_2NH_2.Their agreement with the experimental data on the characteristic frequency vibrations for the normal paraffins and methylamine[5] confirms the relevance of the harmonic strain constanta from the close molecular fragments system system transfer scheme used in the work of [5] to solve the reverse problems in classical theory of molecular vibrations.

The interpretation of spectrum fundamental vibrations of adrenaline suggested in the table 2 confirms the arguments given above which are related to dopamine spectrum identification. All fundamental vibrations of the compounds could be divided into two groups. The first one is connected with benzol skeleton of $C_6H_3X_3$.The adrenaline symmetry decrease to the group of C_1 does not influence on the bands' position and evaluation of their intensity. The second group is interpreted as the vibrations of the fragment CHOH-CH_2NHCH_3. The hydrogen atom substitution into the methyl group in the fragment NH2 leads to the same patterns in the frequency vibrations behavior that happen while transferring from methylamine to dimethylamine [5]. The bands interpreted as valent (q_{OH}) and nonplane deformational vibration of the bond OH (χ_{OH}) of the fragment CHOH-CH_2NHCH_3 is difficult to use for the conformers compounds spectral identification.

Tablr 2. Interpretation of fundamental adrenaline vibrations

Form of vibration	Model A			Model B			Model _C		
	$\nu_{анг}$	IR	CR	$\nu_{анг}$	IR	CR	$\nu_{анг}$	IR	CR
Q,γ,β	1616	16	80	1619	4.4	68	1615	11	62
Q,γ,β	1604	29	7.7	1602	43	7.4	1601	32	29
β	1517	118	9.2	1506	160	3.0	1520	118	6.4
Q,β,β_{OH}	1439	98	3.7	1454	11	15.0	1436	8.7	5.5
β,β_{OH}	1373	28	16	1372	15	8.5	1364	20	27
Q,β	1279	210	21	1286	59	19	1287	142	19
Q,β	1274	23	1.7	1273	132	6.9	1280	3.9	2.5
Q_{CO},γ,β	1252	84	4.2	1253	39	5.5	1261	30	6.1
β,β_{OH}	1188	77	21	1183	31	3.6	1173	278	2.5
β_{OH}	1146	59	12	1147	100	8.5	1156	57	3.3
β,β_{OH}	1139	42	4.9	1139	26	8.2	1148	4.9	11
γ	1092	53	2.6	1090	65	10	1089	44	5.4
γ	933	23	7.5	938	34	5.8	940	27	6.3
ρ	931	6.7	1.1	910	0.9	1.0	909	0.6	1.6
ρ	824	32	7.1	853	6.1	7.7	822	22	4.6
ρ	816	0.3	1.1	790	29	5.0	793	18	3.9
γ	785	11	11	782	7.1	3.6	784	14	5.4
γ	771	40	18	773	42	23	769	20	23
χ,ρ_{CO}	698	1.8	0.9	697	1.7	1.0	698	1.3	0.8
χ,ρ_{CO}	621	9.6	3.1	627	6.1	3.2	627	9.3	3.2
γ,β_{CO}	617	8.7	1.3	616	25	1.5	616	15	1.5
γ	582	13	3.9	581	7.0	6.8	575	8.0	4.4
ρ_{CO},β_{OH},χ	456	4.5	0.4	456	0.1	0.3	457	6.6	0.1
ρ_{CO},β_{OH}	441	17	1.0	441	5.8	0.9	444	4.7	0.7
χ	390	5.8	1.4	390	1.7	1.4	352	101	5.8
β_{CO},β_{OH}	309	14	1.2	308	43	1.1	310	23	1.7
β_{CO}	295	19	1.4	296	11	1.3	307	35	3.1
χ	246	0.7	2.6	247	9.2	1.6	279	12	0.4

β_{CC},β_{OH}	201	24	1.1	199	33	1.6	201	0.7	0.8
χ	170	6.7	1.3	171	7.5	1.1	167	0.3	0.7
Vibrations of fragment CHOH-CH_2NHCH_3									
q_{OH}	3714	69	97	3715	86	131	3698	30	173
q_{OH}	3690	25	89	3691	25	89	3697	90	56
q_{OH}	3658	105	78	3655	105	77	3690	25	88
q_{NH}	3408	3.7	71	3407	3.4	72	3408	3.6	71
α_{HNC},α_{HCH}	1488	25	10	1488	26	10	1488	24	10
Q_{CN},β_{NCH}	1158	28	1.4	1159	10	2.0	1163	80	10
Q_{NC}	1107	23	7.6	1107	39	8.5	1107	32	7.9
β_{CNH}	740	128	2.7	742	133	2.7	741	131	2.8
Q_{CO}	1037	69	3.2	1035	38	4.1	1037	62	3.3
α_{OCC}	519	15	1.6	518	13	1.5	517	13	1.6
χ_{OH}	419	72	1.3	418	65	0.9	393	20	1.5
χ_{OH}	352	107	6.6	350	89	6.1	325	138	1.4
χ_{OH}	205	110	1.5	207	101	2.2	242	5.8	2.2

4. Conclusions

The carried modeling calculations of the vibrational state of rank 1,2,4 replaced benzol particulary dopamine and adrenaline allow us to confirm that the method of the density functional DFT/B3LYP can be used to construct structural and dynamical models of the complex molecule compounds of the given type. The given fundamental vibrations could be relatively divided into two groups. The first group is related to the vibrations of the benzol fragment $C_6H_3X_3$. If to carry out the mass compare (X=F, CH_3, OH) the vibrations could be considered as characteristic according to the frequency and form. Another group is related to the fundamental vibrations of a molecular fragment which is obtained by 4 varied molecular fragments .substitution of a hydrogen atom of a methyl group in a position.

References

[1] Gunasekaran S., Kumar T.R., Ponnusamy S. 2007, Vibrational spectra and normal coordinate analysis of adrenaline and dopamine // *Indian journal of Pure & applied Physics*.**45**. p.881-892.

[2] Lagutshenkov A., Langer J., Berden G., Oomens J., Dopfer O. 2011 Infrared spectra of protonated neotransmitter: dopamine / *Phys. Chem. Chem. Phys.* **13** 2815-2823.

[3] Park S., Lee N, Lee S. 2000.Vibrational analysis of dopamine based of density functional force field // *Bull. Korean cmec. Sc.* **21**, N.10. p.959-968.

[4] Pulin V.F., El'kin M.D., Berezin V.I. 2002. Investigation of the dynamics of molecular compounds of various classes // Saratov, SGTU. 548 p.

[5] Krasnoshekov S.V., Stepanov N.F. 2007. Scaling factors as the effective parameters for correcting the non-empirical force field // *Journal Physics Chemistry*. **81**, N 4. p.680-689.

[6] Sverdlov L. M., Kovner M. A., Krajnov E. P. Vibrational spectra of multiatomic molecules. Moscow: Science, 1970, 559 p.

[7] Green J.H.S., Harrison D.J., Kynaston W. Vibrational spectra of benzene derivatives-XII. 1,2,4-trisubstituted compounds. // *Spectrochim.Acta.* 1971 **27A**, N.6, p.807-815.

[8] Frisch M. J., Trucks G. W., Schlegel H.B. and ath. Gaussian 03, Revision B.03, Gaussian.Inc., Pittsburgh PA(2003).

[9] Hoy A.R., Mills I.M., G.Strey. Anharmonic force constants calculation //*J. Mol. Phys.* **21**, N.6 (1972) 1265-1290.

[10] Elkin M. D., Babkov L.M. 2011 Accounting anharmonic fringe shift in model calculations of the vibrational spectra of hydrogen-bonded dimers // *Izvestiya of SSU. Series "Physics"* V 11, N 1. - p.20-25.

[11] Pulin V.F., El'kin P.M., Erman M.A. 2011 Simulation of the adiabatic potentials of hydroxy benzene // *Vestnik of Saratov State Technical University.* N 3(57). - p.91-95.

[12] Elkin M. D., Smirnov V.V., Dzhalmuhambetova E. A., Grechukhina O. N., Alykova O.M., Gaisina A. R., Ravcheeva N. A. 2013 Structural and dynamic models dimerov monogidroksiuratsily with weak hydrogen communication // *Caspian Journal: Management and High Technology* . - N 2 (22) .- P. 56-63.

[13] Elkin M. D., Smirnov V.V., Alykova O.M. 2015 Comparative analysis of model calculations the frequency of the fundamental oscillation of the hydroxyl moiety in flavonoids // *Izvestiya of Volgograd State University. Series Actual problems of management, computer engineering and computer science in technical systems* May N 2 (157) p. 33-39.

[14] Evans J.C. 1960, The vibrational spectra phenol and phenol-OD //*Spectrochim. Acta* , 26A, № 16, p.1913-1938.

Mobile PET Center Project

O Ryzhikova[1], N Naumov[2], V Sergienko[3] and V Kostylev[4]

[1] Russian Cardiology Research and Production Complex, Moscow, Russia
[2] NGO "Mobile Clinic", Moscow, Russia
[3] Russian Cardiology Research and Production Complex, Moscow, Russia
[4] Association of Medical Physicists in Russia (AMPR), Moscow

E-mail: medphys@mail.ru

Abstract. Positron emission tomography is the most promising technology to monitor cancer and heart disease treatment. Stationary PET center requires substantial financial resources and time for construction and equipping. The developed mobile solution will allow introducing PET technology quickly without major investments.

1. Introduction

Cancers incidence rate in the last decades has continued to grow slowly (1.2 % per annum). Cancers still come second after cardiovascular diseases in the list of causes of death. In the Russian Federation, 2.5 million people are registered as cancer patients in medical institutions. About 450 thousand new cancer cases are registered annually, and the annual mortality rate from cancer is 300 thousand people.

Skin, mammary gland, cervical, gastric, pulmonary and intestinal malignant tumors are the most frequent causes of mortality. Due to modernization programs and the oncology program of the Ministry of Healthcare, mortality rate has been slowed down by 3%, treatment efficiency has been increased and diagnostic procedures improved. However, the mortality rate here is still very high, as not in all regions patients have sufficient access to modern equipment and highly qualified professional medical staff to diagnose and treat oncology diseases, which could help to identify when the disease started and its stage early, evaluate the effect of radiation treatment, radionuclide therapy and chemotherapy.

Compared to advanced European countries and USA, Russia has the lowest five-year survival rate, less than 45%. This level is due to late diagnostics, lack of necessary treatment with lack of qualified professional staff and using outdated equipment for treatment, which has a margin of treatment efficiency as broad as 35%, when the global standard is about 3%. The most frequent male patient cancers are pulmonary, gastric, skin, prostatic gland, large intestine and rectum cancers. The most frequent female cancers are malignant tumors of mammary gland, uterine body and cervix, skin, stomach and large intestine [1].

Due to the level of urgency of this problem, particular attention is paid to such factors as early diagnostics, diagnosis verification and treatment monitoring. One of the methods to act on these incentives is positron emission tomography (PET/CT).

2. Positron emission tomography

Positron emission tomography is a group of radiological diagnostic methods that involves using ultra-short-lived radionuclides to visualize structural and functional condition of organs and systems of the body at cellular level.

Introduction of PET methods in clinical practice has determined the role and importance of non-invasive pathology diagnostics at early stage in treatment of disease entities in oncology, cardiology, neurology and other fields.

Presently there are about 60 PET centers in the Russian Federation. All the centers are at different stages of operation, some of them are only being launched and some are in construction. 34 centers actually work.

Current population of Russia is 146 million people. According to the WHO, 1 PET/CT scanner's operation capacity is 1 million people, so the demand is about 112 PET scanners/centers.

A PET center consists of three units: a cyclotron for generating the USL radionuclides, a radiochemical laboratory for receiving and dispensing a particular radiopharmaceutical, and a PET/CT or a PET/MRI scanner [2]. PET centers may be structurally organized in different ways depending on the location, local population and incidence rate:

1. Research PET centers. These are created as parts of a federal medical science center and may be used to conduct medical studies, as well as to develop new radiopharmaceuticals and pursue following experimental and preclinical research to register new pharmaceuticals, and also for radiopharmaceuticals sales.
2. Clinical PET centers. Creation of such center may be recommended in the regional clinical hospital or a large oncology treatment center, in oncologic dispensaries, territorial and district hospitals. Such PET center consists of a 16-19 MeV cyclotron. It can produce a standard selection of radiopharmaceuticals for PET examinations (fludeoxyglucose, carbon, ammonium, oxygen); equipment: 2 PET/CT scanners.
3. Clinical mini PET centers. These are installed in oncologic dispensaries, in communities of about 500-900 thousand people, or in a remote area with high incidence rate. The center includes a 7.5-10 MeV cyclotron, mini-laboratory and 1 PET/CT scanner.
4. Individual PET scanners installed in hospitals require radiopharmaceuticals supplies. Today the only radiopharmaceutical that can be supplied is [18F]fludeoxyglucose (half-life 120 min).
5. The newest solution for remote areas with high incidence rates is a mobile PET center, which may consist of a complete unit: a cyclotron up to 8 MeV, a mini-laboratory and a PET/CT scanner, or just of a PET/CT scanner.

3. Project: Mobile PET center

This solution is developed by Unix Medical Company together with NGO Mobile Clinic, under supervision of Federal State Budget Institution Russian Cardiology Research and Production Complex of the Ministry of Health of the Russian Federation, Federal State Budget Institution N. N. Petrov Research Institute of Oncology of the Ministry of Health of the Russian Federation and the Association of Medical Physicists in Russia (AMPR).

The project includes 4 mobile modules. Figure 1 shows equipment location diagram in mobile modules. Module 1 – is a reception section, the "cold" waiting area and a doctors room (to prepare reports). Module 2 – is a PET/CT scanner. Module 3 – for radiopharmaceuticals dispensing, "hot" patient waiting area. Module 4 – is a cyclotron with a radiochemical laboratory.

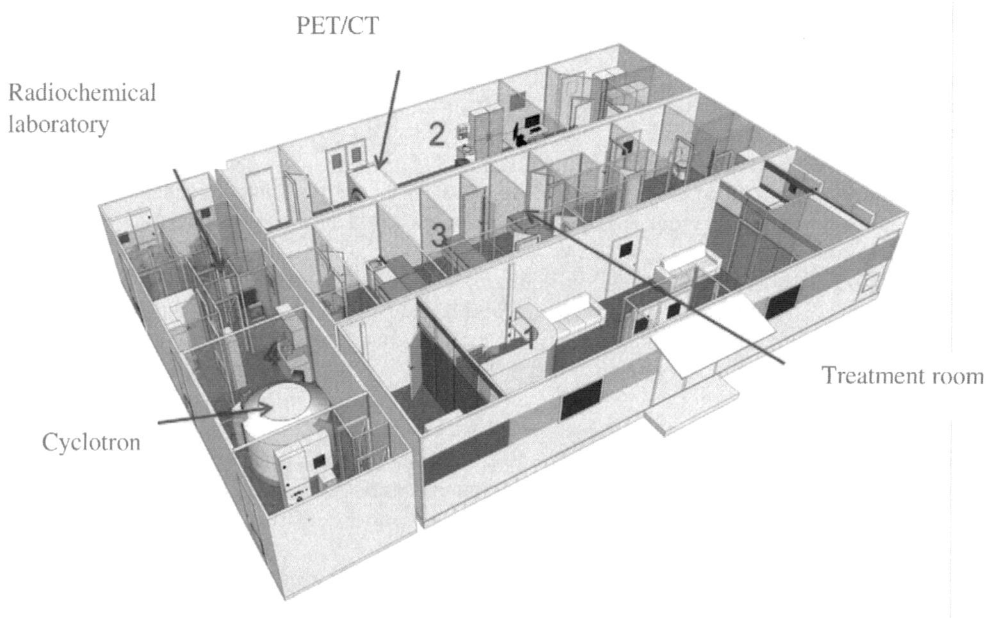

Figure 1. Mobile PET station diagram.

Module 4 radiopharmaceuticals generation technology includes following steps:

- target preparation (raw material for the target for ^{18}F radionuclide production is water of appropriate stable radionuclide composition);
- target matter radiation;
- transportation of radiated target matter to the synthesis module of the safe extraction hood of the synthesis and quality control laboratory;
- radiopharmaceutical synthesis from the generated radionuclide;
- radiopharmaceutical quality control (measuring the pharmaceutical potency, specific potency, chemical and radiochemical purity);
- filling a syringe with a diagnostic volume (dose) of radiopharmaceuticals, dose for each patient is specified individually;
- transferring radiopharmaceuticals to module 2 for radionuclide diagnostics.

Radiopharmaceutical production is performed in clean rooms (areas) with air-locks to provide access for personnel and/or equipment and materials transportation. In the clean rooms (areas), the cleanliness level compliant with the relevant standard is maintained, and the air is supplied through filters of required efficiency.

Each production operation requires measuring the purity of environment in operation condition. In order to ensure compliance with the requirements for clean rooms it is necessary that:

- the concept of air passing from cleaner to "dirtier" rooms is observed in the clean rooms area.
- raw materials enter the area and finished radiopharmaceuticals leave the area through sluice pass-through windows.
- personnel working in clean rooms must arrive at workplace through air-locks.

All the clean rooms of the station are class D purity rooms, since in this module the radiopharmaceuticals are produced in a closed system, radiopharmaceutical dispensing is performed in the closed module and a sterile syringe filling and sterilizing filtration of the finished product are performed on the line.

Radioactive exposure and contamination protection is ensured with barrier technology and air-locks operating by pressure spool principle.

Radiopharmaceutical synthesis, dispensing and quality control take place in one safety box. Linear and cross contamination do not take place, because it is a closed loop automated system.

Main factors stipulating harmful exposure when using sources of ionizing radiation are:

- photon and neutron radiation of cyclotron;
- gamma radiation from radiated targets, solutions, radiopharmaceuticals, waste;
- air contamination with radionuclides during work with bare radioactive sources;
- superficial contamination of equipment and rooms with beta radionuclides;
- radioactive waste;
- radionuclide emissions with vented air into atmosphere.

Radiological safety as a whole is based on the following technical and organizational measures:

- room zoning by level of radiological hazard, personnel access limitation;
- arrangements for radiological and ionizing radiation protection of personnel;
- arrangements for special ventilation in rooms with sources of ionizing radiation;
- arranging an interlock and alarm system that will not allow personnel to enter the cyclotron room when it's working;
- arranging a radiation control system, including an automated radiation control system.

4. Conclusion

Radiopharmaceutical production with an installation at a clinical institution allows solving many problems and offers numerous benefits:

1. There will be possibility to do positron emission tomography more frequently, so-called "dose on demand". As a result, planning will be optimized, and more patients will be served.
2. Financial expenses and time spent on PET/CT technology introduction at the regional level will be reduced as no special bunkers and rooms for radiochemical laboratory are needed, and as a result the number or PET/CT scanning procedures can be increased with no expense increase.
3. This mobile PET center solution produces fluorine-based radiopharmaceuticals, such as fludeoxyglucose, NAF, FMISO, carbon-based pharmaceuticals production is under development.

References

[1] Kostylev V, Ryzhikova O, Sergienko V. Status and perspectives for positron emission tomography methods development in Russia. Med. physics. 2015. No.**2**. pp. 5-16.

[2] Ryzhikova O, Kostylev V. Feasibility study of positron emission tomography centers establishment. Med. Physics. 2016. No.**1**. pp. 78-93.

Medical-biological aspects of radiation effects in Daphnia magna

E Sarapultseva[1,2], D Uskalova[2]. N Savina[2] and K Ustenko[1]

[1] National Research Nuclear University "MEPhI", Kashirskoe Highway, 31, Moscow 115409, Russian Federation
[2] Department of Biology, Institute of Nuclear Power Engineering NRNU MEPhI, Studgorodok,1, Obninsk, Kaluga Region 249040, Russian Federation

E-mail: helen-bio@yandex.ru

Abstract. We have shown that γ-irradiation at doses of 100 and 1000 mGy significantly compromised fecundity and reproductive success of the directly exposed *D. magna*. These effects were also observed among the non-exposed first-generation progeny of irradiated parents, thus implying the manifestation of transgenerational effects in *Daphnia*. We have also shown that compromised viability of irradiated *D. magna* can be attributed cytotoxic effects of irradiation. It would therefore appear that the compromised viability may be attributed to the cytotoxic effects resulted from epigenetic changes affecting some metabolic pathways involved in detoxification of free-radicals. Additionally we have analyzed more distant progeny of irradiated at doses of 10, 100 and 1000 mGy *Daphnia*. Our data demonstrated that multicellular crustacean *D. magna* represent a very useful experimental model for analyse of long-term effects of ionising radiation at the organismal level.

1. Introduction

Over recent years nuclear medicine has become one of the most important and fast-growing component of modern clinical practice. The widespread use of radiopharmaceutical (RFP) is an integral part of nuclear-medical technologies. The basic requirements for application include minimization radiation exposure for both patient and medical staff and safety for patient both RFP and products of its biotransformation in the body. The key characteristics of RFP such as activity per body weight, time and the mode of administration of them, as well as doses of exposure to tumor have been established in experiments with laboratory mice and rats, the species currently regarded in bio pharmacology as gold standard. However, the development of simple and express tests is clearly warranted, as they can provide new insights into the mechanisms underlying the efficiency and toxicity of RFP and help establishing the optimal therapy regimens and improving the outcome of treatment. In our opinion, some alternative experimental models used in radiobiology may also be employed in bio pharmacology.

Over recent years, the work of our research group has been focused on the analysis of low-doses long-term effects of ionising radiation. *Daphnia* represent a very useful experimental model, which allows the very efficient and quick analysis of many aspects of long-term effects of ionising radiation. The life span of these animals is relatively short and seldom exceeds 10-11 weeks; they are also characterised by quite short durations of embryonic and postembryonic development (per 3-4 days each) and gestation (5-8 days). Most importantly, *Daphnia magna* 'enjoy' parthenogenetic reproduction, which quickly allows establishing genetically identical strains of these animals. *D. magna* has been used in numerous studies aimed to analyse the effects of ionising radiation on mortality and fecundity [1]. The aim of our study was to analyse the long-term effects of low-LET irradiation on the survival, life span, fecundity and metabolic activity of directly exposed at low doses test-organism crustacean *D. magna* and their non-exposed generations' offspring.

2. Long-term effects on survival and fecundity in *Daphnia*

The strain of *Daphnia magna* Straus used in our experiments was originally collected in the pond of the Moscow Zoo and maintained for several years at the laboratory in continuous parthenogenetic reproduction following the OECD guideline 211 [2] with modifications. *D. magna* were reared at density of 20 animals per 500 mL in aerated dechlorinated filtered tap-water (pH 7.5 – 8.2, O_2 ~9.0 mg/l; hardness: 128 mg L^{-1}, Ca:Mg 4:1, Fe 0.3 mg/l, Mn 0.1 mg/l) renewed a week. *Daphnia* were fed with green algae suspension (*Chlorella vulgaris*) at daily ration of 90 – 100 μg C *Daphnia*$^{-1}$. *D. magna* were incubated at 20 °C (±0.5 °C) on a 12 h / 12 h light/dark cycle photoperiod at light intensity 700-1200 lux (Climate Control model R2, Russia). Neonates were removed every weekday.

In our previous studies we evaluated the effects of high-dose acute exposure to gamma-rays on the survival of irradiated *Daphnia* [3]. According to our data, the survival rate of *Daphnia* exposed to 20 Gy is significantly compromised. Exposure to a higher dose of 100 Gy substantially compromises their survival, whereas the doses of 250 and 600 Gy can be regarded as lethal. According to our estimates the value semi-lethal dose for *Daphnia* is 63 Gy.

In our recent studies we exposed one-day-old *Daphnia* at doses of 10, 100, 1000 and 10000 mGy (28-1000 mGy min^{-1}) [4, 5]. Control and irradiated *Daphnia* were maintained individually with 50 mL of water in glass vials. Experimental units were checked in a day during 21 days for survival and neonatal release. On days of release, neonates were removed and counted as well as all dead daphnids.

To analyse the effects of parental exposure to the successive generation (F_1), one-day-old neonates from the third broods of generation F_0 were randomly taken from at least three females of irradiated or control groups and transferred to glass vials with 50 mL of water (one *Daphnia* per vial). Using the same protocol, a group of second-generation offspring (F_2) was also established. Following generations F_1 and F_2 were maintained as the original samples but without exposure to γ-irradiation. The survival and fecundity of the parthenogenetic progeny from these generations were measured on a daily basis over 21 days.

2.1 Long-term effect generations on survival in Daphnia

Using the Kaplan Meier approach, we estimated the mean life span for controls and irradiated *Daphnia* [4]. According to our results, the survival of *Daphnia* exposed to 10 mGy of gamma-rays does not significantly differ from that in controls. Thus, the mean life span of this group is approximately 50 days, which does not differ from the mean value in controls of 54 days. However, it is significantly compromised in the two groups exposed to 100 and 1000 mGy. The mean life span of these exposed groups was 43 and 36 days, respectively. Quite surprisingly, the effects of exposure to the two different doses of 100 and 1000 mGy are practically similar. It should also be noted that the results of our study were obtained within the dose-range far below the semi-lethal doses for Crustacean.

2.2 Long-term effect generations on fecundity in Daphnia

According to our data, acute parental exposure to γ-rays equally compromised the total fertility of irradiated F_0 *Daphnia* and their F_1 progeny (Figure 1A; [5]). The fertility of F_0 and F_1 *Daphnia* declined with the dose of parental exposure and significantly decreased at dose of 100 mGy and at higher doses. In contrast, the F_2 total fertility was compromised only among progeny of parents that received the highest dose of 10,000 mGy.

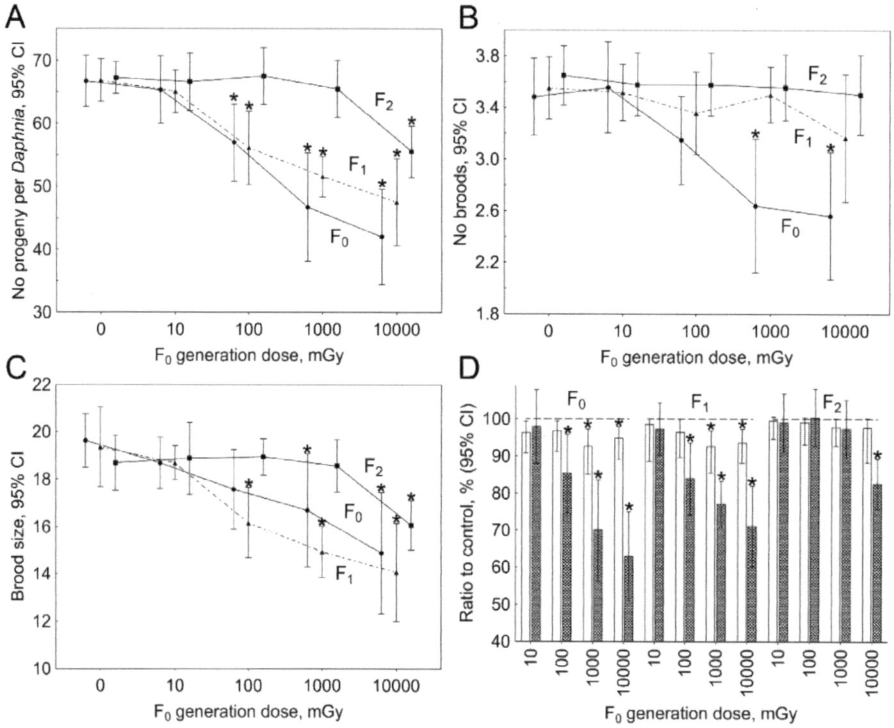

Figure 1. Fertility and life-span of irradiated Daphnia and their progeny. (A) Mean number of progeny per *Daphnia*. (B) Mean number of broods. (C) Mean brood size. (D) Comparison of the effects of parental irradiation on mean life span (open bars) and total fertility (mean number of progeny, shaded bars). 95% confidence intervals (CI) are shown [5]

We also analysed the two main components of the total fertility – the brood size and number of broods. The effects of parental irradiation on the brood size were only observed among the F_0 *Daphnia* exposed to 1000 and 10,000 mGy (Figure 1B). On the other hand, the effects of parental γ-irradiation on the brood size were very close to those for the integrated measure of fertility estimated as the number of progeny per *Daphnia* (Figure 1C).

As we analysed parthenogenic strain of *D. magna*, even early F_0 irradiation resulted in exposure of primordial diploid eggs. In this respect, the design of our study is close to those aimed to analyse the effects of F_0 *in utero* irradiation on the development of F_1 embryos in mammals [6]. According to their results, the developing mammalian embryos show extremely high sensitivity to irradiation, the lethal effects of which substantially exceed those following exposure during adulthood. Our data, summarised on Figure 1D also show that the effects of parental irradiation on the F_0 fertility, which are partially attributed to the early embryonic mortality, considerably surpass those on the life span of exposed adult *Daphnia* and their F_1 offspring.

As far as the results of our study on the transgenerational effects of parental irradiation on viability are concerned, it should be noted that we did not observe a significant shortening of the life span among the F_2 progeny. In this respect it is worth mentioning the results of transgenerational study by Luning and co-authors, showing that paternal irradiation can significantly compromise the viability of F_2 embryos [7].

3. Transgenerational alterations in biochemical activity in *Daphnia*

3.1. Long-term effect generations in free radical reactions in Daphnia

A violation of free radical reactions has been found by the method of free radical copolymerization in the two groups of *D. magna* exposed to 100 and 1000 mGy of γ-rays and their non-exposed first generation progeny [8]. The method is based on a quantitative radiometric registration of the polymerization process that develops in the cells of a multicellular organism in tissue in proportion to the existing number of free radicals. The fact of the increased level of free radicals may be indicative of the hereditary transgenerational effect and the cause of lower survival of *Daphnia* in both generations. However, the effect of a reduced survival rate does not persist in the second generation, which can be explained by the effect of radiation on the embryos of the first generation after acute irradiation of parents.

3.2. Long-term effect generations on mitochondrial activity in Daphnia

The modified *in vitro* MTT-assay [9] was used in our recent study. Four-days-old daphnids from at least first broods were used in the MTT assay. 200 µl of MTT solution (Sigma, St. Louis, USA) (0.5 mg/ml) was added to an Eppendorf tube containing 50 *Daphnia*. After 10-min incubation, *Daphnia* were homogenized. The optical density was measured at 492 and 630 nm on tablet immunoassay analyzer StatFax 2100 (Awareness Technology, USA, VIS-model) and corrected for background noise. The 630 nm OD background were subtracted from the 492 nm OD total signals. The average OD of the blank control wells (homogenized *Daphnia* in DMSO without MTT) was subtracted.

Metabolic test is one of the important diagnostic parameter for understanding the mechanisms underlying the effects of irradiation on mortality. It provides an opportunity to evaluate the oxidative stress that occurs in the body as a result of exposure. Metabolic test have been widely used for the analysis of cytotoxic *in vitro* effects of anticancer chemotherapeutic agents, as well as other drugs. They allow to evaluate the effect of test compounds on cellular metabolism, including the generation of reactive oxygen species (ROS), cells and DNA structure, determination of viability and cell proliferative activity [10; 11].

The methyl thiazol-diphenyl-tetrazolium bromide (MTT) cell viability assay is widely used in determining drug sensitivity profiles and in primary screening of potential efficacy of chemotherapeutic drugs in cell lines [9; 12; 13]. The MTT-assay is based on the conversion of MTT into formazan crystals by living cells, which determines mitochondrial activity. Since for most cell populations the total mitochondrial activity is related to the number of viable cells, this assay is broadly used to measure the in vitro cytotoxic effects of drugs on cell lines or primary patient cells. Thus, any increase or decrease in viable cell number can be detected by measuring formazan concentration reflected in optical density (OD) using a plate reader at 540 and 720 nm.

In the cited articles was show that the *in vitro* MTT assay has been extensively applied to predict drug sensitivity *in vivo*. However, to date this test has seldom been used for the analysis of *in vivo* effects. Earlier MTT-assay was first applied for the investigation *in vivo* of the mechanisms of non-targeted effects of radiation and development of stress in multicellular crustaceans *D. magna*. In our recent study we analyzed the radiation-induced transgenerational alterations of mitochondrial activity at acute exposed in *D. magna* and their non-exposed offspring [14; 15].

According to our data, the cytotoxic effect is significantly elevated in the three groups exposed to 100, 1000 and 10,000 mGy of gamma-rays. Figure 2 shows that the optical density, reflecting changes in mitochondrial activity and expressed as a percentage of the total control, was lower in exposed *Daphnia* from F_0 generation than in control samples with the exception of irradiated at a dose of 10 mGy. The same result was observed in the F1 generation of daphnids.

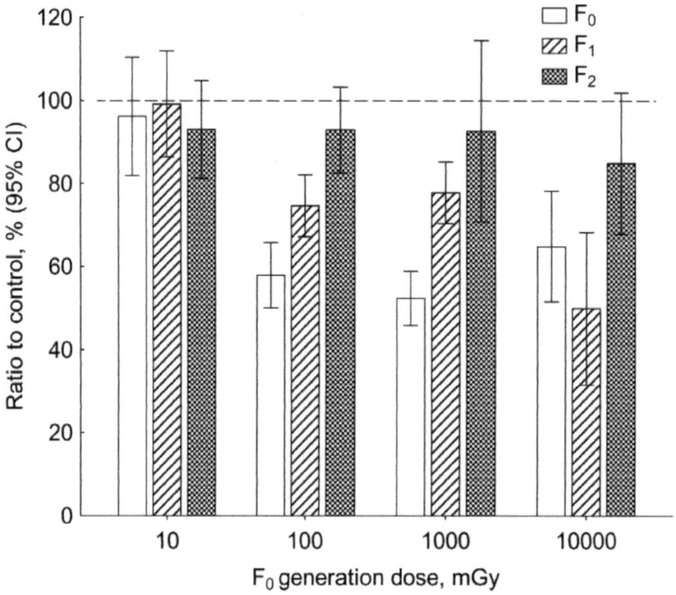

Figure 2. The direct and transgenerational effects of parental irradiation on mitochondrial activity in two generations of *D. magna.*

Thus, the MTT test showed that irradiation at doses of 100, 1000 and 10 000 mGy causes the cytotoxic effect in the directly exposed *Daphnia* and their non-exposed offspring of the first generation. The toxic effect is relatively weak after exposed *Daphnia* at a dose of 10 mGy. Apparently, some of the cells are killed, but the viability of *Daphnia* was not affected.

4. Conclusions

In this study we have analysed the long-term effects of low-LET irradiation at doses from 10 mGy to 10 Gy on functional and metabolic activity of directly exposed *D. magna* and their non-exposed progeny. We observed biological effect at doses at least 100 mGy and dose independence in a wide range of doses. The viability of non-exposed first-generation offspring of irradiated parents is also significantly compromised, thus implying the existence of transgenerational effects of parental exposed in *Daphnia*. Our data are consistent with the results of some previous studies on the transgenerational effects of paternal exposure in mammals.

Our results also imply that the increased level of free radicals can explain compromised viability of irradiated *Daphnia* and their non-exposed progeny. Given that the total activity of dehydrogenases steadily decreases following acute exposure to at least of 100 mGy of γ-rays, the toxic effects of such an exposure should manifest within this dose range. Our data also demonstrate that *Daphnia* represent a very useful and sensitive experimental model, which has a number of potential advantages for express analysis of the long-term effects of exposure ionizing radiation.

Acknowledgments

This work was partially supported by the Russian Foundation for Basic Research (14-48-03002) to E.I.S.

References

[1] Fuller N, Lerebours A, Smith J T and Ford A T 2015 *Aquat. Toxico.* **167** 55-67.
[2] *OECD* 2012. *Organisation for Economic Co-operation and Development*. In: Test No. 202: Daphnia Sp. Acute Immobilisation Test. OECD Publishing, Paris.

[3] Sarapultseva E and Malina J 2009 *Radiats Biol Radioecol* **49**(1) 82−4.
[4] Sarapultseva E I and Gorski A I 2013 *Dose Response* **4** 460–8.
[5] Sarapultseva E I and Dubrova Y E 2016 *Environmental Research* **150** 138−43.
[6] Streffer C, Shore R, Konerman G, Meadows A, Uma Devi P, Preston W J, Holm L E, Stathe J and Mabuchi K 2003 *A report of the International Commission on Radiological Protection* Ann. ICRP **33** 5-206.
[7] Luning K G, Frolen H and Nilsson A 1976 *Mutat. Res* **34** 539-42.
[8] Sarapultseva EI, Melekhova OP, Kossova GV, Igolkina JV and Alexeeva NV 2014 *Radiats Biol Radioecol* **54** (3) 305−8.
[9] *Cancer Cell Culture. Methods and Protocols* 2011, ed. I A Cree (Springer New York Dordrecht Heidelberg London: Human Press) pp 237−44.
[10] Repetto G, del Peso A and Zurita J L 2008 *Nature Protocols* **3** (7) 1125–31.
[11] Shaukat A, Harald G J van Mil and Michael K R 2011 *PLOS ONE* **6** (6) 1−117.
[12] Hayon T, Dvilansky A, Shpilberg O and Nathan I 2003 *Leukemia & Lymphoma* **44** (11) 1957–62.
[13] Hatok J, Babusikova E, Matakova T, Mistuna D, Dobrota D and Racay P 2009 *Clin Exp Med* **9** 1–7.
[14] Sarapul'tseva E I, Ryabchenko N I, Igolkina J V and Ivannik B P 2013 *Radiats Biol Radioecol* **53** (6) 634−8.
[15] Savina N B, Uskalova D V and Sarapultseva E I 2016 *Bulletin Radiation and Risk* **25**(4) 32-6.

Spectrophotometric method for quantitative measuring essential oil in aromatic water and distillate with rose smell

E Semenova[1], V Presnyakova[2], D Goncharov[3], M Goncharov[3], E Presnyakova[4], S Presnyakov[5], I Moiseeva[1], S Kolesnikova[6]

[1] Penza State University, Ministry of Education and Science of the Russian Federation, 440026, Penza, Russian Federation
[2] Academic gymnasium № 1534, 117036, Moscow, Russian Federation
[3] Province boarding school for gifted children, 440046, Penza, Russian Federation
[4] State Commission of Russian Federation for Selection Achievements Test and Protection, 107139, Moscow, Russian Federation
[5] Bauman Moscow State Technical University, 105005, Moscow, Russia
[6] National Research Nuclear University MEPhI (Moscow Engineering Physics Institute), 115409, Moscow, Russian Federation

E-mail: sef1957@mail.ru

Abstract. In this connection, we improved the express methods of determining the mixture of volatile aromatic substances by the spectrophotometry of aromatic water and steam distillate of essential oil raw materials (traditional or biotechnological with rose smell). Direct spectrophotometry of distillation water is impossible because it is a colloid of liquid oil and law is not observed. Therefore, it is necessary to dissolve 1 ml of distillate in ethanol in the ratio 1:4, in this case we take real solution with no lipophilic fall-out on the walls of cuvette, also the light absorption law is observed. There are stable maximums in spectrums of studied oils. Optical density of these maximums is a result of summary absorption of terpenoid components (aromatic and monoterpene alcohols, its ethers). Optical density of tested and standard solutions is measured in appropriate wavelengths. Spectrophotometric method of determination of essential oil quantity in aromatic water with rose smell differs with high sensitivity (10^{-5}-10^{-6} gmol/l) and allows to determine oil concentration from 0,900 to 0,008 mg with an error less than 1%. At that, 1 ml is enough for analysis. It's expedient to apply this method while operating with small quantity of water distillate in biochemical and biotechnological researches and also as express control for extraction and hydrodistillation of essential oil raw material (rose petals and flowers from different origin, eremothecium cultural liquid etc.).

1. Introduction

The production rose of oil is extremely popularly, as more than 50% of world perfumery brands are fabricated on its basis. Its output is an average of 0.025%, and to produce 1 kg of oil is necessary to manually collect and process about 4 tons of petals. After distillation of the oil remains is rose water. In it the share of oil constitutes about 0,2%. A prime vendor of pink water on the world market is Iran, at the same time oil in this country does not produce [1]. On the world market supply rose oil of the highest quality, the amount of which is currently around 600 kg/year, total 4 countries: Saudi Arabia,

Bulgaria, Turkey and Uzbekistan [2-6]. Until 1992 the production of rose oil by steam distillation in the Soviet republics (Ukraine, Moldova, etc.) was about 4 tons/year. Now it has decreased due to the economic crisis in the CIS countries [7]. In 2005 in the Crimea had produced 600 kg of essential oil (extract) of rose, which is 2 times less than the maximum achieved by the essential oil industry in this region of parameters [8].

The world produces several types of valuable aromatic products pink directions: smell of rose, geranium, palmarosa, rosewood [9]. Currently, on the basis of fungal synthesis developed by biotechnology eremothecium oil, similar in composition with rose oil [10, 11]. As a rule, quantitative determination of essential oils and individual compounds aroma-forming conduct gravimetric and chromatography (using a gas-liquid analyzer) methods [12, 13]. They differ in complexity and duration. In this regard, we have attempted to improve the Express method of determining the mixture of volatile aromatic substances by spectrophotometry flavored waters and hydrodistillation essential raw materials with the smell of roses.

2. Materials and methods

As material for a research served 14 samples of aromaprodukt, including 6 – pink water (the aromatic water received from petals fines herbes), 2 – hydrodistillates of leaves of aromatic wild roses; 2 biotechnological semi-products – cultural liquids an eremothecium. For preparation of serial double cultivations of standard solutions used 2 samples of the Crimean rose attar (essential oil and absolutes), a sample of rose oil of the Egyptian origin, 2 samples of eremothecium oil (biotechnological "pink" oil), 2 samples received from chemically synthesized components: pink essence and aromakompozition "A tea rose".

The process of hydrodistillation of essential oil raw (air-dried and fresh petals and leaves of roses and wild roses, the culture fluid of eremothecium) was used for receiving fragrant water. The flask was placed petals or leaves, filled with water, the culture fluid (2/3 volume) and boiled for 2-6 hours. Steam through a tube came to the refrigerator which was cooled with flowing water. The cooled condensate skapyvat in an accumulative flask [12]. Stored water distillate (fragrant water) for the subsequent researches in the refrigerator at 4 ±1 °C.

Distillation (fragrant) waters with the pink direction of a smell represent colloidal solution of oil, their direct spectrophotometry is impossible as at the same time Ber's law isn't observed [14]. Therefore, dissolved 1 ml of the distillate in ethanol in the ratio 1:4, while it is a true solution, eliminates the deposition of lipophilic compounds on the walls of the cuvette and observed the law of light absorption.

The oil concentration was determined in 2 ways: from the calibration graph and by comparison with a known sample. To construct the calibration curve prepared a series of standard solutions, and then 1 ml each of them was dissolved in 4 ml of ethanol and spectrophotometrically.

Since it is known that individual components of the rose essential oil (phenylethanol, monoterpene alcohols, their ethers) have different absorption rates, for the preparation of standard solutions used samples of oils having a different origin and the same composition as the investigated samples, as the mismatch of the component composition of the oils in the studied and standard solutions may lead to inaccurate measurement results.

The optical density of test and standard solutions was measured on the spectrophotometer SF-103 at the respective wavelengths were experimentally determined in the range of 200-300 nm, cuvette thickness of 1 cm, was used as control 80 % ethanol. After measuring the optical density of the investigated solution, according to the schedule found the concentration of oil in 1 ml of distillate.

Determination of the concentration of oil is an ideal method was carried out by measuring the optical density of the investigated solution (D_x) and compare it with the standard optical density of a saturated solution of oil (Dst) the content of 0,4-0,5 mg/ml. Further calculated by the formula: $C_x = C_{st} * D_x / D_{st}$ [15]. The statistical analysis of results of a research was carried out with use of programs of Excel (Microsoft, the USA) and a Statistica 6 package [16, 17].

Journal of Physics: Conference Series **784** (2017) 012053 doi:10.1088/1742-6596/784/1/012053

3. Results and discussion

In the spectrum of studied oils are stable maximum in the region of 250-280 nm, the optical density which is the result of the total absorption of the terpenoid components of nature (aromatic and monoterpene alcohols, their ethers). The absorption spectra of pink and eremothecium natural oils, synthetic aroma compositions (Figure 1) and the dependence of optical density of samples from the oil content (Figure 2) in the sample shown in the figures below.

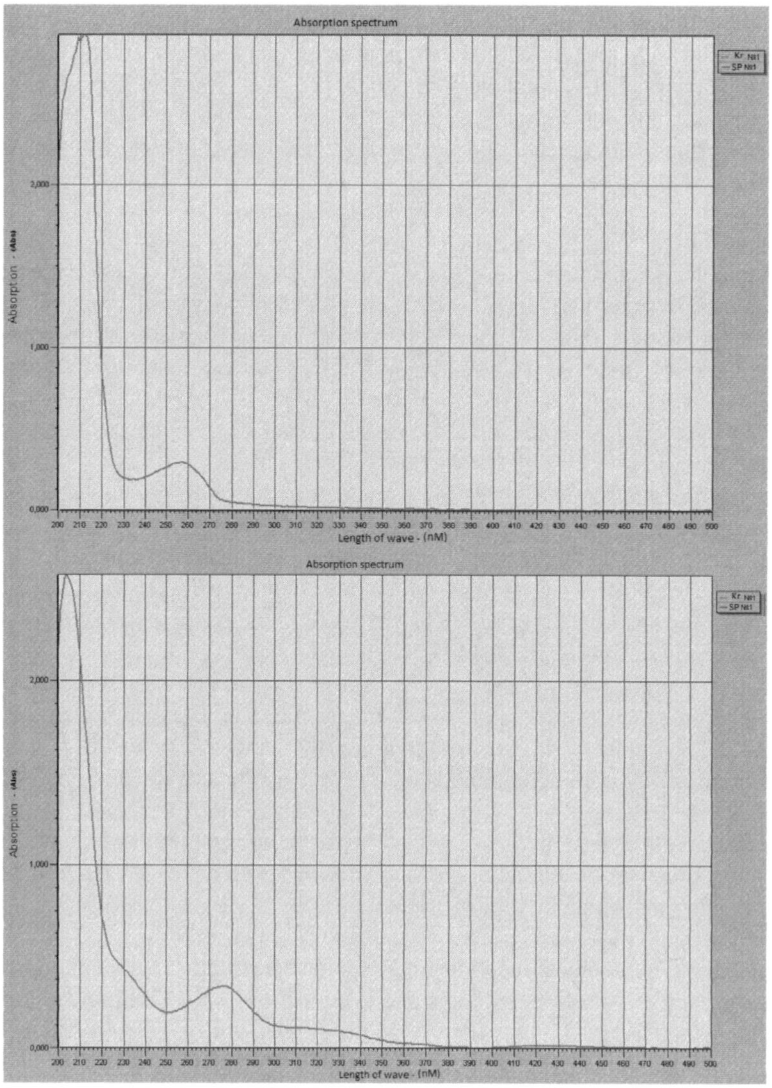

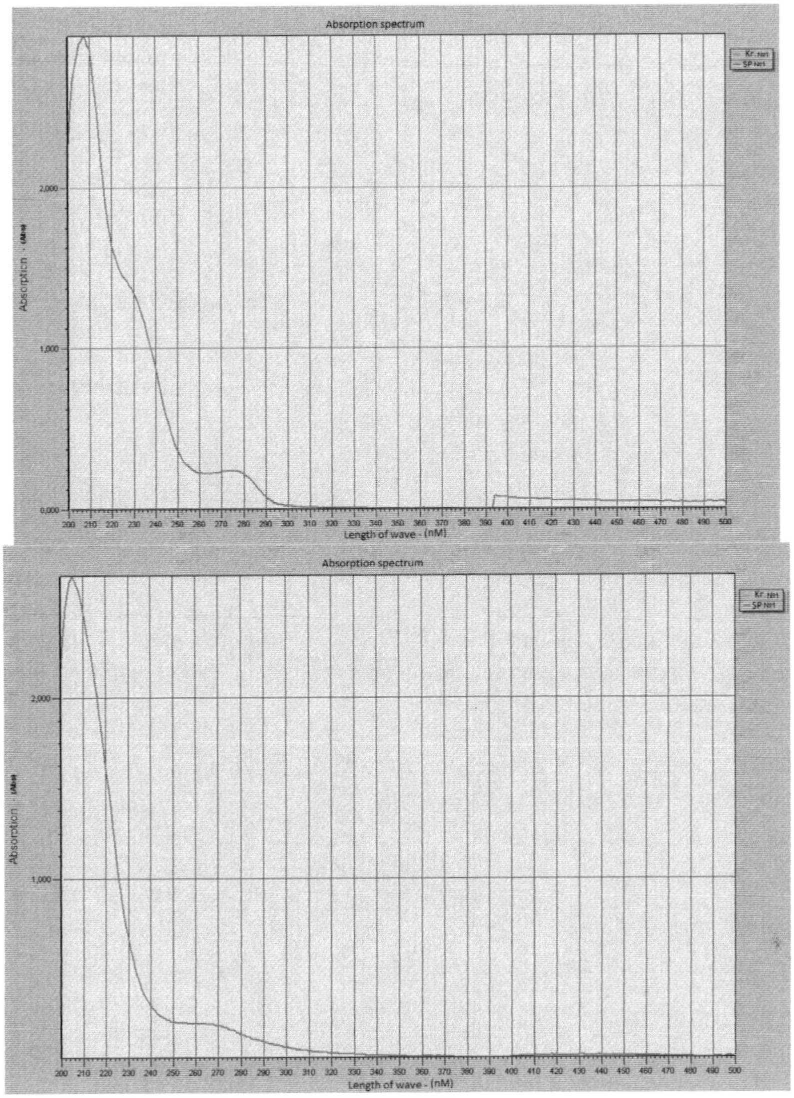

Figure 1. Absorption spectrum aromaproduct And Crimean rose oil (0.125 mg/ml), B - pink essence (1 mg/ml), In - aromacomposition "Tea rose" (0.125 mg/ml), G - eremothecium oil (0.25 mg/ml).

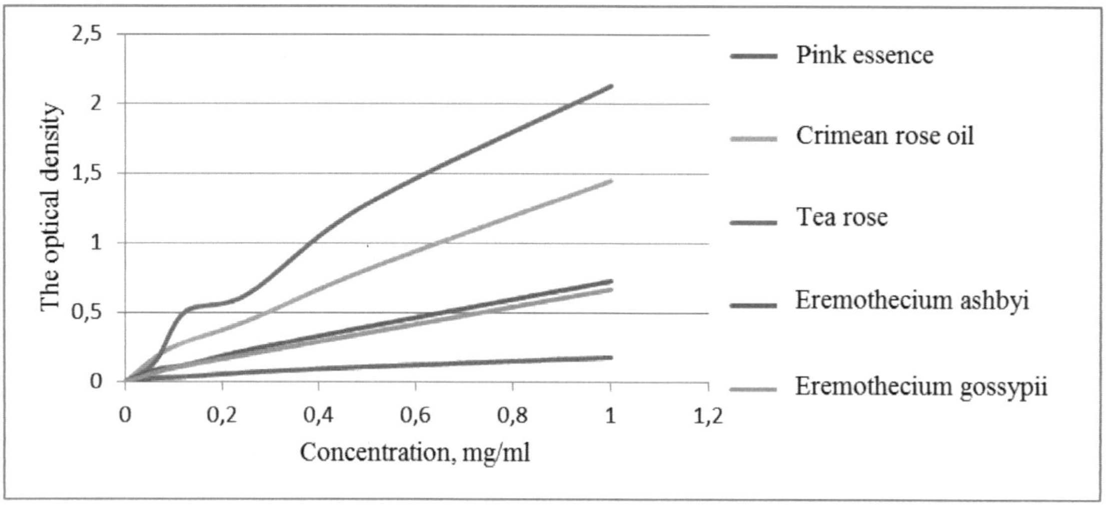

Figure 2. The dependence of optical density of samples from the oil content in the sample.

The carried-out comparative determination of amount of oil in samples of various origin (table 1) spektrofotometrik and gravimetric by methods has shown that metrological characteristics are reproduced and the relative error of a technique doesn't exceed 1%, in this regard it is enough to carry out the analysis no more than in three repetition.

Table 1. Comparative determination of the amount of oil gravimetric and spectrophotometric methods of aromatic waters and distillates of different origin.

The name of the sample	The amount of oil specified (%)		Relative deviation, %
	gravimetric	spectrophotometrik	
Air-dried aromatic rose petals (Baikal, 2016)	0,081	0,082	1,2
Fresh petals decorative fragrant rose (Penza, 2016)	0,117	0,118	0,63
Air-dry the petals essential oil of roses (Crimea, 2013)	0,207	0,208	0,38
Air-dry the petals essential oil of roses (Penza, 2014)	0,185	0,187	0,74
Air-dry the petals essential oil of roses (Iran, 2013)	0,220	0,221	0,36
Air-dry the petals of r essential oil roses of the variety Lada (Russia, 2013)	0,241	0,243	0,59
Fresh leaves of *Rosa pendulina* (Penza, 2016)	0,029	0,029	0,98
Air-dry leaves of the aromatic rose (Baikal, 2016)	0,061	0,062	0,85
Culture liquid *Eremothecium ashbyi*	0,085	0,086	1,0
Culture liquid *Eremothecium gossypii*	0,056	0,057	0,95

The spectrophotometric method for the determination of the number aromaproduct with the smell of roses has a number of advantages. In particular, the use of the absorption of monochromatic light, and this greatly increases the accuracy of the analysis. The method is applicable to mixtures and colorless solutions. Measurements are taken with sufficient speed and with negligible instrument error; the analysis requires a small amount of the solution.

4. Conclusion

Advanced us technique has a high sensitivity (10-5-10-6 gmol/l) and allows to determine the concentration of oil from 0,900 to 0.008 mg with an error of less than 1 %, while for the analysis of 1 ml. is sufficient It is expedient to use when working with small quantities of aqueous distillate in biochemical and biotechnological studies, and for Express control of processes of extraction and hydrodistillation Attar of raw materials (petals of rose flowers of different origin culture fluid of eremothecium, etc.).

Acknowledgments

This work was supported by Competitiveness Program of National Research Nuclear University MEPhI.

References

[1] Haghighi M, Tehranifar A, Nikbakht A, Kafi M 2008 Research and current profile of Iranian production of damask rose (*Rosa damascena* Mill.) *Acta horticulturae* **769** 499-455

[2] Erdogan G 2005 Turkey rose oil production and marketing: a review on problem and opportunities **5(10)** 1871-1875

[3] Rusanov K, Kovacheva N, Stefanova K, Atanassov A, Atanassov I 2009 *Rosa damascena* – genetic resources and capacity building for molecular breeding *Biotechnol.&Biotechnol. Eq.* **23** 1436-1439

[4] Dubal S A, Tilkari Y P, Momin S A, Borkar I V 2008 Biotechnological routes in flavour production *Advanced Biotech March* **3** 20-31

[5] Serra S, Fuganti C, Brenna E 2005 Biocatalytic preparation of natural flavours and fragrances *TRENDS in Biotechnology* **23(4)** 193-198

[6] Kovacheva N, Rusanov K, Atanassov I 2010 Industrial cultivation of oil bearing rose and rose oil production in Bulgaria during 21st century, directions and challenges *Biotechnol. &Bioutechnol. Eq.* **2** 1793-1798

[7] Voytkevich S A 1999 Essential oils for perfumery and aromatherapy M.: Food industry 284 p.

[8] Shlyapnikov V A, Afonin A V, Pehova O A, Suchkova V M 2006 Conception of Essential oils development Crimea industry *Essential oil bearing and herb plants / Scientific reports Essential oil bearing and herb plants institute UAAS* **26** 12-18

[9] Semenova E F, Shpichka A I, Moiseeva I Y 2012 About essential oils biotechnology on the base of microbial synthesis *European Journal of Natural History* **4** 29-31

[10] Semenova E F, Shpichka A I, Moiseeva I Y 2012 About explanation of elaboration of essential Eremothecium oil biotechnology *International journal of experimental education* **3** 35-36

[11] Semenova E F, Shpichka A I, Presnyakova E V, Mezhennaya N A 2016 Processes of essential oil accumulation in petals of Rosa (Rosaceae) and Mycelium Eremothecium (Eremotheciaceae) *Bull. of the State Nikit. Botan. Gard* **118** 25-33

[12] State pharmacopeia RF, XIII edition 2015 (http://femb.ru/feml, Ministry of Health RF) 1-3

[13] ISO 9842 2003 Rose oil

[14] Bershteyn I Y, Kamensky Y L 1975 Spectrophotometric analise in organic chemistry *Chemistry* **7** 16

[15] Melnikov V N, Bugorsky P S 1975 Spectrophotometric method for quantitative measuring rose oil in distillates *Bull.VNII essential oil plants* **8** 69-72

[16] Truhachova N V 2012 Mathematics statistics in biological-medical researchs with use Statistica

M.: GEOTAR-Media 384 p.
[17] Yakovlev V B 2005 Statistics. Microsoft Excel calculations M.: KolosS 350 p.

IOP Conf. Series: Journal of Physics: Conf. Series **784** (2017) 012054 doi:10.1088/1742-6596/784/1/012054

Possibilities of the method of irreversible electroporation in treatment of the local and widespread pancreatic cancer

S G Anaskin[1], Yu V Ivanov[1], D N Panchenkov[2], V B Chertyuk[3], D A Astakhov[4], A A Nechunayev[4], V S Geraskin[1] and T Y Fedotova[1]

[1]National Research Nuclear University MEPhI, Moscow, Russia
[2]Department of endoscopic surgery of SEI HPE MSMSU them. A. I. Evdokimov, Moscow, Russia
[3]Institute of Emergency children's surgery and traumatology, Mockow, Russia
[4]Federal scientific clinical center of FMBA of Russia, Moscow, Russia

E-mail: asg72@list.ru

Abstract. Methods of a local destruction of tumors didn't find till today wide circulation in treatment of patients with formations of the pancreas (P). It is bound to features of blood supply, anatomical and histological structure of PZh, and also a large number of complications and a recurrence.

The technique of the irreversible elektroporation (IE) represents a new unique method of not thermal ablyation at which impact on tissues is carried out by short electric impulses of high voltage (to 3 kV) that involves irreversible rising of permeability of cellular membranes and death of cells. The optimum mechanism of rising of permeability of a cellular membrane electric impulses concerning their frequency or repetitions is yet not up to the end clear; it is recognized that outcomes depend on amplitude, duration and number of impulses. Influence has to be synchronized with a cordial rhythm in order to avoid development of an arrhythmia. Existence at the patient of an arrhythmia and the artificial driver of a rhythm is contraindication to carrying out an elektroporation.

We have experience of use of a technique for 18 patients with a locally-spread pancreatic cancer. It was succeeded to reflect the main advantages of a new technique in our experience, such as comparative simplicity of workmanship, safety for vascular and pro-current structures, the minimum impact on function of a pancreas.

Efficiency of a destruction is proved to ultrasonic, computer and morphological researches in the postoperative period. The maximum term of observation made 19 months.

Though exact indications to use of NE still accurately aren't defined, the international experience and the first own data allow to state efficiency of new technology in palliative surgical treatment of perivascular tumors of a pancreas, and also its safety.

1. Introduction

Thirteen thousand people die of a pancreatic cancer in Russia annually. Among men mortality from pancreatic cancer — 10 out of 100 thousand men, and among women — 8,1 out of 100 thousand women. Statistically speaking, mortality of the population of Russia caused by new malignant growths, the relative frequency of a pancreatic cancer amongst men — 8,2% - 6th place; amongst women — 4,1% - 8th place. The pro-current adenocarcinoma of pancreas enters the top ten of the most frequent causes of death by oncological diseases in Western Europe and America: 40 thousand deceased annually in Europe and 30 thousand in the USA [1]. The growth of pancreas cancer rate is noted in all countries where there is a high life expectancy rate and the so-called "western way" of living taken by the population. According to D.G.Haller [2], 40% of patients with pancreatic cancer to the time of diagnosis already have locally advanced common process, and only less than 20% of cases it is limited by the capsule of the gland. Early forms of pancreatic cancer are found in 3,8% of cases. Nevertheless, an important fact is that new malignant growths of pancreas in general are characterized by low sensitivity to medicinal therapy. However, productivity of some local ways of influence as interventive technologies with embolization of vessels of a tumor, local destruction applying the means of cryo-frequency, thermo-frequency and radio-frequency ablation, which force us to conduct an active research for new methods.The method of irreversible electroporation of tumors originates from a way of injecting cytotoxic substances into a tumorous tissue by means of reversible electroporation known as electrochemotherapy. The conducted research showed [3] that the characteristics of electric fields used at electroporation did not lead to a critical temperature increase ensuing the energy dispersion, which permits to find this data as a method of non-thermal ablation. Despite the medical technology of reversible electroporation used in the past 20 years, an interest in applying this technique has highly increased within the usage of irreversible electroporation. Likewise-reversible, at irreversible electroporation the powerful external electric field influences cells that leads to an increase in transmembrane potential and induces formation of pores in a membrane. Forming of a permeable pore becomes possible only on reaching the critical threshold level which depends on a type of cell and can usually be found in range from 200 to 300 mV/cm. The power of an impulse and its duration with exceeding the threshold level of stability of a cellular membrane leads to formation of multiple nano-pores. The nano-pores created in a membrane under the influence of electric field remains is constantly open. These changes conduct to a cellular death because of inability of a membrane to return to an initial homeostatic state.

Cells are densely compartmented by the bi-lipid plasma membranes, providing internal regulation - and extracellular transportation of solutes. When external electric field influences a cell, it spreads around a cell and not directly through it. It creates a difference in flow density on a surface of a cell and inside, and leads to an onset of potential difference on the surface of membrane. When trans-membrane's potential reaches a certain critical level, the cell becomes unstable and plasmatic membrane is subject to destruction or restructuring with the formation of the nano-pores, providing a passive incoming ionic flow. Because of the membrane becoming permeable, there is an increase of energy demands for maintenance of trans-membrane's difference for concentration of ions cellular increase. Conductivity of a plasmatic membrane sharply increases and, if ATP-dependent ionic pumps are incapable of compensating diffusion of ions through pores in a plasmatic membrane, power malnutrition and termination of biochemical processes take place, the so-called "biochemical arrest" which leads to the death of a cell. Irreversibly-damaged cells are eliminated by immune system. Several authors showed that ultrashort (more than a 100 in a nano-second) and high-intensity (200-300 sc) impulses influence directly the intracellular organelles without causing an irreversible damage to cellular membranes. The cellular death is caused in this case by the apoptosis induced by mitochondrions and connected to the release of calcium from a cytoplasmic reticulum, and the damage of DNA [4,5].

Precision mechanisms of a cellular death at NE are still subject to a debate. However, it is considered that both mechanisms necrosis and apoptosis can play a role in cellular destruction. The latest data indicates an increase of markers of apoptosis in the zone of tissue that has undergone NE. Apoptosis initiates inborn cellular regeneration, thus less time for recovery of tissue is required, than by means of necrosis. Moreover, at apoptosis fibrous changes are found to a lesser extent in areas undergone ablation than at a necrosis which prevents further damage of organs which have undergone ablation, and that is confirmed by data of conducted research.

2. Materials and methods

Patient A., 56 years old, has been hospitalized to the surgery department of FNKTs on April 11, 2013, complaining at the continuous pains in the epigastric area increasing after consuming food, mouth dryness, nausea, weigh-loss, weakness. From the anamnesis, it is known that since March 2012, there have been signs of weakness and fast fatigue. Since April 2012 pains in epigastric area with back irradiation, air regurgitation, vomiting after consumed food began to disturb. An x-ray examination of a stomach performed on 24.04.2012 - depot of barium in peripyloric department of stomach. Throughout May 2012 the specified symptoms remained with a growing tendency. Pains became constant. Patient has undergone treatment on an outpatient basis and in hospital with no visible effect. At EGDS from 17.05.2012-a massive ulcer of subcardial department of a stomach. A biopsy - data for neoplastic process of stomach has not been obtained. On the other hand, the carried-out therapy has reached a positive effect. EGDS of 14.06.2012 - an ulcer of an upper part of a third of a stomach is in a stage of full scarring. EGDS of 20.07.2012: diffusion gastritis. Symptoms of chronic pancreatitis. Retrogression since January 2013 when pains in epigastric area renewed, 03.03.2013 noted vomiting with coffee-ground. In emergency, patient hospitalized to the surgery ward of a local hospital. At EGDS - the ulcer of a bulb in a range from 0,4-0,5 cm covered in fibrin, hemosiderin. Esophagitis. Haemostatic, infusion, anti-ulcerous therapy has been carried out. EGDS from.14.03.13 - hypostasis, a bulb hyperemia, punctate bleeding noted. Patient discharged from hospital on an out-patient stage of treatment. Due to the continuous pain syndrome in the upper parts of the stomach, patient hospitalized for inspection and treatment at the department of gastroenterology of FNKTs of FMBA of Russia.
10.04.2013 of ultrasonography of inards. Conclusion. Liver is condensed, is diffusively-changed. The pancreas has enlarged (head 3,9x4,5; body 1,9; tail 2,4 see), focal education in a pancreas head projection, 3,9kh2,7sm cystous and solid structure, in other departments a parenchyma diffusively heterogeneous, echogenicity is reduced. Virsungov Canal is expanded to 0,4 cm in a body. A bilious bubble banner in the neck and body, and dense walls.

By the results of video-esophagogastroduodenoscopy from 11.04.2013. There is an apparent infiltration of a mucous membrane of a post-bulbar department. A bulb of a duodenum is of normal form, starting from the distal part of a bulb in the post-bulbar department and up to the level of a big duodenal nipple a mucous membrane is apparently hyperemic, edematous a gleam circumferentially is narrowed to 0,9 cm. The compensated circular stenosis of a gleam of a duodenum. The biopsy/cytology is executed.

11.04.2013 KT of abdominal organs with a contrast agent: The liver has not increased in its size, intrahepatic bile ducts has not expanded. Porta is differentiated. The portal vein has not expanded. Vascular drawing is not deformed. Holedokh is not expanded.

The pancreas is positioned as usual, body and tail without visible anomalies. The head is visualized in the form of a conglomerate with the descending part of a duodenum surface of which is maleficiated. The conglomerate is intimately adhering to the frontal wall of the lower vena cava, flattening its gleam at this level, to the frontal surface of the right kidney, to a medial contour of the liver, to the renal leg. Contours of the conglomerate are indistinct with a local infiltration surrounding fatty tissue, undergoing contrasting strengthening, there is a noted active erratic conglomeration of it. Virsungov Canal is not expanded, in structure of education is not traced. The celiac trunk, the general hepatic artery and splenic arteries are traced without signs of involvement in process.

24.04.2013 under an endotracheal anesthesia an operative measure is being executed: Laparotomy. An irreversible electroporation of tumor of the head of pancreas - the "Nano-Knife" system, under intraoperative ultrasonic targeting. The upper median laparotomy has been carried out. In an abdominal cavity, the ascetic exudate up to 500 ml, it's taken for cytology testing. Audit of an abdominal cavity is carried out. In a liver, on innards without focal pathology. Tissue of liver of grayish color, condensed. There is no carcinomatous of a peritoneum. Retroperitoneal and regional lymph nodes are not enlarged. The pancreas' head is hilly, dense and has increased to 5 cm in the diameter. The stomach is mobile. Mobilization of a pancreatoduodenal complex by the method of Kokher. It has been found that there is a pancreas head tumor with involvement of a uniform conglomerate duodenum, it is intimately soldered from the lower vena cava to 3x3,0*3 cm.

Considering the findings, the decision is to limit the operation's volume to cytoreductive, and to execute a tumor electroporation under ultrasonography control has been carried out. The tumor biopsy by the biopsy gun - 1 fragment to 0.6sm is taken, material is sent for planned histologic research. Fabric bleeding is noted. For hemostasis in the tumor i.e. a bleeding part the device, RITA Habib 4 x Kit REF 4401 is inserted, the hemostasis is carried out. Further 4 monopolar electrodes of the "Nano-Knife" system at the distance of 1.5 cm are inserted into the tumor, irreversible electroporation of the tumor in the mode of 80 impulses at 70 microseconds, with a current of 30 amperes, time of exposure - 8 minutes is carried out. Essential changes to the characteristics in homeostasis after electroporation are noted. A vital distinctive feature of the technique is that for the set tumor measurements the software itself defines the mode and time of exposure considering the quantity of electrodes.

3. Results

The current of the after-surgery period on the 7th night has been complicated by development of acute pancreatitis, a focal pancreatectomy. The diagnostic laparoscopy, sanitation and drainage of an abdominal cavity concerning the existence of an exudate in a free abdominal cavity was carried out. Conservative infusion, antibacterial, anti-secretory, analgesic therapy, nutritive support, prevention of thromboembolic complications were carried out. Against the carried-out treatment - positive dynamics in the form of subjective improvement of health, reduction of pain syndrome, permanent decrease in the level of pancreatic amylase of blood, decrease in a leukocytosis, according to tool inspections there is absence of liquid accumulations in peripancreatic cellulose. Patient released in a satisfactory condition for the continuation of treatment on an out-patient stage.

Terms of supervision for this patient make-up 7 months as of today. Since the follow-up examination in November, 2013. Based on ultrasonic, an X-ray, computer and magnetic-resonant inspections, data has not been received for the recurrence of the disease.

4. Discussion

One of the unique characteristics of NE that distinguishes it from other types of ablation - is a lack of damage of tubular structures (arteries, veins, channels) and nerves. For other types of ablation of tumors, such structures are completely obliterated under the influence of the extreme temperatures leading to protein denaturation in an ablation zone. Thus, thermal ablation leads to non-selective destruction of proteins and DNA, thus at NE the cellular death is caused by the apoptosis induced by the formation of the multiple pores in a membrane as mention above. At the same time, the formations created by mainly elastic, collagen structures and proteins of a paracellular matrix are not exposed to damage at NE [6].

The main restriction for the efficient monitoring of hyper thermal as of cold types of ablation, is the long-time supervision necessary for an assessment of an effectiveness of the carried-out therapy. Following the experiments, it is shown that damage caused by NE quickly allowed with the minimal hem in approximately two weeks' time [9]. Most likely, it can be explained by preserving of the micro vessels up to the level of arterioles in all areas of influence of electric field at NE while cryo-and radio-frequency ablation of educations leads to almost full revascularization of a zone of influence and the subsequent long recovery of a microvascular bed from edges of a zone of a necrosis deep into tissues that increase terms of the formation of the devitalized tissue and reparation in the field of thermal influence [10]. The profile of safety of the technique of NE allows to use it widely at various localizations of tumorous processes [11,12].

5. Conclusion

The experiment on the use of NE of a tumor of a pancreas has shown primarily the main advantages of a new technique which are expressed in the minimum invasiveness of intervention on the volume of operational reception, technical simplicity of execution in difficult zones in the anatomic relation, by satisfactory immediate results of a recurrence-free survival. The important characteristic is the safety of NE for pro-current structures and vascular educations provided as at a stage of introduction of electrodes of UZI/KT with targeting, and the direct effects of electroporation concerning tissues of components of the specified structures. This method of ablation and a possibility of its repeated usage for one patient will allow to expand volumes of assistance to initially irresectable patients with pancreas' tumors. A more in-depth indication to the application of this method and frequency rates of its use is in demand of further research.

Acknowledgments
This work was supported by Competitiveness Program of National Research Nuclear University MEPhI.

References

[1] Patyutko Yu. I., Kotelnikov A. G., Abgaryan M. G. 2004 The modern surgical and combined treatment of patients with exocrine cancer of a head of a pancreas and bodies of a periampullary zone. *Practical Oncology* **2**: 94-107.
[2] Haller D.G. Future directions in treatment of pancreatic cancer. Semin. Oncol. 2002; 29 (6): 31-39.
[3] Davalos R.V., Mir L.M., Rubinsky B. 2005 Tissue Ablation with Irreversible Electroporation *Annals of Biomedical Engineering* **33** (2): 223-231.
[4] Meaking, W.S., Edgerton, J., Wharton, C.W., Meldrum, R.A. 1995 Electroporation induced damage in mammalian cell DNA. *Biochim. Biophys. Acta.* **1264** 357–362.
[5] Beebe, S.J., Blackmore, P.F., White, J., Joshi, R.P., Schoenbach, K.H. 2004 Nanosecond pulsed electric fields modulate cell function through intracellular signal transduction mechanisms. *Physiol. Meas.* **25**: 1077–1093.
[6] Livraghi T, SolbiatiL, Meloni MF. 2003 Treatment of focal liver tumors with percutaneous radiofrequency ablation: complications encountered in a multicenter study. *Radiology* **226**: 441-451.
[7] Rhim H, Yoon KH, Lee JM, et al. 2003 Major complications after radiofrequency thermal ablation of hepatic tumors: spectrum of imaging findings. *Radiographics* **23**:123-134.
[8] Livraghi T, Meloni F, Solbiati L, et al. 2012 Complications of microwave ablation for liver tumors: results of a multicenter study. *Cardiovasc Interv. Radiol.* **35**(4):868-874.
[9] Rubinsky, B., Onik, G., Mikus, P. 2007 Irreversible electroporation: a new ablation modality - clinical implications *Technol Cancer Res Treat.* **6**: 37-48.
[10] Lee EW, Chen C, Prieto VE, Dry SM, Loh CT, Kee ST. 2010 Advanced hepatic ablation technique for creating complete cell death: irreversible electroporation *Radiology* **255**: 426-433.
[11] Martin R.C. 2nd, McFarland K, Ellis S, Velanovich V. 2012 Irreversible electroporation therapy in the management of locally advanced pancreatic adenocarcinoma *J Am Coll Surg. Sep* **215**(3):361-369.
[12] Neal R.E. 2nd, Davalos R.V. 2009 The feasibility of irreversible electroporation for the treatment of breast cancer and other heterogeneous systems. *Ann Biomed Eng.* **37**(12): 2615-25.

Search of a solution correction of a lipidic metabolism at acute pancreatitis

S G Anaskin[1], A P Vlasov[2] and I D Korniletsky[1]

[1]National Research Nuclear University MEPhI, Moscow, Russia
[2]N.P.Ogarev Mordovia State University, Saransk, Russia

E-mail: asg72@list.ru

Abstract. Following the results of a pilot study on studying of influence of an emoksipin, verapamil and a reamberin at acute pancreatitis it is possible to say that under the influence of these drugs in fabric structures of the inflamed pancreas there is a decrease in intensity of free radical processes of a lipopereokisleniye, activity of phospholipases, hypoxia phenomena. Level of antioxidant protection of fabric of body increases. In the first three days of supervision the accurate tendency to normalization of the transformed lipidic structure of fabric structures of a pancreas is revealed. At the same time these positive effects are noted in all experienced groups. It demonstrates that though pharmacological drugs are used multidirectional action (antioxidant emoksipin, an antigipoksant reamberin, inhibitor of calcium channels verapamil), they in a varying degree influence on studied pathological (membranodestruktivny, hypoxemic) processes, leading finally to reduction of their expressiveness. So, emoksipin found big ability to increase stability of membranes of pankreatotsit to pathological influence of molecular products the FLOOR, verapamil – to stopping of the activated phospholipases, reamberin – to a hypoxia.
The question solution on the key (prevailing) mechanism in trigger processes of sharp pancreatitis of a definite answer has no. Undoubtedly only the fact that efficiency of antioxidant and inhibitor of calcic channels was rather higher. It suggests that free radical processes of a lipopereokisleniye and activity of fosfolipazny systems predetermine the level and nature of defeat of a cellular biomembrane of pankreatotsit already on the earliest terms of inflammatory process.

1. Introduction

The problem of acute pancreatitis remains to one of actual in modern medicine. Now on occurrence frequency acute pancreatitis takes a leading place among acute diseases of abdominal organs [2]. At the same time at 15–20% of patients inflammatory changes of a pancreas have difficult destructive character [1]. Relevance of this problem is caused by the considerable duration of treatment, high level of a postoperative lethality at destructive forms of pancreatitis [3, 4].

Considerable difficulties in the choice of optimum medical tactics are created by a variety of tactical approaches in treatment of acute pancreatitis. In recent years most of surgeons agree in an opinion that carrying out conservative therapy is shown to most of patients with an acute inflammation of a pancreas [5]. Therefore now optimization of the existing schemes of therapeutic influence at acute pancreatitis represents an important task.

The work purpose – to study influence of an emoksipin, reamberin and verapamil on a functional and metabolic condition of fabric structures of a pancreas and membranodestabiliziruyushchy process of pankreatotsit in the conditions of an acute inflammation.

2. Materials and methods

Basis of work were pilot studies on 40 adult not purebred dogs with an of both sexes weight from 7,3 to 12,8 kg divided into the following groups: the first (n = 8) – control; the second (n = 8) – experienced (with emoksipiny), the third (n = 8) – experienced (with reamberiny), the fourth (n = 8) – experienced (with verapamil). For data acquisition, taken for a norm, researches at 8 healthy animals are made.

Acute pancreatitis was modelled on V. M. Buyanov's way et al. (1989). To adult not purebred dogs carried out a median laparotomy, carried out a puncture of a gall bladder with an intake of bile and the subsequent alloying of the place of a puncture. Further bile was entered into a parenchyma of a vertical part of a pancreas on 0,5 ml in 8 points. Thus reproduced a destructive form of acute pancreatitis. In experienced groups on model of acute pancreatitis estimated a number of pharmakodinamichesky effects of an emoksipin (daily intravenous administrations of 1% of solution (10 mg/kg)) of a reamberin (daily intravenous administrations in a dose of 15 ml/kg), verapamil (daily intravenous administrations of 0,25% of solution of drug (0,2 mkg/kg)).

In control terms of research (the 1st and 3rd days) an animal made relaparotomy, estimated a pancreas condition, defined nature of its damages, and also made a biopsy of its fabrics, a fence of a venous blood.

Experiments were made under an intravenous anesthesia with use of Thiopentalum-natrium at the rate of 0,04 mg/kg of body weight of an animal. After carrying out researches of animals brought out of experiment by introduction of a lethal dose of Thiopentalum-natrium. Researches are executed according to moral requirements to work with experimental animals, approved by local ethical committee.

Lipids from tissue of a pancreas extracted hloroformmetanolovy mix (Higgins J. A., 1990). Lipids fractioned by method of a thin-layer chromatography (Higgins J. A., 1990; Vaskovsky V.E. et al., 1975). The molecular analysis was carried out on the densitometer Model GS-670 (BIO-RAD, the USA) with the corresponding software (Phosphor Analyst/PS Sowtware).

Indicators of intensity of the peroxide oxidation of lipids (POL): the diene conjugates (DC) determined by a spectrophotometric method at the wavelength of 232-233 nanometers (Ganston F. D., 1986); level of low-new dial – a spectrophotometric method in reaction with thiobarbituric acid (TBK) (Sigma), activity of a superoxide scavenger (SOD) – in reaction with nitroblue tetrazoliy (Gurevich V. S., etc., 1990). Activity of a phospholipase of A2 150 mmol a triton of H-100 investigated in the environment containing 10 mmol the tris-HCL-buffer (pH 8,0), 10 mmol of CaCl2 and 1,2 mmol of substrate as which used phosphatidylsincalines of an egg yolk (Trofimov V. A., 1999).

Measurement of content of pyruvic acid (PVK) was performed when carrying out reaction with 2,4 dinitrophenylhydrazine (Kushmanova O. D., Ivchenko G. M., 1983). Content of the lactic acid (LA) was determined by reaction with paraoxodiphenyl (Menshikov V. V., 1987).

Statistical processing of the obtained data was made by the standard methods of statistics with determination of reliability of distinctions between data in experienced and control groups on the basis of calculation of a criterion of Styyudent. The revealed patterns and bonds of the studied parameters between groups and signs were significant at a probability of the faultless forecast r = 95% and more.

3. Results and discussion

Experiences showed that the chosen model of pancreatitis was quite adequate for the solution of the purpose and research problems. In all cases animals had an acute biliary pancreatitis of a destructive form that was confirmed macro - and is microscopic.

The data obtained in control group of experienced animals demonstrate that at acute experimental pancreatitis the forming of disturbances of a lipidic metabolism which is shown increase in tissues of a pancreas of the recreation center level and TBK-aktivnykh products, and also fosfolipazny activity takes place. At the same time the antioxidant potential of the studied fabrics considerably decreased that was expressed by falling of activity of SOD in comparison with an outcome for 44,4–63,8% (p <0,05) in 1–3 days of research.

Table 1. Influence of some drugs on structure of lipids (% of the general maintenance of lipids) in pancreas tissue at acute pancreatitis (M ± m, n = 8 in each group)

Indicator	Norm	Group	Supervision stages	
			1st days	3rd days
Total phospholipids	28,88 ± 1,01	I	16,71 ± 0,53 *	18,87 + 0,49 *
		II	20,79 ± 0,52 * P1 <0,05 P2> 0,05	24,85 + 0,57 * P1 <0,05 P2 <0,05
		III	18,42 ± 0,38 * P3 <0,05	21,46 + 0,21 * P3> 0,05
		IV	19,69 ± 0,34 *	22,34 + 0,69 *
Cholesterol	28,41 ± 1,09	I	27,16 ± 0,77	13,29 * + 0,43 *
		II	28,94 ± 0,74 P1 <0,05 P2> 0,05	17,09 + 0,64 * P1 <0,05 P2> 0,05
		III	26,27 ± 0,59 P3> 0,05	15,07 + 0,31 * P3> 0,05
		IV	27,19 ± 0,80	15,58 + 0,58 *
Free fatty acids	5,97 ± 0,44	I	14,27 ± 0,45 *	16,38 + 0,54 *
		II	11,43 ± 0,36 * P1 <0,05 P2 <0,05	12,44 + 0,42 * P1 <0,05 P2 <0,05
		III	13,62 ± 0,22 * P3 <0,05	14,57 + 0,25 * P3> 0,05
		IV	12,58 ± 0,31 *	13,75 + 0,31 *
Triacylglycerols	18,34 ± 0,78	I	28,06 ± 0,76 *	10,07 + 0,47 *
		II	24,42 ± 0,62 * P1> 0,05 P2> 0,05	13,35 + 0,19 * P1 <0,05 P2> 0,05
		III	26,49 ± 0,77 * P3> 0,05	11,62 + 0,63 * P3> 0,05
		IV	25,15 ± 0,47 *	12,84 + 0,26 *
Cholesterol ethers	14,31 ± 0,89	I	7,43 ± 0,38 *	18,91 + 0,41 *
		II	9,62 ± 0,36 * P1 <0,05 P2> 0,05	16,26 + 0,13 P1 <0,05 P2> 0,05
		III	8,47 ± 0,20 * P3> 0,05	17,37 + 0,34 * P3> 0,05
		IV	8,97 ± 0,30 *	16,79 + 0,35 *

*Notes: hereinafter: I – control group, II – experienced group with emoksipiny, III – experienced group with verapamil; IV – experienced group with reamberiny; * – reliability of changes of indicators in relation to a norm at p <0,05; a bold print – reliability of a difference between data of control and experienced group at p <0,05; P1 <0,05 – reliability of a difference between data II and III groups; P2 <0,05 – reliability of a difference between data II and IV groups; P3 <0,05 – reliability of a difference between data III and IV groups.*

Qualitative and quantitative changes of structure of lipids of tissues of pancreas were characterized by decrease in level of total phospholipids, cholesterol at increase in a quantity monoacylglycerofishing (MAGICIAN), diacylglycerols (DAG) and the free fatty acids (FFA). Content of triacylglycerols (TAG) changed ambiguously: in the first days growth of an indicator, to third day – its reliable decrease took place. The amount of ethers of cholesterol changed exactly the opposite (table 1).

The composition of phospholipids when modeling acute pancreatitis differed in higher rates of lizofosfolipid (LFL), sphingomyelin (CM), phosphatidylsincaline (FH) and smaller values of phosphatidylserine (FS), a phospha-tidylinositol (FI) and phosphatidylethanolamine (FEA) in comparison with a norm (table 2).

Table 2. Influence of some drugs on composition of phospholipids (% of the general maintenance of lipids) in pancreas tissue at acute pancreatitis (M ± m, n = 8 in each group)

Indicator	Norm	Group	Supervision stages	
			1st days	3rd days
Lizofosfolipida	1,55 ± 0,04	I	7,65 ± 0,32 *	14,67 ± 0,46 *
		II	4,97 ± 0,22 * P1> 0,05 P2 <0,05	11,91 + 0,30 * P1> 0,05 P2 <0,05
		III	5,42 ± 0,13 * P3 <0,05	12,73 + 0,40 * P3> 0,05
		IV	5,83 ± 0,10 *	13,02 + 0,30 *
Sphingomyelin	4,84 ± 0,17	I	8,51 ± 0,18 *	7,46 ± 0,08 *
		II	6,83 ± 0,22 * P1 <0,05 P2 <0,05	5,74 + 0,25 * P1> 0,05 P2 <0,05
		III	7,57 ± 0,22 * P3> 0,05	6,38 + 0,18 * P3 <0,05
		IV	7,82 ± 0,23 *	7,05 + 0,21 *
Fosfatidilinozit	19,06 ± 0,52	I	16,55 ± 0,30 *	18,06 ± 0,52
		II	17,92 ± 0,21 P1> 0,05 P2 <0,05	18,96 + 0,45 P1> 0,05 P2> 0,05
		III	17,42 ± 0,22 * P3> 0,05	18,63 + 0,77 P3> 0,05
		IV	16,99 ± 0,27 *	18,27 + 0,35
Fosfatidiletanolamin	41,73 ± 1,03	I	33,88 ± 0,78 *	29,12 ± 1,08 *
		II	36,07 ± 0,44 * P1 <0,05 P2> 0,05	33,90 + 0,68 * P1 <0,05 P2> 0,05
		III	34,66 ± 0,64 * P3> 0,05	34,11 + 0,47 * P3 <0,05
		IV	34,85 ± 0,79 *	32,65 + 0,42 *

Destabilization of lipidic structure of a pancreas was combined with development of the hypoxemic phenomena in it that was shown by growth of the MK, PVK level and coefficient of a hypoxia in control group.

Thus, the results received in control group of animals demonstrate that at acute experimental pancreatitis in fabric structures of a pancreas there is an essential modification of a lipidic metabolism and the hypoxia of body forms that demonstrates development of the membranodestruktivny phenomena in it.

In view of the fact of the major role of frustration of a lipidic exchange in a pathogeny of acute pancreatitis, research of a pharmacodynamics of drugs of metabolic type of action is represented reasonable, one of objects of influence of which are the lipids or processes regulating their exchange. As us are chosen from their numerous quantity emoksipin – drug of antioxidant type of action, reamberin – an antigipoksant and verapamil – inhibitor of calcium channels.

Researches showed that use of the studied drugs allowed to influence therapeutic intensity the FLOOR in pancreas tissues at experimental pancreatitis. Under the influence of the studied drugs the quantity of a recreation center and TBK-aktivnykh products in the 1st and 3 days after modeling of biliary pancreatitis in tissue of a pancreas was reliable less concerning control.

At comparative analysis of experienced groups it turned out that the FLOOR in tissue of a pancreas have the greatest influence on intensity emoksipin and verapamil, a smaller effect – reamberin.

Research of influence of the studied drugs on activity of a phospholipase of A2 in fabric structures of a pancreas at its acute inflammation showed that under the influence of an emoksipin activity of enzyme on the first – third day of experiment authentically decreased in comparison with control by 21,41–28,58%, verapamil – for 37,34–37,48%, a reamberin – for 18,07–19,59% ($p < 0,05$).

Comparative analysis of impact of the studied drugs on fosfolipazny activity in fabric structures of a pancreas at biliary pancreatitis showed that the greatest effect verapamil, smaller – emoksipin and the smallest – reamberin has.

Activity of SOD in fabric structures of the body inflamed experimentally under the influence of an emoksipin on the first and third day increased concerning control respectively by 15,31 and 40,92% ($p < 0,05$). Reliable distinctions of indicators of activity of SOD in pancreas tissue in experienced groups with use of verapamil and a reamberin in comparison with control group appeared only for the 3rd days of a pilot study – activity of enzyme was respectively 25,25 higher and than 22,17% ($p < 0,05$).

The studied drugs influenced also change of structure of lipids in pancreas tissue. So the amount of total phospholipids in a day of experiment authentically increased in comparison with control at inclusion in the scheme of treatment of all approved drugs. On third day of experiment similar dynamics remained. At the same time reduction of the MAGICIAN, DAG and SZhK level was observed. Dynamics of other indicators was less expressed (tab. 1).

Influence of the studied drugs is noted also on composition of phospholipids in pancreas tissue at biliary pancreatitis. As showed research results the most essential changes were observed concerning indicators of LFL and CM which level decreased from first days of experiment in comparison with control data. Reliable dynamics of indicators of FH, FS and FEA is noted mainly only on third day of research, and FI – for the first days of experiment (tab. 2).

When studying values of the indicators allowing to judge lipidic composition of tissue of pancreas at pancreatitis in a comparative aspect it is established that connection of an emoksipin in therapy of this disease rendered the greatest effect for normalization patholologically of modified structure of lipids and phospholipids in fabric structures of body. The smaller effect is gained when using verapamil and a reamberin.

Pilot studies revealed impact of the studied drugs on the arising hypoxemic processes in pancreas tissue at experimental pancreatitis that was shown by decrease in the MK, PVK level and coefficient of a hypoxia from first days of experiment.

Comparative analysis of indicators of the phenomena of a hypoxia in fabric structures of a pancreas at acute pancreatitis showed that most hypoxemic processes in fabrics of the inflamed body were influenced by an antigipoksant reamberin, to a lesser extent – emoksipin and verapamil.

4. Conclusion

Following the results of a pilot study on studying of influence of an emoksipin, verapamil and a reamberin at acute pancreatitis it is possible to say that under the influence of these drugs in fabric structures of the inflamed pancreas there is a decrease in intensity of free radical processes of a lipopereokisleniye, activity of phospholipases, hypoxia phenomena. Level of antioxidant protection of fabric of body increases. In the first three days of supervision the accurate tendency to normalization of the transformed lipidic structure of fabric structures of a pancreas is revealed. At the same time these positive effects are noted in all experienced groups. It demonstrates that though pharmacological drugs are used multidirectional action (antioxidant emoksipin, an antigipoksant reamberin, inhibitor of calcium channels verapamil), they in a varying degree influence on studied pathological (membranodestruktivny, hypoxemic) processes, leading finally to reduction of their expressiveness. So, emoksipin found big ability to increase stability of membranes of pankreatotsit to pathological influence of molecular products the FLOOR, verapamil – to stopping of the activated phospholipases, reamberin – to a hypoxia.

The solution of a question of the key (prevailing) mechanism in trigger processes of acute pancreatitis of a definite answer has no. Undoubtedly only the fact that efficiency of antioxidant and inhibitor of calcium channels was rather higher. It suggests that free radical processes of a lipopereokisleniye and activity of fosfolipazny systems predetermine the level and nature of defeat of a cellular biomembrane of pankreatotsit already on the earliest terms of inflammatory process.

Acknowledgments
This work was supported by Competitiveness Program of National Research Nuclear University MEPhI.

References

[1] Beriashvili Z. A., Parsadanyan A. M., Amiragyan D. M. 2001 Experience of treatment of destructive forms of pancreatitis//*Materials of the Third congress of Association of surgeons of N. I. Pirogov* (M, on October 15-17), 103-104.
[2] Vlasov A. P., Trofimov V. A., Krylov V. G. 2009 A system lipidic distress syndrome in surgery (M.: Science) 224 p.
[3] Gubergrits N. B., Hristich T.N. 2000 Clinical pankreatologiya (Donetsk: Swan) 416 p.
[4] Beger H. G., Rau B. 1995 Acute pancreatitis//*Ann. Ital. Chir.* vol. **66,** p. 209–215.
[5] Benchimol D., Firtion O., Bereder J.M. 1996 Acute pancreatitis treated in a surgery ward. Apropos of 57 cases//*J. Chir.* vol. 133, No. 5, p. 208–213.

Journal of Physics: Conference Series **784** (2017) 012056 doi:10.1088/1742-6596/784/1/012056

Quantum therapy in correction of the lipidic metabolism at acute pancreatitis

S G Anaskin[1], A P Vlasov[2], M A Spirina[2], T I Vlasova[2], T A Muratova[2], I D Korniletsky[1], V S Geraskin[1]

[1] National Research Nuclear University MEPhI, Mockow, Russia
[2] N.P.Ogarev Mordovia State University, Saransk, Russia

E-mail: asg72@list.ru

Abstract. Attempt to establish efficiency of laser therapy in correction of a lipid metabolism at patients with acute pancreatitis was the purpose of work. There were clinical laboratory researches of 48 patients with acute heavy pancreatitis. To the first clinical group (comparison) standard therapy was carried out. To patients of the second clinical group (main) in addition to basic therapy within 10 days daily sessions of laser therapy by the device "Matrix" were held later. Radiation with the wavelength of 635 nanometers, 2 MW was used. Percutaneous laser radiation of blood was carried out to projections of a cubital vein within 30 minutes daily.
Inclusion of laser therapy in complex treatment of patients with pancreatitis led to more significant positive dynamics. Reduction of weight of endotoxemia in the main group is set that was verified by decrease in level of both hydrophilic, and hydrophobic toxins.
The analysis of the data obtained as a result of research in the main group revealed decrease in concentration of products of free radical oxidation of lipids in comparison with group of comparison for 12,1 – 17,3% of % (p. <. 0,05). Laser radiation of blood as a part of complex treatment led to reliable inhibition of activity of enzymes of phospholipase system in blood plasma, in particular activity of a phospholipase of A2 fell for 13,2 – 34,4% (p <0,05).
Thus, inclusion of laser therapy in structure of complex treatment of sharp pancreatitis allowed to reduce significantly expressiveness of endogenous intoxication, intensity of processes of free radical oxidation of membrane lipids and activity of phospholipase systems.

1. Introduction

In modern emergency abdominal surgery the most complex problem is acute pancreatitis. It is caused not only by growth of incidence and development of heavy complications, but also the remaining high lethality which in general makes 4,3 – 5,5%, and at destructive forms reaches 28 – 80% (Andreyev A. V., etc., 2015; Kubyshkin V. A., etc., 2015 Aranda-Narvaez J.M. et al., 2014; FengY.C. et al., 2014).

At acute pancreatitis against the expressed endotoxemia reduction of speed of a blood-groove in capillaries, deterioration in rheological properties of blood and forming of a circulator and metabolic hypoxia with activation of processes of peroxide oxidation of lipids is observed that as a result leads to the membrane destructive phenomena (Vlasov A. P., etc., 2009; Dibirov M. D., etc., 2014; Bradley E.L. et al., 2010). Proceeding from specified, one of possible pathogenic medical agents can act such which are capable to reduce expressiveness of endotoxemia, to improve microcirculation, to effectively correct pathological transformations of a lipidic matrix biomebran (Tsedrik N. I., etc., 2008; Kudelich O. A., 2014). In the last decades the special attention is drawn by laser therapy which exerts positive impact on a number of pathogenic links at various diseases (Geynits A. V., etc., 2013; Musikhin L. V., etc., 2014). However effects of its action at acute pancreatitis on the relation of a lipidic exchange are studied insufficiently fully.

Work purpose. To establish a role of disturbances of a lipidic metabolism in a pathogeny of acute pancreatitis; to estimate pathogenic effects of laser therapy.

2. Materials and methods

Clinical laboratory researches of 48 patients with acute heavy pancreatitis are conducted. To the first clinical group (comparison) standard therapy was carried out. To patients of the second clinical group (main) in addition to basic therapy within 10 days daily sessions of laser therapy were held by the device "Matrix" (the registration certificate No. FSR 2007/00589, the certificate of conformity of ROSS RU.AV35.D00082). The head of "KLO-3" was used (radiation with the wavelength of 635 nanometers, 2 MW). Percutaneous laser radiation of blood was carried out to projections of a cubital vein within 30 minutes daily. Blood sampling was carried out to control terms (the 1, 2, 4 and 6 days).

In work the following methods of research are used. Chromatographic research of a lipidic range of a blood plasma and erythrocytes (Higginsdzh. And., 1990); determination of level of spontaneous low-new dial (Egorov of Yu., Kozlov A. V., 1988); studying of activity of a phospholipase of A2 (Trofimov V. A., 1999); definition maintenance of molecules of average weight (Pikuza O. I., Shakirova L. Z., 1994); studying of the general and effective concentration of albumin (Gryzunov Yu. A., Dobretsov G. E., 1994); a laser Doppler flowmetry by means of the LAKK-02 analyzer (LLC NPP LAZMA, Russia).

The obtained digital data processed by method of variation statistics with use of a criterion of Styyudent. Calculations made on CPU3,10GHz "Intel® Core™ i3-2100" by means of the software package of MicrosoftOffice 2007. Dynamics of indicators is reflected in the schedules constructed with use of the program of the spreadsheets MicrosoftExcel 2007.

3. Results and discussion

Distribution of patients on a sex, age, a prescription and disease severity, laboratory data and data of tool researches in the studied groups was comparable.

The syndrome of endogenous intoxication is diagnosed for patients of the first group at receipt in clinic and during the period of supervision that was expressed in increase in a blood plasma of level of toxins of the hydrophilic and hydrophobic nature. So, increase of maintenance of molecules of average weight with lengths of waves of 254 and 280 nanometers at 27,3 is registered –.142,9 % ($p < 0,05$) and 39,5 –.96,8 % ($p < 0,05$) respectively. The general concentration of albumine decreased on 13,0 – 34,6 % ($p < 0,05$), effective – on 32,4 –58,3 % ($p < 0,05$) that caused its low ability to binding of toxic ligands. When calculating falling of a reserve of binding of an albuminan 17,3 is established –.65,2 % ($p < 0,05$), increase of an index of toxicity of plasma – on 168,3 –.378,1 % ($p < 0,05$), and the maximum growth of an indicator is recorded for the fourth days of the postoperative period. Despite some positive shifts by the end of supervision, in group of comparison it is not recorded normalization of the studied indicators. Thus, the carried-out treatment of acute pancreatitis did not interfere with progressing of endogenous intoxication and did not eliminate an imbalance in system of a homeostasis.

For the purpose of studying of a condition of microcirculation at patients with acute pancreatitis record doppler-gramm was carried out in points Zhang-Men (the free end of the XI edge) - an alarm point of a meridian of a pancreas (Briskin B. S., Bukatko V. N., 2002). Researches showed a reliable deviation of indicators from a norm. So, indicators of microcirculation and shunting increased in comparison with a norm on 25,3 – 78,3 and 20,4 – 83,2 % ($p < 0,05$) respectively, the Index of efficiency of microcirculation increased on 15,2 – 40,8 % ($p < 0,05$). The relation of the maximum respiratory and low frequencies exceeded a norm on 31,3 – 121,7 % ($p < 0,05$) with a maximum for the first days of research. The relation of cardiac and low frequencies was 60,2 - 98,4 norms higher % ($p < 0,05$) on – the sixth days of supervision. The obtained data confirmed development of disturbance of microcirculation of the I-II degree with turning on of respiratory and cardial compensatory mechanisms (existence of R-and C-wave on different terms of supervision). Use of traditional therapeutic actions did not bring to elimination of disorders of microcirculation.

Definition of indicators of quantitative and qualitative spectral structure of lipids, intensity of free radical reactions of a lipoperoxidation, activity of a phospholipase of A2 and catalase in a blood plasma at patients with acute pancreatitis showed an intensification of processes of peroxide oxidation of lipids and increase in activity of a phospholipase of A2. Concentration of primary products the FLOOR in a blood plasma was 46,3 - 211,3 norms higher % (p < 0,05). Level of low-new dial exceeded normal value for 87,3 – 158,2% (p <0,05). Activity of a phospholipase of A2 was increased on 320,2 – 647,3 % (p < 0,05). Increase in activity of the catalase participating in an inactivation of free radicals on 20,3 was registered –.79,7 % (p < 0,05) that, probably, is connected with compensatory and adaptive reactions in response to an intensification of free radical oxidation of lipids.

Disturbances of a lipidic exchange in a blood plasma remained throughout research. So, the amount of total phospholipids in a blood plasma was reduced concerning a norm by 15,1 – 20,4% (p <0,05), sphingomyelin – for 10,2 – 16,7% (p <0,05), phosphatidylsincaline – for 15,1 – 20,3% (p <0,05), a phosphatidylinositol – for 19,6 – 31,4% (p <0,05), phosphatidylethanolamine – for 15,3–16,1% (p <0,05). Content of free fatty acids was authentically raised in comparison with a norm for 26,2 – 50,1%, cholesterol ethers – for 38,1 – 54,2%, and lizophospholipid – for 350,4 – 514,2% (p <0,05).

Research of indicators of a lipidic metabolism of erythrocytes similar changes in a type of change of maintenance of the main membranebased fractions of lipids, growth of level of chaotropic fractions against increase in intensity of free radical processes of oxidation of membrane lipids and a superactivity of a phospholipase of A2 were revealed.

Carrying out traditional medical actions promoted reduction of the revealed changes of a lipidic metabolism at patients acute pancreatitis, but did not lead to full elimination of the dislipid phenomena in blood.

Laser radiation of blood as a part of complex therapy led to a reliable detoxification effect. Against use of laser therapy decrease in level of middlemolecular peptides on 24,1 was registered –.50, 3% (p < 0,05) concerning group of comparison; increase of the general and effective concentration of albumine on 15,1 –.25,6% (p < 0,05) and 28,6 –.80,7% (p < 0,05) respectively, the index of toxicity decreased on 35,9 –.64,5 % (p < 0,05) concerning group of comparison. Clinical trials confirmed reduction of expressiveness of endogenous intoxication and normalization of the arising homeostatic disturbances at acute pancreatitis under the influence of laser therapy in earlier terms, than at the patients receiving only basic treatment.

At patients of the main group against use of laser therapy positive dynamics of indicators of microcirculation was revealed that was shown on early terms of a disease in transition of extent of disturbance of microcirculation from II in I in a look decrease in amplitude of a R-wave, emergence of the S-wave, decrease in an indicator of microcirculation by 10,3–25,2% (p <0,05) and a shunting indicator – for 9,7 – 15,2% (p <0,05) concerning data of group of comparison (figure 1).

Studying of concentration of products of a lipoperoxidation in the main group showed the following: content of low-new dial decreased on 11,6 –.29,0 % (p < 0,05). Also reduction of enzymatic activity in a blood plasma is registered: catalases for 19,1% (p <0,05), A2 phospholipases for 21,8 – 64,3% (p <0,05) (figure 2).

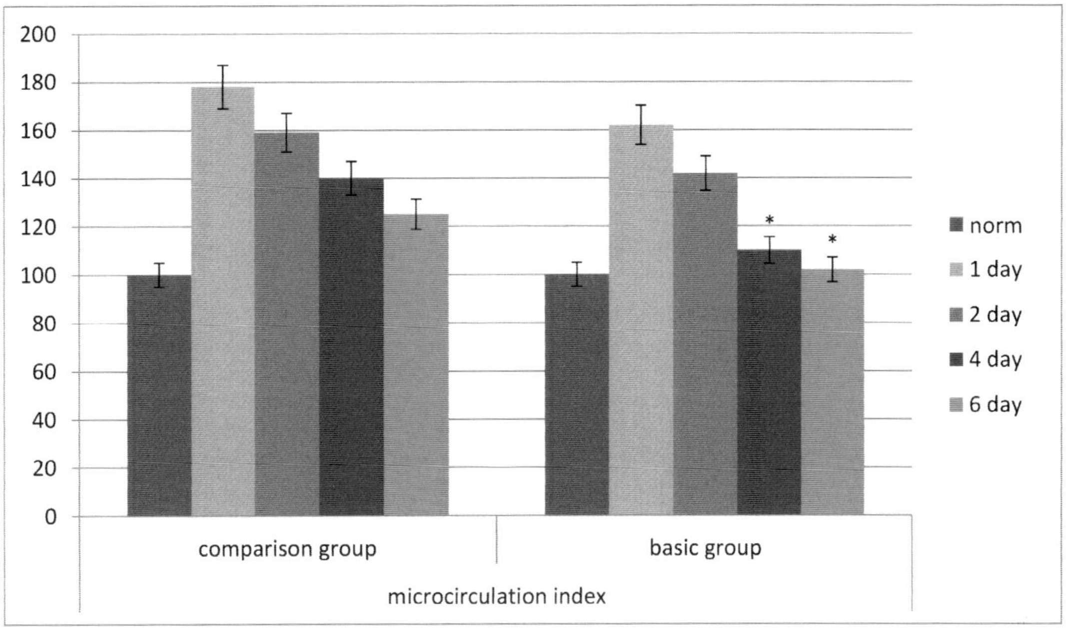

Figure 1. An indicator of microcirculation of patients with acute pancreatitis against laser therapy use. The established norm is taken for 100%, * – reliability of difference from data of group of comparison (p <0,05); I – comparisons; II – the main group.

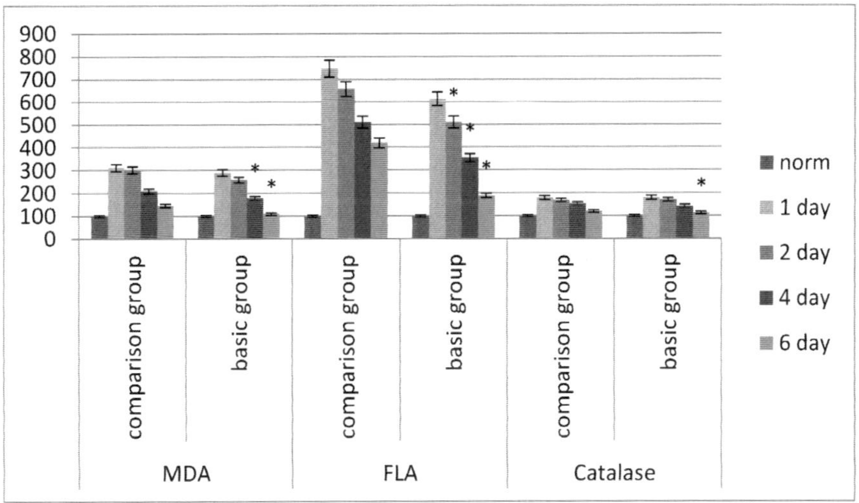

Figure 2. Concentration of by-products the FLOOR and activity of enzymes in plasma of patients with acute pancreatitis against laser therapy use. The established norm is taken for 100%, * – reliability of difference from data of group of comparison (p <0,05); I – comparisons; II – the main group.

At comparative analysis of fractional structure of lipids of a blood plasma at patients with acute pancreatitis in group of comparison and the main group it turned out that the content of total phospholipids in a blood plasma of the patients receiving laser therapy was above data of group of comparison for 10,3 – 16,8%, phosphatidylsincaline – for 10,1 – 15,2% (p <0,05), the indicator of triacylglycerols was below control for 12,0 – 20,3%, lizophospholipid – for 25,1 – 30,1%, free fatty acids – for 12,8 – 17,3% (p <0,05). Studying of indicators of a lipidic metabolism of erythrocytes in the main group showed decrease in contents primary and by-products the FLOOR in erythrocytes of patients of the main group there was less than that in control group for 21,1 – 17,0% (p <0,05) and reduction of activity of a phospholipase of A2 by 53,2% (p <0,05). Authentically the lipidic structure of erythrocyte membranes concerning data of group of comparison (figure 3) improved.

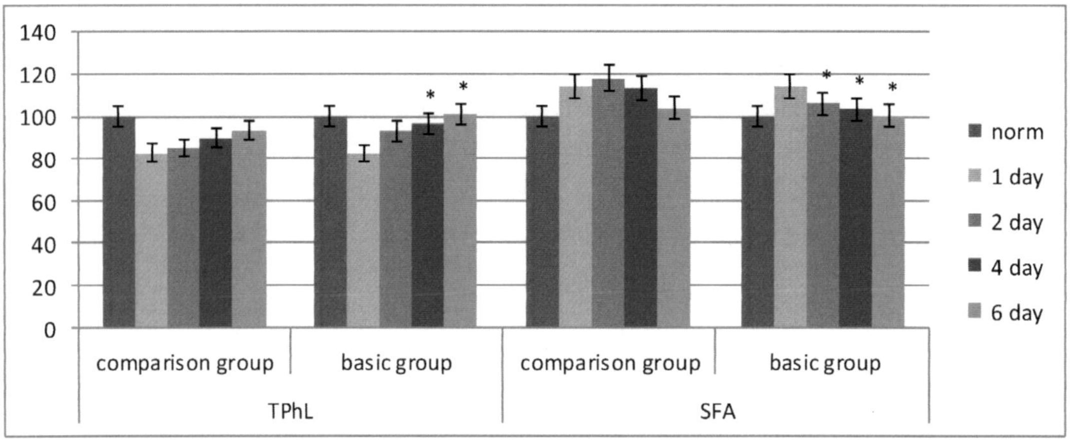

Figure. 3. Indicators of lipidic structure of erythrocytes of patients with acute pancreatitis against laser therapy use (TPhL – total phospholipids; SFA – free fatty acids). The established norm is taken for 100%, * – reliability of difference from data of group of comparison (p <0,05); I – comparisons; II – the main group.

In conclusion it should be noted that use of laser therapy at acute pancreatitis showed high performance in correction of disturbances of a lipidic metabolism – one of the leading pathogenetic components of this serious illness. lipid modulating potential of laser radiation of blood is verified by increase in a blood plasma and cellular membranes of erythrocytes of a share of total phospholipids, phosphatidylsincaline, phosphatidylserine and decrease in concentration of free fatty acids, acylglycerofishing and lizophospholipid. The specified positive action of the approved way of therapy is implemented through inhibition of processes of peroxide oxidation of lipids and reduction of activity of phospholipazny systems. Possibly, the above-stated effects of laser therapy are defined by its ability to improve microcirculation, to activate enzymes of own antioxidant protection, to increase synthesis of ATP, nucleic acids and proteins, and also to intensify cellular respiration (Nechipurenko N. I., 2014). Certainly, reduction of the membrane destabilizing phenomena was one of significant factors of correction of a syndrome of endogenous intoxication at acute pancreatitis. Use of laser radiation of blood allowed to reduce quickly concentration in a blood plasma of toxic metabolites of both the hydrophilic, and hydrophobic nature.

4. Conclusion

Let's note that productivity laser radiation of blood in recovery of indicators of a homeostasis significantly exceeded possibilities of standard therapy both on expressiveness, and on manifestation time.

Thus, use of laser therapy in complex treatment of acute pancreatitis allows to level dislipid changes in an organism and to prevent progressing of the membrane destabilizing phenomena, and also to quickly correct a syndrome of endogenous intoxication, significantly improving a current and the forecast of a disease in general.

Acknowledgments
This work was supported by Competitiveness Program of National Research Nuclear University MEPhI.

References

[1] Andreyev A. V. 2015 Treatment of the infected pancreatonecrosis by means of miniinvasive interventions / A.V. Andreyev, V. G. Ivshin, V. R. Goltsov//*Annals of surgical hepathology* vol. 20, No. 3, p. 110-116.
[2] Briskin B. S., Bukatko V. N. 2002 Otsenka of a condition of microcirculation at acute pancreatitis by method of a laser Doppler floumetriya//*the Siberian medical review* No. 4.
[3] Vlasov A. P. 2009 A system lipidic distress syndrome in surgery / A.P. Vlasov, V. A. Trofimov, V. G. Krylov (Moscow: Science) 224 p.
[4] Geynits A. V. 2012 Intravenous laser radiation of blood (wavelength – 405 nanometers) in complex treatment of patients with peritonitis / A.V. Geynits, R. D. Mustafayev, A. M. Mamedov, etc.//*Laser medicine* vol 16, No. 1, p. 13-16.
[5] Dibirov M. D. 2014 Modern approaches to complex treatment of a biliary pancreatonecrosis / M. D. Dibirov, T. A. Atayev, V. N. Yushchuk, etc.//*Magazine of N. V. Sklifosovsky Emergency medical service* No. 1, p. 17-19.
[6] Kubyshkin V. A. 2015 Modern approach to treatment it is purulent - necrotic complications of destructive pancreatitis/VA. Kubyshkin, O. V. Moroz, Yu. V. Kulezneva, etc.//*plenum Materials pravl. associations of gepatopankreatobiliarny surgeons of the CIS countries.* (Samara) p. 81-82.
[7] Kudelich O. A. 2014 A syndrome of endogenous intoxication and a possibility of its correction at heavy acute pancreatitis / O. A. Kudelich//*Surgery. Eastern Europe* vol. 11, No. 3 p. 99-111.
[8] Musikhin L. V. 2013 VLOK – modern opportunities of not pharmacological solution of problems of an intensive care and reanimation / L.V. Musikhin, P. V. Smolnikov, V. S. Shiryaev, etc.//*Laser medicine* vol 17 Issue 3 p. 4-10.
[9] Nechipurenko N. I. 2014 Influence of a laser hematotherapy on some biochemical indicators of blood in the acute period of lacunary heart attacks of a brain / N. I. Nechipurenko, L. N. Anatskaya, L. I. Matusevich//*Magazine of neurology and psychiatry of S. S. Korsakov* No. 7, p. 43-48.
[10] Tsedrik N. I. 2008 An oxidizing homeostasis at various forms of acute pancreatitis: features of diagnosis and way Correction / Tsedrik N. I., Vinnik Yu. S., Savchenko A. A., etc.//*Achievements of modern natural sciences* No. 9, p. 44-51.
[11] Aranda-Narváez J.M. Acute necrotizing pancreatitis: Surgical indications and technical procedures/J.M. Aranda-Narváez, A.J. González-Sánchez, M.C. Montiel-Casado et al.//World. J. Clin. Cases. – 2014. – Dec. 16. - Vol. 2(12). – P. 840–845.
[12] Bradley E.L. 2010 Management of Severe Acute Pancreatitis: A Surgical Odyss/E.L. Bradley, N.D. Dexter//*Ann. Surg.* vol. 251, p. 6–17.
[13] Feng Y.C. 2014 Study on acute recent stage pancreatitis/Y.C. Feng, M. Wang, F. Zhu et al.//*World. J. Gastroenterol.* No. 21. vol. 20(43) p. 16138–16145.

Allium-test as a tool for toxicity testing of environmental radioactive-chemical mixtures

A A Oudalova[1,2], S A Geras'kin[2], N S Dikareva[2] and S V Pyatkova[1]

[1]Obninsk Institute for Nuclear Power Engineering, National Research Nuclear University MEPhI, Obninsk, Russia
[2]Russian Institute of Radiology and Agroecology, Obninsk, Russia

E-mail: oudalova@mail.ru

Abstract. Bioassay-based approaches have been propagated to assess toxicity of unknown mixtures of environmental contaminants, but it was rarely applied in cases of chemicals with radionuclides combinations. Two *Allium*-test studies were performed to assess environmental impact from potential sources of combined radioactive-chemical pollution. Study sites were located at nuclear waste storage facilities in European and in Far-Eastern parts of Russia. As environmental media under impact, waters from monitor wells and nearby water bodies were tested. Concentrations of some chemicals and radionuclides in the samples collected enhanced the permitted limits. Cytogenetic and cytotoxic effects were used as biological endpoints, namely, frequency and spectrum of chromosome aberrations and mitotic abnormalities in ana-telophase cells as well as mitotic activity in *Allium* root tips. Sample points were revealed where waters have an enhanced mutagenic potential. The findings obtained could be used to optimize monitoring system and advance decision making on management and rehabilitation of industrial sites. The *Allium*-test could be recommended and applied as an effective tool for toxicity testing in case of combined contamination of environmental compartments with radionuclides and chemical compounds.

1. Introduction

Environmental exposure levels could be substantially increased through anthropogenic nuclear and industrial activities. Many surveys have been carried out to specify and justify constraints for radionuclide discharges and chemical substances releases into environment compartments, but a development of strategies for regulatory decision-making is still a challenge for specialists in radiation and chemical protection of the environment and the public. In real-time cases when territories appear contaminated through human civil and military activities such as nuclear facilities operation, mining sites, nuclear weapon testing, there is an essentially complex impact so that a mixture of radionuclides is often supplemented by other potentially hazardous substances (e.g. heavy or alkali metals, organic compounds, etc.). An actual hazard of such combined contaminations is difficult to estimate and predict correctly because of a lack of knowledge about relationships between various types and combinations of environmental pollutants and consequences for biota and human health. To mitigate the problem, bioassay-based approaches have been propagated to assess toxicity of unknown mixtures of environmental contaminants [1-3], but it was rarely applied in cases of chemicals with radionuclides combination. One of the standardized bioassays is the *Allium*-test that has been developed as a laboratory-based, single-species method for rapid screening of chemicals and environmental samples to assess their toxicity and genotoxicity [4-6].

In this paper findings are presented from the *Allium*-test application at an assessment of environmental impact from potential sources of combined radioactive-chemical pollution at two different study-sites.

2. Materials and methods

2.1. RWSF study-site

The first study was implemented in a vicinity of the radioactive waste storage facility (RWSF) in the European part of Russia [7]. The storage was established in the beginning of 1960th, radioactive wastes disposal had continued 8 years. Since 1998, there had been detected a leakage of ^{90}Sr from one of the storage tanks, and there were found a significant increase of ^{90}Sr specific activity (up to 40 Bq/l) in one of the monitor wells [8]. Maximum concentration of radioactivity occurred in a vicinity of the leaking storage tank (up to 180-210 Bq/l). To estimate consequences for the environment and population, a comprehensive radioecological survey was completed, and a bioassay-based study was one of its components. In a few years, the leakage was fixed, and a full-scale countermeasure program was implemented. By now, the RWSF has been decommissioned.

Here bioassay data are presented from testing water samples collected at and around the RWSF in 2004. Samples were taken from 7 monitor wells (Samples 5-7) and nearby waterbodies (Samples 3, 4, 8, 9). Water from a forest brook was considered as control (Sample 1), also distilled water was tested (Sample 2). In the most contaminated waters, ^{90}Sr specific activities were about 10 times of the intervention level (IL = 5.0 Bk/kg) adopted in Russia for drinking water. In all samples except control heavy metals concentrations exceeded the Russian guideline values for maximum permissible content (MPC) in waterbodies, especially for Zn and Mn (70-250 times of MPCs), Ni (18-30 times of MPC).

2.2. STS study-site

The second study was carried out within the site for temporary storage (STS) of radioactive wastes and spent nuclear fuel in the Far-East of Russia where a military coastal technical base of the nuclear fleet of the former USSR used to operate. The STS is now a branch of the regional RW management center under supervision of the Federal State Unitary Enterprise 'DalRAO'. From previous studies [9], gamma dose rate within the STS amounted from 0.1 to 65 μSv/h, at regional natural background level of 0.10-0.15 μSv/h. The main source of exposure is radiation penetrating walls of specialized containers and buildings where radioactive materials are stored. Radioactive contamination of territory is mainly formed by ^{137}Cs and ^{90}Sr; their specific activities in soil vary significantly within the site.

Water samples were collected in 2013 from 7 monitor wells (Samples 6, 12-16, 21) located within the STS [10]. Tap water was taken as control. Specific activities of ^{137}Cs in some samples reached up to 30% of the IL; for ^{90}Sr the corresponding IL was enhanced 2-3 times. Samples were of low mineralization, but for some chemicals the MPCs were broken in many cases.

2.3. Bioassay with Allium-test

To test samples cyto- and genotoxicity, the *Allium*-test was applied. General procedure is described in [4, 11], detailed description is also given in [7, 10]. Onion, *Allium cepa* L., was taken as a test-species. Bulbs (10-30 per every water sample) were germinated on water samples under testing. Onion roots of the 15 mm length were randomly sampled and fixed in an alcohol/acetic acid mixture (3:1). To test samples cyto- and geno-toxicity, squashed slides of *A. cepa* root tips were prepared with aceto-orcein as a dye and examined by optical microscopy. All slides were blindly and randomly coded before cytogenetic analysis. Sample cytotoxicity was estimated based on root proliferation that was quantified in parallel for all samples as mitotic index. Geno-toxicity was assessed with a frequency of aberrant cells in root meristem of *A. cepa*.

3. Results and discussion

3.1. Cyto- and genotoxicity of waters from the RWSF study-site

The bioassay results obtained at the onion bulbs germination in the water sampled from the RWSF and nearby territory are shown in figure 1. The mitotic activity in root meristem changed from 55 to 190% of control level (figure 1a). The inhibitory effect was observed for water collected from a small bog (Sample 3) and distilled water (Sample 2); both these samples were obviously unfavorable for higher plant ontogenesis. Sample 4 was taken from a brook flowing close to the RWSF fence; this sample contained high amounts of ^{90}Sr, Zn and Cu. Comparing to Sample 1 (control), an increase of mitotic activity was found in Samples 8, 5 and 6. Interesting that these samples had got rather high concentrations of chemicals, and it was difficult to relate increased mitotic index and chemical contamination.

For cytogenetic disturbances (figure 1b), a transparent effect of water contamination was shown for all water samples taken from and nearby the territory of the RWSF since the aberrant cells occurrence in root meristem of *Allim cepa* was significantly over the control, which indicated a potential adverse impact on the natural environment in a vicinity of the storage. The main contribution to the genotoxic impact was given by laggard chromosome occurrence, which frequency significantly enhanced the control level in many samples [7]. Remarkable role was also played by such severe disturbances as chromosome bridges and tripolar mitoses.

Thus, in water samples from the territory affected by the RW storage in European part of Russia, combined effect of different toxicants induced a wide spectrum of responses (from inhibition to stimulation) according to the bioassay of "proliferative activity in root meristem of *Allium cepa* L.". Genotoxic effect was revealed for all water samples collected from the impacted territory which showed an efficacy and high sensitivity of the bioassay of "aberrant cells frequency in root meristem of *Allium cepa* L." in a case of combined contamination of environment with ^{90}Sr and heavy metals.

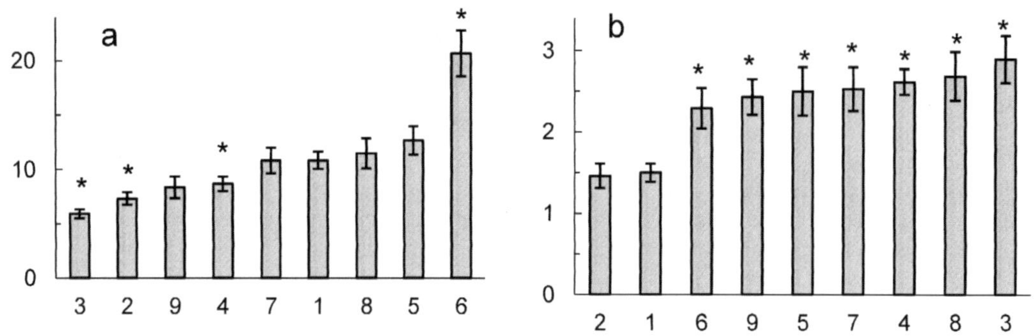

Figure 1. a - Mitotic index (%), b - aberrant cell frequency (%) in root tip cells of *Allium cepa* bulbs germinated in waters sampled from the RWSF and nearby territory, (mean ± se).

* Difference from the control (Sample 1), $p < 0.05$.

3.2. Cyto- and genotoxicity of waters from the STS study-site

At onion bulbs germination in waters sampled from the monitor wells at the STS territory there were found changes in mitotic activity of root meristem cells of both types: an inhibition and stimulation (table 1). Mitotic index decreased below the control in Samples 13, 16 and 21. These alterations were statistically significant, but in absolute value they amounted only to 10-13% of the control level; it could be considered as a small cytotoxic effect. In other water samples (6, 12, 14, 15) a significant increase of mitotic activity up to 50-60% of the control level was observed.

Mutagenicity estimated via the aberrant cells frequency in root meristem was high in the same samples (13, 16, 21) which showed the inhibition effect on the proliferative activity (table 1). Moreover, there was a close correlation between mitotic index and aberrant cell occurrence ($r = -0.88$, $p = 0.008$). In other samples, genotoxicity was not different from the control level, and in two samples

(14 and 15) a significant, 3-fold decrease was detected. Chromatid aberrations recognized as damage of moderate severity were prevailing type in aberration spectrum in most samples [10]. However, in cyto- and genotoxic waters in Samples 13, 16 and 21, there were found an appreciable fraction of genomic disturbances such as laggings and conglutinations.

Thus, the *Allium*-test application revealed that onion bulbs germination in waters from the STS monitor wells (far-Eastern part of Russia) resulted in statistically significant alterations of both the mitotic activity in root meristems in all impacted samples and aberrant cells frequency in most water samples, and these two biological indices correlate to each other at high level of significance.

Table 1. Mitotic index and aberrant cell frequency in root tip cells of *Allium* bulbs germinated in waters sampled from the STS territory

Sample (No. monitor well)	Mitotic index (%)	Aberrant cells (%)
Tap water	11.0±0.2	2.04±0.42
6	16.2±0.3 [a]	1.08±0.24
12	16.6±0.3 [a]	1.75±0.35
13	10.1±0.2 [a]	3.72±0.58 [a]
14	16.8±0.3 [a]	0.69±0.21 [a]
15	16.7±0.3 [a]	0.78±0.23 [a]
16	10.0±0.2 [a]	3.96±0.58 [a]
21	9.8±0.2 [a]	5.01±0.66 [a]

[a] Difference from tap water (control), $p < 0.05$.

4. Conclusions

Biological monitoring data obtained with the plant bioassay – the *Allium*-test – at testing the environmental media from territories impacted by two different facilities for radioactive material management are presented in this paper. In both study-sites, an industrial intake of chemicals and radionuclides into the environment appeared during the facilities operation. As a rule, the enhanced concentrations of some contaminants were found in technical or nearby areas around the RWSF and the STS that could pose a risk both to human health and to biotic components of natural ecosystems.

The identification and quantification of single chemicals considered as harmful for man or the environment are routine procedures in a system of environmental monitoring. Measurements of doses or contaminant concentrations could give us a detailed description of pollution levels, but they hardly indicate or forecast potential biological consequences from mixed pollution of the environment. Bioassays, in turn, quantify biological impact of mixtures. To have a realistic estimation of risk to the environment caused by any contamination, the simultaneous application of pollutant chemical control methods followed by biological tests is necessary.

The *Allium*-based root tips assay of chromosome aberrations in anaphase-telophase has been used in many studies to test genotoxicity of wastewater, contaminated soils, remediation techniques, new substances, etc [4, 6, 11, 12]. It was not so extensively exploited at testing of media contaminated with radionuclides or radionuclide-chemical mixtures, and this gap has been partly fulfilled by the findings presented here. Indeed, in both studies carried out in different radioecological conditions, water samples bearing high genotoxicity potential were revealed, with the important contribution from severe types of cell damage. The positive results obtained in the *Allium*-test should be considered as a warning and also an indication that the tested environments may be of risk to both human health and biological components of natural ecosystems.

Thus, the *Allium*-test is shown to be an effective tool for toxicity testing in case of combined contamination of environmental compartments with radionuclides and chemical compounds.

Information gained from the bioassay can help in optimizing monitoring and rehabilitation activities. An effectively linking of bioassay-screening assays to the well-established environmental pollution monitoring approach is a way of improving and upgrading an existing system of the public and the environment protection.

Acknowledgements
This work was supported by Russian Foundation for Basic Research (№ 05-04-96721 and № 16-48-400837) and Federal Target Program "Industrial utilization of weapons and military equipment of the nuclear complex for 2011-2015 and for the period till 2020" (contracts № 10-3-14-2013 and № 31401098199).

References

[1] Sharma C B S R 1983 *Current Science* **52** 1000–2
[2] Grant W F 1999 *Mutat. Res.* **426** 107–12
[3] Geras'kin S A, Evseeva T I and Oudalova A A 2011 *Encyclopedia of Environmental Health* vol 4, ed Nriagu J O (Burlington: Elsevier) pp 571–9
[4] Grant W F 1982 *Mutat. Res.* **99** 273–91
[5] Fiskesjo G 1985 *Hereditas* **102** 99–112
[6] Rank J and Nielsen M H 1993 *Hereditas* **118** 49–53
[7] Oudalova A A, Geras'kin S A, Dikarev V G and Dikareva N S 2014 *Radiat. Biol. Radioecol.* **54** 97–106 (in Russian)
[8] Starkov O V, Vaizer V I, Bogdanovich N G and Kozmin G V 2003 *Nuclear Energetics* **2** 67–73 (in Russian)
[9] Shandala N K, Kiselev S M, Lucyanec A I, Titov A V, Seregin V A, Isaev D V and Akhromeev S V 2011 *Radiat. Prot. Dosimetry* **146** 129–32
[10] Oudalova A A, Pyatkova S V, Geras'kin S A, Kiselev S M and Akhromeev S V 2016 *Radiat. Biol. Radioecol.* **56** 208–19 (in Russian)
[11] Fiskesjo G 1988 *Mutat. Res.* **197** 243–60
[12] Leme D M and Marin-Moralis M A 2009 *Mutat. Res.* **682** 71–81

The assessment of human skin biomatrixes using raman spectroscopy method

E V Timchenko[1], P E Timchenko[1], L T Volova[2], D A Dolgushkin[2], P Y Shalkovskaya[1], S V Pershutkina[1], I F Nefedova[2]

[1]Samara National Research University, 443086 Russia, Samara, st. Moscow highway, 34.
[2]Institute of Experimental Medicine and Biotechnology (IEMB) Samara State Medical University, 443079 Russia, Samara, st. Gagarin, 20.

E-mail: laser-optics.timchenko@mail.ru

Abstract. There are presented the results of the analysis of the implants made of human skin by Raman scattering method. The main spectral distinctions of bioimplants by using various methods for their manufacture are shown at wavenumbers 1062 cm^{-1}, 1645 cm^{-1}, 1260 cm^{-1}, 850 cm^{-1}, 863 cm^{-1}, corresponding to components that are important for the quality of implant: glycosaminoglycans, amide type I, amide type III, asymmetrical association C-O-S of vibration of glycosaminoglycans GAGs, tyrosine and a C-C stretching of proline ring, ribose. Has been carried out two-dimensional analysis of optical coefficients providing an opportunity to control the quality of cutaneous implants in the process of manufacturing it, and detailed analysis of Raman scattering spectroscopy.

Keywords: Raman scattering spectroscopy, optical coefficients, cutaneous implant, quality control.

1. Introduction

Over the last years in different branches of medicine widely have been using the materials of biogenicity for tissue defects replacement. The full recovery of skin integument after thermal injury and information research for this purpose are the important problems of Combustiology. 60 thousand people die each year from burn injuries worldwide [1, 2]. Unfortunately, the optimum plastic materials for tissue defects replacement have not been found yet. The biotechnological methods for obtaining and using skin equivalents in clinical practice are actively developing. However, there is a problem of quality control of such implants.

In order for use the implants effectively for tissue defects replacement during the manufacturing process must be carefully removed all immunogenic factors that can lead to rejection of donor material [3, 4]. At the same time in the process of manufacturing of implants must maintain their native structure and composition. The methods used to achieve these goals are usually having the opposite effects: the extremely aggressive removal of immunogenic components can destroy the structure and composition of the tissue, while the more sparing techniques can save the enhanced immunogenicity of implants [5, 6].

For quality control of implants is used biochemical, cytological and histological research methods [7, 8, 9]. However, these methods are time-consuming and often lead to changes in the structure and composition of the material quality in the process of their analyzing.

On the current level of science and technology development to evaluate the implants made of human skin may be used physical and optical research methods that allow people to examine complex biological objects quickly without damaging their structure. Raman scattering spectroscopy (RS) [10, 11, 12] because of its noninvasiveness is widely used in biological objects for quality control of materials of tissue engineering [13, 14, 15]. It is known work [12], in which the Raman spectroscopy method RS is claimed to be the best method for analyzing the chemical composition of human skin. During the analysis, it was found that the absence of Raman peaks at wave numbers 717 cm-1, 1157 cm-1 and 1526 cm-1, indicates the absence of lipids in the horny layer of skin. Also in works [15, 16] was used Raman spectroscopy method to evaluate the collagen type I and glycosaminoglycans (GAGs) in composition of biomaterials, corresponding to lines 937 cm-1 and 1062 cm-1.

The mission is to use Raman spectroscopy to evaluate the quality of a person's skin implants manufactured in different ways.

2. Materials of research

As the objects of research were used 8 lyophilized bioimplants with preserved epidermis (the samples manufactured according to the protocol number 1), 8 lyophilized bioimplants without epidermis (the samples manufactured according to the protocol number 2), and 2 control samples of untreated human skin (Figure 1). Lyophilized implants were manufactured by the technology "Lioplast» ® (TU-9398-001-01963143-2004), which is based on the use of physical factors processing of biomaterials (ultrasonic, lyophilization) [16, 17].

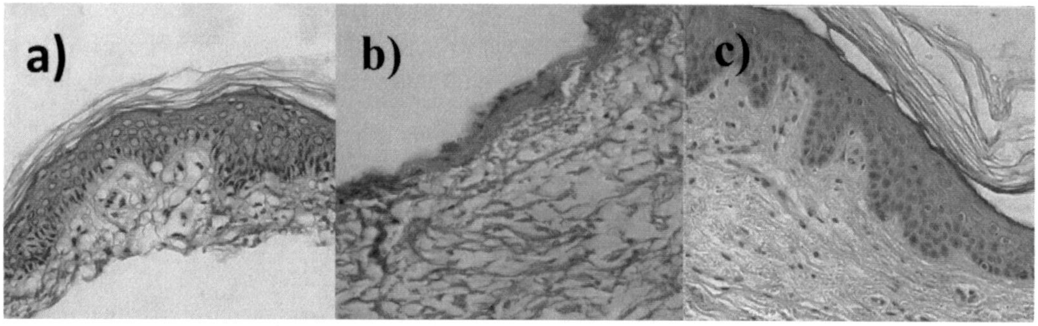

Figure 1. Smears of the samples: a) a sample with the epidermis treated with the protocol number 1; b) the sample without the epidermis treated with the protocol number 2; c) A control sample of untreated human skin. H & E stain. The increase of 400.

The skin has typical histological structure in control samples: the epidermis was presented with 5 layers; dermis papillary and reticular layers. The samples manufactured according to the protocol №1, were keeping arrangement of layers of the epidermis; the cells of basal layer were more elongated, the nuclei were flattening assuming drop shape. In the cells of spinous and granular layers was found empty vacuoles on the place of the nuclei. Dermal papillae were smoothed. The dermis was represented as loose areolar connective tissue with fibers of different thickness. Under basal layer of the epidermis was observed focal changes in the form of lacunae of various sizes. On some fibers were located single cells. The sample manufactured according to the protocol №2, did not contain the epidermis. The boundary between the papillary and reticular layers of dermis has not been visualized. Dermis was presented with fragmented fibers of loose areolar connective tissue and contained capillary loop (pic. 1, a, b, c). Control of the qualitative composition of skin surface of implants was carried out using an experimental stand, including a high-resolution digital spectrometer Shamrock sr-303i with built-in cooling chamber DV420A-OE, a fiber optic probe the RPB-785 for Raman spectroscopy, combined with laser module LuxxMaster LML-785.0RB -04 [16]. The stand has the following characteristics: scanning range of wavelengths from 190 nm to 1200 nm; time of exposure from 0.04 to 600s; camera resolution of 1024 *255 pixels, laser power 50-450 mW. Processing of the spectra was carried out in the program Wolfram Mathematica 8. Error of method at determining of used coefficients was <7% (GOST 8.207-76).

3. Results and discussion

In Figure 2a presented average Roman spectrum obtained from the surface of lyophilized bioimplants and control skin samples. In view of the fact that the skin is a multi-component object, in the spectrum of which many lines overlap, it had to carry out the separation of the obtained spectra into spectral lines using deconvolution of Lorentz-Gauss functions in MagicPlotPro software environment (Figure 2b). Average value of the coefficient of determination of the resulting spectrum, generated by the spectral profiles from the original Raman spectrum for area 300-2199 cm^{-1} was R^2=0,98, and average standard deviation of analysis σ = 8,81. Registered Raman bands correspond to oscillation modes, which are shown in Table 1.

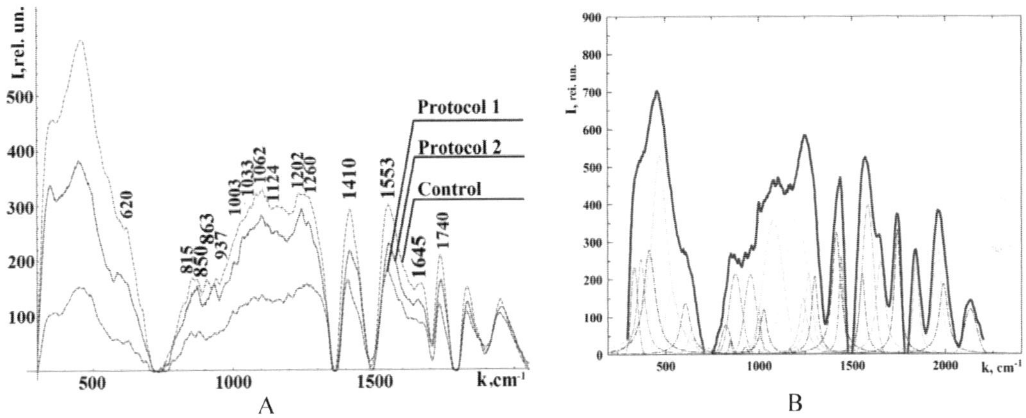

Figure 2. The results of research: A) Averaged Raman spectrum of bio-implants manufactured by Protocols 1, 2 and control samples; B) The software division lines in the spectrum.

Table 1. Decoding of the Raman spectra of bioimplants made of human skin

Wavenumber, cm^{-1}	Substance, oscillation
620	γt C–C phenylalanine (protein) [18]
815	RNA phosphodiester bands [18]
850	Asymmetrical C-O-S association of vibration GAGs, tyrosine and C-C stretching of proline ring (protein) [18]
863	Ribose vibration (RNA) [18]
937	Glycosidic bond C-O-C vibration in the GAGs, C-C stretching of proline / valine (protein) [18]
1003	Phenylalanine (C-C-aromatic ring) (protein) [11,14]
1033	ν (C-H) Phenylalanine Breath (protein) [18]
1062	Glycosaminoglycans (α- glycans) [18]
1124	C-C stretching of lipids, C-H stretching of protein , CH и C-OH deformation [14]
1196	CH3 lipid ring
1202	Hydroxyproline, tyrosine [18]
1260	Amide III [18]
1383	NH2 vibration
1410	ν(C-O) from COO (amino acids of glutamine acid) [18]
1553	NH deformation, Amide II [18]
1645	Amide I [18]

Analysis of the spectra from the surface of skin implants, obtained by different methods showed differences between them at wavenumbers 1062 cm^{-1}, 1645 cm^{-1}, 1260 cm^{-1}, 850 cm^{-1}, 863 cm^{-1}, corresponding to components that are important for the quality of implant: glycosaminoglycans, amide I, amide III, asymmetrical C-O-S association of vibration GAGs, tyrosine and C-C stretching of proline ring and ribose [17, 18, 19]. Therefore, these wavenumbers can be used as criteria for the quality of processing of skin bioimplants. Relatively permanent component in skin implants is an amide II [15, 16], corresponding to the wave number 1553 cm^{-1}, so it was used as the denominator (I_{1553}) in the input optical coefficient (k):

$$k = \frac{I_i}{I_{1553}},$$

where I_i – the values of intensity on the wavenumbers for the analyzed components. Recovery properties of implants are identified by their structure and composition. Thus, the saved glycosaminoglycans, proline and hydroxyproline affect its architectonics and elasticity while the residual RNA cells are capable of eliciting a reaction of immunological response and rejection of the implants. The proposed coefficients allow evaluating these criteria. In the Figure 3 are presented two-dimensional diagrams of the input optical coefficients k.

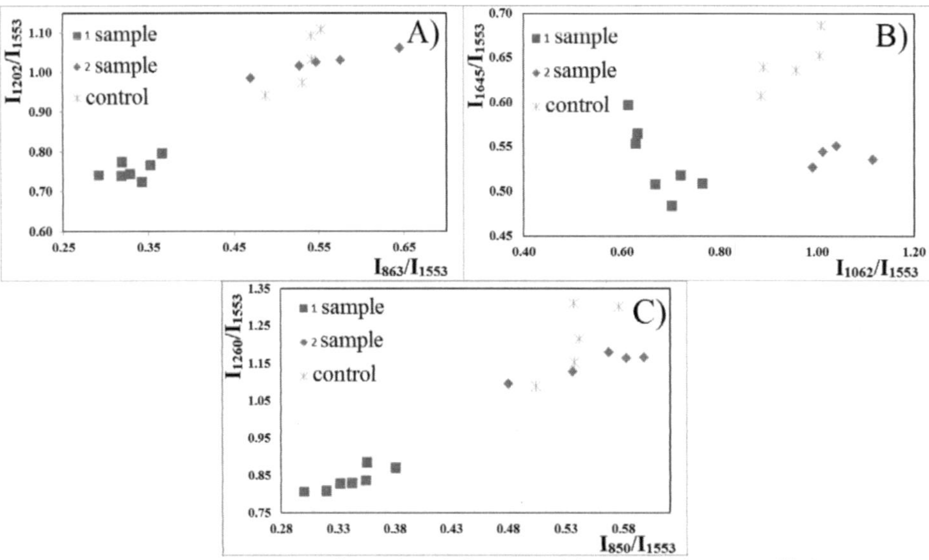

Figure 3. Two-dimensional diagram of the optical input coefficients.

Analysis of two-dimensional diagrams in Figure 3 shows that the treated samples of skin implants manufactured according to the protocol №1 have values of optical coefficients $I_{1645} / I_{1553} > 0,49$, and $0,73 < I_{1202} / I_{1553} < 0,81$, and characterized by lower content of amide I and hydroxyproline than the sample manufactured according to the protocol №2. Quite the opposite, the content of glycosaminoglycan is greater than in the samples manufactured according to the protocol №2, which has the values of optical coefficient $0,91 < I_{1062} / I_{1553} < 1,13$. Also Figure 3 shows that the value of the lines intensity of amide I and amide III in control sample has a maximum value and corresponds to optical coefficients $0,6 < I_{1645} / I_{1553} < 0,71$, $1,1 < I_{1260} / I_{1553} < 1,34$.

Based on the carried out analysis, diagrams were constructed relatively represented the maximum content of substances found by Raman spectroscopy method on the surface of samples of human skin bioimplants manufactured in different ways (Figure 4).

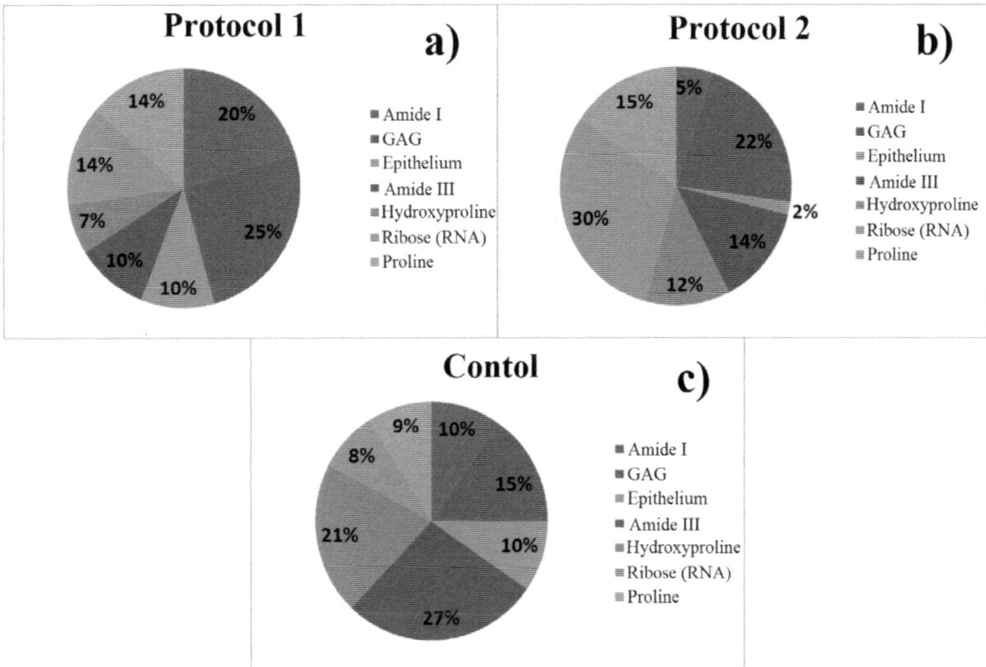

Figure 4. Diagrams substances as defined by the surface of the skin implants: a) a sample of the epidermis treated with the protocol number 1; b) the sample without the epidermis treated with the protocol number 2; c) A control sample of untreated human skin.

Analysis of the diagrams, shown in Figure 4, showed that with proposed coefficients and using Raman spectroscopy can be compared the relative content of the components of surfaces of the skin implants, also choose the optimal methods for their processing and control their effectiveness.

4. Conclusion

The spectra's analysis from the surface of skin implants obtained by different methods showed differences between them at wavenumbers 1062 cm^{-1}, 1645 cm^{-1}, 1260 cm^{-1}, 850 cm^{-1}, 863 cm^{-1}, corresponding to components that are important for the quality of implant: glycosaminoglycans, amide I, amide III, asymmetrical C-O-S association of vibration GAGs, tyrosine and a C-C stretch ring proline, ribose.

Raman spectroscopy allows to carry out quick noninvasive control of the structure of the skin bioimplants; obtain an composition evaluation of components of surfaces bioactive matrixes; choose optimal bio-carriers and control the effectiveness of methods of their processing; optimize the creation of cellular-tissue products for regenerative medicine.

5. References:

[1] Ledovskoy S N, Burda Y E, Lazarenko V A 2008 Analysis of the clinical efficacy of fetal and adult allogeneic diploid fibroblasts in the treatment of burns border *Scientific J.: Advances modern science* chapter 9 pp 92-94

[2] Gerasimov L 2000 The problem of burns on the threshold of XXI century *Scientific J. Kombustologiya* chapter 2 pp 36-42

[3] Livesey S A, Herndon D N, Hollyoak M A, Atkinson Y H, Nag A 1995 Transplanted acellular allograft dermal matrix. Potential as a template for the reconstruction of viable de rmis *J. Transplantation* chapter 60 pp 1-9

[4] Gubareva E A, Sotnichenko A, Gilewicz I, Macchiarini P 2012 Morphological evaluation of the quality decellularization heart and diaphragm of rats *J. Cell Implant and tissue engineering* chapter 3 pp 22-27

[5] Sergeyevich D S, Podhvatilina N A, Vasilyeva M B 2012 Morphological and functional features of the aortic graft after decellularization *J.Complex problems of cardiovascular disease* pp 3-6

[6] Kurapeev D I, Lavreshin A V, Anisimov S V 2012 Tissue engineering of heart valves: decellularization allo- and xenografts *J.Cell Implant and tissue engineering* chapter 1 pp 34-39

[7] Isayev A D 2014 Skin cancer. Causes, symptoms and signs, diagnosis and treatment of cancer *J. Polismed* chapter 1 pp 33-39

[8] Afanasiev Y I 2012 *Histology, embryology, cytology* M.: Medicine p.800

[9] Bykov V L 2002 *Handbook on the Cytology and general histology* vol 1, ed (St. Petersburg: SOTIS) p 254

[10] Caspers P J, Lucassen G W, Puppels G J 2001 Combined In Vivo Confocal Raman Spectroscopy and Confocal Microscopy of Human *Skin J. of Investigative Dermatology* chapter 116 pp 434-442.

[11] Matousek P, Draper E R, Goodship A E, Clark I P, Ronayne K L, Parker A W 2006 Non-invasive Raman spectroscopy of human tissue in *J. Central Laser Facility Annual Report* chapter 4 pp 133-135

[12] Ali S M, Bonnier F, Tfayli A, Lambkin H, Flynn K 2013 *J. of Biomedical Optics* p 8

[13] Khmaladze A, Ganguly A, Kuo S 2013 Tissue-Engineered Constructs of Human Oral Mucosa Examined by Raman Spectroscopy *J.Tissue Engineering Part* chapter 19 pp 299-306

[14] Zhao J Y, Chai J K, Song H F, Sun T J, Li D J, Liu L Y, Gao Q W, Liang L M 2011 Effects on collagen I and III after transplantation of porcine acellular dermal matrix with hyaluronic acid *J. Zhonghua Yi XueZaZhi* chapter 91 pp 1276-1280

[15] Kunstar A A, Leferink A M, Okagbare P I 2013 *J. of the Royal Society Interface* p 10

[16] Timchenko E V, Timchenko P E, Frolova L T, Milyakova M N, Maksimenko N A, Taskina L A 2015 *J. Optical Society of America* p 30

[17] Timchenko E V, Timchenko P E, Volova L T et al. 2014 *J. Quantum Electronics* p 696

[18] Tfaili S, Gobinet C, Josse G, Angiboust J, Manfait M, Piot O 2012 Confocal Raman microspectroscopy for skin characterization: a comparative study between human skin and pig skin *J. Analyst* chapter 137 pp 3673–3682

[19] Zanyar M, Shazza R, Ihtesham U 2007 Raman Spectroscopy of Biological Tissues *J. Applied Spectroscopy Reviews* chapter 42 pp 493-541

Generators and automated generator systems for production and on-line injections of pet radiopharmaceuticals

G Shimchuk, Gr Shimchuk, G Pakhomov, G Avalishvili, G Zavrazhnov, I Polonsky-Byslaev, A Fedotov and P Polozov

POZITOM-PRO RPC LLC, Moscow, Russia

E-mail: medphys@mail.ru

Abstract. One of the prospective directions of PET development is using generator positron radiating nuclides [1,2]. Introduction of this technology is financially promising, since it does not require expensive special accelerator and radiochemical laboratory in the medical institution, which considerably reduces costs of PET diagnostics and makes it available to more patients. POZITOM-PRO RPC LLC developed and produced an ^{82}Sr-^{82}Rb generator, an automated injection system, designed for automatic and fully-controlled injections of ^{82}RbCl produced by this generator, automated radiopharmaceutical synthesis units based on generated ^{68}Ga produced using a domestically-manufactured ^{68}Ge-^{68}Ga generator for preparing two pharmaceuticals: Ga-68-DOTA-TATE and Vascular Ga-68.

1. Introduction

Presently, an efficient offer is equipping fully functional PET centers and medical institutions that have PET scanners with radionuclide generator systems, in particular, ^{68}Ge/^{68}Ga and ^{82}Sr/^{82}Rb, to generate positron radiating radionuclides and radiopharmaceuticals based on them which notably expands the range of PET method application in clinical practice.

Radionuclide generator is a rather compact device with own protection from ionizing radiation. Its operation does not require to specially design and then provide maintenance for complex specialized rooms systems, like with accelerator equipment.

2. Radionuclide generator system construction

Radionuclide medical generator is an installation in which the long-lived (mother) nuclides are used to generate other (daughter) nuclides with shorter half-life, in the form of pharmaceutically acceptable (or compatible with it) in vivo medical preparation.

By method of nuclide pair separation extraction and chromatographic generators are distinguished. In extraction generators nuclide pair separation is performed by extracting the required nuclide into an appropriate solvent.

However, in practical terms chromatographic generators, where parent radionuclide is fixed on mineral or organic sorbent, are more process-efficient. Daughter radionuclide is extracted by eluting from the column with an appropriate solvent (eluent).

Number of possible extractions (elutions) depends on the life duration of mother radionuclide. Daughter radionuclide extraction rate depends on how quickly the equilibrium is achieved. If the daughter radionuclide half-life is negligibly short compared to the mother radionuclide, then 50% of the equilibrium amount is accumulated by the end of the first half-life, 75% by the end of the second half-life and 99% by the end of 6 half-lives. Consequently, eluate from a generator with short-lived daughter radionuclide may be obtained as frequently as needed.

3. Radionuclide generator system $^{82}Sr/^{82}Rb$

In cardiovascular disease diagnostics, myocardial blood flow condition evaluation is of great importance [3].

Absorption of $^{82}Rb^+$ is evaluated by blood flow, the condition of Na^+–K^+ATP pump and integrity of cellular membrane. For this reason, already from the first attempts of using it as an alternative for $[^{13}N]NH_3$, it was introduced in non-invasive myocardial blood flow measurements by PET, as well as for visualization of functional changes connected with cerebral blood circulation condition evaluation in various cerebral diseases diagnostics.

$^{82}Rb^+$ can be used for qualitative and quantitative evaluation of cerebral cells' integrity, which is particularly important to obtain clearer outline of cerebral tumors distribution and to distinguish tumors and to evaluate radiation necrosis when radiation treatments are introduced.

$[^{82}Rb]RbCl$, apart from the simplicity of generation process, has a number of other advantages, as it has short half-life (76 seconds). Besides this, not only is the patient radiation exposure reduced, but several stress tests can be performed consecutively, which leads to shorter test duration and getting a more objective picture of the condition.

Procedure duration at rest and with a pharmaceutical sample is not over 30 minutes, while a full-scale myocardial perfusion exam with any other known method of diagnostic nuclear medicine would require more time.

4. Radionuclide generator system $^{68}Ge/^{68}Ga$

Perspectives for ^{68}Ga ($[^{68}Ga]$DOTATOC)-labeled pharmaceuticals for neuroendocrine malignant tumor diagnostics may be estimated by analyzing an earlier preparation version based on ^{111}In. ($[^{111}In]$-DTPA-D-Phe^1-Tyr^3-octreotide) [4].

This pharmaceutical helps to diagnose a broad range of diseases not only of neuroendocrine nature, which increases interest in similar ^{68}Ga-labeled predecessors. In evaluating the disease diagnostics efficiency on the equivalent objects, positron emitting radionuclide ^{68}Ga-labeled pharmaceuticals have demonstrated higher sensitivity due to conceptually different detection scheme. Detection principle is different from that of SPECT, it involves lead collimators, and it accounts for higher sensitivity of PET.

Radionuclide generator system has optimal characteristics of the mother and daughter radionuclides. Mother radionuclide half-life (271 days) ensures efficient generator performance for about 1 year, with the possibility to obtain daughter radionuclide each 4 hours (70% from the initial potency).

^{68}Ga radionuclide may be used to label more proteins, peptides and small molecules. Half-life is 68 minutes, which ensured higher diagnostics efficiency without additional radiation exposure for patients.

5. Automated injection system

A pharmaceutical obtained from the generators described above may be injected directly intravenously (generator system 82Sr/82Rb) or after additional treatment (generator system 68Ge/68Ga). For these procedures in clinical practice, there has been developed and manufactured an automated injection system [5] (see figure 1), which is an automated system of solution infusion, including: means to produce a solution; means for infusion; means for radioactive solution dosing based on potency level or volume; means of automatic monitoring for current potency, occlusion, weight and solution radioactivity in the waste container; own protection from ionizing radiation. Infusion process is fully automated.

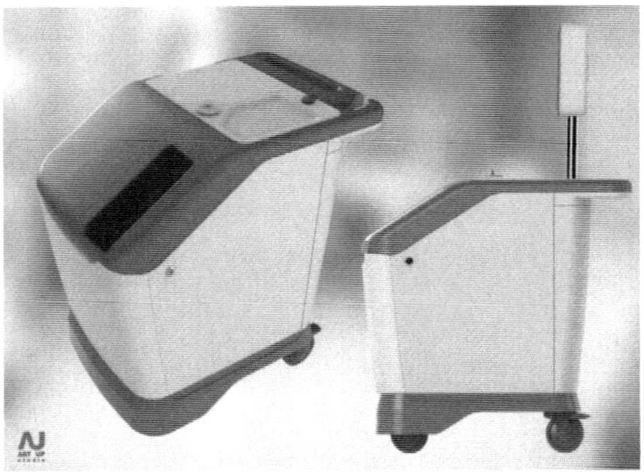

Figure 1. Domestically produced injection system. POZITOM-PRO RPC LLC Upgraded.

In order to enhance the potential of the automated injection system, POZITOM-PRO RPC LLC has developed and pilot tested an automated radiopharmaceutical synthesis module based on generated Ga-68.

The designed automated radiopharmaceutical synthesis module based on generated Ga-68 produces preparations from domestically produced lyophilizates, to produce two pharmaceuticals: Ga-68-DOTA-TATE and Vascular Ga-68. Preclinical studies of these pharmaceuticals are conducted together with the Federal State Institution A. I. Burnazyan Federal Medical and Biophysical Center.

6. Conclusions

Radionuclide generator system is a rather compact device with individual protection from ionizing radiation. Its main advantage is that it does not depend on cyclotron and can be installed in medical institutions with only PET scanners available. Introduction of the generator technology for radiopharmaceutical production is also financially attractive, as costly accelerator equipment and a radiochemical laboratory are not needed in this case, which considerably reduces cost of PET diagnostics and makes it affordable to a big part of the population of the Russian Federation.

The described automated injection system helps facilitate generator systems usage, ensure precision and automation of needed activity selection, reduce personnel and patient radiation exposure, and ensure control over all operation processes.

References

[1] M Phelps. Positron emission tomography provides molecular imaging of biological processes. Nat. Acad. Sci. USA, 2000, **97, 16,** P. 9226-9233.

[2] J Czernin and M Phelps. POSITRON EMISSION TOMOGRAPHY scanning: Current and Future. Applications Annual Rev. Med., 2002, **53**, P. 89-112.

[3] J Knuuti and F Bengel. Positron emission tomography and molecular imaging. Heart, 2008, 94, P. 360-367.

[4] E Krenning, D Kwekkeboom, W Bakker et al. Somatostatin receptor scintigraphy with [111In-DTPA- D-Phel] – and [123I-Tyr3-octreotide: the Rotterdam experience with more than 1000 patients. Eur. J. Nucl. Med.; 1993, 20, P. 716-731.

[5] G Shimchuk, N Kogay. Study of elution curve stabilization methods in real-time 82Sr/82Rb generator systems, in PET // MEPhI science session, 2008, vol. 3.

Spectral analysis of allogeneic hydroxyapatite powders

P E Timchenko, E V Timchenko, E V Pisareva, M Yu Vlasov, N A Red'kin and O O Frolov

Samara National Research University named after academician S.P. Korolev, 443086, Samara, Moskovskoye shosse 34, Russia.

e-mail: laser-optics.timchenko@mail.ru

Abstract. In this paper we discuss the application of Raman spectroscopy to the *in vitro* analysis of the hydroxyapatite powder samples produced from different types of animal bone tissue during demineralization process at various acid concentrations and exposure durations. The derivation of the Raman spectrum of hydroxyapatite is attempted by the analysis of the pure powders of its known constituents. Were experimentally found spectral features of hydroxyapatite, based on analysis of the line amplitude at wave numbers 950-965 cm^{-1} ($(PO_4)^{3-}$ (ν_1) vibration) and 1065-1075 cm^{-1} ($(CO_3)^{2-}$(ν_1) B-type replacement). Control of physicochemical properties of hydroxyapatite was carried out by Raman spectroscopy. Research results are compared with an infrared Fourier spectroscopy.

1. Introduction

At the present stage of medicine development in the field of surgery and dentistry for the restoration of human bone tissue increasingly are widely used biomaterials based on hydroxyapatite (HA). According to its physical and chemical properties and physiological reactions in the body, a finely divided crystalline synthetic hydroxyapatite can be compared with the inorganic component of bone tissue [1]. However, osseointegration properties and biological activity of these materials can vary significantly depending on method of their preparation, and consequently, there are ways to increase their effectiveness.

Injection of the hydroxyapatite to the bone implant increases the physico-mechanical properties of the composite: increases the flexural strength, elastic modulus and hardness, approaching to the lower property values of natural bone [2].

Naturally occurring calcium hydroxyapatite is also interesting because of the large number of substituent elements (Na, Mg, K, Fe, Zn, Cu, Ba, F, Cl, C, S) as part of the HA crystals. The impurity concentration is not more than 3-5%, but affects the chemical-biological and physico-mechanical properties of biogenic HA due to distortions in the crystal lattice caused by the presence of vacancies, interstitial and substitutional impurities. Due to the large number of isomorphic substitution, it is possible to change the properties of hydroxyapatite directed and getting materials in composition close to the native bone [2, 3]. Since the bone matrix serves as a buffer system for the atoms of different metals, synthetic biomaterials based on HA must have different concentration of impurities for best correspondence to the replaceable area of bone [4-7].

At work with complex objects such as calcium phosphate and hydroxyapatite, as well as composites applying in dentistry and medicine, it is necessary to evaluate their composition. Control of physicochemical properties can be implemented by Raman spectroscopy method.

In work [8] was carried out determination of biocompatibility and efficacy of hydroxyapatite in clinical application to replace bone tissue using multislice computed tomography (MSCT). In a comparative analysis of tomograms, authors of Ref concluded that the application of hydroxyapatite to fill defects in bone tissue during surgery accelerates bone repair processes. However, the CT method is time consuming and expensive method.

Infrared spectroscopy method is actively used in the analysis of biomaterials based on hydroxyapatite. Through this technique, it is possible to make conclusion about composition of the molecular groups in the sample, to detect distortion in the crystal structure of HA and to determine the influence of defects on the molecular composition of hydroxyapatite. The advantage of this method is the relative simplicity of the experiment and a high selectivity. In addition, advantage of the infrared method is the possibility to use in the analysis of multi-component mixtures [9, 10].

For example, in ref [11-14] points to related changes in the frequency shifts corresponding PO_4^{3-} relations, during the formation of hydroxyapatite. The intensity redistribution of the main peak PO_4^{3-} group corresponds to a broadening of the X-ray diffraction lines, which confirms the influence of crystallite size on the IR spectra calculated theoretically. However, this method can not distinguish between types of minerals in the composition of the implant [10].

In Raman spectroscopy realized high spatial resolution (up to 1 micron), combined with ease of sample preparation, the lack of destruction after analysis and reduced demands to the sample (which may be opaque, hydrated, etc.) [15].

In the works of authors [16, 17] have shown that the line of symmetric stretching vibration PO_4^{3-} (ν_1) has a high sensitivity to the mineral encirclement: the frequency and form of this line depends on the local surroundings and change as a result of substitution of anionic groups and changes in the degree of crystallinity. The substituted carbonate-apatite B-type phosphate.ν_1 line appears in the range of 955–959 cm^{-1}. In a crystalline unsubstituted hydroxyapatite line ν_1 shifted to 962–965 cm^{-1}. Finally, at a frequency line 945–950 cm^{-1} indicates the presence of a disordered lattice phosphate apatite. In general, the phosphate band ν_1 is a superposition of three components; it usually has an asymmetric shape due to the disordered vibrations contribution of phosphate and unsubstituted hydroxyapatite.

In addition to phosphate mineral component, in the Raman spectra also reflects the region 1065-1075 cm^{-1}, corresponding to fluctuations in the carbonate ion B-type [18, 19].

In the region 1400-1800 cm^{-1}, the most pronounced peaks of protein bonds are amide groups -CO-NH-. Following bands recorded in the spectrum: Amide I (1655–1675 cm^{-1}), which is mainly expressed by carbonyl groups C=O; Amide II (1555-1565 cm^{-1}); Amide III (1230–1289 cm^{-1}), which includes C-N stretching and N-H deformation vibrations; deformational twisting oscillation CH_2 (1455-1465 cm^{-1}). Low-intensity peak at 1620 cm^{-1}, authors of the article [15] correlated with fluctuations in the side chain of tyrosine (Y8a).

In the papers [20-22] have also used Raman spectroscopy for the study of changes in the composition of hydroxyapatite obtained from dental hard tissues in various pathological processes. The results of experimental studies indicate the information content of the method of Raman spectroscopy to study the crystallinity, structural features of biomaterials based on HA.

The aim of this study was to analyze the spectral composition of the hydroxyapatite powders obtained from different bone tissue types of various animal species by Raman spectroscopy.

2. Materials and methods of research

Objects of study were 30 samples of hydroxyapatite powder obtained from a mixture of compact and cancellous bone tissue of different animal species. To obtain a demineralized biomaterial, bone was placed in a solution of hydrochloric acid, the concentration of which depends on the type of bone (for birds' bone tissue samples 1.8H HCl solution was chosen, as 2.4H and 3.6H hydrochloric acid solution respectively used for bone tissue of sheep and cows). The obtained demineralized solution is used to produce hydroxyapatite. Difference thus obtained against synthetic hydroxyapatite powder lies in the fact that the final addition of hydroxyapatite powder contains organic components, and small concentrations of other apatite in the bone tissue. Measurements were taken of each sample at three different points.

The main method of research hydroxyapatite powder was used Raman spectroscopy method (Figure 1), implemented by the high-resolution digital spectrometer Andor Shamrock sr-303i with built-in cooling camera DV420A-OE, which provides spectral resolution of 0.15 nm, and fiber optic probe for Raman spectroscopy RPB785, combined with the laser module LuxxMaster LML-785.0RB - 04 (up to 500 mW, wavelength 785 nm). This wavelength well proven in Raman spectroscopy in biological applications, due to the lower level of autofluorescence.

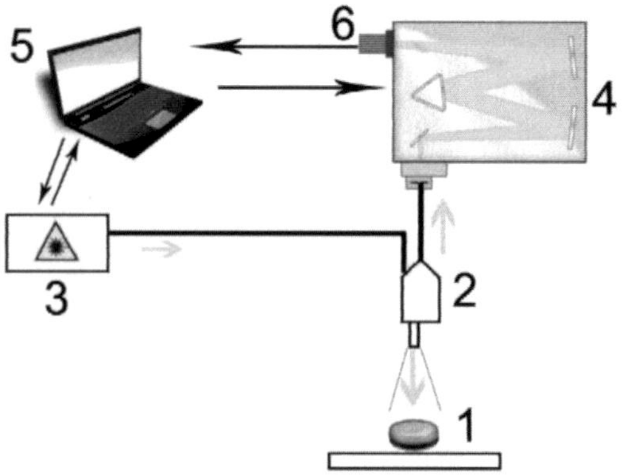

Figure 1. Experimental stand:
1 – studied object;
2 – Raman probe RPB785;
3 – laser module LuxxMaster LML-785.0RB-04;
4 - spectrometer Shamrock sr-303i;
5 – computer;
6 – cooling camera DV420A-OE.

Error of method does not exceed 4%. Processing spectra carried out in a software environment Wolfram Mathematica 9 and consisted in the removal of noise by the smoothing median filter. Then at the selected range of 400-2200 cm-1 using an iterative algorithm determined approximating line (a polynomial of the fifth degree) of autofluorescence component and subtracted it, yielding a dedicated Raman spectrum.

An additional method of research method was FTIR spectroscopy. IR - spectra of the samples were obtained on the console of frustrated total internal reflection (FTIR) from Pike Technologies Inc., which connected to the FTIR spectrometer SPECTRUM 100 (PerkinElmer, USA) in the range of wave numbers 550-4000 cm-1 with spectral resolution of 4 cm-1 and the signal accumulation of 20 scans. To record the spectra of hydroxyapatite powders used, which was applied directly onto the FTIR crystal consoles.

3. Results and discussion

Figure 2 shows the typical Raman spectra of hydroxyapatite powder samples obtained from biological materials with varying degrees of processing.

Registered Raman bands correspond to vibrations shown in Table 1. It is seen that these Raman peaks are presented in all samples.

IOP Conf. Series: Journal of Physics: Conf. Series **784** (2017) 012060 doi:10.1088/1742-6596/784/1/012060

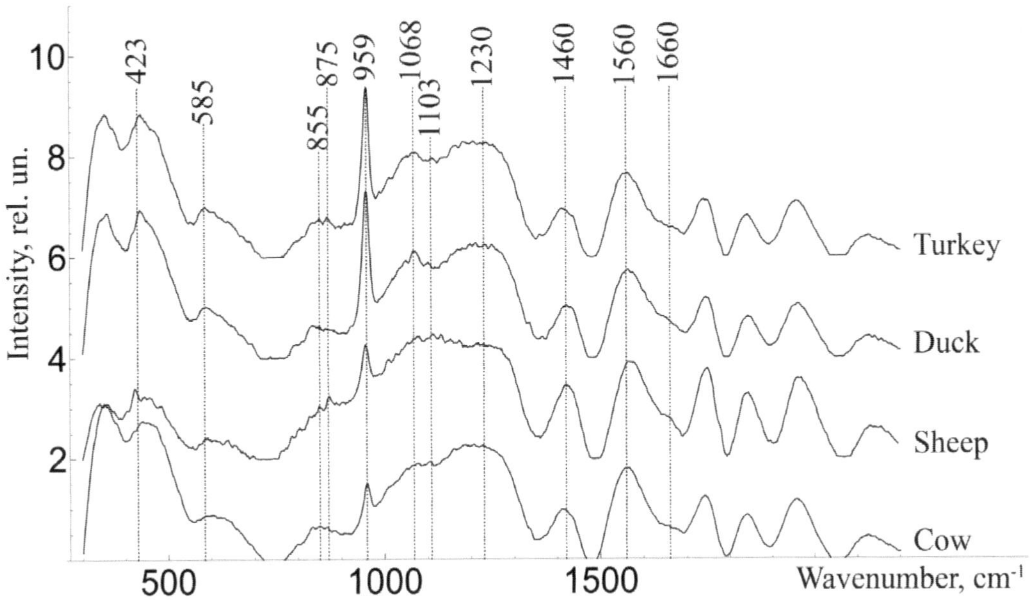

Figure 2. Averaged Raman spectra of hydroxyapatite powder samples.

The degree of substitution of the atoms and ions in the crystal of hydroxyapatite carbonate ions during the life changes, particularly there is an intensive substitution of the hydroxyl group with age corresponding hydrated CO_3^{2-} A-type (1095-1103 cm^{-1}). Therefore, in the production of composite materials based on HA seek to obtain mainly product with a predominant content of carbonate-B-type. [27, 28].

Figure 2 shows that the line corresponding to the intensity carbonate inclusion in the crystal lattice of hydroxyapatite (1065-1075 cm^{-1}, $(CO_3)^{2-}$ B-type substitution), predominates over Raman line, reflecting the concentration of the hydrated carbonate ion-type A. Hence the predominant content of carbonate-apatite B-type in studied HA samples is correspond to biomaterial close in chemical composition to the young bone tissue [18, 19].

Also, there is a relatively high content of organic components. Apart from Raman lines of proline and hydroxyproline, collagen component is represented by the Amide III groups (in region 1230-1289 cm^{-1}), Amide II (1555-1565 cm^{-1}) and Amide I (1655-1675 cm^{-1}) [17].

The observed differences in the intensity of the lines at wavenumbers of 950-965 cm^{-1} (vibrations PO_4^{3-} (v_1)) are the result of processing samples in hydrochloric acid solution with a normality varying depending on the animal species. Also, the differences can be explained by significant fluctuations in the content of amorphous calcium phosphate, depending on the individual and the age of the donor bone tissue. Amorphous calcium phosphate prevails at an early age; in the mature bone crystalline hydroxyapatite becomes predominant. Accordingly, on the composition of material affects both the type of bone tissue, and the age of the organism [24].

IR spectra of hydroxyapatite samples obtained from cancellous bone tissue of sheep, cows, turkeys and ducks are presented in Figure 3.

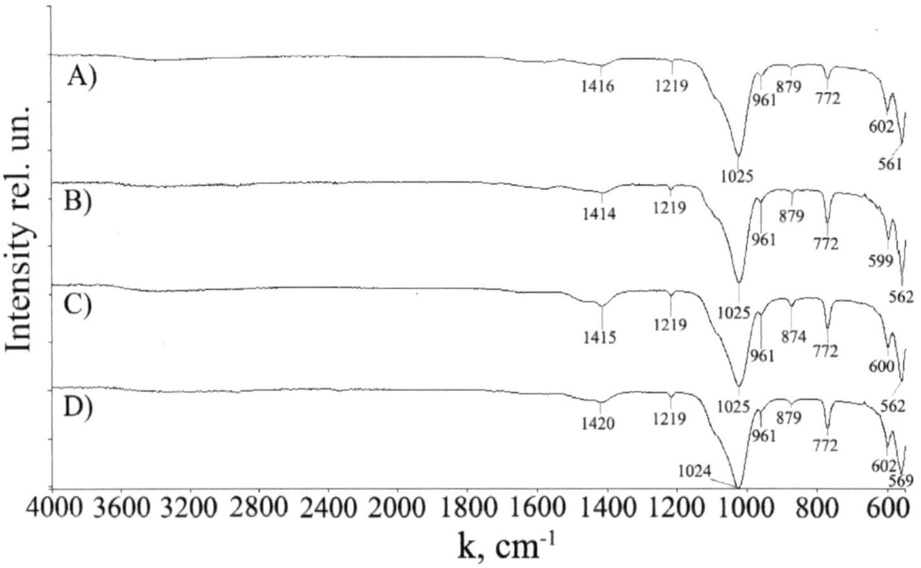

Figure 3. IR absorption spectra of hydroxyapatite for sample extracted from sheep cancellous bone tissue (A), cow (B), turkey (C) duck (D).

No significant difference in IR spectra of HA, obtained from different animal cancellous bone were found. Main intervals of wave numbers 950-965, 1020-1080, 520-660 cm^{-1} correspond to the vibrations of the phosphate ion [25]. The hydroxyl groups are shown in regions 3000-3600 and 1550-1700 cm^{-1}, the frequencies are overlapped with the frequencies of bound water. Frequencies 1415-1420, 1065-1075 and 870-880 cm^{-1} correspond to the presence of carbonate ions, which precipitates together with phosphates [26].

For all absorption spectra of hydroxyapatite, bands associated with organic components (Amide III, Amide II, Amide I) is significantly weaker than the line associated with the mineral component.

Table 1. Raman bands for hydroxyapatite powders.

Raman shift, cm^{-1}	Assignments
423	PO_4^{3-} (v_2) (P-O stretching)
561, 585, 602	PO_4^{3-} (v_4) (P-O bending)
855	Benzene ring of proline
875	Benzene ring hydroxyproline
880	CO_3^{2-} (v_2) (C-O stretching)
950–965	PO_4^{3-} (v_1) (P-O symmetric stretch)
1023,1030,1045,1052	PO_4^{3-} (v_3) (P-O asymmetric stretch)
1065-1075	CO_3^{2-} (v_1) B-type substitution (C-O in plane stretch)
1079-1080	Amorphous potassium carbonate (v_1)
1095-1103	CO_3^{2-} (v_1) A-type substitution (C-O in plane stretch)
1230-1289	Amide III (C-N-H stretch) (bending N–H)
1455-1465	CH_2 wag
1555-1565	Amide II (C-N-H stretch) (bending N–H)
1610-1620	Tyrosine Y8a (resonance-enhanced side-chain vibrations)
1655-1675	Amide I (C=O stretch)

The composition of the hydroxyapatite powders depends on the method of their preparation. HA, obtained from bone tissue, contains mineral components and the relative concentration of which depends on the degree of processing in the hydrochloric acid solution. Hence, for the most convenient control of changes in hydroxyapatite composition, analyze its main components, corresponding to mineral structure, by intensity of characteristic Raman lines: 950-965 cm^{-1} (symmetric stretching vibration PO_4^{3-} (v_1)) and 1065-1075 cm^{-1} (planar stretching vibration CO_3^{2-} (v_1)) [16-19].

For a comparative analysis of the results obtained using methods of Raman spectroscopy and FTIR - spectroscopy, we were built according shown in Figures 4-5.

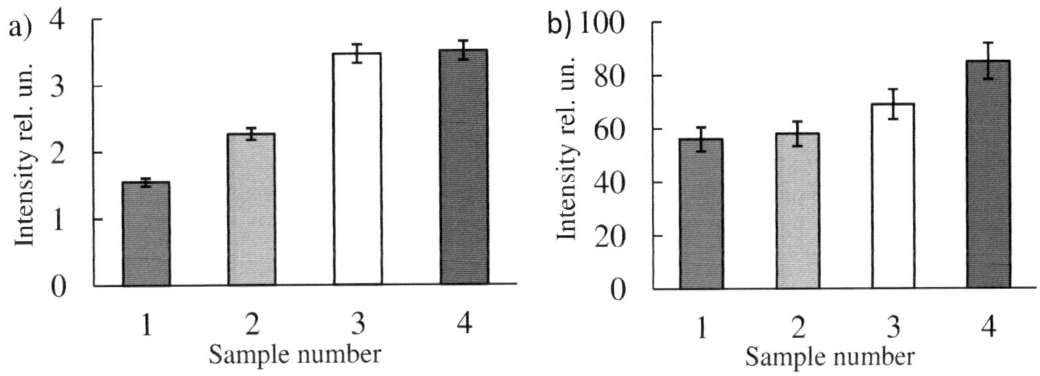

Figure 4. The intensity of the lines at the wavenumber of phosphate component for samples with different degrees of processing based on the data: a) Raman spectroscopy; b) FTIR spectroscopy.

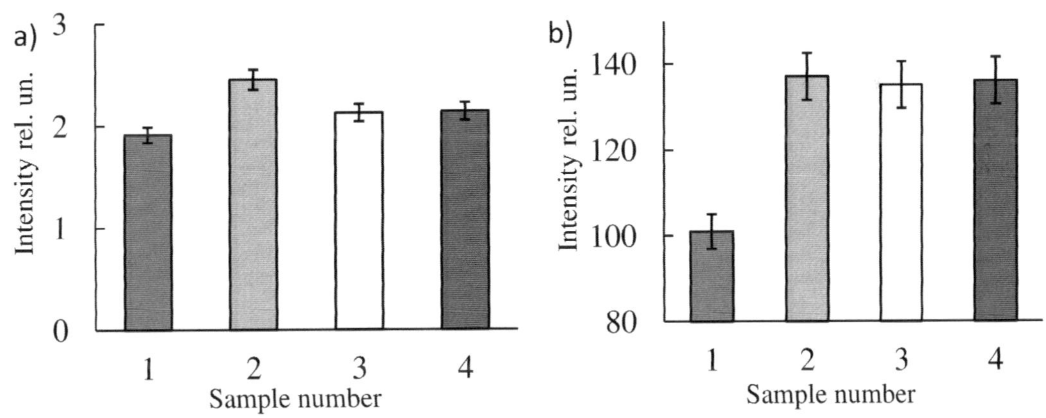

Figure 5. The intensity of the lines at the wavenumber of carbonate component for samples with different degrees of processing based on the data: a) Raman spectroscopy; b) FTIR spectroscopy; where HA samples obtained from the bone, are designated: 1. Cow, 2. Sheep, 3. Duck, 4. Turkey.

From the histograms presented in figures 4-5 it is seen that the intensity ratio of the amplitudes of the peaks Phosphate and carbonate ions at wave numbers 950-965 cm^{-1} and 1065-1075 cm^{-1} obtained by the method of Raman spectroscopy have a positive correlation with the results FTIR spectroscopy and reflects similar relative concentration dependence of substances.

The calculated data correlation coefficient of the two methods for of phosphate and of carbonate component is R_{PO4} = 0,86 and R_{CO3} = 0,76 respectively.

4. Conclusions

Were experimentally detected features of Raman spectra for hydroxyapatite powder samples obtained from a mixture of compact and cancellous bone tissue of different donors during demineralization, based on analysis of amplitude of $(PO_4)^{3-}$ line, integral line intensity $(PO_4)^{3-}$, integral line intensity $(CO_3)^{2-}$.

It was found that the degree of processing is characterized by change in the spectrum at wavenumbers of 950-965 cm^{-1} and 1065-1075 cm^{-1} corresponding phosphate and carbonate ions.

Research results are compared with Fourier transform infrared spectroscopy and represent a similar dependence of the relative concentrations of substances from the intensity amplitudes ratio for the major Raman peaks and the absorption bands in the infrared spectra. The calculated data correlation coefficient of the two methods for of phosphate and of carbonate component is R_{PO4} = 0,86 and R_{CO3} = 0,76 respectively.

5. Acknowledgments

This work was supported by the Ministry of Education and Science of the Russian Federation.

References

[1] Seredin P V, Al-Zubaidi A A, Ippolitov Y A 2013 *Research hydroxyapatite crystals and organic components of hard tooth tissues affected by dental caries using FTIR-microspectroscopy and XRD microdiffraction* (Kondensirovannye sredy i mezhfaznye granitsy) vol 3 pp 224 - 231

[2] Volozhin A I, Popov V K, Krasnov A P, Matveichuk I V, Doktorov A A, Roginskii V V, Nemeryuk D A 1999 *Physical-mechanical and morphological characteristics of the new composites based on ultrahigh molecular weight polyethylene and of hydroxyapatite* (New in Dentistry) vol 8. pp 35-43

[3] Pisareva E V, Vlasov M Yu, Volova L T 2015 *Structure and features of biomaterials based on mineral component of bone tissue* (Modern problems of science and education) vol 5

[4] Savel'ev V I, Kalinin A V 2001 *Experience of production and use of demineralized bone tissue in the experiment and clinic* (Biomedicinskie tehnologii: Sb. nauch. tr. NIC BMT vol 17) pp. 17–24

[5] Salgado C P, Sathler P C, Castro H C, Alves G G, Oliveira A M, Oliveira R C, Maia M D C, Rodrigues C R, Coelh P G, Fuly A *et al.* 2011 *Bone Remodeling, Biomaterials and Technological* (Journal of Biomaterials and Nanobiotechnology vol 1) pp 318-328

[6] Rey C, Combes C, Drouet C, Sfihi H, Barroug A 2007 *Physico-chemical properties of nanocrystalline apatites: Implications for biominerals and biomaterials* (Material science and engineering vol 27) pp 198-205

[7] Luong N D and Nam J-D 2001 *Synthesis of hydroxyapatite and application in bone and dental regeneration in human body* (Sungkyunkwan University) p 106.

[8] Druzhinin T V, Kamenchuk J A, Trofimov K V, Rostovtsev A V, Zhulyabin A, Isupov A A, Verevin A V 2014 *Qualitative and quantitative assessment of the application of of hydroxyapatite osteoregeneration in surgical practice* (BULLETIN of the East Siberian Scientific Center of the Academy of Medical Sciences) p 131

[9] Boskey A and Mendelsohn R 2005 *Infrared analysis of bone in health and disease* (Journal of Biomedical Optics vol 10) p 031102-1-9

[10] Tolstoy V P Chernyshova I V, Skryshevsky V A 2003 *Handbook of infrared spectroscopy of ultrathin films* (John Wiley & Sons, Inc., Hoboken. -New Jersey) p 738

[11] Gadaleta S J, Paschalis E P, Betts F, Mendelsohn R 1996 *Fourier transform infrared spectroscopy of the solution-mediated conversion of amorphous calcium phosphate to hydroxyapatite: new correlations between x-ray diffraction and infrared data* (Calcif. Tissue. Int vol 58) pp 9-16

[12] Pleshko N, Boskey A, Mendelsohn R 1991 *Novel infrared spectroscopic method for the determination of crystallinity of hydroxyapatite minerals* (Biophysical Journal vol 60) pp 786-793

[13] Nakano T, Tokumura A, Umakoshi Y 2002 *Variation in Crystallinity of Hydroxyapatite and the Related Calcium Phosphates by Mechanical Grinding and Subsequent Heat Treatment* (Metallurgical and materials transactions) pp 521-528
[14] Ippolitov Yu A, Lukin A N, Seredin P V 2012 *Research by IR spectroscopy using synchrotron radiation intact and infected carious process of enamel and dentin of human teeth* (Journal of New Medical Technologies vol 2) pp 343-346
[15] Ager III J W, Nalla R K, Balooch G, Kim G, Pugach M, Habelitz S, Marshall G W, Kinney J H, O'Ritchie R 2006 *On the Increasing Fragility of Human Teeth With Age:A Deep-UV Resonance Raman Study* (Journal of Bone and Mineral Research vol 21) pp 1879-1887
[16] Kiseleva D V 2010 *Application of Raman microspectroscopy to study structural features of biogenic apatite* (YEARBOOK 2009 vol 157) pp 332–335
[17] Tarnowski C P, Ignelzi Jr M A, Morris M D 2002 *Mineralization of Developing Mouse Calvaria as Revealed by Raman Microspectroscopy* // (Journal of Bone and Mineral Research).. 17, № 6. pp 1118–1126
[18] Penel G, Leroy G, Rey C, Bres E 1998 *Micro-Raman spectral study of the PO_4 and CO_3 vibrational modes in synthetic and biological apatites* (Calcified Tissue International 63) pp 475-481
[19] Awonusi A, Morris M D, Tecklenburg M M J 2007, *Carbonate assignment and calibration in the Raman spectrum of apatite* (Calcified Tissue International 81) pp 46-52
[20] Ionita I 2009 *Diagnosis of tooth decay using polarized micro-Raman confocal spectroscopy* (Romanian Rep. Phys. vol 61) pp 567-574
[21] Pezzotti G 2005 *Raman piezo-spectroscopic analysis of natural and synthetic biomaterials* (Anal. Bioanal. Chem., 381) pp 577-590
[22] Kirchner M T, Edwards H G M, Lucy D and Pollard A M *Ancient and Modern Specimens of Human Teeth: a Fourier Transform Raman Spectroscopic Study* (J. Raman Spec., 28) pp 171–178
[23] Boskey A L and Mendelsohn R 2005 *Infrared spectroscopic characterization of mineralized tissue* (Vib. Spect., 38) pp 107-114
[24] Baryshev E S, Baranova O V, Hamburg T V 2011 *Theoretical Foundations of Biochemistry: Tutorial* /; (Orenburg State. Univ.) 360
[25] Rasskazova L A, Korochenko N M, Guzeev V V 2013 *Compare the properties of unmodified and modified silicon hydroxyapatite, obtained by microwave exposure* (Polzunovsky Vestnik) № 1 pp 176-179
[26] Knyazev A V, Bulanov E N, Aleynik D Y, Charykova I N, Zemskov A E, Kalentev A V 2012 *Study of nano-sized hydroxyapatite models in vitro* (Bulletin of the Nizhny Novgorod University) № 5 (1) pp 88–94
[27] Landi E, Tampieri A, Celotti G, Langenati R, Sandri M, Sprio S, 2005 *Influence of Synthesis and Sintering Parameters on the Characteristics of Calcium Phosphate* (Biomaterials vol 26) p 2835.
[28] Starikov V V and Rudchenko S O 2010 *Optimization composite properties based on chitosan and of hydroxyapatite by varying its composition and heat treatment regimes* (Harkivsky national University) vol 14 pp 35-39